2016 29th Symposium on Integrated Circuits and Systems Design (SBCCI 2016)

Belo Horizonte, Brazil
29 August – 3 September 2016

IEEE Catalog Number: CFP16237-POD
ISBN: 978-1-5090-2737-8

Copyright © 2016 by the Institute of Electrical and Electronics Engineers, Inc
All Rights Reserved

Copyright and Reprint Permissions: Abstracting is permitted with credit to the source. Libraries are permitted to photocopy beyond the limit of U.S. copyright law for private use of patrons those articles in this volume that carry a code at the bottom of the first page, provided the per-copy fee indicated in the code is paid through Copyright Clearance Center, 222 Rosewood Drive, Danvers, MA 01923.

For other copying, reprint or republication permission, write to IEEE Copyrights Manager, IEEE Service Center, 445 Hoes Lane, Piscataway, NJ 08854. All rights reserved.

***This publication is a representation of what appears in the IEEE Digital Libraries. Some format issues inherent in the e-media version may also appear in this print version.**

IEEE Catalog Number: CFP16237-POD
ISBN (Print-On-Demand):   978-1-5090-2737-8
ISBN (Online):   978-1-5090-2736-1

Additional Copies of This Publication Are Available From:

Curran Associates, Inc
57 Morehouse Lane
Red Hook, NY 12571 USA
Phone: (845) 758-0400
Fax: (845) 758-2633
E-mail: curran@proceedings.com
Web: www.proceedings.com

2016 29th Symposium on Integrated Circuits and Systems Design (SBCCI 2016)

Belo Horizonte, Brazil
29 August – 3 September 2016

IEEE Catalog Number: CFP16237-POD
ISBN: 978-1-5090-2737-8

SBCCI 2016 List of Papers

Development Process for MEMS Pressure Sensors for Standarized CMOS Read-Out Circuitry.....1
Wolfgang Schreiber-Prillwitz, and Reinhart Job

Energy-Aware Light-Weight DMM-1 Patterns Decoders with Efficiently Storage in 3D-HEVC.....7
Gustavo Sanchez, Luciano Agostini, and César Marcon

A Mutual Rectification-Interference Avoidance Technique with Cascade Filters for Both Downward-Direction Tailed-RDF Deconvolution.....13
Hiroyuki Yamauchi, and Worawit Somha

A Lightweight Software-based Runtime Temperature Monitoring Model for Multiprocessor Embedded Systems.....19
Guilherme Castilhos, Fernando Gehm Moraes, and Luciano Ost

Evaluating the Impact of Circuit Legalization on Incremental Optimization Techniques.....25
Renan Netto, Vinicius Livramento, Chrystian Guth, Luiz C.V. dos Santos, and José Luís Güntzel

Modeling and Design of High-Efficiency Power Amplifiers Fed by Limited Power Sources.....31
Arturo Fajardo Jaimes, and Fernando Rangel de Sousa

Integrated CMOS Class-E Power Amplifier for Self-Sustaining Wireless Power Transfer system.....37
Arturo Fajardo Jaimes and Fernando Rangel de Sousa

Low-Power Hardware Design for the HEVC Binary Arithmetic Encoder Targeting 8K Videos.....43
Fábio Luís Livi Ramos, Jones Goebel, Bruno Zatt, Marcelo Porto, and Sergio Bampi

A 0.3 V, High-PSRR, Picowatt NMOS-Only Voltage Reference using zero-V_T Active Loads.....49
David Cordova, Arthur C. de Oliveira, Pedro Toledo, Hamilton Klimach, Sergio Bampi, and Eric Fabris

A Standard Cell Characterization Flow for Non-Standard Voltage Supplies.....55
Matheus Gibiluka, Matheus Trevisan Moreira, Walter Lau Neto, and Ney Laert Vilar Calazans

Design and Analysis of the HF-RISC Processor Targeting Voltage Scaling Applications.....61
Felipe Todeschini Bortolon, Sergio Johann Filho, Matheus Gibiluka,Sergio Bampi, Ney Laert Vilar Calazans, Fabiano Passuelo Hessel, and Matheus Trevisan Moreira

A Placement and Routing Algorithm for Quantum-dot Cellular Automata.....67
Alyson Trindade, Ricardo Ferreira, José Augusto M. Nacif, Douglas Sales and Omar P. Vilela Neto

A Novel Pruned-Based Algorithm for Energy-Efficient SATD Operation in the HEVC Coding.....73
Leonardo Bandeira Soares, Cláudio Machado Diniz, Eduardo Antonio César da Costa, and Sergio Bampi

Architectural Exploration of Last-Level Caches targeting Homogeneous Multicore Systems.....79
Rodrigo Cataldo, Guilherme Korol, Ramon Fernandes, Debora Matos, and César Marcon

Side Channel Attack on NoC-based MPSoCs are practical: NoC Prime+Probe Attack.....85
Cezar Reinbrecht, Altamiro Susin, Lilian Bossuet, Georg Sigl, and Johanna Sepúlveda

Cache Sizing for Low-Energy Elliptic Curve Cryptography.....91
Felipe Piovezan, Tarcísio E. M. Crocomo, Luiz C. V. dos Santos

Cluster-based Architecture Relying on Optical Integrated Networks with the Provision Of a Low-latency Arbiter.....97
Felipe Göhring de Magalhães, Fabiano Hessel, Odile Liboiron-Ladouceur, and Gabriela Nicolescu

A Security Aware Routing Approach for NoC-based MPSoCs.....103
Ramon Fernandes, César Marcon, Rodrigo Cataldo, Jarbas Silveira, Georg Sigl, and Johanna Sepúlveda

MagPDK: an Open-Source Process Design Kit for Circuit Design with Magnetic Tunnel Junctions.....109
Raphael M. Brum, and Gilson I. Wirth

Efficient Hardware Implementation of the Richardson-Lucy Algorithm for Restoring Motion-Blurred Image on Reconfigurable Digital System.....115
Oscar Anacona-Mosquera, Janier Arias-García, Daniel M. Muñoz, and Carlos H. Llanos

A systematic design approach for nanoscale inductor-less regulated cascode stages.....121
C. Talarico, G. D'Amato, G. Avitabile, G. Piccinni, and G. Coviello

Characterization and Nonlinear Modeling of MASMOS® Transistor in Order to Design Power Amplifiers for LTE applications.....126
Frédérique Simbélie, Sylvain Laurent, Pierre Medrel, Michel Prigent, Raymond Quere, Myrianne Regis, and Yann Creveuil

Focal-Plane Image Encoder with Cascode Current Mirrors and Increased Vector Quantization Bit Rate.....132
Fernanda D. V. R. Oliveira, Tiago M. de F. Lopes, José Gabriel R. C. Gomes, Fernando A. P. Barúqui and Antonio Petraglia

An Ultra Wide Band Analog-to-Digital Converter based on a Delta-Riemann architecture.....138
Francois Rivet, Elina Fiawoo, Richard Montigny, Patrick Garrec, and Yann Deval

A balanced logic routing block for Factorial-DLL based frequency generation.....142
Yann Deval and François Rivet

Energy-aware Scheduling in Transactional Memory Systems.....146
Ademir Marques Junior and Alexandro Baldassin

A Parallel Motion Estimation Solution for Heterogeneous System on Chip.....152
Mateus Melo, Gustavo Smaniotto, Henrique Maich, Luciano Agostini, Bruno Zatt, Leomar Rosa Jr, and Marcelo Porto

A Digitally Tunable 4th-order Gm-C Low-Pass Filter for Multi-Standards Receivers.....158
Mateus S. Oliveira, Paulo C. de Aguirre, Lucas C. Severo, Alessandro G. Girardi, and Altamiro A. Susin

A new two-step ΣΔ architecture column-parallel ADC for CMOS image sensor.....164
Pierre Bisiaux, Caroline Lelandais-Perrault, Anthony Kolar, Filipe Vinci Dos Santos, and Philippe Benabes

New Asynchronous Protocols for Enhancing Area and Throughput in Bundled-Data Pipelines.....170
Jean Simatic, Abdelkarim Cherkaoui, Rodrigo Possamai Bastos, and Laurent Fesquet

A Design Methodology for Low-Noise CMOS Transimpedance Amplifiers Based on Shunt-Shunt Feedback Topology.....176
A. F. Ponchet, J. W. Swart, E. M. Bastida, C. A. Finardi, R. R. Panepucci, S. Tenenbaum and S. Finco

A 450 mV Supply Self-biased Wideband Inductorless Balun LNA for sub-GHz Applications.....182
Arthur Liraneto Torres Costa, Hamilton Klimach, and Sergio Bampi

A Hardware Accelerator for the Alignment of Multiple DNA Sequences.....188
Antonyus P. A. Ferreira, Joao G. M. Silva, Jefferson R. L. Anjos, Luiz H. A. Figueiroa, Victor W. C. Medeiros, Edna N. S. Barros, and Manoel E. Lima

Inserting permanent fault input dependence on PTM to improve robustness evaluation.....194
Rafael B. Schivittz, Rafaél Fritz, Denis T. Franco, Lirida Naviner, Cristina Meinhardt, and Paulo F. Butzen

An FPGA-based accelerator for multiple real-time template matching.....200
Erika S. Albuquerque, Antonyus P. A. Ferreira, João G. M. Silva, João P. F Barbosa, Renato L. M. Carlos, Djeefther S. Albuquerque, and Edna N. S. Barros

Successful Prototyping of Complex Integrated Circuits with Focused Ion Beam.....206
E. Petitprez, D. M. Colombo, F. M. Henes, L. Courcelle, R. Tararam, S. Jacobsen, R. Soares, C. Krug, and M. Lubaszewski

Automatic Layout Integration of Bulk Built-In Current Sensors for Detection of Soft Errors.....211
Mário Vinícius Guimarães and Frank Sill Torres

Analytic Boundaries for 6T-SRAM Design in Standby Mode.....217
Fabián Olivera and Antonio Petraglia

A 0.7V Fully Differential First Order GZTC-C Filter.....223
Pedro Toledo, Renê Timbo, David Cordova, Hamilton Klimach, Sergio Bampi, and Eric Fabris

Software-Defined Radio Design based on GALS Architecture for FPGAs.....229
Eduardo Lussari, Duarte L. Oliveira, Lester A. Faria, and Orlando Verducci

A Digital Offset Correction Method for High Speed Analog Front-Ends.....235
Andres Amaya, Hector Gomez, and Elkim Roa

SBCCI 2016 Foreword

The Symposium on Integrated Circuits and Systems Design (SBCCI) is an international forum dedicated to Integrated Circuits and Systems Design, Test and Electronic Design Automation (EDA), held annually in Brazil.

The **SBCCI** has been established as an important international forum for presentation of advanced research results on leading edge aspects of integrated circuits and systems design, such as Analog circuits, Mixed-signal and Digital Integrated Circuits Design, Dedicated and Reconfigurable Architectures, EDA tools, Design Methods, Embedded Systems, Nanoarchitectures and Nanocomputing, as well as Verification and Test Methods. In the Systems-on-Chip (SoC) and Internet-of-things (IoT) era, all these technical fields contribute to the advancement of computing, communication and information systems.

The SBCCI 2016 Program Committee has made a major effort to thoroughly review 98 papers, which were electronically submitted to five technical tracks. After double-blind reviewing by the 106 members of the PC, 40 papers have been selected for the final program, distributed over 9 technical sessions. This year SBCCI received submissions coming from 12 countries, namely Austria, Brazil, Canada, China, Colombia, France, Germany, India, Italy, Japan, Great Britain and United States.

We would like to thank all the members of the Program Committee for their invaluable contributions, and in particular the chairs of each of the five tracks. Further, we would like to thank all the members of the Organization Committee, the Finance Chairs, the Local Arrangement Chair, the Tutorial chair, the Publication Chair and the Industry and Regional Liaisons. Their work and organizational efforts have resulted in an excellent infrastructure and a stimulating environment. The event counts on a variety of corporate and institutional sponsors, including national and international technical societies, and is co-organized by the Brazilian Microelectronics Society (SBMicro), the Brazilian Computer Society (SBC), the IEEE Circuits and Systems Society (CASS), and the ACM Special Interest Group on Design Automation (SIGDA). Our acknowledgments are naturally extended to the funding agencies, especially BNDES, CNPq, CAPES and FAPEMIG, as well as to the corporate supporters, whose financial support made this conference possible.

Davies William de Lima Monteiro	Frank Sill Torres	Leandro Soares Indrusiak
SBCCI'16 General Chair	SBCCI'16 Program Co-Chair	SBCCI'16 Program Co-Chair
Federal University of Minas Gerais	Federal University of Minas Gerais	University of York
Brazil	Brazil	United Kingdom

SBCCI 2016 Symposium Organization

SBCCI 2016 Committee

General Chair	Davies William de Lima Monteiro, DEE UFMG, Brazil
Program Chairs	Frank Sill Torres, DELT, UFMG, Brazil franksill @ufmg.br
	Leandro Soares Indrusiak, University of York, United Kingdom
Local Arrangements Chair	Omar Paranaiba Vilela Neto, DCC, UFMG, Brazil
Tutorials Chair	Gilson Inácio Wirth, UFRGS, Brazil
Panels Chair	Hercules Pereira Neves
	UNITEC Semiconductors, Brazil
Finance Chairs	Ado Jório, DFIS, UFMG, Brazil
	Luciana P. Salles, DEE, UFMG, Brazil
Publicity Chair	Ricardo Reis, UFRGS, Brazil
Publication Chair	Carlos Augusto de Moraes Cruz, UFAM, Brazil
Industry Liaison	Marcelo Lubaszewski, CEITEC, Brazil
North America Liaison	Kaushik Roy, Purdue University, USA
Europe Liaison	François Blanchard-Rivet, IMS-Bordeaux, France
Asia Liaison	Rui P. Martins, University of Macau, China
Latin America Liaison	Victor Champac, INAOE, Mexico

SBCCI 2016 Program Committee

Achim Rettberg, Uni-Oldenburg
Altamiro Susin, UFRGS
André Reis, UFRGS
Antoine Frappé, ISEN
Antonio Carlos Cavalcanti, POLIMI
Antonio Lopez-Martin, UNAVARRA
Antonio Miele, UFPB
Benjamin Carrion Schafer, PolyU
Bernardo Leite, UFPR
Bruno Zatt, UFPEL
Carlos Cruz, UFAM
Carlos Llanos
Cesar Prior, UFSM
Chao Wang, USTC
Cleonilson Souza, UFPB
Costas Psychalinos, UPATRAS
Cristina Meinhardt, FURG
Daniel Mauricio Muñoz Arboleda, UNB
Denis Masliah, ACCO Semiconductor
Diana Goehringer, RUB
Didier Belot, CEA
Dirk Timmermann, Uni-Rostock
Djones Vinicius Lettnin, UFSC
Dominique Dallet, IMS-Bordeaux
Edna Barros, UFPE
Eduardo Lima, UFPR
Edward David Moreno, UFS
Eric Kerherve, IMS-Bordeaux
Erika Cota, UFRGS
Felipe Marques, UFPEL

Fernanda Kastensmidt, UFRGS
Fernando Moraes, PUCRS
Fernando Rangel, UFSC
Franck Badets, CEA
Gilson Wirth, UFRGS
Hans Joachim Wunderlich, Uni-Stuttgart
Heider Marconi, DFCHIP
Ian Harris, UCI
Jean Baptiste Bégueret, IMS-Bordeaux
Jean-Max Dutertre, EMSE
Jeronimo Castrillon, TU Dresden
Jones Yudi Mori da Silva, UNB
José Augusto Nacif, UFV
José Monteiro, ULISBOA
Laurent Fesquet, TIMA
Leandro Buss Becker, UFSC
Letícia Maria Bolzani Pöhls, PUCRS
Luciano Agostini, UFPEL
Luciano Ost, Univ. of Leicester
Luis Lolis, UFPR
Luiz Filipe Menezes Vieira, UFMG
Luiz Franca-Neto, HGST
Marcelo Pavanello, FEI
Márcio Kreutz, UFRN
Márcio Schneider, UFSC
Marco Aurélio Wehrmeister, UTFPR
Marie-Lise Flottes, LIRMM
Mário Lúcio Cortes, UNICAMP
Martha Johanna Florez, TUM
Michael Hübner, RUB
Mounir Benabdenbi, TIMA
Nathalie Deltimple, IMS-Bordeaux
Ney Calazans, PUCRS
Nobuo Oki, UNESP
Norian Marranghello, UNESP
Olivier Sentieys, IRISA
Omar Paranaiba Vilela Neto, UFMG
Oscar Gouveia, UFPR
Paulo Augusto Dal Fabbro, Chipus
Paulo César Crepaldi, UNIFEI
Paulo Flores, INESC-ID
Paulo Francisco Butzen, FURG
Peter Marwedel, TU Dortmund
Raafat Lababidi, ENSTA
Renato Ribas, UFRGS
Ricardo de Oliveira Duarte, UFMG
Ricardo Jacobi, UNB
Rodolfo Azevedo, UNICAMP
Sara Vinco, Polit. Di Torino
Sergio Bampi, UFRGS
Shuai Li
Tales Pimenta, UNIFEI
Tsung-Chuan Ma, NTU
Volnei Pedroni, UTFPR
William Prodanov, Chipus
Yoan Veyrac, IMS-Bordeaux

SBCCI 2016 Sponsors & Supporters

SBCCI 2016 Organization

SBCCI 2016 Master Sponsors

SBCCI 2016 Sponsoring Societies

SBCCI 2016 Funding Agencies

SBCCI 2016 Silicon Sponsor

SBCCI 2016 Corporate Supporters

Development Process for MEMS Pressure Sensors for Standarized CMOS Read-Out Circuitry

Wolfgang Schreiber-Prillwitz

Sensor Business Group
TDK-EPC AG & Co. KG
14532 Stansdorf, Germany
wolfgang.schreiber-prillwitz@epcos.com

Reinhart Job

Dept. of Electrical Engineering and Computer Science
Muenster University of Applied Sciences
48565 Steinfurt, Germany
reinhart.job@fh-muenster.de

Abstract—Pressure sensor systems can be realized by the combination of two different technologies: CMOS-technology for the electrical part and MEMS-technology for the pressure sensor part. The challenge is to design the pressure sensor cell in such a way that a wide pressure range can be covered with one fix lateral membrane size, as any change in this lateral dimensions would lead to a new CMOS design, resulting in high costs. The authors describe an approach how to realize pressure sensor systems that on one hand are able to measure a nominal pressure range of several bar, but also to measure in the low pressure range of only several tenth of mbar by keeping a sufficient signal to noise ratio, and without changing the original CMOS circuitry. For the calculation of the sensor signals, an analytical approach, based on the theory of piezoresistivity, is combined it with structural mechanical simulation of mechanical stress, utilizing finite element analysis.

Keywords—MEMS; pressure sensor system; piezoresistive pressure sensor; signal conditioning; calibration

I. INTRODUCTION

Today, piezoresistive pressure sensor systems are wide spread in a lot of different applications. A pressure sensor system is the combination of the pressure sensor cell for the conversion of the physical pressure into an electrical signal, and the subsequent signal conditioning. To be usable within an application, the raw signal from the sensor cell has to be amplified and corrected for offset, linearity and temperature effects via an electrical circuitry into a calibrated output that gives a true electrical value proportional to the pressure, without any dependencies of the ambient temperature.

In contrary to standard CMOS integrated circuits (ICs) like analog-digital converters, microcontrollers, or operational amplifiers, to name a few, there are no standard products available in the area of pressure sensors. Anyway, for cost saving reasons, the challenge for a pressure sensor supplier in designing a pressure sensor system is to find a solution that in principle could cover a wide range of applications, which means pressure ranges. This is the key factor for high volume production, and by this to set up a stable and cost effective production process.

Two approaches for pressure sensor systems are common: first, to integrate the pressure sensor cell into the signal conditioning circuit (co-integrated pressure sensors), i.e. into one chip, and second, to choose a two-chip solution, where the pressure sensor cell and the readout IC are separated and interconnected by wires. Both options need a defined range of the raw signal from the sensor cell, since a single readout circuitry cannot operate at any input signal range.

This paper describes important aspects to be considered for designing a pressure sensor cell, and gives an overview of how the signal conditioning leads to a calibrated output signal.

II. PRESSURE SENSOR CELL DESIGN

The pressure sensor cell converts the physical pressure into an electrical signal.   a)
b)

Fig. 1 Schematics (a) and principle sketch (b) of a piezoresistive pressure senso shows the typical arrangement (a) and the schematic (b) of the piezoresistors on a silicon membrane. In   a)
b)

Fig. 1a the resistors are named after the North, South, East, West, respectively, according to their position related to the top view of a membrane (this notation is later used for calculating the change in resistivity). The direction of the current flow through the resistors and the applied directional stress is given by dotted arrows for the current and solid arrows for the directional stress. In case of stress and current are oriented in the same direction, the applied stress is called 'longitudinal', and in case of stress and current flow are perpendicular to each other, the applied stress is called 'transverse'.

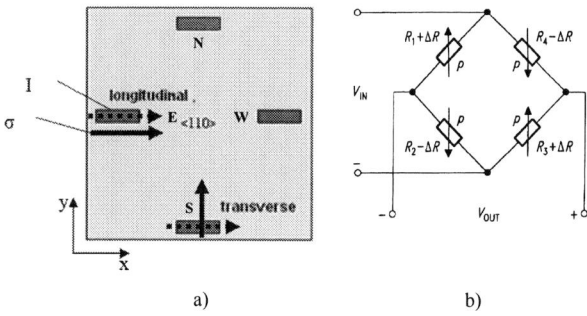

a)   b)

Fig. 1 Schematics (a) and principle sketch (b) of a piezoresistive pressure sensor

When pressure causes mechanical stress on the membrane, the resistivity of the longitudinally stressed resistors is rising, while the resistance of the transversely stressed resistors decreases. The change in resistivity is caused by the piezoresistive effect, which can be expressed according to Tufte [1] by the combination of the mechanical stress and the piezocoefficients, i.e.

$$\frac{R}{R_0} = \sigma_l \pi_l + \sigma_t \pi_t \qquad (1)$$

where σ_l and σ_t are the longitudinal (current and stress are parallel) and transverse (current and stress are perpendicular) stress values along the according resistors, and π_l and π_t are the so called effective piezocoefficients, respectively. The relation between the fundamental piezocoefficients and the effective piezocoefficients after [2] are given in Table I.

TABLE I.   EFFECTIVE PIEZOCOEFFICIENTS

Effective piezocoefficients	
Longitudinal direction π_l	Transverse direction π_t
$\frac{1}{2}(\pi_{11} + \pi_{21} + \pi_{44})$	$\frac{1}{2}(\pi_{11} + \pi_{21} - \pi_{44})$

The fundamental piezocoefficients for n-doped and p-doped silicon after [3] are given in Table II.

TABLE II.   FUNDAMENTAL PIEZOCOEFFICIENTS

Material	Fundamental piezocoefficients		
	$\pi_{11} \cdot 10^{-11}\ Pa^{-1}$	$\pi_{12} \cdot 10^{-11}\ Pa^{-1}$	$\pi_{44} \cdot 10^{-11}\ Pa^{-1}$
p-Si	+6.6	-1.1	+138.1
n-Si	-102.2	+53.4	-13.6

By the different changes of the single resistors (longitudinal: increasing, transverse: decreasing) the Wheatstone bridge gets imbalanced and provides a voltage deviation at the output. This voltage signal can be calculated as shown here:

$$V = \frac{R_1 \cdot R_2 - R_3 \cdot R_4}{R_1 + R_2 + R_3 + R_4} \cdot I \qquad (2)$$

According to [4] the piezoresistive coefficients are dependent on the carrier concentration of the resistors (i.e. the doping profile) and from the temperature.

$$P = \frac{300}{T} * \left(\frac{1}{1 + e^{\frac{-ef}{kB*T/q}}} \right) \Big/ ln\left(1 + e^{\frac{ef}{kB*\frac{T}{q}}} \right) \qquad (3)$$

P is the so called the piezoresistance factor, and is shown exemplarily in Fig. 2 for a p-type specimen [5].

The fundamental piezocoefficients have to be corrected by the piezoresistance factor P to take into account the carrier concentration and the temperature characteristics. Hence, they are scaling the fundamental piezocoefficients. For the design of a piezoresistive pressure sensor, this empirical model has two consequences:

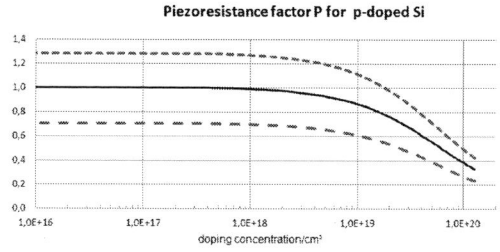

Fig. 2: Trend of the relative piezoresistance factor P for p-doped Si for -40°C, 25°C, 125°C

1. the lower the doping concentration, the higher the piezoresistive effect (and with this the gained signal), and

2. the higher the doping concentration, the lower the temperature dependency of the piezoresistive effect (and with this a more stable signal over temperature).

For the sensor design, it has to be considered what would be the trade off by choosing option 1 or option 2. The characteristics of the piezoresistance factor allows for higher signals up to a factor of two for relatively low doping of the resistor structures in the range of 10^{18} cm^{-3} compared to concentrations of about 10^{20} cm^{-3}. A more detailed analysis of these options is given in [6].

III.   CALCULATION OF SENSOR SIGNALS

As an example, the signal of a pressure sensor cell for a pressure range of 600 mbar should be calculated. The membrane geometry is 800 µm by 800 µm, with a thickness of nominal 12 µm. The target values of the sensor bridge signal should be higher than 3 mV/V and lower than 50 mV/V for the full scale range of the pressure. Application specific integrated circuits (ASSP) for piezoresistive sensor bridges typically can operate with such signals [7]. Although this is only one single case study, the following methodology could be applied to any other membrane design or different used doping profiles forming the piezoresistors.

A. Stress distribution

Signal evaluations starts with structural simulations of the deflection of the silicon membrane under pressure load. For this purpose the finite element analysis (FEA) tool ANSYS® [8] is used, which delivers membrane deflection, normal stress values σ_{xx}, σ_{yy}, and the shear stress σ_{xy}. A stress distribution over the membrane of 800 µm side length and 12 µm thickness for a pressure load of 600 mbar is shown in Fig. 4.

ANSYS offers a tool to define a result path by coordinates to extract results from the structural analysis at certain regions of the geometry. Here this option is used to define a path across the membrane. The resulting path is extending the membrane side length of 800 µm by 100 µm on each side leading to a resulting path length of 1000 µm. As the used coordinate system has its origin at the topside in the center of the membrane, the coordinates of the path range from -500 µm to +500 µm (Fig. 3b). The result path is indicated by the bold white arrow.

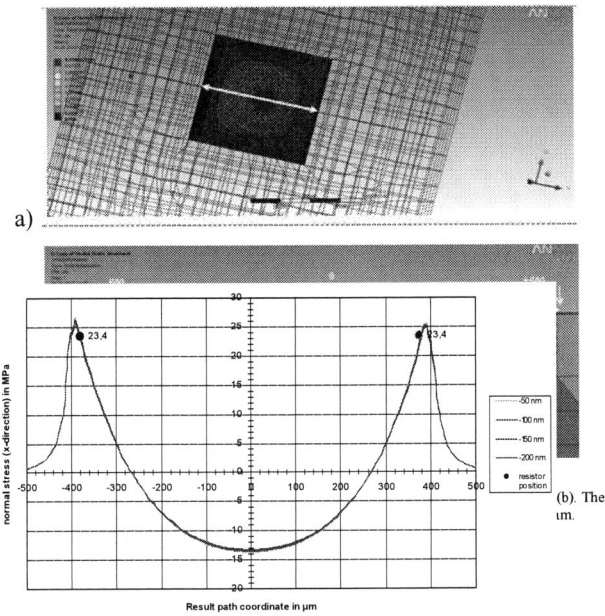

Fig. 4: Mechanical stress along the defined result path for different depths within the membrane

Applying the nominal pressure of 600 mbar to the model provides the stress profile across the middle of the membrane as shown in Fig. 4.

The x-axis represents the resulting path and gives the position along the cross section of the membrane (see also Fig. 3b). At the membrane rims, i.e. at x = −400 µm and x = +400 µm, the stress reaches its maximum. The black dots show the position of the geometric center of the resistors relatively to the rim, i.e. on both sides 13 µm inwards on the membrane. The y-axis shows the directional stress σ_{xx} in MPa. The stress value for the 12 µm thick membrane at 600 mbar is found to be 23.4 MPa at the geometric center of the resistors.

B. Doping profiles

Beside the mechanical stress – according to equation (3) – the piezoresistive coefficients are dependent from the carrier concentration of the resistors, i.e. on the doping profile. The doping is done by the technology of ion implantation that brings the doping specimen (e.g. boron) into the intrinsic silicon substrate. The effect of the additional carriers is that more free states (described by the effective density of states) could be occupied in case of a change of the band structure within the solid state, caused by mechanical stress.

For typical piezoresistive pressure sensors, mainly p-type resistors are used [9], [10]. The reason lies in the larger piezocoefficients π_{44} for p-type silicon of 138.1, as compared to −13.6 of n-type silicon [11] (see also **Error! Reference source not found.**II). The p-type resistor is formed by boron implantation, within a phosphorous doped n-well.

The doping concentration of the n-well is typically in the range of 10^{16} cm^{-3}. Compared to that, the boron concentration of the p-doping for the resistor structures is about 10000 times higher. By means of simulations, the implant profile can be calculated, taking into account the dose, implant energy and process temperature load. A simulated doping profile for p-doped piezoresistors is shown in Fig. 5.

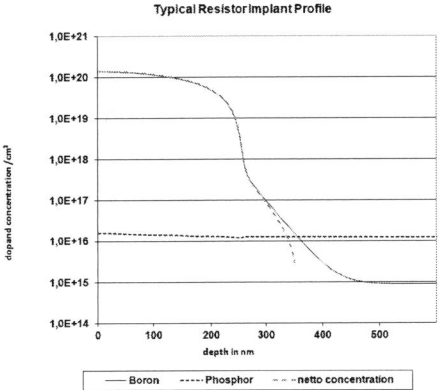

Fig. 5: Typical doping profile for piezoresitors

The x-axis represents the depth into the silicon wafer in nm, with zero representing the surface. The y-axis shows the carrier concentration per cm^3. The solid line shows the boron concentration, the horizontal dotted line the n-well. At a certain depth both concentrations are equal (junction depth), and the net concentration for boron and phosphorous is zero at this point (not shown in **Error! Reference source not found.**).

After implantation and annealing, the maximum doping concentration of the profile is located up to around 100 nm to 200 nm below the silicon wafer surface.

According to [6] it might be beneficial to decrease the doping concentration to raise the bridge signal in certain cases (see also Fig. 2).

C. Signal evaluation methode

Utilizing the stress values at the location of the resistors, and the carrier concentration within the resistors, the evaluation of the sensor bridge signal can be done.

978-1-5090-2737-8/16 $31.00 © 2016 IEEE

The conversion of stress distributions to electrical signals was done by subsequently calculation of the physical relations for semiconductors after the following the simplified flow chart shown in Fig. 6.

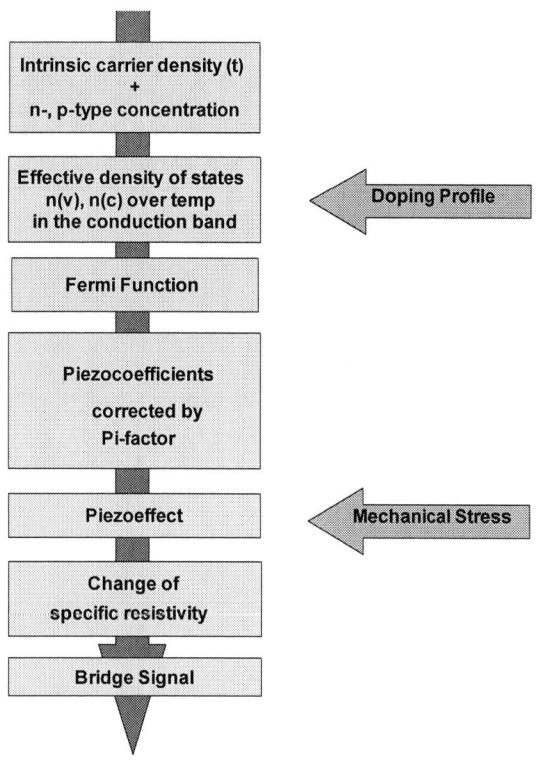

Fig. 6: Flow chart for calculating the sensor signal from semiconductor fundamentals, utilizing numerical stress simulations and real doping profiles

The conversion of stress distributions to electrical signals was done by a MATLAB® program, which interpolates the nodal stress values from FEM simulation and matches them to the resistor geometries. The routines take into account process related parameters like doping concentration or p-n junction depth from the doping profile. Resistivity changes for each resistor are calculated from the difference of the stress values for the unloaded membrane (zero pressure) and the full scale pressure of 600 mbar.

The bridge signal is generated by a change of the specific resistivity ρ, caused by the mechanical stress within the membrane by the applied pressure. According to [10], only the mechanical stress σ_{xx} is affecting the resistor pair E-W, σ_{yy} the resistor pair N-S (a)
b)

Fig. 1**Error! Reference source not found.**) and σ_{xy} (shear stress) could be neglected in an ideal case. The calculation of the relative change in resistivity of the E-W resistors after [10], using Pixy (with xy = 11, 12, 44) as the notation for the corrected piezocoefficients π_{xy} by the piezoresistance factors P, i.e. Pixy = P·π_{xy}, we obtain:

$$\Delta\rho\,(E-W) =$$

$$\left(\frac{\sigma_{xx}}{2}\right) * (Pi11 + Pi12 + (Pi11 - Pi12) + Pi44) \qquad (4)$$

Taking the simulated directional stress in x-direction σ_{xx} of 23.4 MPa and the piezocoefficients from Table II, the variation of ρ of the E-W resistors delivers $4.77 \cdot 10^{-3}$ or 0.477%.

For symmetry reasons, the y-directional stress σ_{yy} is equal and the change of resistivity for the N-W resistors after equation (4):

$$\Delta\rho\,(N-S) =$$

$$\left(\frac{\sigma_{yy}}{2}\right) * (Pi11 + Pi12) + (-Pi11 + Pi12) - Pi44) \quad (5)$$

The variation of ρ of the N-S resistors delivers $-4.40 \cdot 10^{-3}$ or -0.44 %.

Both values describe the relative change of the resistivity. Assuming a base resistivity of 5000 Ω per resistor and an excitation voltage of 1 V of the whole Wheatstone bridge, as arranged according to Fig. 1, this value will change for the E-W resistors R_E, R_W to 5023.84 Ω, and for the N-S resistors R_N, R_S to 4977.99 Ω. As the bridge signal consists of the difference of the two pairs of equally orientated resistors, the output signal can be calculated after

$$\frac{(R_N * R_S) - (R_E * R_W)}{(R_E + R_S) * (R_N + R_W)} * 1000mV = 4.59\ mV \qquad (6)$$

This specific example can be generalized. As the output signal is dependent from the combination of mechanical stress and the doping concentration after equations (3), (4) and (5), there is a certain degree of freedom to increase the piezoresistive signal, even if the mechanical stress is low. With this, for a fixed lateral membrane size, it is possible to adjust piezoresistive pressure sensor cell signals for wide pressure ranges of several mbar up to 10th of bars to available commercial signal conditioning ASSPs.

IV. READ OUT CIRCUITRY

In general the raw sensor bridge signal cannot be used within an application. First of all, it is too small to be used directly, and it is depending on the membrane thickness and the temperature behavior of the piezoresistors after [4] and [5]. The sensitivity, and with this the signal height, shows a certain distribution because the membrane thickness is deviating due to process inhomogeneity from the etching process of the membrane. For a standard time controlled KOH-etching process the deviation from the target thickness typically is around 25% over one wafer. Additionally, the quality of the assembly technique plays a major role, because the mismatch of the temperature coefficient of expansion (TCE) between the silicon and the substrate leads to a high temperature coefficient of the signal.

978-1-5090-2737-8/16 $31.00 © 2016 IEEE

Fig. 7 shows the principal architecture of a commercial application specific integrated circuit (ASSP) for piezoresistive pressure sensors.

Sensor signals of about 3 mV/V can be processed by a typical standard signal readout circuitry [7].

The sensor bridge signal is amplified by a switched capacitor amplifier (SC-amp.). The multiplexer provides switching between the pressure sensor signal and the signal of the temperature sensor that is needed for the temperature calibration. The following analog-to-digital converter (ADC) converts the analog temperature signal and the amplified pressure sensor signal into a digital word. During calibration, a multi-order correction over pressure and temperature of the sensor characteristics concerning offset, linearity, temperature behavior of offset and sensitivity, and supply voltage deviations is done. The therefore needed correction coefficients of the transfer function are stored in the EEPROM. The digital signal processor (DSP) then uses the calibration coefficients for linearization of the sensor transfer function and the correction of the offset and sensitivity with regard to the temperature. The calibrated signal can either be read as a bit-stream via I²C interface, or it can be converted back into a calibrated and amplified analog signal by a digital-to-analog converter (DAC).

V. SIGNAL CALIBRATION

The signal calibration is needed to always provide a defined signal for a certain pressure sensor system, which is proportional to the applied pressure on the sensor membrane. As mentioned before, the sensor cell signal is mainly dependent on the temperature, due to the temperature dependency of the piezoresistive effect and the interaction of sensor cell and sensor substrate. Further on, the signal at zero pressure (the offset) might be different from sensor to sensor due to intrinsic mechanical stress of the sensor cell, and the sensitivity might be different because of process deviations during the etching process of the membrane.

Fig. 88 shows the typical difference between the ideal transfer function (target value) of a sensor (straight line) and the real characteristic (true value) of the sensor cell (curved).

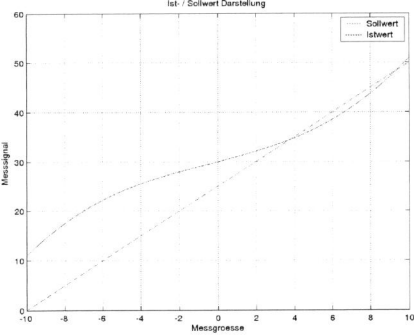

Fig. 8: Target and true transfer function of a pressure sensor cell

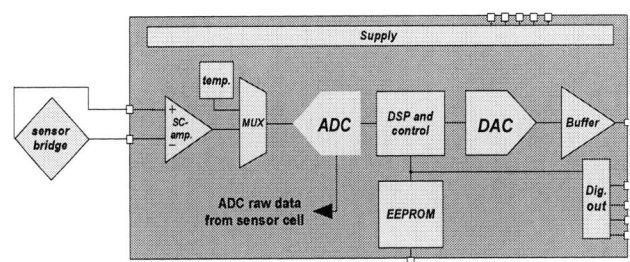

Fig. 7: Block diagram of an ASSP for piezoresistive pressure sensors

Utilizing the least-square method, correction coefficients are derived, which calculate the target values from the actual values. This could be expressed by (7), where the value of φ should become minimal.

$$\varphi = \sum_{i=1}^{n}(s_i - p_{grad}(c, m_i))^2 \qquad (7)$$

In this expression, s_i represents the target values, m_i the true values, c the correction coefficients, $p_{grad}(c, m_i)$ the expansion in series, and grad the polynomial order.

The expansion in series is represented by

$$p_{grad}(c, m_i) = \sum_{j=0}^{grad} c_j \cdot m_i^{j} \qquad (8)$$

The index i indicates the number of true values, which are used for the calibration.

To find the minimum of the function φ, according to curve sketching, the first deviation has to be zero, and the second deviation has to be > zero. In case the grade of the polynomial is one (representing a linear function which needs two measurements m_i), $p_{grad}(c, m_i)$ gives the expression:

$$p_{grad}(c, m_i) = c_0 \cdot m_i^{0} + c_1 \cdot m_i^{1} = c_0 + c_1 \cdot m_i \quad (9)$$

and the following equation has to be solved:

$$\varphi = \sum_{i=1}^{n}(s_i - p_{grad}(c, m_i))^2$$
$$= \sum_{i=1}^{n}(s_i - c_0 - c_1 \cdot m_i)^2 \qquad (10)$$

In this equation, only the correction coefficients c are variable, as the target values s_i are defined, and the true values m_i are measured by the sensor cell.

For a linear polynomial with two measurement values, equation (10) reduces to

$$\varphi = (s_1 - c_0 - c_1 \cdot m_1)^2 + (s_2 - c_0 - c_1 \cdot m_2)^2 \quad (11)$$

As the first derivation of this function has to be zero and the second derivation should be minimal, partial derivations of the coefficients have to be applied, i.e.

$$\frac{\partial \varphi}{\partial c_i} = 0 \quad (12)$$

Concerning the partial derivation with respect to c_0 this leads to

$$\frac{\partial \varphi}{\partial c_0} = s_1 + s_2 - 2 \cdot c_0 - c_1 \cdot (m_1 + m_2) = 0 \quad (13)$$

And for the partial derivation with respect to c_1 we get

$$\frac{\partial \varphi}{\partial c_1} = s_1 \cdot m_1 + s_2 \cdot m_2 - $$
$$c_0 \cdot (m_1 + m_2) - c_1 \cdot (m_1^2 + m_2^2) = 0 \quad (14)$$

The solution follows via a matrix formulation utilizing the inverse matrix:

$$\vec{A} \cdot \vec{c} = \vec{x} \Leftrightarrow \vec{A}^{-1} \cdot \vec{x} = \vec{c} \quad (15)$$

with

$$\vec{A} \cdot \vec{A}^{-1} = \vec{1} \quad (16)$$

Equations (13) and (14) now can be written in matrix form:

$$\begin{pmatrix} 2 & m_1 + m_2 \\ m_1 + m_2 & m_1^2 + m_2^2 \end{pmatrix} \cdot \begin{pmatrix} c_0 \\ c_1 \end{pmatrix} = \begin{pmatrix} s_1 + s_2 \\ s_1 \cdot m_1 + s_2 \cdot m_2 \end{pmatrix} \quad (17)$$

and according to equation (15) the coefficients c_0 and c_1 can be calculated by

$$\begin{pmatrix} 2 & m_1 + m_2 \\ m_1 + m_2 & m_1^2 + m_2^2 \end{pmatrix}^{-1} \begin{pmatrix} s_1 + s_2 \\ s_1 \cdot m_1 + s_2 \cdot m_2 \end{pmatrix} = \begin{pmatrix} c_0 \\ c_1 \end{pmatrix} \quad (18)$$

The inverse matrix of the left term in equation (18) can be calculated by solving the condition given by equation (16). This can be done for instance by a calculation tool like MATLAB.

For higher order correction coefficients, additional true values have to be measured to solve the according equation systems.

VI. CONCLUSION

The design of a pressure sensor system demands a deep knowledge of the transformation of the physical unit 'pressure' into an electrical signal, that can be processed further on by a commercial ASSP into a calibrated output signal. The combination of numerical calculation for the mechanical stress values (FEA) combined with the analytical calculation of the electrical signal utilizing this FEA results enables to find the appropriate solution for a sensor geometry for a given readout circuitry. This allows for an optimal adjustment of the sensor cell output range to the ASSP, especially to its input stage, that has to first amplify this signal and convert it into digital values. The individual sensor parameters like zero offset and sensitivity, which always show a certain distribution caused by process deviations, have to be determined during the calibration process that allows to calculate individual correction coefficients for each sensor. It has to be noticed that this calibration procedure in general is necessary for every single sensor system, and consumes a significant fraction of the production cost of each system.

REFERENCES

[1] O. N. Tufte, E. L. Stelzer. Piezoresistive Properties of Silicon Diffused Layers. J. Appl. Phys. 1963, Vol. 34, Februar 1963, p. 313

[2] W. G. Pfann, R. N. Thurston. Semiconducting stress transducers utilizing the transverse and shear piezoresistance effects. J. Appl. Phys. 1961, Vol. 32, pp. 2008–2019

[3] C. S. Smith. Piezoresistance effect in germanium and silicon. Phys. Rev. 1954, Vol. 94, pp. 42–49

[4] Y. Kanda. The Principle of the Piezoresistive Effect. [ed.] Lausanne Elsevier Sequoia. January 1991

[5] Y. KANDA. A Graphical Representation of the Piezoresistance. IEEE Transactions on Electron devices. ED-29, 1982, Vol. NO. 1

[6] "Improvement of Integrated Pressure Sensor Systems Fabricated by a Combined CMOS- and MEMS-Technology with regard to Low Pressure Ranges", W. Schreiber-Prillwitz, R. Job; "Microelectronics Technology and Devices – SBMicro 2012", Editors: G. Wirth, N. Morimoto, D. Vasileska, SBMicro 2012, Aug. 30th – Sept. 2nd, 2012, Brasília, Brazil, ECS Transactions U49U (1), 417 (2012)

[7] Elmos Semiconductor. E520.18: Advanced Versatile Sensor Signal Processor. [Data Sheet]. June 6, 2014

[8] ANSYS, Inc. [Online] www.ANSYS.com

[9] A. Senturia. Piezoresistive pressure sensor. Chapter 18

[10] Sugiyama, Toshiyuki, Toriyama, Susumu. Analysis of Piezoresistance in p-Type Silicon for Mechanical Sensors. Journal of Microelectromechanical Systems. October 2002, Vol. 11, pp. 598 - 604.

[11] E. V. Thomsen, J. Richter. Piezo Resistive MEMS Devices: Theory and Applications. May 18, 2005

Energy-Aware Light-Weight DMM-1 Patterns Decoders with Efficiently Storage in 3D-HEVC

[1,2]Gustavo Sanchez, [3]Luciano Agostini, [1]César Marcon

[1] PPGCC – Pontifícia Universidade Católica do Rio Grande do Sul, Porto Alegre, Brazil
[2] IFFarroupilha – Instituto Federal Farroupilha, Brazil
[3] GACI/PPGC – Universidade Federal de Pelotas, Pelotas, Brazil
gustavo.sanchez@acad.pucrs.br, cesar.marcon@pucrs.br, agostini@inf.ufpel.edu.br

Abstract—One of the major problems in Depth Modeling Mode 1 (DMM-1) hardware design is the large memory area required for storing all wedgelet patterns. This work presents four energy-aware light-weight coding techniques to reduce the memory usage and power dissipation. Experimental results show that the proposed techniques are capable of reaching a compression rate of 76.9%, representing a reduction of 140 Kbits. In the DMM-1 hardware design, this compression rate results in smaller memory, fewer memory accesses and, thus a power dissipation decrease, reaching 678 mW for ST 65nm standard cells technology.

Keywords—3D-HEVC; DMM-1; Pattern Compression; Memory Size Reduction

I. INTRODUCTION

Due to the current increase in 3D video coding usage, the Joint Collaborative Team on 3D Video Coding Extension Development (JCT-3V) [1] spends significant effort in research and development to extend the High-Efficiency Video Coding (HEVC) standard [2] to 3D video applications, aiming a bandwidth reduction for 3D video transmission/storage, while maintaining its quality. By February 2015, the JCT-3V has finalized the 3D-HEVC standard, which uses the most advanced features provided by HEVC and inserts many 3D-specific features aiming to obtain higher efficiency.

The adoption of Multiview Video plus Depth (MVD) [3] representation is one of the key factors for 3D-HEVC being capable of achieving high levels of compression rate. In MVD, each texture view is associated with a depth map. The same camera captures the texture images and depth maps, which provides geometrical information according to the distance between the objects and the camera. These maps are composed of 8-bits samples, where the closer the object is from the camera, the lighter is the shade of gray the object is represented. Fig. 1 presents a (a) texture view and its associated (b) depth map extracted from *Newspaper_CC* video sequence.

The motivation for MVD usage is to reduce the bandwidth for a 3D video transmission. This reduction is possible because, in the decoder, virtual views can be synthesized by interpolating texture and depth data with the use of techniques such as Depth Image Based Rendering (DIBR) [3]. These techniques allow synthesizing a dense set of texture views of the scene [4] and, consequently, to reduce the number of transmitted texture views.

The quality of the depth maps is crucial to allow generating high-quality virtual views, which is one of the main challenges in this scenario. Fig. 1 shows that depth maps contain characteristics that contrast with texture frames; i.e., they contain large areas of constant values (background or body of objects) and sharp edges (border of objects) while texture frames have smooth transitions between its pixels [5].

(a)   (b)
Fig. 1. (a) Texture view and its associated (b) depth map extracted from *Newspaper_CC* video sequence.

It is important to emphasize that distortions in the depth maps indirectly impact on the video quality since they are used to synthesize new texture views of the same scene. Then, it is important to encode the depth maps as precisely as possible, preserving these edges (i.e., without smoothing them) and avoiding errors in the video synthesis process.

The standard HEVC intra prediction [2] was designed considering the smooth transition between texture pixels, and this prediction is not efficient in the presence of depth maps sharp edges. Taking into account the importance of edge preservation, the JCT-3V has designed a coding tool called bipartition modes, which should be applied in the 3D-HEVC intra prediction to encode better depth maps edges [6]. The bipartition modes are composed by two Depth Modeling Modes (DMMs) – DMM-1 and DMM-4. Notice that DMM-2 and DMM-3 were removed in 3D-HEVC standard since they present low coding efficiency [7].

One of the major problem for DMM-1 hardware design is related to the huge memory area required to store the wedgelet patterns [8]. This work proposes four light-weight approaches that reduce the area, quantity of accesses and energy consumption of the memory used for DMM-1 wedgelet patterns storage. The remainder of this paper is organized as follows: Section II presents the 3D-HEVC depth intra prediction algorithm. Section III describes the proposed solutions. Section IV presents the simulation results comparing with the standard HEVC storage requirements. Finally, Section V renders the conclusions of this work.

II. DEPTH MAPS INTRA PREDICTION

Fig. 2 illustrates a high-level block diagram of the depth maps intra prediction implemented in 3D-HEVC reference software [9]. In the intra prediction, depth maps can be encoded by two modules: the (i) **HEVC intra prediction** that implements the same intra algorithms used for texture videos, and the (ii) **bipartition modes**, which explore depth maps properties that are not used on HEVC intra prediction.

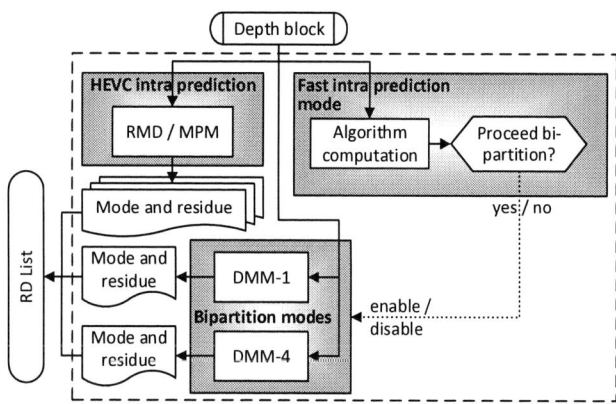

Fig. 2. Main blocks and flow of 3D-HEVC depth maps intra prediction.

The bipartition modes should be evaluated in parallel with the basic HEVC intra prediction step. However, since depth coding blocks are very flat or smooth, in general, and bipartition modes obtain better results in edges of sharp transitions, the **Fast intra prediction mode** proposed in [10] is applied to enable/disable the use of bipartition modes, reducing the computational complexity of the 3D-HEVC depth intra prediction.

The **Fast intra prediction mode** only enables the use of the bipartition modes when the first mode obtained by HEVC intra prediction in the Rate-Distortion list (RD-list) is not the planar mode, and the encoding block variance is higher than a pre-defined threshold [10], meaning that the encoding block has a high probability of containing an edge. When this technique fails, it means that the encoding block represents a flat or a smoothing area of the scene, which is the case where the HEVC intra prediction obtains better results. This procedure reduces the computational complexity of the intra prediction process without reducing encoding efficiency, because it reduces the unnecessary bipartition modes evaluation significantly. When the bipartition modes are enabled, the DMM-1 and DMM-4 are processed, and their results are added in the RD-list. In the following steps inside the encoder, the RD-cost is computed for all results of HEVC intra modes and bipartition modes added to this list.

The bipartition modes divide a given depth block into two regions. The encoded block encompasses a pattern containing the segmentation information, which specifies the region each sample belongs to, and a constant value that represents each region [11]. This pattern is composed of an array with N×N elements (N is the block width), containing 0 or 1 when the element belongs to region 0 or 1, respectively.

The DMM-4 algorithm dynamically creates the segmentation pattern from the texture data information, while the DMM-1 algorithm, which is the focus of this work, requires an evaluation over a predefined pattern set, which is a memory consuming approach.

A. Depth Modeling Mode 1 (DMM-1)

The DMM-1 algorithm segments blocks in two regions using a straight line named wedgelet. Regarding a continuous space, as exemplified in Fig. 3(a), the wedgelet could be represented and stored by the straight line equation. However, taking into account the discrete samples space, the wedgelet is stored in an N×N-array containing the binary pattern that defines the regions 0 and 1, as exemplified in the 8×8 block of the Fig. 3(b).

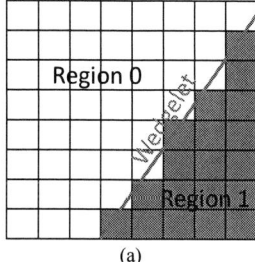

Fig. 3. Wedgelet segmentation model of a depth block: (a) the straight line dividing Region 0 and Region 1, and (b) discretization with constant values.

Fig. 4 presents a high-level diagram of the DMM-1 encoding algorithm, which is composed of three stages: (i) **Main Stage**, (ii) **Refinement Stage**, and (iii) **Residue Stage**. The **Main Stage** evaluates the entire initial wedgelet set (i.e., all wedgelets that must be assessed before the refinement) and finds identifies the best wedgelet pattern among the available ones.

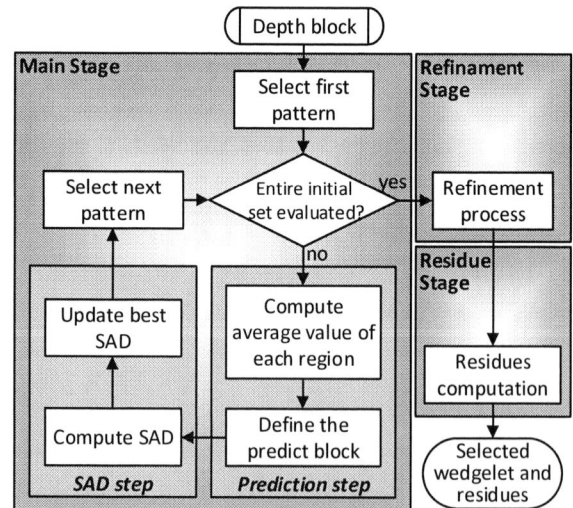

Fig. 4. The main blocks of the DMM-1 encoding algorithm.

978-1-5090-2737-8/16 $31.00 © 2016 IEEE

For each wedgelet pattern, Fig. 5 shows the pseudo-code applied to obtain the predicted block, where the $block_{i,j}$ represents the samples of the encoding block, $wedge\_pattern_{i,j}$ is the binary array of the wedgelet pattern, and $pred\_block_{i,j}$ is the resulting predicted block.

The DMM-1 algorithm starts computing avg_0 and avg_1, which are the average values of all samples mapped into regions 0 and 1, respectively (lines 1-11). Subsequently, the algorithm generates the predicted block inserting the avg_0 or avg_1 in the pred_block matrix, according to the wedgelet pattern (**Prediction step**). Next, Equation (1) computes the Sum of Absolute Differences (SAD) for each wedgelet pattern, where $|P_{i,j} - O_{i,j}|$ is the absolute value of the residue between the predicted and original depth samples at position (i, j). Finally, all SADs are compared, and the pattern with the lowest SAD defines the best wedgelet (**SAD step**).

1. avg_0 ← 0, avg_1 ← 0, num_0 ← 0, num_1 ← 0
2. for i ← 1 to N
3.   for j ← 1 to N
4.   if wedge_pattern$_{i,j}$ = 1
5.   avg_0 ← avg_0 + block$_{i,j}$
6.   num_0 ← num_0 + 1
7.   else
8.   avg_1 ← avg_1 + block$_{i,j}$
9.   num_1 ← num_1 + 1
10. avg_0 ← avg_0 / num_0
11. avg_1 ← avg_1 / num_1
12. for i ← 1 to N
13.   for j ← 1 to N
14.   if wedge_pattern$_{i,j}$ = 1
15.   pred_block$_{i,j}$ ← avg_1
16.   else
17.   pred_block$_{i,j}$ ← avg_0

Fig. 5.   Pseudo-code for DMM-1 prediction block generation.

$$SAD = \sum_{i=1}^{N} \sum_{j=1}^{N} |P_{i,j} - O_{i,j}| \qquad (1)$$

The **Refinement Stage** evaluates up to eight wedgelets around of the selected one (i.e., with a similar pattern) in the previous operation. Again, the wedgelet that obtained the lowest SAD among these eight possibilities, along with the first wedgelet selected in **Main Stage,** is elected as the best one. Finally, the **Residue Stage** subtracts the predicted block of the elected wedgelet from the original one and adds this wedgelet into the RD-list.

Fig. 6 exemplifies the encoding of a 4×4 depth block along with the evaluation of three wedgelet patterns. DMM-1 prediction process encodes the depth block sample according to the evaluated wedgelet (i.e., patterns **a**, **b**, and **c**). This procedure maps the pixels of the block sample in one of the two regions. Subsequently, the predicted block step computes the average value of all pixels in the region (e.g., the average value of regions 0 and 1 of pattern **a** are 64 and 76, respectively). The residue step annotates the position corresponding of each pixel with the difference between the

predicted and original depth sample. The SAD of all patterns is attained and finally, the pattern **b** is selected since it has the lowest SAD.

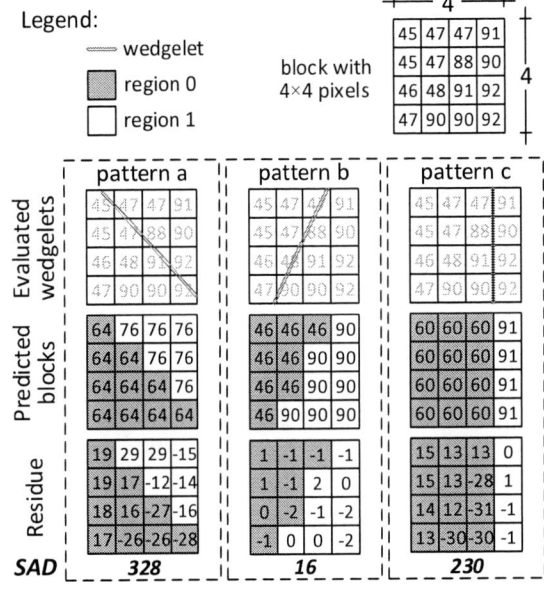

Fig. 6.   Example of a 4×4 depth block encoding with the DMM-1 algorithm.

One of the major problems in DMM-1 hardware design is efficiently storing all patterns because there are a lot of possible wedgelets [8]. Table I shows the number of wedgelets and its storage requirements, considering that the memory stores all patterns using N×N bits; i.e., without any compression, which is done in 3D-HEVC Test Model (HTM) version 16.0 reference software due to the required performance.

TABLE I.   NUMBER OF EVALUATED WEDGELETS IN DMM-1

Block size	Wedgelets	Storage requirements (bits)
4×4	86	1,376
8×8	802	51,328
16×16	510	130,560
32×32	510	-
Total		183,264

Table I depicts that the DMM-1 hardware requires 183,264 bits to store all wedgelet patterns for all available block sizes. Notice that the DMM-1 wedgelet patterns of 32×32 blocks are not stored because they are obtained up-scaling 16×16 wedgelets. Furthermore, the standard DMM-1 algorithm requires frequent access to the wedgelet patterns implying large energy consumption. This work explores and compares light-weight techniques for wedgelet patterns compression to reduce the storage area and energy consumption.

III. PROPOSED COMPRESSION TECHNIQUES

This section describes four techniques proposed to reduce the wedgelets pattern memory size: (i) First Bit and Change (FB&C); (ii) Huffman Code; (iii) Block Change Map (BCM); and (iv) Line Change Map (LCM).

A. First Bit and Change (FB&C)

FB&C algorithm is grounded in the fact that DMM-1 is a bipartition approach that divides each block into two and only two regions. Therefore, the capacity of the standard DMM-1 to represent interlaced ones and zeros in a single line is unnecessary producing an inefficient storage model. For example, regarding a 4×4 block, each line should never contain the 0110 pattern because any wedgelet is capable of describing such pattern. All forbidden lines patterns for 4x4 blocks are presented in Table II along with the allowed ones and its proposed FB&C representation. For an N×N block, there are 2×N allowed lines in DMM-1 patterns.

TABLE II.   FB&C REPRESENTATION FOR 4×4 PATTERN LINES

Forbidden lines	Allowed lines	Proposed representation
0010	0111	000
0100	0011	001
0101	0001	010
0110	0000	011
1001	1000	100
1010	1100	101
1011	1110	110
1101	1111	111

Fig. 7(a) illustrates the FB&C technique that codifies each line of the block copying the 1st bit content of the uncoded line to the coded one. Let RB be the remaining bits required for codifying a line of the N×N block, which can be obtained applying Equation (2). Then, the remaining RB bits of the coded line indicate how many subsequent positions, the line will change to the other region.

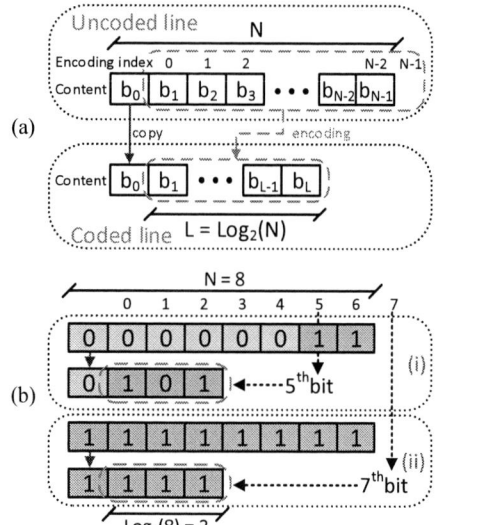

Fig. 7.   (a) FB&C encoding model; and (b) two examples with 8×8 blocks.

$$RB = log_2(N) \tag{2}$$

Note in the examples of Fig. 7(b) that the encoded value of N-1 means that the line contains only a single region. Moreover, one can notice that the amount of bits required by the FB&C algorithm grows logarithmically, while the same storage requirement increases linearly for the standard DMM-1

technique. The Equation (3) analytically computes the compression rate obtained with FB&C technique.

$$Compression\ Rate = 1 - \frac{log_2(N) + 1}{N} \tag{3}$$

B. Huffman Code

Huffman code is a traditional algorithm in data compression field, which creates a prefix binary code tree with minimum expected code-word length. This algorithm works reducing the representation of the most frequent lines and increasing the representation of the less frequent lines. The statistical analysis to create Huffman code tree is obtained offline because wedgelets are predefined before the encoding execution.

Table III shows the original samples with all available lines, its probability and the Huffman code for all available 4×4 pattern lines. One can notice that the samples 0000 and 1111, which are the most often samples, are represented by only 2 bits, while 0011 and 0001, which are the less often samples, are represented by 5 bits, resulting in a reduction in storage requirements.

TABLE III.   PROBABILITY AND HUFFMAN CODE FOR 4×4 PATTERN LINES

Original Sample	Probability	Huffman Code
0111	6.10%	0001
0011	5.81%	00001
0001	6.10%	00000
0000	21.52%	10
1000	12.50%	010
1100	15.12%	001
1110	12.50%	011
1111	20.35%	11

The Average Line Size (ALS) used to represent each line when applying Huffman Code is obtained using Equation (4), where $p(i)$ is the probability of the sample i appear.

$$ALS = \sum_{i=1}^{2 \times N} p(i) * length(HuffmanCode(i)) \tag{4}$$

The compression rate can be analytically obtained applying the Equation (5). For the 4×4 block example, the ALS is 2.884 bits, and the compression rate is 27.98%.

$$Compression\ Rate = 1 - \frac{ALS}{N} \tag{5}$$

C. Block Change Map (BCM) Algorithm

In BCM algorithm, the first wedgelet pattern can be encoded with any algorithm or even be stored without compression. Next, for each wedgelet pattern, a matrix called *Block Change Map* is created indicating when this wedgelet pattern has a bit change comparing with the previous pattern. The positions that contain a change are signalized with 1 while the remaining positions are signalized with 0.

This Block Change Map creation helps to increase the entropy of the wedgelet patterns and then, Huffman algorithm is applied per line, reducing the storage requirements considerably. An example of its application for two 4×4 wedgelets patterns is presented in Fig. 8.

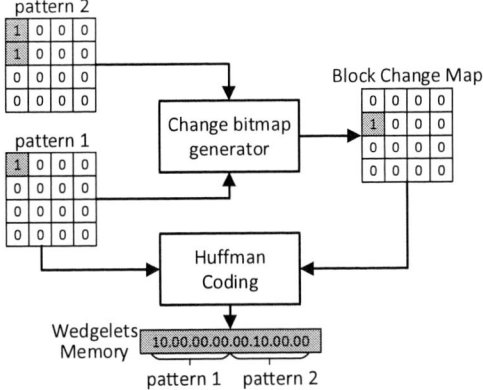

Fig. 8. Example of BCM application for two 4×4 block patterns.

In this example, the Huffman Coding algorithm has been applied to encode the first wedgelet pattern. A comparison is performed bit by bit in Change bitmap generator block to encode the second pattern, where it resulted in 0 in the positions that identically are 1 in the positions that presented a change. The same Huffman Coding block that encoded the first pattern is used to encode the change bitmap. Finally, all encoded data are inserted into the wedgelets memory.

D. Line Change Map (LCM) Algorithm

The LCM algorithm presents high similarity with the BCM, however, instead of generating a block change map, it generates a line change bitmap for each line. Similar to the previous solution, the first line can be encoded using any technique or even be stored uncompressed. When next blocks are being encoded, the LCM between its first line and the last line of previous block is generated.

Fig. 9 presents an example of this technique, where the first line has been stored using the same Huffman coding tree employed to store the line change bitmap. The following lines are compared against the previous line, where every bit change is indicated with 1 and non-bit change is indicated with 0. This resulting line change map is encoded using Huffman code algorithm, and all data is stored in wedgelet memory.

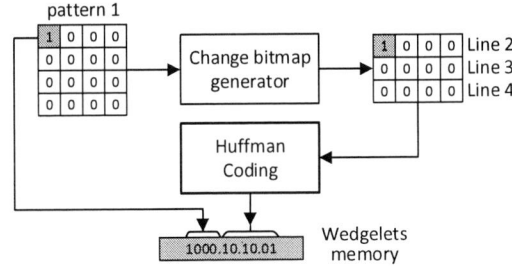

Fig. 9. Example of LCM application for a 4×4 block pattern.

IV. EXPERIMENTAL RESULTS AND DISCUSSION

A. Simulation Results

All wedgelets patterns were extracted from 3D-HTM 16.0 and the proposed solutions were implemented using Matlab. Table IV presents the quantity of bits required to store all

available wedgelets of all block sizes regarding standard DMM-1 and the four proposed solution.

TABLE IV. QUANTITY OF BITS REQUIRED TO STORE ALL WEDGELETS

Algorithm	4×4	8×8	16×16	Total
Standard	1,376	51,328	130,560	183,264
FB&C	1,032	25,664	40,800	67,496
Huffman	991	23,503	34,298	58,792
BCM	761	14,428	27,175	42,364
LCM	1,086	22,301	27,108	50,495

Fig. 10 shows the impact of the compression techniques having as reference the Standard DMM-1. The FB&C and Huffman techniques, which are the simpler solutions, are capable of achieving a compression rate of 63.17% and 67.92%, respectively. The highest compression rates are achieved applying BCM and LCM algorithms, where memory reduction of 76.88% and 72.45% are obtained, respectively.

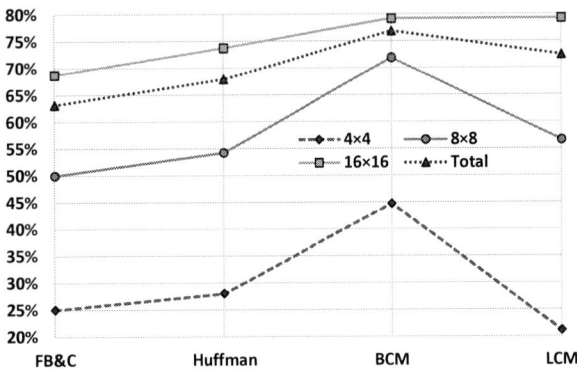

Fig. 10. Compression rates for all stored patterns of all proposed technique.

B. Implementation Impact on the Target Architecture

Memory reduction techniques usage should care about the impact on the target architecture. In this direction, it is important to evaluate the area and power consumption overhead. We synthesized the architectures that implement the decoder of the proposed solutions using ST 65nm standard cells technology for a frequency capable to achieve real-time passing by the entire initial wedgelet set for HD 1080p video processing. Table V depicts the area of the additional hardware that is necessary to implement each technique.

TABLE V. AREA OF THE TARGET ARCHITECTURE OF EACH PROPOSED TECHNIQUE (IN GATES)

Technique	4×4	8×8	16×16	32×32	Total
FB&C	19	25	67	65	176
Huffman	17	47	92	93	249
BCM	210	1089	6257	6257	13813
LCM	39	244	406	324	1013

The standard DMM-1 storage does not require additional hardware resources in its implementation while all of the proposed solutions requires an increase in hardware usage with a considerable amount reduction in memory size. The hardware implementation of FB&C and Huffman Code are the smaller, requiring only a small lookup table capable of converting the encoded lines into the uncoded line.

BCM is the solution that requires the highest area among our solutions, however, it is the only solution that delivers an entire NxN block per cycle. It requires N lookup tables to convert the encoded Huffman data to the change bitmap and NxN 1-bit registers to store the previous block pattern. Finally, an N×N-array of X-ORs should convert the previous pattern to the current pattern.

The LCM hardware design presents high similarity compared to BCM. However, it delivers a patterns line per cycle, requiring only one lookup table, N 1-bit registers and, N-array of X-ORs to convert the previous line to the current line.

Along with area evaluation, energy consumption results for memory access of each proposed technique have been analyzed, considering that 100pJ are consumed each time a byte is accessed in a DDR3 memory, according to [12]. Table VI presents some results of this evaluation.

TABLE VI.   POWER DISSIPATION OF THE STANDARD DMM-1 AND THE PROPOSED TECHNIQUES (IN mW)

Technique	Resource	4×4	8×8	16×16	32×32	Total
Standard	Memory	49.77	250.39	304.82	304.82	909.79
FB&C	Memory	37.32	125.19	95.26	95.26	353.03
	Hardware	0.39	1.60	2.16	0.68	4.83
	Total	37.71	126.79	97.41	95.94	357.86
Huffman	Memory	35.84	114.65	80.08	80.08	310.65
	Hardware	0.14	1.19	1.08	0.43	2.84
	Total	35.99	115.84	81.16	80.51	313.49
BCM	Memory	27.53	70.38	63.43	63.43	224.78
	Hardware	0.42	2.30	3.23	1.05	7.00
	Total	27.95	72.69	66.66	64.49	231.78
LCM	Memory	39.28	108.79	63.28	63.28	274.63
	Hardware	0.47	3.42	3.82	0.63	8.34
	Total	39.75	112.21	67.10	63.91	282.97

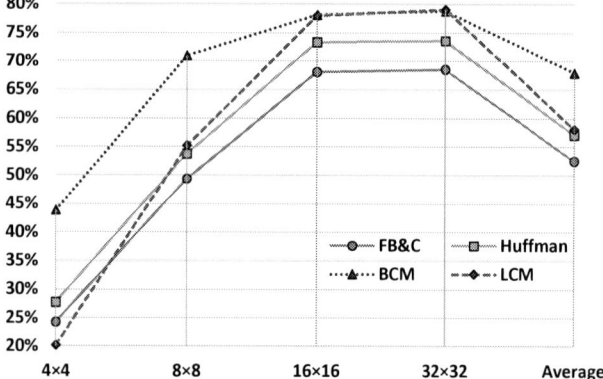

Fig. 11. Power dissipation decrease when using the proposed techniques.

The power dissipated from hardware overhead required by the proposed techniques is small, when compared to the power dissipated by the memory accesses. Fig. 11 illustrates the how much each proposed technique enable to reduce power when compared with the standard DMM-1 implementation.

As one can notice, the proposed techniques are capable of reducing from 20% to 45% when taking account small blocks, and from 68% to 80% for the larger blocks implying a total power dissipation decrease between 551 mW and 678 mW.

C. Comparison with Related Work

We only find the work [13] that proposed simplifications on the wedgelet set. In [13], a reduction of 27.8% in the memory size is obtained with a loss in the compression rate of 0.03%. This result is obtained storing the entire 16×16 wedgelet patterns and dynamically down-sampling these patterns for other block sizes. No energy information is given in [13]. The techniques proposed here are capable of providing more memory area reduction without impacting the compression rate of the encoding videos. Moreover, the proposed techniques provide a high decrease in power consumption, which is not demonstrated in [13].

V. CONCLUSIONS

This work presented four techniques to reduce the storage requirements of DMM-1 wedgelets. The standard technique store all patterns bit by bit, without considering the entropy of this information, resulting more than 180 Kbits to store the entire wedgelet set. The proposed techniques are capable of achieving a compression rate ranging between 63.17% and 76.88%, which represents a reduction between 112 and 140 Kbits using only simple and light-weight techniques. Additionally, the implementations of the proposed techniques result in a smaller area, fewer memory accesses and, consequently, a power consumption decrease.

REFERENCES

[1] JCT-3V. Available at //phenix.int-evry.fr/jct2, access in Ago. 2015.

[2] G. Sullivan et al. "Overview of the high efficiency video coding (HEVC) standard," Transactions on circuits and systems for video technology, v. 22, n. 12, pp. 1649-1668, Dec. 2012.

[3] P. Kauff, et al. "Depth map creation and image based rendering for advanced 3DTV services providing interoperability and scalability," Image Communication, v. 22, n. 2, pp. 217-234, Feb. 2007.

[4] C. Fehn. "Depth-image-based rendering (DIBR), compression, and transmission for a new approach on 3D-TV," SPIE, Stereo-scopic Displays and Virtual Reality Syst, v. 5291, pp. 93-104, May 2004.

[5] A. Smolic et al. "Intermediate view interpolation based on multiview video plus depth for advanced 3D video systems," IEEE International Conference on Image Processing (ICIP), pp. 2448-2451, 2008.

[6] G. Sanchez et al. "Complexity reduction for 3D-HEVC depth maps intra-frame prediction using simplified edge detector algorithm", International Conference on Image Processing, pp. 3209-3213, 2014.

[7] G. Tech et al. "Overview of the Multiview and 3D extensions of High Efficiency Video Coding," IEEE Transactions on Cicuits and Systems for Video Technology (TCSVT), v.26, n.1, pp. 35-49, Sep. 2015.

[8] G. Sanchez et al. "Real-time scalable hardware architecture for 3D-HEVC bipartition modes," Journal of Real-Time Image Processing, 2016.

[9] L. Zhang et al. "3D HEVC Test Model 6," ISO/IEC JTC1/SC29/WG11, Geneva, Oct. 2013.

[10] Z. Gu et al. "Fast Intra Prediction Mode Selection for Intra Depth Map Coding," ISO/IEC JTC1/SC29/WG11, Vienna, Aug. 2013.

[11] P. Merkle et al. "3D video: Depth coding based on inter-component prediction of block partitions," Picture Coding Symposium (PCS), pp. 149-152, 2012.

[12] T. Vogelsang. "Understanding the Energy Consumption of Dynamic Random Access Memories," IEEE/ACM International Symposium on Microarchitecture (MICRO), pp. 363-374, 2010.

[13] S. Ma et al. "Reducing Wedgelet lookup table size with down-sampling for depth map coding in 3D-HEVC," IEEE International Workshop on Multimedia Signal Processing (MMSP), pp. 1-5, 2015.

A Mutual Rectification-Interference Avoidance Technique with Cascade Filters for Both Downward-Direction Tailed-RDF Deconvolution

Hiroyuki Yamauchi[†]

[†]Faculty of Information Engineering
Fukuoka Institute of Technology
Wajiro-Higashi, Higashi-ku, Fukuoka, Japan,
E-mail: yamauchi@fit.ac.jp[†]

Worawit Somha[††]

[††]Faculty of Industrial Education,
King Mongkut's Institute of Technology
Ladkrabang, Bangkok, Thailand,
E-mail: ksworawi@kmitl.ac.th[††]

Abstract— **This paper proposes a cascade filter design technique to avoid mutual interference in the error rectification process of the Lucy-Richardson (LR) deconvolution. The deconvolution is designed to adequately retrieve the variation factors caused by the Random Dopant Fluctuation (RDF) from the complexly coupled SRAM margin variations. The proposed filter avoids the mutual interferences between the two rectification processes for the both downward-direction-tails of the RDF caused margin distribution. The proposed filter contributes to suppress the relative errors to 7-orders of magnitude smaller than that for the conventional single filter. A 400-fold larger error reduction and a 5-times faster speed of the error reduction are achieved compared with the parallel filter.**

Keywords—Random dopant fluctuation, Lucy-Richardson deconvolution, Cascade filter, Mutual interference

I. INTRODUCTION

The reverse engineering to retrieve the device-history information such as the screening point (SP) from the shipped static random access memory (SRAM) is expected to become more challenging problem as the device size is scaled down to a sub-10-nm range.

This is because: (1) the Random Telegraph Noise (RTN)-caused fail-probability is expected to become larger than that for the Random Dopant Fluctuation (RDF) and (2) the RTN tail distribution is largely deviated from the Gaussian [1-3].

Thus, ordinary Gaussian-based forward/inverse analyses can no longer be used for the reverse engineering, instead we have to heuristically deconvolute the two factors (**f** and **g**) for RDF and RTN from the tail distribution of **h**, namely "$g_{DEC}=h\otimes^{-1}f$ and $f_{DEC}=h\otimes^{-1}g$". Where, **h** is the convolution result ($h=f\otimes g$) of **f** and **g**, and the two symbols ($\otimes$ and $\otimes^{-1}$) denote the convolution and deconvolution, respectively. The g_{DEC} and f_{DEC} represent the deconvolution results of **g** and **f**, respectively.

These analyses are needed for reverse engineering to retrieve the device history such as the screening point (SP), as shown in Fig. 1. And furthermore, we have to solve the "Blind deconvolution (**Blind-deconv**)", *i.e.*, we need to retrieve the information not only for RTN but also RDF simultaneously, given solely the information of **h**, as shown in Fig. 1.

The Lucy-Richardson (LR) deconvolution (**LR-deconv**) is one of the most widely used techniques and has been available

MEXT/JSPS KAKENHI Grant Number of 23560424 and 26420326 and grant from Information Sceience Laboratory of Fukuoka Institute of Technology

to use in the MATLAB® tool box. Recently, some demonstrations of the RTN-**deconv** ($g_{DEC}=h\otimes^{-1}f$) were presented [6,7] but have not been reported yet for the RDF-**deconv** ($f_{DEC}=h\otimes^{-1}g$). This is because the RDF-**deconv** is more difficult than the RTN-**deconv**.

The main reason is due to the need of more complex rectification position assignment in the process of the dual direction down sloped RDF-**deconv**. This increases the possibility of wrong rectification, resulting in an intolerably large deviations from the expected **f** line. As a result, and the SP information cannot be precisely retrieved.

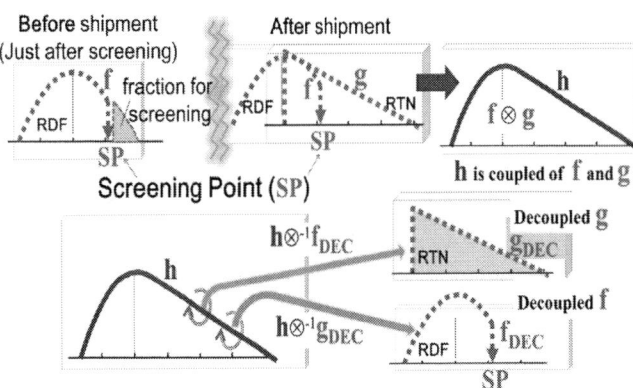

Fig. 1 The relationships among f, g, and h. After screening, g is coupled with f. g and f have to decouple from h.

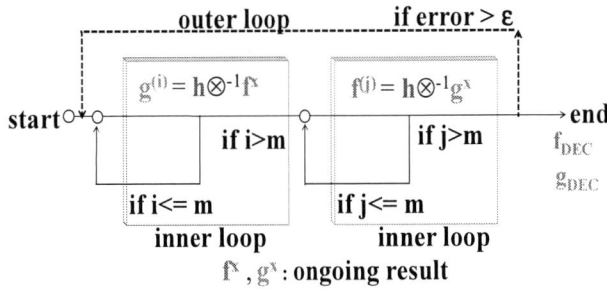

Fig. 2 Outer-loop and dual inner loops for g_{DEC} and f_{DEC} Blind deconvolutions.

978-1-5090-2737-8/16 $31.00 © 2016 IEEE

The biggest obstacle to success of the **Blind-deconv** [7] is strong mutual interference between the two different rectifications in the loops of the operations ($\mathbf{g}^{(i)}$=$\mathbf{h}\otimes^{-1}\mathbf{f}^X$ and $\mathbf{f}^{(i)}$=$\mathbf{h}\otimes^{-1}\mathbf{g}^X$), as shown in Fig. 2. Where, $\mathbf{f}^X$ and $\mathbf{g}^X$ represent the ongoing deconvolution results for $\mathbf{f}_{DEC}$ and $\mathbf{g}_{DEC}$, respectively and the superscripts (i) and (j) of $\mathbf{f}$ and $\mathbf{g}$ represent the number of iterations in each inner loop, respectively.

Thus, the avoidance of the mutual interferences is prerequisite to realize the faster convergence of the two error rectification loops for not only $\mathbf{g}_{DEC}$ but also more troublesome $\mathbf{f}_{DEC}$.

The reason why the $\mathbf{f}_{DEC}$ deconvolution is more difficult than the case for $\mathbf{g}_{DEC}$ will be described. TABLE-I shows the comparison with the previous works [6-7] in terms of the two: (1) the downward directions of the distribution tails of $\mathbf{f}_{DEC}$ and $\mathbf{g}_{DEC}$, and (2) filter structures for alleviating ringing behavior.

For example, the $\mathbf{g}_{DEC}$ can be approximated by only right side downward slopes. On the other hand, the tails of $\mathbf{f}_{DEC}$ consist of both right and left downward slopes, as shown in TABLE-II.

Since the number of downward directions for $\mathbf{f}_{DEC}$ is two, the filters for $\mathbf{f}_{DEC}$ have to be designed for the two for right and left tails, respectively. Where, the mean values of the two filters for the right and left tails are μ_L and μ_R, respectively.

The μ_R for $\mathbf{f}_{DEC}$ takes some positive value, unlike that the μ_L for the $\mathbf{g}_{DEC}$ filter has some positive value.

TABLE-I Comparisons between This Work and Previous Filters

	Deconvoluted -Object	Downward direction of tails	Filter structure
This work	$\mathbf{f}_{DEC}$ for RDF	Both right and left downward- sloped tails	Cascade dual
Previous works Ref. [6-7]	$\mathbf{g}_{DEC}$ for RTN	Only right downward-sloped tail	Single/parallel*

* Parallel filter is used for complex RTN but the polarity of μ_L is negative only

TABLE-II Filter Designs for Single and Dual Downward Sloped Tails

Shape	This Work for RDF		Filter for RTN Ref. [6-7]
	Symmetric dual tails (right and left downward slopes)		Single right downward slope
Concept of filter structure v.s.	Dual filters with zero and plus for right, for left μ_L=0, μ_R		Single filter with only minus for right μ_L, 0
Shape of deconvolution-object	left, right $\mathbf{f}_{DEC}$ RDF		right $\mathbf{g}_{DEC}$ RTN

The most important role of the filter is to align the x-position of the rectification on the tail. Since the right x-position for the rectification depends on the downward-direction of the tail, the variety of the slope direction increases, the filter design becomes more complex.

For example, the two different cases of the x-positions of the error rectification between A-A' and B-B' are compared, as shown in Figs. 3(a) and 3(b). Where, A' and B' corresponds to the x-positions of error rectification for the deconvolution error at A and B on $\mathbf{h}$.

If the x-position difference of A-A' ($=\varphi$) is same as that for B-B' (like the case of $\mathbf{g}_{DEC}$), the simple parallel shifting by φ can be accepted for $\mathbf{g}_{DEC}$ error rectification (Fig. 3(a)). Thus, a single filter can be accepted. Contrary, if the φ_A and φ_B for the right and left tails of $\mathbf{f}_{DEC}$ are largely different, multiple filters

are needed, as shown in TABLE-II. This is caused by the opposite downward direction of the tails.

Thus, the required filter designs for the RDF-**deconv** are completely different from the conventional RTN-**deconv**, as summarized in TABLE-I and TABLE-II. This work is for the first time to demonstrate the **LR-deconv** for the RDF caused $\mathbf{f}_{DEC}$.

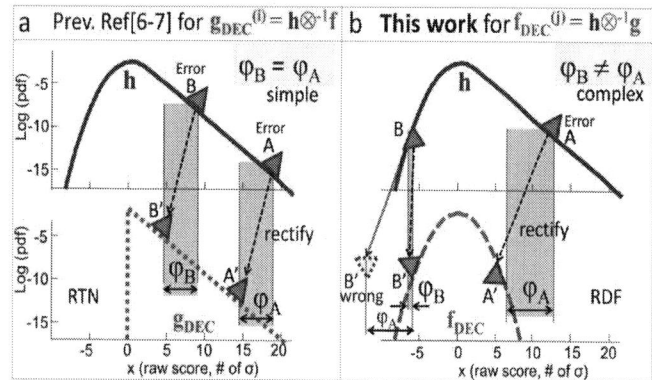

Fig. 3 Comparisons of x-position alignment complexity for error rectification between $\mathbf{g}_{DEC}$ (RTN) and $\mathbf{f}_{DEC}$ (RDF) deconvolution.

The rest of the paper is organized as follows. The filter design for **LR** RDF-**deconv** is proposed in Section III following discussed its backgrounds in Section II. The advantages over the conventional ones are demonstrated in section IV, followed by conclusion in section V.

II. Background of RDF Deconvolution

A. Algorithm for LR-RDF-deconv

The iterations of **LR-deconv** are obeyed to the equation (1).

$$\mathbf{f}_{DEC}^{(i+1)}= \mathbf{f}_{DEC}^{(i)}\times[\mathbf{h}/(\mathbf{f}_{DEC}^{(i)}\otimes\mathbf{g})] \otimes ALG \quad ------- (1)$$

where the superscripts (i) and (i+1) of $\mathbf{f}_{DEC}$ represent the number of iterations. $\mathbf{f}_{DEC}^{(i)}$ is updated to $\mathbf{f}_{DEC}^{(i+1)}$ every iteration cycle depending on the amount of the error determined by the ratio of $\mathbf{h}/(\mathbf{f}_{DEC}^{(i)}\otimes\mathbf{g})$.

ALG denotes the filter for x-position alignment of the error rectification that is designed with the Gaussian distribution. The concept of this error rectification is illustrated in Fig. 4. Note that x-axis in Figs. 3 and 4 shows the raw score, *i.e.*, # of σ for the Gaussian distribution $\mathbf{f}_{DEC}$. This corresponds to the SRAM margins such as the minimum operating voltage (**VCCmin**)[6,7]. The non-Gaussian distributions for both $\mathbf{g}$ and $\mathbf{h}$ are placed on the relative x-position to that of $\mathbf{f}_{DEC}$. Y-axis shows the probability density function (pdf).

B. Errors Caused by Ringing Behaviors

This subsection explains the mechanism of ringing generation and amplification in the iteration cycles of **LR-deconv**.

A certain error of $\mathbf{h}_{DEC}^{(i)}$ at $x=x_O$ caused by the error of $\mathbf{f}_{DEC}^{(i)}$ is assumed at iter#=0 (see Fig. 4(a)).

Based on the equation (1), the updated $\mathbf{f}_{DEC}^{(i+1)}$ is rectified by $\mathbf{h}/(\mathbf{f}_{DEC}^{(i)}\otimes\mathbf{g})\otimes ALG$ so that the error of $\mathbf{h}_{DEC}^{(i+1)}$ at $x=x_O$ can be smaller. However, if the error rectification (A) is done at the

978-1-5090-2737-8/16 $31.00 © 2016 IEEE

wrong x-point (=x_A), additional error (A') is generated at x=x_A, as shown in Fig. 4(b).

Thus, the curve of $f_{DEC}^{(i)}$ are modulated and exhibits an initial symptom of the ringing. As this phenomenon continues in the successive cycles (iter#=2,3,..) the ringing is expanded to (A'+B'+C'+...), as shown in Figs. 4(c) and 4(d).

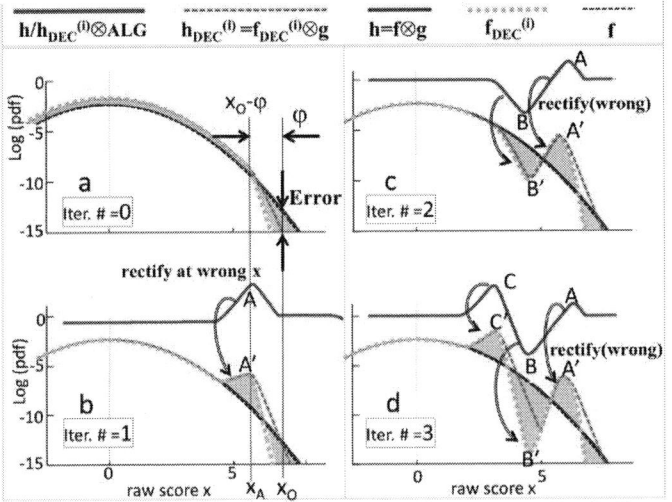

Fig. 4 Ringing amplification due to accumulations of error rectification at wrong x-points during iteration # of 0 to 3.

C. Error Rectification at Right Point

Figures 5(a) and 5(b) show the comparisons of the error rectifications at the wrong and right x-points, respectively.

The curves for $f_{DEC}^{(i-1)}$ (solid-line) and $f_{DEC}^{(i)}$ (dotted-line) denote the deconvolution results before and after the error rectification at the iteration cycle #=1, respectively.

If the x-position is wrong due to an inadequate ALG design, the ringing behavior can be exhibited, as shown in Fig. 5(a). On the other hand, if the x-position of the rectification (=$h/(f_{DEC}^{(i)}\otimes g)\otimes ALG$) can be set on the right position with the adequate filter ALG, no unwanted deviations from expected line can be seen, as shown in Fig. 5(b).

Thus, the important phase alignment filter design technique will be proposed and discussed in the next section.

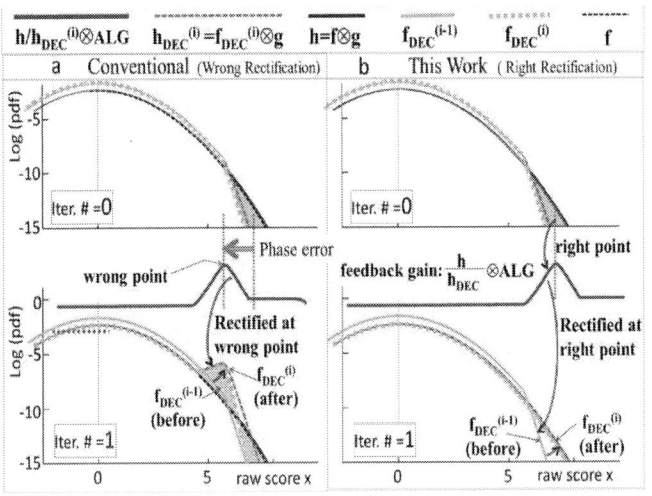

Fig. 5 Comparisons of error rectification at wrong and right positions.

III. PHASE ALIGNMENT FILTER FOR RDF LUCY-RICHARDSON DECONVOLUTION

A. Filters for Both Side Directions Downward Tails

The distribution of the RDF-caused variation (**f**) follows the Gaussian that has both right and left side downward tails. Thus, the complexity of x-point alignment on both tails for the rectification becomes doubled at least, as explained earlier.

In order to elucidate the filter characteristics-sensitivity to the rectification convergence, the two extreme cases of using only ALG_L (μ_L=0) and only ALG_R (μ_R=20) are compared in Figs. 6(a) and 6(b), respectively.

Where, f_{EXP} represents the expected (golden) value for **f**. |f_{DEC} – f_{EXP}| corresponds to the f_{DEC} deconvolution error, as shown in Fig. 6(b).

Where ALG_L and ALG_R denote the filters which consist of single Gaussian with μ_L and μ_R for σ_f=0.05, respectively.

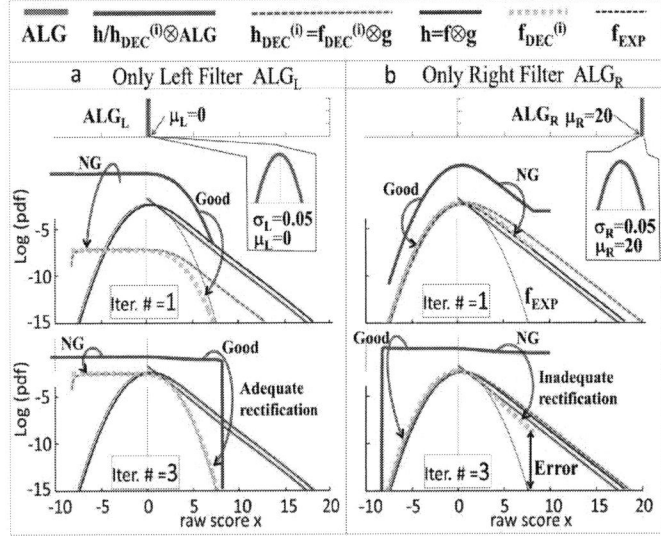

Fig. 6 Comparisons of error rectification between using either ALG_L or ALG_R.

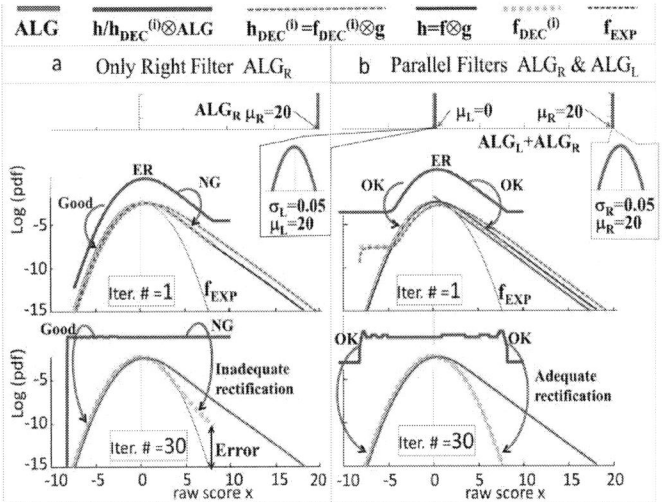

Fig. 7 Comparisons of error rectification between using single ALG_L and both ALG_L and ALG_R in parallel.

Obtained results and remarks are as follows: (1) single use of either ALG_L or ALG_R provides a better result for only right or left tail of $\mathbf{f_{DEC}}^{(i)}$, respectively, while leaving the other side unchanged, as shown in Figs. 6(a) and 6(b), and (2) the error rectification on the right and left tails with ALG_L and ALG_R is almost completed within only 3-iteration cycles.

This indicates that the filter using a combination of ALG_L and ALG_R may work well for both side error rectifications.

To elucidate the effects of using the combination of ALG_L and ALG_R, the comparisons of $\mathbf{f_{DEC}}^{(i)}$ deconvolution result at iteration cycle #=1 and 30 between the two cases : (1) using only single ALG_R and (2) both ALG_L and ALG_R in parallel manner [8] are shown in Figs. 7(a) and 7(b), respectively.

The difference in the shape of the error rectification curve (**ER**) defined by $\mathbf{h}/(\mathbf{f_{DEC}}^{(i)} \otimes \mathbf{g}) \otimes ALG$ (see brown solid line) between the two cases suggests as follows: (1) the filter using only single ALG_R makes the shape of **ER** asymmetric and this provides better fitting on the left tail only, Contrary, (2) the filter using both ALG_L and ALG_R in parallel makes the shape of **ER** symmetric that is similar to the shape of **f** and this contributes to deconvolute the both tails, and (3) the progress of deconvolution of the left tail at iteration cycle #=1 slows down, as can be seen in Figs. 7(a) and 7(b).

The means to remove this detrimental effect will be discussed in the following subsection.

B. Parallel Filter versus Proposed Cascade Filter

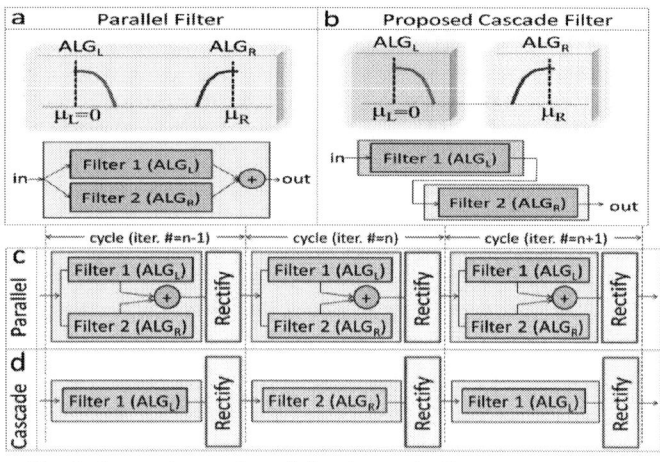

Fig. 8 Comparisons of parallel and cascade filters.

To remove the detrimental effects caused by the parallel filters of ALG_L and ALG_R, the cycle time division multiplexing between the two is proposed in this subsection.

Note that the parallel and cascade filters in this paper are determined as shown in Fig. 8. Both filters consist of ALG_L and ALG_R. Main difference between the two is which type of filter configuration is used in each iteration cycle, as shown in Figs. 8(c) and 8(d). The parallel filter uses both ALG_L and ALG_R in parallel manner for the error rectification.

Thus, there is some overlapping between the distributions of the characteristics of ALG_L and ALG_R.

This can cause some detrimental effects. On the other hand, the cascade filter uses one of either ALG_L or ALG_R alternatively. Thus, no overlapping happens between the two.

This allows the deconvolution on the left and right tails independently without any mutual interference noises. As a result, a slow-down of the rectification progress on the left tail can be avoided unlike the case of the parallel.

It is found that the cascade filter allows a speed-up of the progress of deconvolution of both left and right tails, as shown in Fig. 9(d). Compared with the parallel case (see Fig. 9(b)), much better fitting can be completed within number of iteration cycle #=5, as shown in Fig. 9(d).

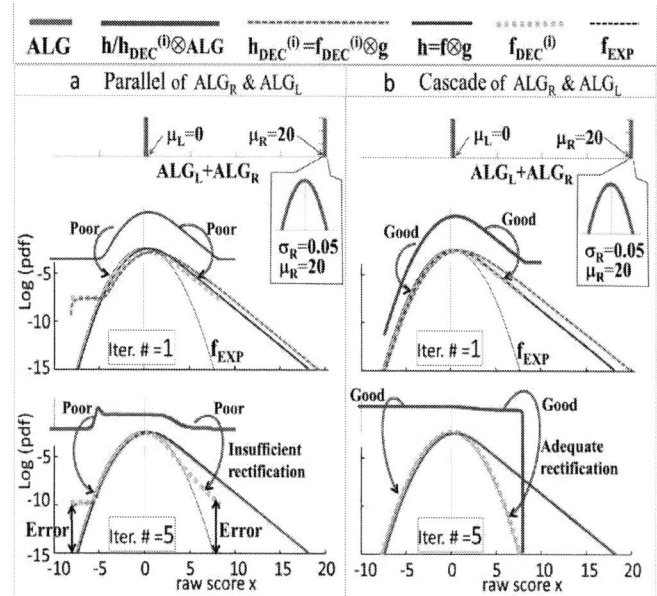

Fig. 9 Comparisons of error rectification with parallel and cascade filters.

C. Gaussian Filter Design for LR-RDF-deconv

We have found that the required amount μ_L and μ_R can be predicted based on the given **h** as follows: 1) the average gradients α_L and α_R of the left and right tails of **h** ($\mathbf{h_L}$ and $\mathbf{h_R}$), respectively, are measured and 2) μ_L and μ_R are picked up from the look up table, as shown in Fig. 10.

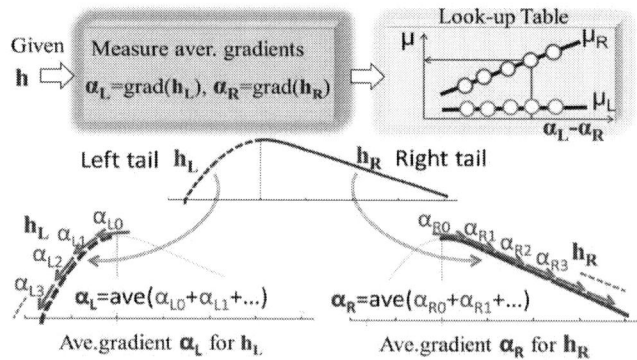

Fig. 10 Look-up table whose input is h and output is μ_R.

The relationships of μ_R and μ_L with ($\alpha_L - \alpha_R$) for the different tails of $\mathbf{h_1}, \mathbf{h_2},$ and $\mathbf{h_3}$ are shown in Fig. 11.

The tail length of $\mathbf{h_i}$ depends on the tail of $\mathbf{g_i}$ because of the relationship of $\mathbf{h_i} = \mathbf{g_i} \otimes \mathbf{f}$ (i=1,2,3). The three different cases for

g_i (g_1, g_2, g_3) are assumed, whose tail length are normalized by f, respectively (See Fig. 11(a)) in this paper. The tail-length depends on the generation of the device scaling, e.g., g_3 corresponds to a sub-10-nanometer region [6].

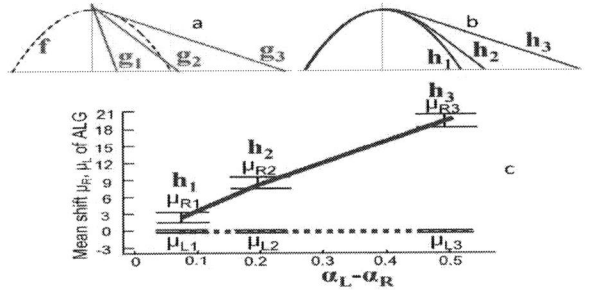

Fig. 11 Relationships of μ_R with (α_L - α_R) for different RTN slopes g_i and its convolution result h_i (i=1,2,3).

The look-up table whose input=h and output=(μ_R, μ_L), can be made based on the relationships shown in Fig. 11.

Allowable ranges for μ_R and μ_L (corresponding to the error bar) are large enough across overall the range of (α_L - α_R). The best sets of (μ_R, μ_L) and (σ_R, σ_L) where the error becomes minimum are found based on the 3-D mesh plots, as shown in Fig. 12. When seeking the best μ_R and μ_L, σ_R=σ_L=0.05 are assumed. Based on these results, μ_L=0 and σ_R=σ_L=0.05 are used for all cases in this paper. Contrary, since μ_R has a strong dependency on the tail length of RTN, μ_R=2.6, 9.0, and 20.0 are used for RTN1, RTN2, and RTN3, respectively.

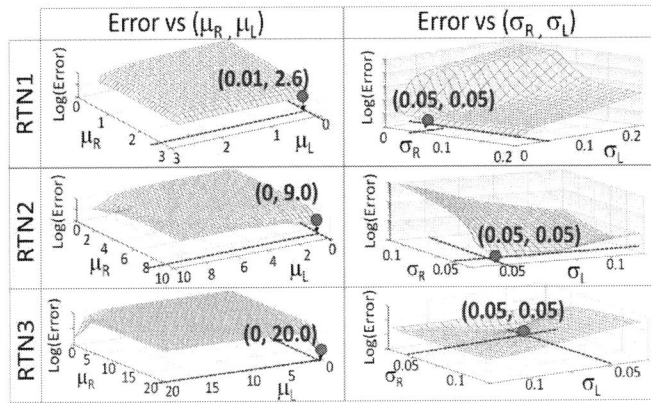

Fig. 12 3-D mesh plots for the sets of (μ_R, μ_L) and (σ_R, σ_L) versus the deconvolution errors.

IV. RDF LUCY-RICHARDSON DECONVOLUTION RESULTS

Since there have been no reports about the RDF-caused f_{DEC}-**LR-deconv**, the filter design dependencies of the deconvolution error are demonstrated first in this section. The filter designs are categorized into the three: (1) Cascade, (2) Single, and (3) Parallel, as shown in TABLE-III.

TABLE-III COMPARISONS OF THREE-TYPES OF FILTER CONFIGURATIONS

	Cascade	Single	Parallel
Filter Configuration	Alternatively switch ALG_L and ALG_R every cycle	Only ALG_R	ALG_L+ALG_R connected in parallel

The error reduction speed characteristics are compared among the three types of filter configurations for the **LR-RDF-deconv** for h_3, as shown in Fig. 13.

Due to the limitation of space, only the case for h=h_3 and g=g_3 are shown (See Fig. 11).

In order to compare: (1) the convergence characteristics of the average error across x-range and (2) the two different errors defined by the average-absolute value of (f_{DEC}-f_{EXP})/f_{EXP} and f_{DEC}/f_{EXP} are used in the Y-axis for Figs. 13 and 14, respectively. Where, f_{DEC} and f_{EXP} denote the RDF caused f deconvolution result and the expected value or the golden value, respectively.

Obtained results and remarks are as follows: (1) the cascade filter provides the best reduction speed. The error defined by (f_{DEC}-f_{EXP})/f_{EXP} is suppressed to less than 1% within 6-iteration cycles. This is 5-times smaller than that for the parallel filter.

The cascade filter can enjoy the benefits from the independence of the error rectification between the right and left downward tails. Contrary, (2) the parallel filter suffers from the distortion of the error rectification characteristics caused by the mutual interference in the error rectification for the right and left tails. In the initial cycles within 12, the error is larger than that for the single filter that can focus on the error rectification on the left side tail.

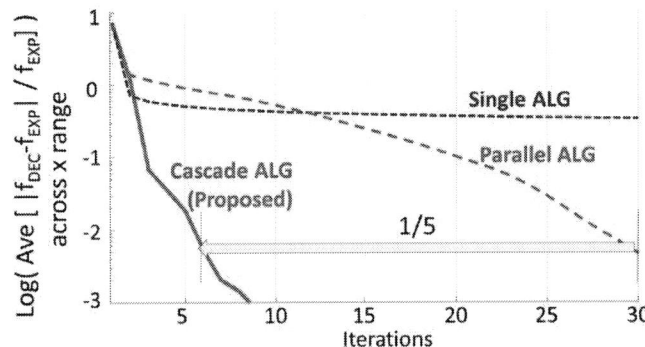

Fig. 13 Reduction speed comparisons of errors defined by the average-absolute value of (f_{DEC}-f_{EXP})/f_{EXP} for h_3.

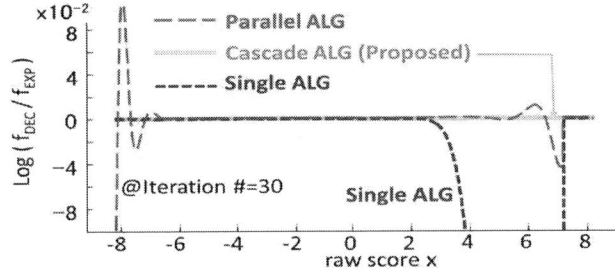

Fig. 14 Comparisons of error variations across x-range for f_{DEC} for h_3 among three types of filter configurations.

Instead of the average values across the x-range that are used in Y-axis of Fig. 13, individual error value at each x-point is used in Fig. 14 to clarify the x-position dependencies. Since the error is defined by log(f_{DEC}/f_{EXP}), the centerline at y=0 corresponds to the case of no error (f_{DEC}=f_{EXP}). The ranges of x<0 and x>0 correspond to the left tail and right tail, respectively.

Obtained results at iteration #=30 and remarks are as follows: (1) errors for the parallel filter largely remain in both edges of the tail due to the ringing behavior, (2) the single filter cannot provide sufficient error rectification to the right tail, and (3) the proposed cascade filter provide the best result in overall x-ranges. The results when the iteration cycles are increased up to 100 are shown in Fig. 15. It is found that the errors for the proposed cascade filter at iteration #=30 and 100 are 400-times and 10^7-times smaller than those for the parallel filter and the single filter, respectively.

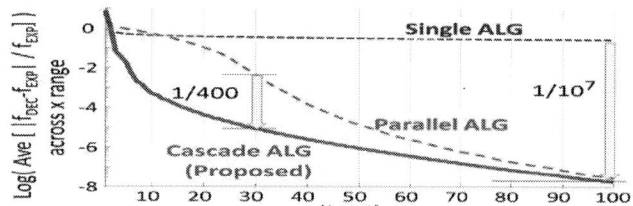

Fig. 15 Error reduction speed comparisons of for h_3 among three types of filter configurations.

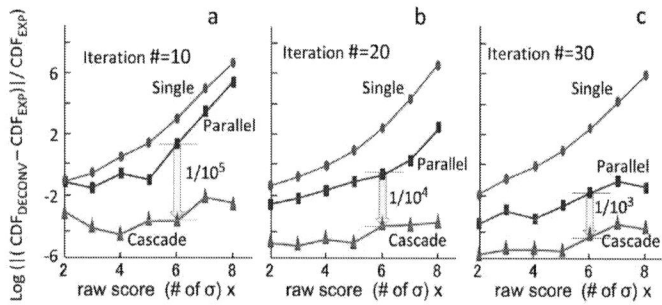

Fig. 16 Comparisons of cdf errors for h_3 among three types of filters at iteration cycles # of (a) 10, (b) 20, and (c)30.

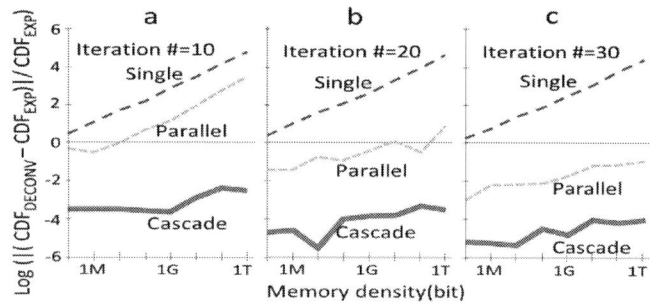

Fig. 17 Comparisons of cdf-error impacts on fail-bit count error for different memory sizes of 1M, 1G, and 1Tbit.

In order to elucidate the error-impacts on the retrieving accuracy of the screening point (SP), the correlation of the deconvolution error with the fail-bit-count (FBC) prediction error will be described.

The cumulative density function (cdf) of f_{DEC} is evaluated. The number of FBC at a certain x-point (i.e., SP point) can be estimated by the cumulative density function (cdf(x)) of f_{DEC}.

For example, if the point of SP is set at x=6, cdf(6)=10^{-9} corresponds to a 1bit fail screening probability for 10^9-bit SRAM, i.e., 1bit screening probability for 1G-bit SRAM [7].

The three lines in Figs. 16(a-c) represent the relative cdf errors when using the filters of single, parallel, and cascade. It is found that the proposed cascade filter can reduce the relative cdf error by 5, 4, and 3-orders of magnitude at x=6 at the iteration cycles #=10, 20, and 30, respectively, compared with the parallel filter, as shown in Figs. 16(a-c).

Since the decision of the screening condition depends on the memory density [7], the relationships of the FBC estimation error versus the memory density are shown in Fig. 17.

Obtained results and remarks are as follows: (1) the errors for single filter are too large to use even for 1Mb memory, (2) the accuracy for the cascade filter is allowable even for 1Tbit. Contrary, the error amounts for the parallel filter at iteration cycle #=20 are not acceptable for the FBC estimations for 1Gbit and 1Tbit memories.

V.   CONCLUSIONS AND FUTURE WORKS

We have successfully demonstrated the LR-RDF-deconvolution for f_{DEC}. The errors for f_{DEC} are suppressed to less than 1% within only 6-iteration cycles. This is 5-times shorter than the case of using parallel filter. The errors at iteration cycle #=30 and 100 for the case of using cascade is 400-times and 10^7-times smaller than those for the parallel filter and the single filter, respectively.

For the next steps, we will combine this time proposed f_{DEC} filter with the previously proposed g_{DEC} filter for the Blind deconvolution [7] and elucidate the effects on the speed and accuracy of the deconvolution.

REFERENCES

[1]   K. Takeuchi, T. Nagumo, and T. Hase, "Comprehensive SRAM Design Methodology for RTN Reliability" , Digest of IEEE Symposium on VLSI Technology, (2011), pp. 130-131

[2]   X. Wang, A.R. Brown, B. Cheng, and A. Asenov, "RTS amplitude distribution in 20nm SOI FinFETs subject to Statistical Variability", SISPAD 2012, (2012) pp.296-299

[3]   N. N.Tega, H.Miki, M.Yamaoka, H.Kume, T.Mine, T.Ishida, Y.Mori, R.Yamada, and K.Torii, "Impact of threshold voltage fluctuation due to random telegraph noise on scaled-down SRAM" , Reliability Physics Symposium, IRPS2008. IEEE International , vol., no., pp.541,546, April 27 2008-May 1 2008

[4]   F.Dell'Acqua1, "A Modified Damped Richardson-Lucy Algorithm to Improve the Estimation of Fiber Orientations in Spherical Deconvolution", Proc. Intl. Soc. Mag. Reson. Med. 16 (2008)

[5]   D. A. Fish, A. M. Brinicombe, and E. R. Pike, "Blind deconvolution by means of the Richardson Lucy algorithm", Journal of the Optical Society of America A 12 (1): pp.58–65

[6]   H.Yamauchi and W.Somha, "Feedback Gain Phase Alignment Effects on Convergence Characteristics in Lucy-Richardson Deconvolution for Inversely Predicting Complex-Shaped RTN Distributions", IEEE 58th MWSCAS 2015, (2015).

[7]   H.Yamauchi and W.Somha, "A Filter Design for Blind Deconvolution to Decouple Unknown RDF/RTN Factors from Complexly Coupled SRAM Margin Variations",   IEEE LASCAS 2016, pp.1-6 (2016).

[8]   H.Yamauchi and W.Somha, "A Parallel Filter Technique to Stabilize Error-Rectification Behavior in RDF Deconvolution Process for SRAM Screening Test", 2016 SAI Computing Conference (SAI) to be presented

A Lightweight Software-based Runtime Temperature Monitoring Model for Multiprocessor Embedded Systems

Guilherme Castilhos, Fernando Gehm Moraes
PUCRS University, Computer Science Department
Porto Alegre, Brazil
guilherme.castilhos@acad.pucrs.br,
fernando.moraes@pucrs.br

Luciano Ost
University of Leicester, Department of Engineering
Leicester, UK
luciano.ost@leicester.ac.uk

Abstract—High-thermal variation and temperature operation can have a noteworthy impact on system performance, power consumption and reliability, which is a major and increasingly critical design metric in emerging multiprocessor embedded systems. Existing thermal management techniques rely on physical sensors to provide them with temperature figures to regulate the system's operating temperature and thermal variation at runtime. However, on-chip thermal sensors present limitations (e.g. extra power and area cost), which may restrict their use in large scale systems. In this regard, this paper proposes a lightweight software-based runtime temperature model, enabling to capture detailed temperature distribution information of multiprocessor systems at a negligible overhead. To validate the proposal, the model is embedded in a distributed-memory MPSoC platform described in RTL. Further, results show that the average absolute error of the temperature estimation, compared to HotSpot is smaller than 4% in systems with up to 36 processing elements.

Keywords—MPSoC; NoC; monitoring; HostSpot; temperature estimation

I. INTRODUCTION

The technology scaling allied with growing processing capability of multiprocessor embedded systems, cause higher on-chip thermal variation and temperature operation. Managing thermal variation and temperature operation is key to accomplish a reliable and efficient system operation [1]. The higher on-chip temperature may lead to overall system performance degradation (e.g. energy efficiency), transient faults (e.g. occurrence of bit-flips) due to timing violations, as well as physical/permeant faults [2][3].

To balance at runtime temperature variation while satisfying power budget constraints of many-core systems, researchers have been considering different techniques like dynamic voltage and frequency scaling (DVFS), task scheduling, mapping, and migration. DVFS reduces high-temperature peaks by lowering the supply voltage and system operating frequency. Effects of application workload allocation on thermal system behavior have been investigated in several works [4][5], showing that load imbalance decisions can generate hotspots zones and consequent thermal implications, which may result in unreliable system operation. While task mapping and migration techniques aim to balance application workloads across multiple processing elements, task scheduling focus on local tasks/threads

assignment optimizations to meet the temperature constraints.

Aforementioned techniques rely on, or assume, the existence of various on-chip temperature sensors to dynamic manage system temperature [6]. Although collecting real on-chip temperature values can be considered ideal, this approach presents several limitations. Physical sensors may be located away from hotspot zones, which impacts on both temperature measurement accuracy and response time [7]. To minimize such limitations, the use of numerous sensors is an option. As reported in [8], to achieve an efficient thermal management the IBM microprocessor combines 44 digital thermal sensors on a chip. Incorporating multiple sensors on the same die may reduce temperature underestimation and response delay, but it incurs extra costs regarding power and area, reducing its use in large scale systems that require detailed temperature distribution information at runtime.

The *goal* of the present work is to propose a lightweight software-based runtime temperature monitoring model, which can be used to collect detailed temperature distribution information of multiprocessor embedded systems without penalizing applications' execution time and system operation.

The main *contributions* of this work are the following:

- proposal of a software-based runtime temperature monitoring model that can be used for temperature management of large scale systems;
- validation of the proposed model by integrating it onto a cycle-accurate SystemC NoC-based MPSoC platform, considering several benchmark scenarios;
- comparison of the proposed model with a well-known temperature tool.

This work is organized as follows. Section II presents related works. Section III describes the proposed temperature evaluation flow. Section IV details the temperature heuristic. Experimental results including model accuracy and cost are given in Section V. Section VI concludes the paper and presents directions for the future works.

II. RELATED WORK

Beyond models and tools, like HotSpot [9], which allows thermal analysis at design time, researchers have been

978-1-5090-2737-8/16 $31.00 © 2016 IEEE

investigating techniques to manage on-chip system temperature and power budget constraints at runtime.

For instance in [4][5], tasks are mapped according to thermal condition of cores, which are collected at runtime by physical sensors. In [10], sensors are also employed to monitor the system temperature and based on collected information task migrations may be triggered, aiming to balance system temperature. A similar task migration scheme is proposed in [11], aiming to reduce hotspots on multi-core systems. Another work that targets multi-core thermal balance is presented in [12]. Wu et al. [13] present a temperature sensor-based DVFS, which allows to fine-tuning the system frequency operation according to its temperature condition. Such techniques rely on monitoring approaches, which provide necessary power/temperature information that is used to invoke a thermal management technique (e.g. task migration) when necessary. Underlying approaches consider the presence of on-chip sensors. Despite increasing chip power and area cost, physical sensors are vulnerable to noise and process variation.

In an attempt to replace or at least reduce the number of on-chip thermal sensors, a software-based approach is described in [14]. This approach extends the HotSpot model and uses physical performance counters available in the Pentium 4 to collect its power activity at runtime. The main limitation of this approach is the performance overhead related to the HotSpot temperature model computation. Results show that the performance overhead of this approach is more than 50% depending on the SPEC2000 benchmark profile.

Different from reviewed works, this paper proposes a purely software-based monitoring model, which has been integrated into a cycle-accurate NoC-based multiprocessor platform. The proposed model enables gathering detailed temperature distribution information of large scale multiprocessor systems with little runtime performance (less than 5%) and power (less than 4%) overheads. Further, due the low memory usage (less than 2 KB), the proposed approach can be easily integrated into other operating systems, which opens up new possibilities for enhancing dynamic temperature management of other multiprocessor platforms.

III. TEMPERATURE ESTIMATION FLOW

This Section describes the two-phase flow used to estimate the MPSoC temperature, at the processing element (PE) level. In this work, each PE includes a processor, a router, and a local memory. Fig. 1 illustrates the design-time and the runtime phases.

The *design time* phase consists in two steps:
(1) Energy calibration ('1' in Fig. 1). This step generates an *Energy Table*, which includes the energy cost of the processor and the NoC routers, for a given technology. Each PE stores this table. Section III.A details this step.
(2) Temperature calibration ('2' in Fig. 1). This step creates a temperature model using the Hotspot tool as the reference. As output, a *Thermal Table* with the thermal influences is generated. Section III.B details this step.

The *runtime phase* also contains two steps:
(1) Periodic power monitoring ('3' in Fig. 1). Each PE monitors its processor and router energy according to a parameterizable monitoring window. Monitored power values are sent to a manager processor (*M* in Figure 1).
(2) Runtime temperature estimation ('4' in Fig. 1). The manager processor receives the power dissipation of each PE and computes at runtime the current temperature of all PEs. Section IV details this step.

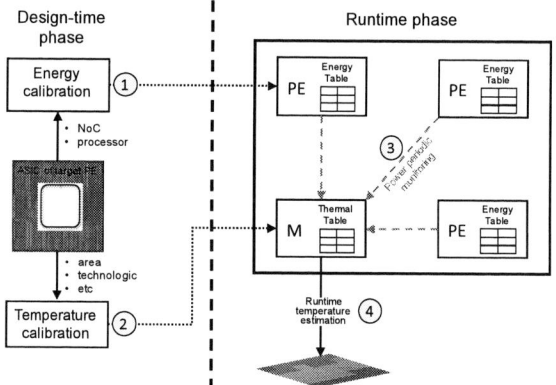

Fig. 1. Temperature Estimation Flow.

A. Processor and NoC Power Estimation

The energy consumption in a NoC-based multiprocessor system is mainly due to three components: processors, NoC (routers and links), and the memory. The current work does not consider the memory energy cost. This paper assumes an MPSoC that adopts scratchpad as local storage memory. In this case, the memory energy consumption per PE is similar, representing an offset of the consumed energy per PE.

As described in the literature [15], the energy consumption (EC) of a processor is defined by its static and dynamic consumption. The processor EC related to the execution of a given task is a function of the number of executed instructions. In our model, the energy cost of each instruction is determined from a gate-level implementation of the processor, as proposed by Rosa et al. [16]. In this model is accounted both static and dynamic consumptions.

Each processor contains an instruction analyzer module, which counts the number of executed instructions for different classes at runtime. Equation 1 computes the processor energy consumption for a given monitoring period.

$$E_{processor} = \sum_{i=0}^{8} energy(c_i) * total\_instructions(c_i) \quad (1)$$

where: $energy(c_i)$, energy to execute a given instruction belonging to the class c_i; $total\_instructions(c_i)$, number of executed instructions belonging to the class c_i.

The NoC EC is proportional to the number of transmitted flits at each router port [17]. A gate level description of the NoC is used to determine the energy consumption of the main router components: buffers, internal crossbar and control logic. Equation 2 gives the energy consumption for a given monitoring period.

$$E_{router} = nb\_flits * E_{buffer} + E_{crossbar} + E_{control\_logic} \quad (2)$$

where: nb_flits corresponds to the number of flits transferred by the router; E_{buffer}, $E_{crossbar}$, and $E_{control\_logic}$ to the energy consumption of the main router components.

978-1-5090-2737-8/16 $31.00 © 2016 IEEE

Each PE monitors the processor and router energy according to a parameterizable *monitoring period*. Equation 3 computes the PE power dissipation (processor and NoC) for a given monitoring period.

$$P_{PE} = \frac{E_{router} + E_{processor}}{monitoring\ period} \qquad (3)$$

B. Temperature Calibration

The temperature calibration uses as the reference the HotSpot tool [9], which provides an accurate thermal model widely used in the computer architecture research community. The worst case error values of HotSpot model for steady-state temperatures and transient temperatures is less than 5% and 7%, respectively.

HotSpot thermal characterization is produced according to a floorplan that represents the target circuit, considering physical properties and power dissipation of components (e.g. processors), modeled as an RC pair, on the die. In our case, HotSpot is used to capture the heat flow into and within the thermal package from each active PE, as well as the lateral heat flow between the PEs.

Using the HotSpot tool, it is applied the maximum power dissipation in a given PE, with a set of surrounding PEs. The goal of executing this simulation is twofold. First, to evaluate how power dissipation affects temperature over time. Secondly, to evaluate how the temperature of the PE under evaluation affects the neighbor PEs.

Fig. 2 illustrates the impact of the *target* PE temperature (i.e. PE under evaluation) on its direct neighbor PEs (*lateral* and *diagonal*), and on PEs far from the target PE (*other neighbors* PEs).

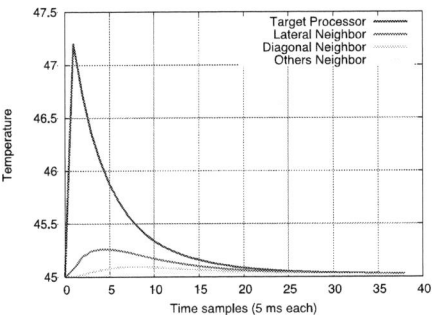

Fig. 2. Effect of temperature (0C) over time in a given PE and its neighbor PEs.

To create a temperature model, a set of assumptions are established to reduce the complexity of the model.
- Assumption 1. It is considered that the thermal influence of a processor affects only its direct neighbors. As Fig. 2 shows, the temperature of a given processor has a small impact in distant PEs.
- Assumption 2. The effect of the temperature decay (*thermal inertia*), as modeled by HotSpot, is too long. This work assumes that at the end of a given period (INERTIA_SIZE) the effect of the applied power ends. In our current model, this period is assumed as 100 ms, corresponding to 20 sampling windows of 5 ms.

- Assumption 3. As the temperature is linear with the power, it is possible to discretize the power values in intervals, assigning a temperature for each power interval.

- Assumption 4. Only integer values are used instead of floating point numbers. This assumption enables faster computation than using floating point numbers.

These assumptions may induce an error on the temperature estimation. The error of the model is evaluated in the Results section.

Using the above assumptions, it is created an LUT-based model. This LUT model comprises 3 *thermal matrices*. Each *thermal matrix* corresponds to the effect of the power over time for PEs in the boundary of the system (*lateral*), in the corners of the system (*diagonal*), or in the center of the system (*central*). Fig. 3 presents the placement of these PEs in a 6x6 system.

diagonal	lateral	lateral	lateral	lateral	diagonal
lateral	central	central	central	central	lateral
lateral	central	central	central	central	lateral
lateral	central	central	central	central	lateral
lateral	central	central	central	central	lateral
diagonal	lateral	lateral	lateral	lateral	diagonal

Fig. 3. Relative position of the PEs for the LUT-based model.

Fig. 4 presents an abstraction of the data structure employed for one *thermal matrix*. Each *thermal matrix* has 3 dimensions:

- *x* coordinate: power index, with the power discretized in 10 ranges according to the maximum power dissipated in the sampling window;

- *y* coordinate: thermal inertia (20 values);

- *z* coordinate: effect of the power on the temperature of the PE under evaluation and its neighbors (3 values). For $z=0$ it is the PE under evaluation, $z=1$ the lateral PEs and $z=2$ the diagonal PEs.

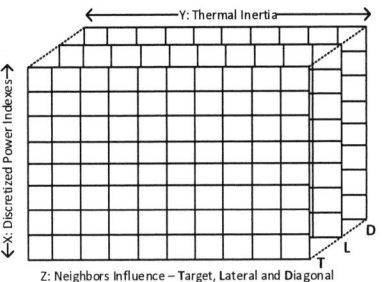

Fig. 4. Example of the data structure adopted for the *thermal matrices*.

Each thermal matrix requires 600 integers (10 * 20 *3), which represents a low memory usage (less than 2 KB) to store the set of tables to estimate the temperature.

IV. TEMPERATURE ESTIMATION HEURISTIC

To compute at runtime the current temperature of all PEs, the heuristic must be as simple as possible, without incurring in inaccuracy. The basis of the temperature computation are the 3 *thermal matrices*, obtained in the offline phase.

The proposed heuristic adopts 3 data structures to compute the temperature of each processing element:

- temperature[|PE|]: vector with the temperature of each PE, all initialized at ambient temperature (45°C).

- power[|PE|]: vector with the average power of each PE, obtained at the end of each monitoring window (5 ms in our current implementation). If a given PE does not send the monitoring packet, the average power in the period is assumed zero (for example, a processor may be in hold state).

- thermal_influence[INERTIA_SIZE][|PE|]: matrix used to store the effect of the temperature over time. It works as a circular FIFO.

To understand as the *thermal_influence* matrix is used, let's assume: INERTIA_SIZE=4, |PE|=1. Also, for one 1 W the effect of the temperature is {10°, -5°, -3°, -2°} and for 0.5 W {6°, -3°, -2°, -1°}. Table I presents a hypothetical example, with at the top of the Table the current temperature (T_{cur}), and the average power obtained from monitoring. Below, it is presented the current state of the *thermal_influence* matrix and the effect of the applied power on the temperature (in fact the value of the current effect corresponds to the addition of both values). The gray cells correspond to the current access position of the *thermal_influence* matrix. After accounting the effect of the power in the temperature, its influence is zeroed for the next measurement (red zeros). The last row is the new temperature considering the effect of the power on the temperature, as well as the effect of the previously sampled power values.

TABLE I. EXAMPLE OF HOW THE MEASURED AVERAGE POWER AFFECTS THE TEMPERATURE USING THE THERMAL_INFLUENCE MATRIX.

T_{cur}	45°		55°		60°		58°		56°	
Power	1W		1W		0.5W		0.5W		1W	
	cur	effect	cur	effect	cur.	effect	cur	effect	cur	effect
thermal_influence_matrix	0	10	0	-2	-2	-2	-4	-3	-7	10
	0	-5	-5	10	0	-1	-1	-2	-3	-5
	0	-3	-3	-5	-8	6	0	-1	-1	-3
	0	-2	-2	-3	-5	-3	-8	6	0	-2
New T	55°		60°		58°		56°		59°	

The first power sample adds 10° in the PE temperature. The second power sample adds 5° in the PE temperature (-5° due to the previous thermal influence plus 10° due to the current power). The third power sample, 0.5W, would add 6° to the PE, but due the thermal influence of the previous power samples, the temperature reduces 2°. The Figure also shows the circular behavior of the *thermal_influence* matrix.

Fig. 6 presents the proposed heuristic to update the current temperature of all PEs. Besides the *temperature* and *power* vectors, the procedure receives the *idx* values, which corresponds to the access position of the *thermal_influence* matrix (gray cells of the example presented in Table I). This heuristic uses 3 functions:

- **get_matrix_type**(PE) (at lines 3 and 7) – according to the position in the system, the PE under evaluation can be diagonal (PEs 1, 3, 7, 9 in Fig. 5), lateral (PEs 2, 4, 6, 8) or central (PE 5). This function returns a pointer to one of the *thermal matrices*.

PE 7	PE 8	PE 9
PE 4	PE 5	PE 6
PE 1	PE 2	PE 3

Fig. 5. PE index to illustrates the concept of diagonal, lateral and central PEs.

- **neighbors** (PE) (line 5) – this function returns a list with the direct neighbors of a given PE. For example (Fig. 5), if the PE under evaluation is 1 the returned set is {2,4,5}, if PE=5 the returned set is {1,2,3,4,6,7,8,9}.

- **get_ng_influence_over_pe**(PE) (line 8) – returns how a given PE (ng) influence the PE under evaluation. In the previous example, if PE=1 and ng=2, the function returns *lateral*. If ng=5, the function returns *diagonal*.

---

void **update_temperature**(*idx, temperature[], power[]*)

    // *idx*: current position in the *thermal_influence* matrix
    // *temperature*[|PE|]: current temperature of all PEs, updated by this procedure
    // *power*[|PE|]: monitored average power of all PEs in the monitoring window

1.   **FOR EACH** PE pe in the system
2.   **FOR** inertia =1 to INERTIA_SIZE
3.   **m* = **get_matrix_type**(pe)   // matrix type (lat/diag/cent)
4.   *thermal_influence(idx, pe)* += m [*power*(pe)] [inertia] [0]
5.   *neighbors_list* ← **neighbors**(pe)
6.   **FOR EACH** PE ng in the *neighbors_list*
7.   **m* = **get_matrix_type**(ng)
8.   pos = **get_ng_influence_over_pe**(pe, ng)   // lat or diag
9.   *thermal_influence (idx, pe)* += m [*power*(ng)] [inertia] [pos]
10.   **END FOR**
11.   *idx* ← (*idx* +1) mod INERTIA_SIZE
12.   **END FOR**
13.   *temperature*(pe) += *thermal_influence(idx, pe)*
14.   *thermal_influence(idx, pe)* ← 0
15. **END FOR**
16. *idx* ← (*idx* +1) mod INERTIA_SIZE

Fig. 6. Heuristic to update the current temperature of all PEs.

The external loop (lines 1 to 15 in Fig. 6) updates the temperature of all PEs. The loop between lines 2 to 12 updates the *thermal_influence* (TI) matrix, as in the example of Table I, now considering the influence of the direct neighbors. Lines 3-4 update the TI matrix considering only the PE under evaluation. Next, lines 5-10 the TI matrix is updated considering the direct neighbors.

Line 11 increments the current index of the TI matrix. Note that the loop 2-12 has, in fact, two indexes: *inertia*, which accounts the effect of the temperature over time; *idx*, used to fill the TI matrix. When the loop 2-12 ends, *idx* returns to its original value. At line 13 the temperature of the PE under evaluation is updated. At line 14 the thermal influence in the current TI matrix is zeroed, as shown in the example of Table I. When all PEs have their temperature updated, the *idx* parameter is incremented (line 16).

This heuristic has a computational complexity O(*n*), where *n* is the number of PEs. In the experiments for an MPSoC running at 100 MHz and 3x3 size, the worst case execution time of the heuristic is 325 µs.

V. RESULTS

The experiments were executed in a public domain NoC-based MPSoC [18], using a clock cycle accurate model described in SystemC. Each PE executes a multi-task operating system (μkernel) and user tasks.

Five benchmarks, described in C language, are used: (*i*) DTW - Digital Time Warping (DTW), with 10 tasks; (*ii*) MPEG decoder, with 5 tasks; (*iii*) DJK - Dijkstra, with 6 tasks; (*iv*) SYN1, synthetic application, with 12 tasks, which emulates the communication behavior of an MPEG4 full decoder; (*v*) SYN2, synthetic application, with 12 tasks, that emulates the communication behavior of VOP (Video Object Plane) decoder application.

Experiments are conducted using the scenarios presented in Table II with three different MPSoC Size: 3x3; 4x4; 6x6. In all MPSoC sizes, one PE is reserved for management purposes (e.g. to execute the proposed heuristic). The experiments limit the size of the system to 36 PEs, since larger systems adopt a cluster-based organization, where the MPSoCs is partitioned in clusters, with one manager per cluster, ensuring scalability [19].

TABLE II. SCENARIOS USED TO EVALUATE THE HEURISTIC TO ESTIMATE THE TEMPERATURE AT RUNTIME.

Scenario	Applications	Number of tasks
1	20 x MPEG, 20 x DJK, 20 x SYN1, 20 x SYN2, 20 x DTW	780
2	10 x MPEG, 10 x DJK, 10 x SYN1, 10 x SYN2, 10 x DTW	390
3	50 x MPEG	250
4	100 x DTW	1000
5	100 x MPEG	500

A. Accuracy of the Proposed Model

This section compares the proposed model w.r.t. HotSpot. Table III summarizes the absolute errors considering the HotSpot as the reference. Two mapping heuristics are used in the evaluation: "*Standard*", whose cost function is the minimization of communication energy in the NoC; "*Energy-Aware*" [20], whose cost function is the workload distribution over time.

TABLE III. ERROR OF THE PROPOSED HEURISTIC W.R.T HOTSPOT.

Scenario / MPSoC size		"Standard" Mapping			Energy-Aware Mapping		
		AVG	STDEV	MAX	AVG	STDEV	MAX
3x3	1	2.33%	2.19%	8.88%	1.34%	1.33%	5.71%
	2	1.98%	2.12%	8.88%	1.14%	1.23%	5.71%
	3	1.46%	1.61%	8.00%	1.08%	1.07%	5.44%
	4	1.21%	0.81%	5.14%	1.13%	1.06%	5.50%
	5	2.42%	2.44%	9.74%	1.44%	1.57%	6.69%
4x4	1	3.52%	3.31%	11.99%	1.87%	1.84%	6.94%
	2	2.53%	2.53%	10.86%	1.38%	1.41%	6.18%
	3	1.92%	1.66%	7.70%	0.87%	0.80%	3.66%
	4	1.79%	1.88%	9.02%	1.14%	0.98%	4.05%
	5	3.13%	3.03%	12.14%	1.43%	1.41%	5.75%
6x6	1	3.18%	3.16%	14.20%	1.91%	1.64%	5.89%
	2	2.17%	2.22%	12.05%	1.05%	0.99%	4.43%
	3	0.95%	0.88%	5.00%	0.52%	0.45%	2.49%
	4	0.57%	0.50%	2.06%	0.68%	0.53%	2.28%
	5	1.66%	1.78%	10.07%	0.94%	0.88%	3.87%

The average absolute error (columns AVG) is below 3.6%, with and standard deviation below 3.2%. The "*Standard*"

mapping creates hotspot zones, inducing a larger temperature in some PEs, revealing the effects induced by the assumptions made in the model. A mapping heuristic that distributes the workload evenly prevents the creation of hotspots, resulting in a smaller average absolute error and standard deviation compared to the "*Standard*" mapping. In most cases, the maximum absolute error (columns MAX) is below 10%. The higher errors are observed with the "*Standard*" mapping due to the hotspot zones. The maximum error, using the "*Energy-Aware*" mapping, is 6.94%.

The proposed software-based model (executed in 0.35 ms@100MHz) enables to estimate with accuracy the temperature of each PE at runtime. A temperature estimation with an error smaller than 10% w.r.t HotSpot enables to manage safely the system temperature at runtime. The temperature is available for each PE (fine-grain data), differently from approaches with thermal sensors, which provide temperature data for the entire system or larger areas.

B. Thermal Maps

Fig. 7 presents the thermal distribution for scenario 4 in a 6x6 MPSoC (similar results are observed for the others scenarios). Each square represents a processor in the MPSoC (x and y-axis), and the time axis represents different moments of the simulation (5%, 25%, 50%, 75% and 100% of execution time). As illustrated in Fig. 7(a), the proposed model produces a similar thermal distribution compared to Hotspot Model (Fig. 7(b)), which validates the proposed model.

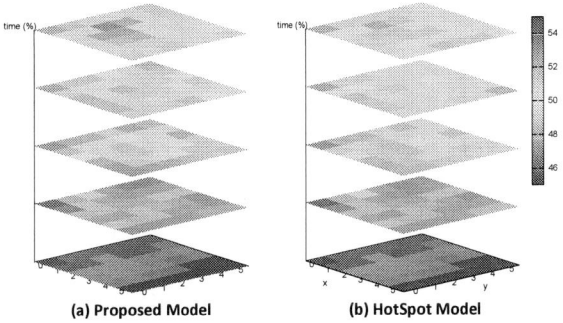

(a) Proposed Model **(b) HotSpot Model**

Fig. 7. 6x6 MPSoC platform temperature distribution considering (a) proposed and (b) HotSpot models.

Fig. 8 compares thermal maps, considering the two mapping heuristics.

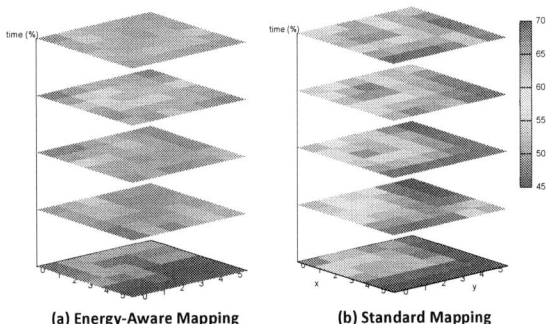

(a) Energy-Aware Mapping **(b) Standard Mapping**

Fig. 8. (a) Energy-Aware Mapping, (b) Standard Mapping.

978-1-5090-2737-8/16 $31.00 © 2016 IEEE

The goal is to present the occurrence of hotspots (Fig. 8(b)). With the presence of hotspots, the influence of hot processors increases, rising the error of the model, as presented in Table III. Even with hotspots the average error is small (<3.6%). Only the processors in the hotspot zones have the estimation error increased (worst case: 14.2%). This thermal comparison aids in the development of heuristics, like mapping, or other power management heuristics (e.g. DVFS).

C. Computational Cost Estimation

This Section evaluates two overheads induced by the inclusion of monitoring and temperature estimation in the system. The graph presented in Fig. 9 evaluates the execution time and energy overheads. All scenarios were executed with and without the proposed method. The total execution time overhead of the proposed model is less than 4.9% in the worst case. As the execution time increases, the energy consumed by MPSoC also increases. Results show that the energy overhead of our model varies from 0.87% to 3.64% depending on the scenario. These results demonstrate the low cost of proposed model.

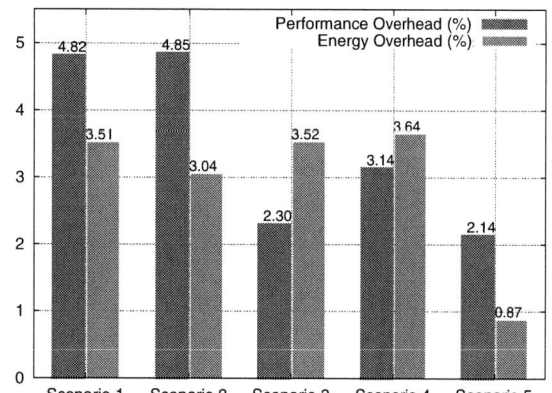

Fig. 9. Overhead of the proposed model (6x6 MPSoC).

VI. CONCLUSIONS AND FUTURE WORKS

This paper proposed a lightweight temperature model for monitoring the system temperature at runtime. The features included in proposed temperature model include scalability, small computation cost, and runtime execution. HotSpot tool is used to generate the system thermal behavior (*design-time phase*), which is used to calculate the system temperature at runtime (*runtime phase*).

The proposed model achieved a low error, with minimal impact on energy consumption (<4%) and runtime performance overhead (<5%). Results show that the average absolute error of the temperature estimation, compared to HotSpot is smaller than 4% in systems with up to 36 processing elements.

Future works include to: (1) integrate of a lifetime model to evaluate MTTF (Mean Time to Failure); (2) include the proposed temperature model to guide mapping heuristics.

VII. ACKNOWLEDGMENTS

The Author Fernando Moraes is supported by CNPq - projects 472126/2013-0 and 302625/2012-7, and FAPERGS - project 2242- 2551/14-8.

REFERENCES

[1] Henkel, J.; Bauer, L.; Dutt, N.; Gupta, P.; Nassif, S.; Shafique, M.; Tahoori, M.; Wehn, N. *"Reliable On-chip Systems in the Nano-era: Lessons Learnt and Future Trends"*. In: DAC, 2013, 10p.

[2] Semenov, O.; Vassighi, A.; Sachdev, M. *"Impact of self-heating effect on long-term reliability and performance degradation in CMOS circuits"*. IEEE Trans. on Device and Materials Reliability, v.6(1), 2006, pp. 17–27.

[3] Hruska, J. *"NVIDIA denies rumors of faulty chips, mass GPU failures"*. http://arstechnica.com/gadgets/2008/07/nvidia-denies-rumors-of-mass-gpu-failures/.

[4] Chantem, T.; Xiang, Y.; Hu, X. S.; Dick, R. P. *"Enhancing multicore reliability through wear compensation in online assignment and scheduling"*. In: DATE, 2013, pp. 1373–1378.

[5] Rudi, A.; Bartolini, A.; Lodi, A.; Benini, L. *"Optimum: Thermal-aware task allocation for heterogeneous many-core devices"*. In: HPCS, 2014, pp. 82–87.

[6] Wanner, L.; Apte, C.; Balani, R.; Gupta, P.; Srivastava, M. *"Hardware Variability-Aware Duty Cycling for Embedded Sensors"*. IEEE Trans. on Very Large Scale Integration Systems, v.21(6), 2013, pp. 1000–1012.

[7] Kong, J.; Chung, S. W.; Skadron, K. *"Recent Thermal Management Techniques for Microprocessors"*. ACM Computing Surveys, v.44(3), 2012, 42 p.

[8] Ware, M.; Rajamani, K.; Floyd, M.; Brock, B.; Rubio, J. C.; Rawson, F.; Carter, J. B. *"Architecting for power management: The IBM POWER7 approach"*. In: HPCA, 2010, 11 p.

[9] Huang, W.; Ghosh, S.; Velusamy, S.; Sankaranarayanan, K.; Skadron, K.; Stan, M. R. *"HotSpot: a compact thermal modeling methodology for early-stage VLSI design"*. IEEE Trans. on Very Large Scale Integration Systems, v.14(5) , 2006, pp. 501–513.

[10] Ge, Y.; Malani, P.; Qiu, Q. *"Distributed task migration for thermal management in many-core systems"*. In: DAC, 2010, pp. 579–584.

[11] Liu, Z.; Tan, S. X. D.; Huang, X.; Wang, H. *"Task Migrations for Distributed Thermal Management Considering Transient Effects"*. IEEE Trans. on Very Large Scale Integration Systems, v.23(2), 2015, pp. 397–401.

[12] Kursun, E.; Cher, C.-Y. *"Variation-aware thermal characterization and management of multi-core architectures"*. In: ICCD, 2008, pp. 280–285.

[13] Wu, Y. K.; Sharifi, S.; Rosing, T. S. *"Distributed thermal management for embedded heterogeneous MPSoCs with dedicated hardware accelerators"*. In: ICCD, 2011, pp. 183–189.

[14] Lee, K. J.; Skadron, K. *"Using performance counters for runtime temperature sensing in high-performance processors"*. In: IPDPS, 2005, pp. 8 -13.

[15] Jejurikar, R.; Pereira, C.; Gupta, R. *"Leakage aware dynamic voltage scaling for real-time embedded systems"*. In: DAC, 2004, pp. 275-280.

[16] Rosa, F.; Ost, L.; Raupp, T.; Moraes, F; Reis, R. *"Fast energy evaluation of embedded applications for many-core systems"*. In: PATMOS, 2014, 6 p.

[17] Martins, A.; Silva, D.; Castilhos, G.; Monteiro, T.; Moraes, F. *"A method for NoC-based MPSoC energy consumption estimation"*. In: ICECS, 2014, pp. 427-430.

[18] Carara, E.; Oliveira, R.; Calazans, N.; Moraes, F. *"HeMPS - a Framework for NoC-based MPSoC Generation"*. In: ISCAS, 2009, pp. 1345-1348.

[19] Castilhos, G.; Mandelli, M.; Madalozzo, G.; Moraes, F. *"Distributed Resource Management in NoC-Based MPSoCs with Dynamic Cluster Sizes"*. In: ISVLSI, 2013, pp. 153-158.

[20] Castilhos, G.; Mandelli, M.; Ost, L.; Moraes, F. *"Hierarchical Energy Monitoring for Task Mapping in Many-core Systems"*. Journal of Systems Architecture, v.63, 2016, pp. 80–92.

Evaluating the Impact of Circuit Legalization on Incremental Optimization Techniques

Renan Netto*, Vinicius Livramento†, Chrystian Guth*,
Luiz C.V. dos Santos* and José Luís Güntzel*

*Dept. of Computer Science, Federal University of Santa Catarina, Brazil
†Dept. of Automation and Systems Engineering, Federal University of Santa Catarina, Brazil
{renan.netto, vinicius.livramento, chrystian.guth}@posgrad.ufsc.br
{luiz.santos, j.guntzel}@ufsc.br

Abstract—During physical synthesis, global placement and incremental optimization steps such as gate sizing, buffer insertion and timing-driven placement, produce placements where cells are overlapped or misaligned with respect to sites and rows predefined in the used standard cell library. Therefore, a legalization procedure must be used to keep the placement legality. In the case of incremental optimization techniques, the legalization step can be applied as a final step, after each optimization iteration or even after each primitive placement transformation. Although different legalization strategies are used by different works on incremental optimization techniques found in the literature, they do not present practical results on how those different strategies impact on the runtime and quality of the optimized placements. This work investigates such issue by comparing the runtime, timing degradation, density profile and wirelength resulted from using three legalization strategies on an incremental timing driven placement technique. Experimental results show that incremental legalization is from 22% to 71% faster than a single final legalization step and achieves similar density and wirelength profiles, but may increase timing violations up to 48% (9% on average). On the other hand, legalizing the circuit after each optimization iteration is the least interesting strategy, since it leads to similar quality results than a single final legalization, at the cost of a runtime about 3.2 times greater on average. Therefore, incremental legalization may be an interesting strategy for incremental optimization techniques due to its shorter runtime. Finally, we point out alternatives to improve the quality of incremental legalization techniques.

I. INTRODUCTION AND RELATED WORK

In the physical synthesis flow the global placement step is responsible for finding locations to circuit standard cells while optimizing an objective function, which is typically the wirelength [1]. In order to place complex circuits that may contain up to a few millions of cells within acceptable runtime, global placement disregards cells' dimensions to find an initial rough solution that does not necessarily respect the valid locations. Therefore, a legalization step must be applied to align all cells to predefined sites and rows as well as to remove overlaps between cells [2].

The legalized initial placement solution produced after global placement must undergo a number of optimization steps so as to make possible the timing closure of the design. This is done by using different incremental techniques such as buffer insertion, gate sizing and incremental timing-driven placement, to optimize the placement considering metrics

such as wirelength, density and timing [2]. Such optimization techniques receive as input a legal solution and are expected to keep placement legality by applying one of the following legalization strategies: 1) final legalization, 2) iterative legalization, and 3) incremental legalization.

The first strategy consists of relocating all cells occupying illegal locations in a single step in the end of the optimization process. It is used by the optimization techniques reported in [3][4]. It is also the strategy employed to legalize the initial (illegal) global placement solution. The iterative legalization, by its turn, relocates all cells that occupy illegal locations after every optimization iteration, being applied by the techniques described in [5][6][7]. As long as in both previous strategies a large number of cells are to be relocated, the solutions legalize the entire circuit by using one of the following methods: greedy heuristics [8], dynamic programming formulation [9][10], diffusion-based techniques [11] or by solving a minimum cost maximum flow problem [12][13].

Incremental legalization relocates cells after every primitive placement transformation (e.g. movement of a cell) that takes place during incremental optimization. Such strategy is employed by a few optimization techniques [14][15][16][17]. Typically, a placement transformation requires a single cell relocation and its legalization. However, if the target location is already occupied, other cells must be moved and those movements must be also legalized. Since those operations are repeated many times during optimization, the legalization must be fast enough not to compromise the whole runtime. Incremental legalization may be accomplished by either adapting an existing algorithm that legalizes the whole circuit so as to legalize only a small portion of it [10] [16] or by using a specialized technique [17]. In particular, all incremental legalization techniques found in the literature limit the legalization region to only a single circuit row to speed up the runtime.

Although several works report optimization results considering the legalization strategies described in the previous paragraphs, so far no one provides a comparison between different legalization strategies applied to the same incremental optimization technique. This work investigates such issue by comparing the three legalization strategies in the frame of an incremental timing-driven placement technique. Therefore, its main contribution is to provide, for the first time, a quantitative comparison between legalization strategies considering total optimization runtime and solution quality, the latter measured in terms of timing violations, density profile and circuit wirelength.

The experimental evaluation, performed using industrial circuits provided by the ICCAD 2015 Incremental Timing-Driven Placement Contest Infrastructure [18], shows that the incremental legalization strategy is faster than a final legalization and achieves similar density and wirelength profiles, but results in more timing violations. The iterative strategy, by its turn, leads to a higher runtime, which is not compensated by its solution quality, as the timing violations, density and wirelength results are similar to running a final legalization step.

The remaining of this paper is organized as follows. Section II presents the problem formulation for the legalization problem and Section III describes the legalization algorithms evaluated in this work. Section IV brings the experimental results and comparisons between the three legalization strategies. Finally, Section V draws the conclusions.

II. PROBLEM FORMULATION

A circuit can be represented by a set of cells $\mathcal{C} = \{c_1, c_2, ..., c_n\}$ where each cell $c_i \in \mathcal{C}$ occupies a location (x_i, y_i) and has dimensions (w_i, h_i). A placement is legal if it satisfies the legality constraints defined by Equations (1) to (5). The constraints in Equations (1) and (2) bound the cells to within chip boundaries, represented by $X_{left}, X_{right}, Y_{bottom}, Y_{top}$. Equations (3) and (4) align them to circuit sites and rows, whose dimensions are given by W_{site} and H_{row}. In addition, each pair of cells in the same row must satisfy Equation 5, which prevents overlaps between them.

$$X_{left} \leq x_i \leq X_{right} - w_i \qquad (1)$$

$$Y_{bottom} \leq y_i \leq Y_{top} - h_i \qquad (2)$$

$$x_i = n \times W_{site}, n \in \mathbb{N} \qquad (3)$$

$$y_i = m \times H_{row}, m \in \mathbb{N} \qquad (4)$$

$$y_j = y_k \Rightarrow x_j > x_k + w_k \vee x_j + w_j < x_k \qquad (5)$$

Given an illegal placement, the legalization problem consists of moving a subset of cells $\mathcal{C}' \subseteq \mathcal{C}$ to satisfy the legality constraints while minimizing the cost function in Equation (6). This cost function measures the placement perturbation as the Manhattan distance between the cell locations before and after the legalization, represented by (x_i, y_i) and (x_i', y_i'), respectively.

$$\sum_{c_i \in \mathcal{C}} |x_i - x_i'| + |y_i - y_i'| \qquad (6)$$

Both the iterative and the final legalization strategies have to operate on a placement that may contain many cell misalignments and overlaps that are spread in the circuit area. Therefore, these strategies legalize the entire circuit by considering the constraints in Equations (1) to (5), while minimizing the cost function from Equation (6).

The incremental legalization strategy differs from the other two strategies in the sense that it receives as input a legal placement and a placement transformation to be performed, such as cell replacement (in the case of cell sizing), buffer insertion or cell relocation. Then, the problem consists in performing the placement transformation while keeping the placement legality given by Equations (1) to (5) and also minimizing the cost function from Equation (6). Although every placement transformation requests a single cell to be relocated and legalized, such operation may probably require that other cells are also relocated and legalized to make room for the incoming cell. In this way, a subset of cells $\mathcal{C}''$ will be moved during incremental legalization. Nevertheless, it is expected that $|\mathcal{C}''| \ll |\mathcal{C}'|$ since only a single placement transformation is performed.

III. LEGALIZATION ALGORITHMS

To evaluate the legalization strategies, we implemented two state-of-the-art legalization algorithms: 1) an algorithm that legalizes the whole circuit, allowing it to be used for the final and iterative legalization strategies and 2) an incremental algorithm, which can be used for the incremental strategy. We implemented the Abacus algorithm from [9] to legalize the whole circuit and its incremental version, proposed by [16]. With them we can fairly compare the different strategies, since both use the same dynamic programming formulation to minimize cell displacement during legalization. This section describes them showing their main differences.

Both algorithms described in this section assume that the circuit is composed of a set of rows $\mathcal{R}$, where each row $r_j \in \mathcal{R}$ has a location (x_j, y_j) and dimensions (w_j, h_j). In addition, we represent the cells located in a given row r_j by its attribute $C(r_j)$. Given the set $C(r_j)$ of a row, its capacity $cap(r_j)$ is calculated by Equation (7) as the row space that is not occupied by cells.

$$cap(r_j) = w_j - \sum_{c_i \in C(r_j)} w_i \qquad (7)$$

A. Abacus

Given an illegal placement, the Abacus algorithm legalizes it by moving all circuit cells ordered by their x coordinates. For each cell, it finds its optimal location by evaluating the cost of placing the cell in each circuit row. The cost is calculated as the cell displacement from its location in the illegal placement to its location after legalization. Abacus uses a dynamic programming formulation to find the optimal location of all cells in a given row while minimizing the cell displacement.

Algorithm 1 shows the steps of the Abacus legalization algorithm. First it sorts the cells according to their x coordinates (line 1). Then it iterates through all sorted cells (lines 2-16) finding the optimal location of each cell. Observe that the algorithm evaluates a row $r_j \in \mathcal{R}$ only if its capacity is greater than the cell width. In this case, the optimal locations of all cells in that row are found using the Place Row function,

978-1-5090-2737-8/16 $31.00 © 2016 IEEE

Algorithm 1: Abacus

Input : Illegal placement and set of rows $\mathcal{R}$
Output: Legal placement

1 $C_{sorted} \leftarrow$ cells sorted according to x coordinate;
2 **foreach** $c_i \in C_{sorted}$ **do**
3    $cost_{best} \leftarrow \infty$;
4   **foreach** $r_j \in \mathcal{R}$ **do**
5   **if** $cap(j) \geq w_i$ **then**
6        $C(r_j) \leftarrow C(r_j) \cup \{c_i\}$;
7   Place Row(r_j);
8        $cost \leftarrow |x_i - x_i'| + |y_i - y_i'|$;
9   **if** $cost \leq cost_{best}$ **then**
10          $cost_{best} \leftarrow cost$;
11          $r_{best} \leftarrow r_j$;
12        $C(r_j) \leftarrow C(r_j) - \{c_i\}$;
13   **end**
14    $C(r_{best}) \leftarrow C(r_{best}) \cup \{c_i\}$;
15   Place Row(r_{best});
16 **end**

Algorithm 2: Incremental Abacus

Input : Set of rows $\mathcal{R}$ and target location (x_i^t, y_i^t) of cell c_i
Output: $true$, if it is possible to legalize, $false$ otherwise

1 $r_{curr} \leftarrow r_j$ such that $c_i \in C(r_j)$;
2 $C(r_{curr}) \leftarrow C(r_{curr}) - c_i$;
3 $r_{target} \leftarrow r_j$ such that $y_j = y_i^t \wedge (x_j \leq x_i^t \leq x_j + w_j)$;
4 **if** $cap(r_{target}) \geq w_i$ **then**
5    $C(r_{target}) \leftarrow C(r_{target}) \cup \{c_i\}$;
6   Place Row(r_j);
7   **return** $true$;
8 **else**
9    $C(r_{curr}) \leftarrow C(r_{curr}) \cup c_i$;
10   **return** $false$;
11 **end**

---

which is described in [9]. It is also important to notice that, when evaluating the cost of each row, the cells are not actually moved in the circuit, the algorithm only finds their optimal locations. After finding the best row, the algorithm places the cell in the best row (lines 14-15), actually performing the movements.

The main advantage of Abacus is that the Place Row function finds the optimal locations for all the cells in a given row, and not only for the cell c_i that was just placed. Although such approach results in longer runtime, it can globally minimize the cell displacement in that row, achieving better results.

B. Incremental Abacus

Instead of moving all cells in order to legalize the whole circuit, the incremental version of Abacus updates the cell locations in a single circuit row in order to perform a single placement transformation. The optimal location of the cells in that row is found by using the same Place Row method from [9]. By legalizing a single row, the incremental algorithm reduces the number of calls to the Place Row function, resulting in a shorter runtime.

Algorithm 2 shows the pseudocode of the incremental version of Abacus. The algorithm receives as input the target location (x_i^t, y_i^t) of a cell c_i resulting from a placement transformation and tries to relocate c_i while keeping the placement legality. If it is not possible to do so, c_i remains in its original location. First, the cell is removed from its current row (lines 1-2). Then the target row where the cell should be placed is found (line 3). If the capacity of row r_{target} is greater than the width of c_i, the cell is added to that row and Place Row is called to find the optimal locations of all cells in r_{target}. Otherwise, c_i is restored to its original row and it is not moved.

Since the incremental version of Abacus does not evaluate all circuit rows in order to find the optimal location of cell c_i, it might not be possible to perform the placement transformation and legalize the circuit (for example, if there is not enough room in r_{target} to place c_i). In this case, the legalization may fail and the cell is kept in its original location.

IV. EXPERIMENTAL RESULTS

We generate experimental results for the 8 circuits available from the ICCAD 2015 Contest (problem C: Incremental Timing-Driven Placement) [18]. Those circuits were derived from industrial designs having from 768k to 1.93M cells. For each circuit, we used the initial placement solution provided by the ICCAD 2015 Contest. We implemented the Abacus algorithm and its incremental version using C++ based on the pseudocodes provided in [9] and [16], respectively. We performed all experiments in a Linux workstation with four Intel®Core®i5-4460 CPUs @ 3.20 GHz and 32GB RAM.

In order to evaluate the three legalization strategies within a real circuit optimization flow we integrated them in the incremental timing-driven placement technique from [7], giving rise to three versions of optimization technique: 1) a version that uses Abacus to legalize the entire circuit only in the end of the optimization (referred to as *final legalization*, or simply FIN) ; 2) an *iterative legalization* version that uses Abacus to legalize the entire circuit after each optimization iteration (named ITE, for shortly); 3) an *incremental legalization* version that employs the incremental Abacus algorithm to legalize the circuit after each cell movement (referred to as INC).

The incremental timing-driven placement technique from [7] receives as input the number of primary outputs (POs) with negative slack to be optimized, where a PO may be either a circuit output pad or a flip flop input pin. Starting from a given PO, the algorithm traverses the critical path that reaches that PO towards a primary input (PI) while optimizing the cells with negative slack. In such algorithm the number of cells to be moved is directly correlated with the number os

POs that are selected for optimization. Therefore, to evaluate the behavior of the legalization strategies in function of the number of cells to be moved we varied the number of POs to be optimized within the range of 200 to 2000. In addition, we defined five as the number of iterations for the incremental timing-driven placement technique, as proposed in [7].

To evaluate the runtime, we measured the required time to execute the whole incremental timing-driven placement technique, so as to encompass the runtimes of both optimization and legalization steps. We evaluate the quality of the solution obtained by each legalization strategy through the timing violations, density and wirelength of the resulting placements. The results obtained for ITE and INC strategies, in terms runtime and quality evaluation metrics, were normalized using as baseline the results obtained for FIN strategy.

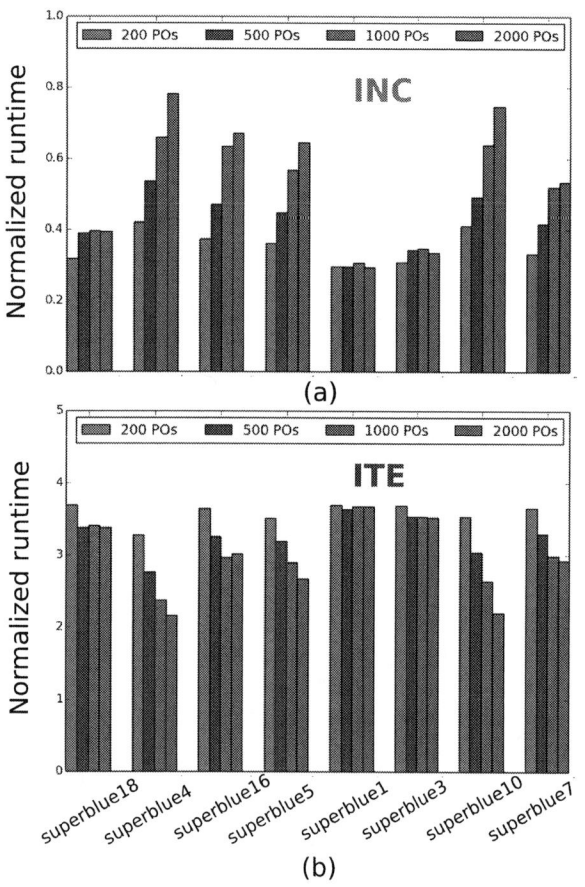

Fig. 1. Normalized runtime for each benchmark w.r.t FIN. (a) Results obtained by INC. (b) Results obtained by ITE.

Figures 1 (a) and (b) show the normalized runtime results for INC and ITE. INC achieves the shortest runtimes for all circuits, being from 22% to 71% faster than FIN. On the other hand, ITE is from 2.16 to 3.69 times slower than FIN, since it has to legalize the whole circuit five times. Observe that the legalization runtime of INC depends on the number of moved cells, while this is not the case for ITE and FIN, which always legalize the whole circuit. As a consequence, the normalized runtime of INC increases with the number

of optimized POs, since more cells tend to be moved and, consequently, legalized. The normalized runtime of ITE, by its turn, decreases as the number of optimized POs increase, because the legalization runtime remains constant while the optimization runtime increases. However, for the circuits superblue18, superblue1 and superblue3 the normalized runtime remains approximately constant when increasing the number of POs. Since those circuits do not have enough POs with negative slack, increasing the number of optimized POs does not raise the number of moved cells.

To evaluate the timing violations in the end of the optimization, we measured the Total Negative Slack (TNS), calculated as the sum of the slacks in all POs with negative slack. Figures 2 (a) and (b) show the normalized TNS results for INC and ITE. INC is the strategy that leads to greater TNS, resulting in 9% more timing violations than FIN, on average. The reason why INC resulted in more timing violations can be explained with the help of the example illustrated in Figure 3.

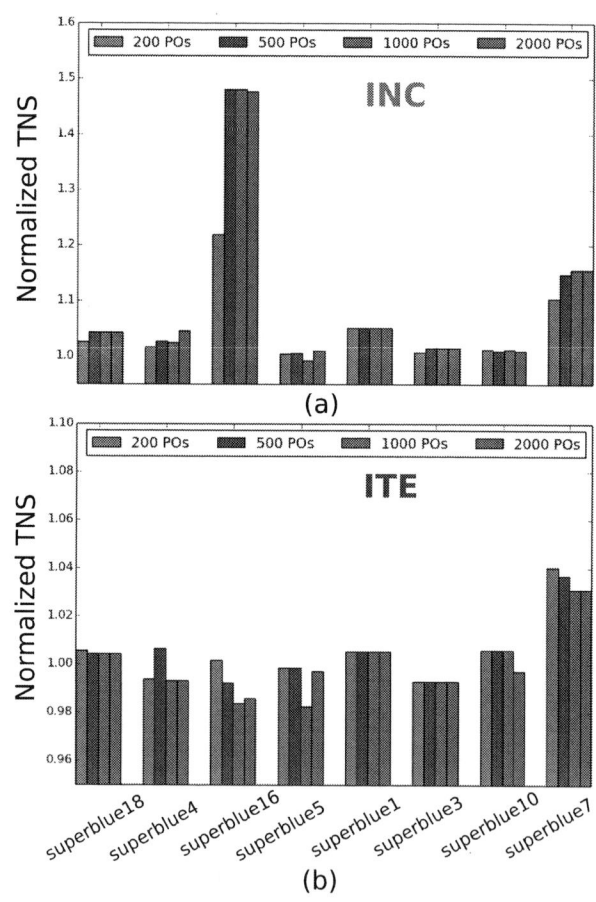

Fig. 2. Normalized TNS for each benchmark w.r.t FIN. (a) Results obtained by INC. (b) Results obtained by ITE.

In Figure 3 (a), cell c_1 is to be moved to the location showed by the dashed rectangle, which lies over a fixed macro block (big black rectangle) in a crowded region. Since the row containing the target location is crowded, INC is not able to legalize this movement and hence, c_1 stays in its original location, as shown in Figure 3 (b). On the other hand, since FIN legalizes the circuit only when all movements have been performed, other cells in the same row might have been moved, allowing c_1 to be placed close to its target location, as shown by Figure 3 (c). The larger displacement achieved by INC results in c_1 being placed far from its target location as found by the incremental timing-driven placement algorithm, leading to more timing violations. In particular, circuit superblue16, which resulted in the largest normalized TNS, has several cells with negative slack lying on crowded regions with many macro blocks, worsening the results obtained from INC.

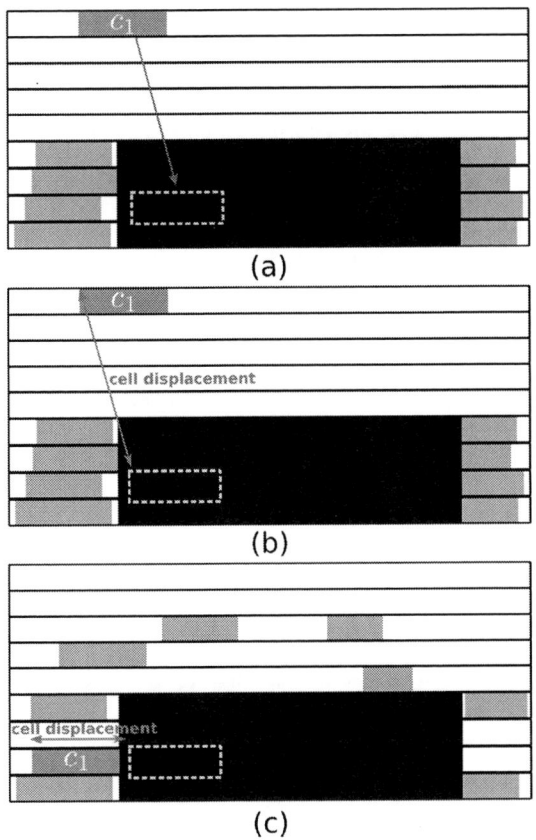

Fig. 3. Example of cell displacement using two different legalization strategies. Cell c_1 (orange rectangle) has a target location (dashed green rectangle) lying over a fixed macro block (big black rectangle). (a) Legalization resulted from using INC. (b) Legalization resulted from using FIN. The red arrow shows the cell displacement.

Since the ITE strategy uses the same algorithm as FIN to legalize the circuit, it shows no significant difference on timing violations as compared to FIN, with the normalized TNS ranging from a 2% decrease to a 4% increase. In addition, for both INC and ITE, the normalized TNS remains similar when increasing the number of POs for most circuits. Such results show that the number of timing violations does not depend on the number of moved cells, but on circuit characteristics.

To measure the density profile, we used the ABU metric as stated in [18]. This metric divides the circuit area in bins and measure the density as the average utilization of overflowing bins. Figures 4 (a) and (b) show the normalized ABU results

for INC and ITE. Observe that there is no significant difference among the results obtained from the three strategies in most of the circuits. Considering INC, circuit superblue5 resulted in the largest ABU difference (13% reduction), while the other circuits resulted in a difference of up to only 6%. On the other hand, for ITE the largest ABU reduction is of 5% for superblue16 with 1000 POs, and up to 3% for all other circuits. In addition, the normalized ABU remains similar when changing the number of POs for 5 out of the 8 circuits. Such results show that the density results are neither highly affected by the adopted legalization strategy used nor by the number of moved cells.

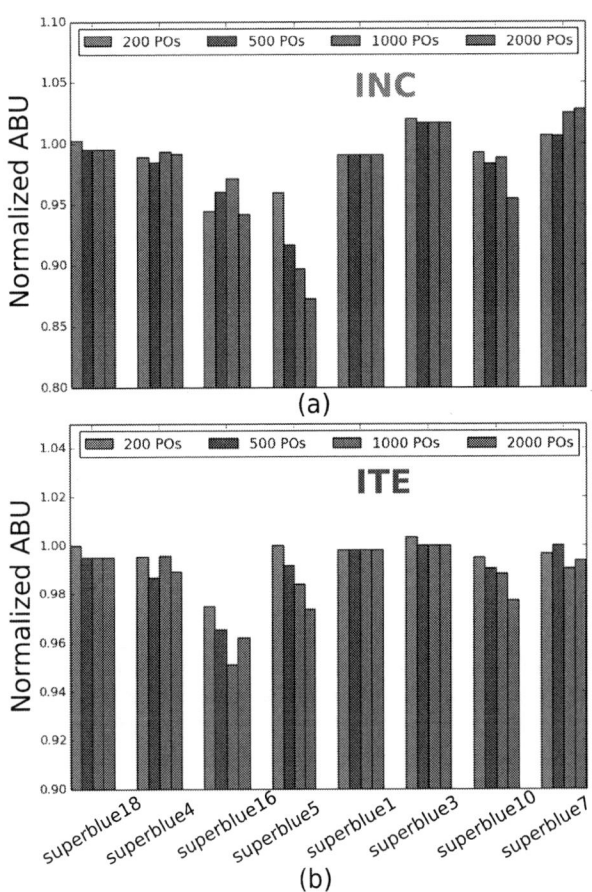

Fig. 4. Normalized ABU for each benchmark w.r.t FIN. (a) Results obtained by INC. (b) Results obtained by ITE.

We used the FLUTE algorithm [19] to assess the circuit interconnections and thus to obtain the wirelengths. We observed that the difference between all the three strategies is less than 1% for all circuits. This happens because only a small number of cells are moved during the incremental optimization. To show such behavior, Figure 5 shows the average percentage of cells that are moved by each of the three versions of optimization technique (each one with a given legalization strategy) when the number of POs to be optimized is varied within the considered range. Those percentages include all cells that are moved, including the ones during legalization. Notice that the number of moved cells is less than 3% of

978-1-5090-2737-8/16 $31.00 © 2016 IEEE

the total number of cells in the circuit even when 2000 POs are considered for optimization. Such results show why no significant difference on wirelength was observed, and similar results are expected for other incremental techniques, since such techniques are designed to move only a small subset of cells. The small number of moved cells also show why the runtime of INC is always shorter than the runtimes of FIN and ITE. Since less than 3% of the cells are moved, the overhead of legalizing the whole circuit, which is the case of FIN and ITE, is much larger than running a small legalization for each cell movement.

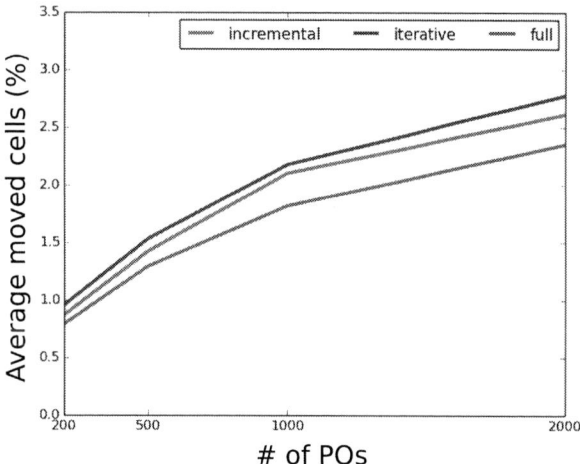

Fig. 5. Average percentage of moved cells for each number of POs

V. CONCLUSIONS

Different legalization strategies are typically employed for incremental optimization techniques: legalizing the entire circuit only in the end of the optimization, legalizing the entire circuit after each optimization and incrementally legalizing only a limited circuit region after each placement transformation. Although these strategies have been reported in the context of different optimization techniques, so far no experimental results were published to quantify the impact of them in terms of runtime and placement quality on any specific optimization technique.

This work bridges this gap by evaluating the impact of those three different legalization strategies on an incremental timing-driven placement flow. To do that we performed experiments considering different numbers of cell movements and changing the number of primary outputs to be optimized by the incremental timing-driven placement technique.

Experimental results show that the incremental strategy may be the most interesting one for incremental optimization techniques, since it is from 22% to 71% faster than the final legalization strategy. However, it resulted in up to 48% more timing violations (9% on average), since its incremental nature may fail to perform some movements, placing cells far from their target locations found by the optimization technique. On the other hand, the iterative strategy results in similar quality

results as the final legalization, which does not pay off for a runtime 3.2 times larger on average.

To improve the quality obtained by incremental legalization algorithms, we envisage some alternatives, such as: 1) evaluating the cost of placing the cell in different rows, while keeping runtime low; 2) constraining the maximum cell displacement during each legalization, to avoid movements that place the cell too far from its target location or 3) new incremental legalization algorithms with a more global view of the circuit to minimize the cell displacement. As future works, we intend to perform experiments with different legalization algorithms and applied on different incremental optimization techniques, such as gate sizing and buffer insertion.

REFERENCES

[1] C. J. Alpert, D. P. Mehta, and S. S. Sapatnekar, *Handbook of algorithms for physical design automation.* CRC press, 2008.
[2] A. B. Kahng, J. Lienig, I. L. Markov, and J. Hu, *VLSI physical design: from graph partitioning to timing closure.* Springer Science & Business Media, 2011.
[3] H. Ren, D. Pan, and D. Kung, "Sensitivity guided net weighting for placement-driven synthesis," *TCAD*, vol. 24, no. 5, pp. 711–721, 2005.
[4] T. Luo, D. Newmark, and D. Pan, "A new lp based incremental timing driven placement for high performance designs," in *DAC*, 2006, pp. 1115–1120.
[5] W. Choi and K. Bazargan, "Incremental placement for timing optimization," in *Proceedings of the 2003 IEEE/ACM international conference on Computer-aided design.* IEEE Computer Society, 2003, p. 463.
[6] A. Chowdhary, K. Rajagopal, S. Venkatesan *et al.*, "How accurately can we model timing in a placement engine?" in *DAC*, 2005, pp. 801–806.
[7] C. Guth, V. Livramento, R. Netto *et al.*, "Timing-driven placement based on dynamic net-weighting for efficient slack histogram compression," in *ISPD*, 2015, pp. 141–148.
[8] D. Hill, "Method and system for high speed detailed placement of cells within an integrated circuit design," Apr. 9 2002, uS Patent 6,370,673.
[9] P. Spindler, U. Schlichtmann, and F. M. Johannes, "Abacus: fast legalization of standard cell circuits with minimal movement," in *ISPD*. ACM, 2008, pp. 47–53.
[10] J. C. Puget, G. Flach, M. Johann, and R. Reis, "Jezz: An effective legalization algorithm for minimum displacement," in *SBCCI*. ACM, 2015.
[11] H. Ren, D. Z. Pan, C. J. Alpert, and P. Villarrubia, "Diffusion-based placement migration," in *DAC*. ACM, 2005, pp. 515–520.
[12] M. Cho, H. Ren, H. Xiang, and R. Puri, "History-based vlsi legalization using network flow," in *DAC*. ACM, 2010, pp. 286–291.
[13] U. Brenner, "Vlsi legalization with minimum perturbation by iterative augmentation," in *DATE*. IEEE, 2012, pp. 1385–1390.
[14] H. Ren, D. Pan, C. Alpert *et al.*, "Hippocrates: first-do-no-harm detailed placement," in *ASP-DAC*, 2007, pp. 141–146.
[15] D. Papa, T. Luo, M. Moffitt *et al.*, "Rumble: an incremental timing-driven physical-synthesis optimization algorithm," *TCAD*, vol. 27, no. 12, pp. 2156–2168, 2008.
[16] S. Popovych, H. Lai, C. Wang *et al.*, "Density-aware detailed placement with instant legalization," in *DAC*, 2014, pp. 1–6.
[17] W. Chow, J. Kuang, X. He *et al.*, "Cell density-driven detailed placement with displacement constraint," in *ISPD*, 2014, pp. 3–10.
[18] M. Kim, J. Hu, J. Li, and N. Viswanathan, "Iccad-2015 cad contest in incremental timing-driven placement and benchmark suite," in *ICCAD*, 2015, pp. 921–926.
[19] C. Chu and Y. Wong, "Flute: Fast lookup table based rectilinear steiner minimal tree algorithm for vlsi design," *TCAD*, vol. 27, no. 1, pp. 70–83, 2008.

Modeling and Design of High-Efficiency Power Amplifiers Fed by Limited Power Sources

Arturo Fajardo Jaimes*†
* Department of Electronics Engineering
Pontifical Xavierian University, Bogota, Colombia
Email: fajardoa@javeriana.edu.co

Fernando Rangel de Sousa †
† Radiofrequency Laboratory
Department of Electrical and Electronics Engineering
Federal University of Santa Catarina (UFSC), Florianpolis, Brazil
Email: rangel@ieee.org

Abstract—Recently the non-electromagnetic energy sources and wireless power transfer (WPT) techniques have been used to obtain electromagnetic (EM) energy sources for providing energy autonomy to electric devices when the energy in the environment is insufficient. Typically, this type of systems are composed by a harvester, a DC/RF converter, a WPT interface and a load. Further, in the power stage of the DC/RF converter the energy flow is processed using a power amplifier (PA). This paper proposes a design methodology for a generic PA fed by a limited power source. This methodology extracts the maximum available power of the EPS with the maximum PA efficiency using impedance matching and a novel PA modeling based on its impedance ports (*DC* and *AC*). Moreover, models of class-A PA and class-D PA were developed as examples. Furthermore, a Class-A PA was designed using its impedance ports model and the proposed methodology as a proof of concept. The results reflect that the designed PA extracts the maximum available power of the source with its maximum efficiency.

Keywords—Class-A power amplifier, class-D power amplifier, efficiency, power amplifier modeling, wireless power transfer.

I. INTRODUCTION

Currently, there is an increasing interest in providing energy autonomy to electric devices (e.g. sensors) in order to implement concepts such as the Internet of Things (IoT), and Wireless Body Area Networks (WBAN) [1]–[3]. Energy-harvesting technologies allow this autonomy, collecting energy from primary energy-sources (i.e. solar, thermal, kinetic, or electromagnetic) and converting it to DC power using a transducer. This system can be modeled using the energy power source (EPS) concept proposed in [4]. Following this modeling approach, the system composed by the primary energy source (e.g. solar) and the harvester (e.g. photovoltaic cell) can be considered as a Power EPS, that means, it supplies a high amount of energy (i.e. self-sustaining), but it has a weak power density and cannot sustain a stable output voltage under loading conditions [5].

When the energy in the environment is insufficient a self-sustaining WPT system can be used as a secondary energy source for powering an self-sustaining device [6]. A typical self-sustaining WPT system is illustrated in Fig. 1, it is composed by a power EPS, DC/RF converter (i.e. oscillator and power amplifier, or power oscillator), a WPT interface (i.e. antenna or inductive link), and a load. For instance, in [7] it is proposed a self-sustaining WPT as an additional EM energy source when there is availability of solar light but limited EM signals in the environment. This system was based on

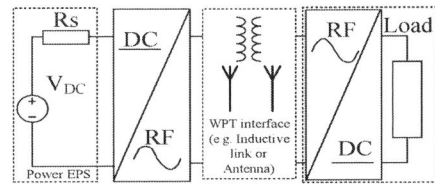

Fig. 1. Simplified block diagram of a self-sustaining WPT system

photovoltaic cells, class-E oscillator (905 MHz), and monopole antenna. The measured efficiency of the DC/RF converter was 43%. In general, given that the system EPS is a power EPS the efficiency of the overall power-chain (i.e. the circuit used for processing the energy flow from primary source to the load) must be very high [8].

In the traditional design approach of self-sustaining devices the interactions among EPSs, converters and loads are reduced to a specification (i.e. a voltage or a current) between the subsystems. Then they are optimized individually using this specification as a constraint. For example, in [5] a voltage specification of 1.9 V is imposed by design, then the power EPS (a regulated voltage source based on multi-harvest EPSs) was optimized for the extraction of the maximum available power. The self-sustaining voltage source, proposed by [9], for powering smart nodes of IoT is another example of this design approach. This EPS achieves low power operation and a high efficiency for the predefined voltage specification (3.3V), and imposes this voltage as a constraint for the load design.

For the self-sustaining WPT system, the traditional design approach of self-sustaining devices is inadequate, because it does not maximize the power delivered to the load. Further, the designer can improve the system performance by optimizing both the trade-offs and the interactions within the circuits. For instance, a non regulated voltage between the EPS and the power amplifier (PA) was explored in [10]. They proposed a regulator-less PA with an adaptive output matching network for maintaining the load power constant while the supplied voltage decreases. With this approach the overall system efficiency was increased and the burst time of a wireless sensor network node was maximized. However, this technique increases the hardware complexity. On the other hand, the harmonic balance optimization, proposed by [7], maximizes the conversion efficiency of the power oscillator without penalty on the system complexity, but it is limited to a specific harvester condition.

This paper proposes a methodology for maximizing the

978-1-5090-2737-8/16 $31.00 © 2016 IEEE

output power of a PA fed by a power EPS without an increase of the circuit complexity. It extracts the maximum available power of the EPS (P_{avs}) with the maximum PA efficiency using impedance matching and a novel PA modeling based on its impedance ports (DC and AC).

II. APPLICATION CONTEXT: THE WPT NODE CONCEPT

The PA analyzed in this paper is part of a self-sustaining WPT system for transferring energy from a WBAN node to an implanted device as shown in the Fig. 2. In order to achieve energy autonomy, the WPT system harvests energy from the body environment (e.g. solar and thermal), then it is transferred through an inductive link to the implanted device. The simplified system block diagram is shown in Fig. 3(a). The inductive link uses the magnetic coupling between two inductors: the primary inductor (L_1) is connected to the source and the secondary (L_2) is embedded into an implanted device. This inductive link is expected to operate under weak coupling regime, for this reason, the load of the DC/RF converter may be approximated as the impedance of primary inductor. A generic linear power EPS was modeled using a resistor (R_s) and a ideal voltage source (V_{DC}), the R_s captures both the limited power characteristic of the EPS and the output voltage drop when it is loaded.

III. ENERGY POWER SUPPLY (EPS)

The EPS energy (E_s) is supplied by the combination of current (i_s) and voltage (v_s) with an energy rate (P_s) given by:

$$P_s(t) = \frac{dE_s(t)}{dt} = v_s(t) i_s(t). \tag{1}$$

An ideal EPS must deliver an infinite energy at a required load power, in a small form factor [4]. On the contrary, any real EPS delivers a limited energy at a limited power with a specific form factor [4]. When an EPS is based on energy harvester, it supplies a high amount of energy, but at a limited low power [8]. Additionally, this maximum power value (P_{avs}) can be extracted only when the power transfer condition is satisfied. On the other hand, when the EPS is based on an energy carrier (e.g. battery), it supplies a high power during a limited time, because it has a limited energy-storage-capacity (E_{st}). In [4] a taxonomy based on the E_{st} and the P_{avs} EPS characteristics was proposed, it is summarized in Table I.

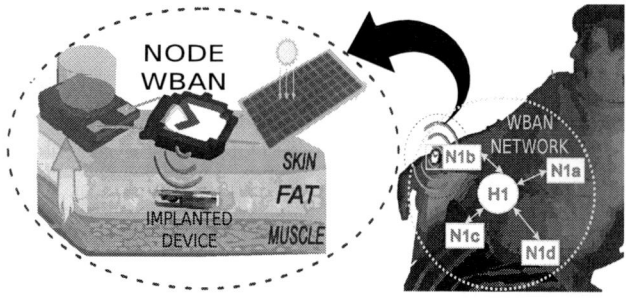

Fig. 2. Self-sustaining WPT system for powering implanted device.

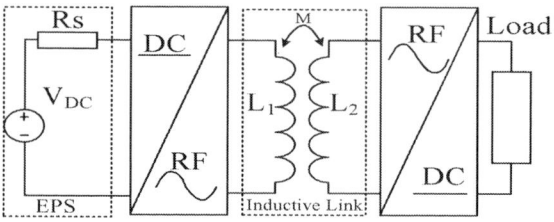

(a) Simplified block diagram

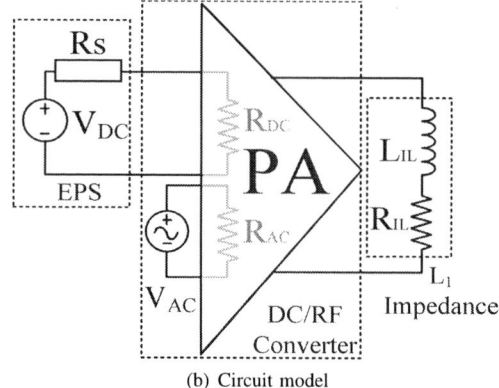

(b) Circuit model

Fig. 3. WPT system for powered an implanted device

TABLE I.   EPS TAXONOMY BASED ON [4]

EPS Name	P_{avs}	E_{st}	Example
Ideal EPS	infinite	infinite	The sun
Real EPS	finite	finite	Electric distribution network
Power EPS	low	high	Solar Energy + PV cell
Energy EPS	high	low	AAA battery

IV. PA MODELING

A. Modeling of the energy-flow process using the PA impedance ports

Assuming a DC/RF converter based on an oscillator that drives a PA as shown in Fig. 3(b), the energy flows from the the DC source (V_{DC}) and the AC source (V_{AC}) to the load ($R_L = R_{IL}$) connected to the RF port. In order to model the energy-flow process in the PA, equivalent resistances in its input ports are proposed: R_{DC} in the PA DC-port and R_{AC} in the PA AC-port, as depicted in Fig.4(b). Some of power "disipated" in R_{DC} and R_{AC} is transfered to the load. Considering resonant load, narrow-band operation, and ideal passive elements, the load power in steady state depends only on the external elements connected to the PA, therefore the load port could be considered as a power source. This circuit element imposes the power on its load [11], the I–V characteristic of the a DC power source of 1W is illustrated in Fig.4(b). Using this model, the current (I_{RF}) and the voltage (V_{RF}) in the RF port are imposed by both the load and the power source, and are given by:

$$I_{RF} = I_m sin(\omega_0 t); \tag{2}$$
$$V_{RF} = V_m sin(\omega_0 t); \tag{3}$$

where, I_m is the peak load current, V_m is the peak load voltage, and ω_0 is the natural frequency of the resonant load (i.e. C_0, L_0

and R_L). In addition, the power delivered to the load (P_{RF}), the power supplied by the DC source (P_{DC}), and the the power supplied by the AC source (P_{AC}) are given by:

$$P_{RF} = \frac{I_m{}^2}{2} R_L = \frac{V_m{}^2}{2 R_L} = \frac{I_m \cdot V_m}{2}; \qquad (4)$$

$$P_{AC} = \frac{\omega_0}{2\pi} \int_0^{\frac{2\pi}{\omega_0}} i_{AC}(t) \cdot v_{AC}(t) \cdot dt; \qquad (5)$$

$$P_{DC} = I_{DC}{}^2 R_{DC} = \frac{V_{DC}{}^2}{R_{DC}} = I_{DC} V_{DC}; \qquad (6)$$

where, R_L is the load impedance at resonance, i_{AC} is the current supplied by the AC source, v_{AC} is the voltage supplied by the AC source, I_{DC} is the DC current supplied by the EPS, V_{DC} is the voltage supplied by the EPS, and R_{DC} is the impedance imposed by the amplifier in its DC-port.

B. PA efficiency predicted by the PA model based on its impedance ports

The power "disipated" in R_{DC} and R_{AC} and transfered to R_L (P_{RF}) can be calculated following the power added efficiency (PAE) definition (7). On the other hand, following

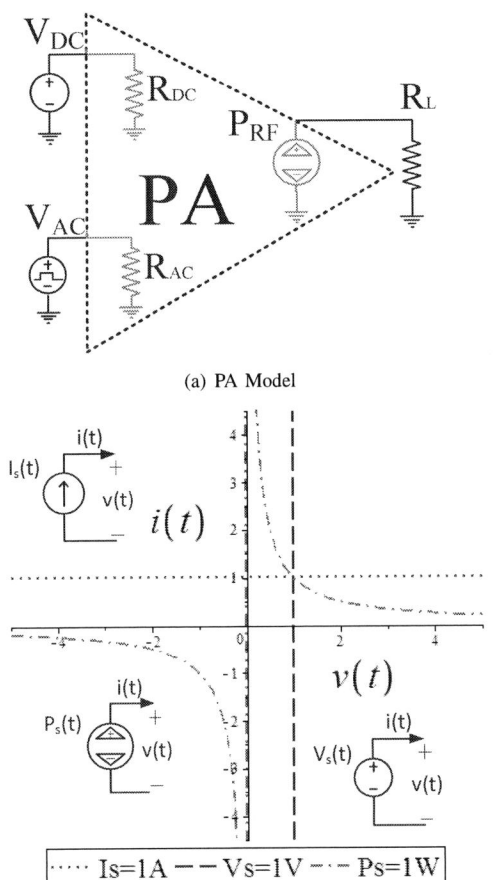

(a) PA Model

(b) Power source example

Fig. 4.   PA Model based in its port resistors and a controled power source.

the drain (or colector) efficiency (η_D) definition, P_{RF} could be calculated as (8). Further, following the PA efficiency (η) definition the P_{RF} is given by (9).

$$P_{RF} = PAE \cdot P_{DC} + P_{AC}; \qquad (7)$$

$$P_{RF} = \eta_D \cdot P_{PA}; \qquad (8)$$

$$P_{RF} = \eta \cdot (P_{DC} + P_{AC}); \qquad (9)$$

where, P_{DC} is supplied by the DC source (V_{DC}), P_{AC} is supplied by the AC source (V_{AC}), and P_{PA} is the power consumed by the PA power stage. When the P_{AC} is negligible compared to P_{DC}, the value of PAE, η_D and η are equal. Hence they can be approximated by (10). Using (4) and (6), the equation (10) can be rewritten as (11).

$$\eta \approx PAE \approx \eta_D \approx \frac{P_{RF}}{P_{DC}}; \qquad (10)$$

$$\eta = \frac{R_L}{2 R_{DC}} \left(\frac{I_m}{I_{DC}} \right)^2. \qquad (11)$$

The power delivered by V_{DC} is consumed by both the bias circuit (or the driver circuit) and the PA power stage, therefore the I_{DC} is given by (12). Using (12) the R_{DC} can be calculated by (13).

$$I_{DC} = I_{bias} + I_{PA} = \frac{R_{PA} \cdot I_{PA}}{R_{bias}} + I_{PA}; \qquad (12)$$

$$R_{DC} = \frac{V_{DC}}{I_{DC}} = R_{bias} \| R_{PA}; \qquad (13)$$

where, I_{bias} is the DC current of the bias circuit, I_{PA} is the DC current of the PA power stage, R_{bias} is the resistance imposed by the bias circuit to the DC source and R_{PA} is the resistance imposed by the PA power stage to the DC source. Hence, using (12) and (13) the equation (11) can be rewritten as (14).

$$\eta = \frac{1}{2} \frac{R_L R_{DC}}{R_{PA}{}^2} \left(\frac{I_m}{I_{PA}} \right)^2 = f(G_R); \qquad (14)$$

where, G_R is the PA impedance factor defined as $G_R = R_{DC}/R_L$, and $f(x)$ is a function dependent on the PA topology. The PA efficiency is maximum for an optimum relationship between its port impedances (i.e. $G_R = G_{R_{opt}}$). In the next subsections we discuss the models of class-A PA and class-D PA based on its impedance ports as examples of the proposed modeling technique.

C. Example 1: BJT class-A PA

Considering the schematic shown in the Fig. 5(a), a negligible energy consumption of the bias circuit (i.e. $R_{DC} \approx R_{PA}$) and (14), the η can be expressed as:

$$\eta \approx \frac{0.5 R_L}{R_{PA}} \left(\frac{I_m}{I_{PA}} \right)^2 \approx \frac{0.5 R_L}{R_{DC}} \left(\frac{I_m}{I_{DC}} \right)^2. \qquad (15)$$

Considering the exponential model of the BJT transistor presented in [12], (15) can be rewritten as:

$$\eta = \frac{1}{2 G_R} h_i{}^2 \left(\frac{|v_{ac}|}{\phi_t} \right); \qquad (16)$$

978-1-5090-2737-8/16 $31.00 © 2016 IEEE

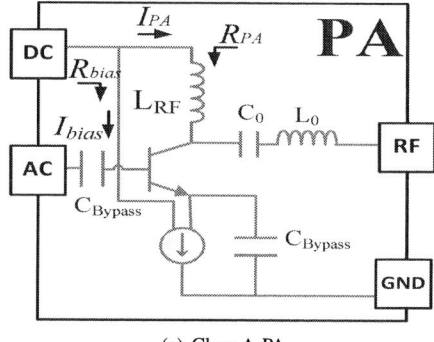

(a) Class-A PA

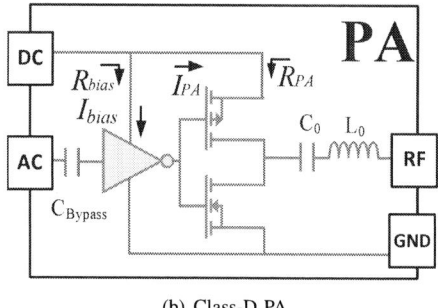

(b) Class-D PA

Fig. 5.   Analyzed PA topologies.

where, $|v_{ac}|$ is the peak voltage value of the sinusoid input signal (V_{AC}) normalized by the thermal voltage (ϕ_t) and the function $h(x)$ is given by:

$$h\left(x\right) = 2\frac{I_1\left(x\right)}{I_0\left(x\right)}; \qquad (17)$$

where, $I_n(x)$ is the modified Bessel function of the first kind, of order n and argument x. The PA operates as class-A PA until I_m is equal to the I_{PA} or V_m is equal to the Collector bias voltage (V_C). Therefore, in order to achieve the class-A operation (i.e. $I_m < I_{PA}$ and $V_m < V_C$) the $|v_{ac}|$ value must be lower than a maximum value given by:

$$|v_{ac}|_{\phi_{t M}} = \begin{cases} h^{-1}\left(1\right) \approx 1.16 & G_R > 1; I_m = I_{PA} \\ h^{-1}\left(G_R\right) & G_R < 1; V_m = V_C \end{cases} ; \quad (18)$$

where, $h^{-1}(x)$ is the inverse function of the $h(x)$. Finally, from (16) and (18) the PA efficiency is given by :

$$\eta = f_{A_{BJT}}\left(G_R\right) = \begin{cases} 0.5 \cdot G_R & G_R < 1 ; \Rightarrow I_m = I_{PA} \\ \frac{1}{2G_R} & G_R \geq 1; \Rightarrow V_m = V_C \end{cases} . \tag{19}$$

This efficiency is maximum when $G_R = G_{R_{opt}} = 1$ is forced by design. In this case, the maximum value of $\eta = f_{BJT}\left(G_{R_{opt}}\right)$ is 50%.

D. Example 2: CMOS class-D PA

Considering the schematic shown in the Fig. 5(b), ideal components, pulse signal of 50% of duty cycle in the PA AC-port, assuming a resonant load with high loaded quality factor, modeling the MOSFET (N and P type) as an ideal switch and a series resistance (R_{on}), I_m can be approximated by (20). Additionally, the current waveform of the PA power stage is a

rectified sinusoid with peak value of I_m, therefore its average value (I_{PA}) is given by (21). Further, using (20) and (21) the impedance imposed by the PA power stage (R_{PA}) can be calculated as (22). Moreover, the power used to charge and discharge the total gate capacitances of the MOSFETs (C_G), dissipated by the driver, could be modeled with an equivalent resistance (R_{bias}) given by (23).

$$I_m = \frac{V_m}{\left(R_L + R_{on}\right)} = \frac{4V_{DC}}{\pi\left(R_L + R_{on}\right)}; \qquad (20)$$

$$I_{PA} = 2I_m/\pi; \qquad (21)$$

$$R_{PA} = \frac{V_{DC}}{I_{PA}} = \frac{\pi^2}{8}\left(R_L + R_{on}\right); \qquad (22)$$

$$R_{bias} = \frac{V_{DC}{}^2}{P_{bias}} \approx \frac{V_{DC}{}^2}{V_{DC}{}^2 \cdot f \cdot \alpha \cdot C_G} = \frac{1}{f \cdot \alpha \cdot C_G}; \quad (23)$$

where, the factor α represents the capacitance excess due to the driver implementation and f is the frequency of the AC input. The $C_G \cdot R_{on}$ product could be considered quasi constant and given by the product of two technology parameters (i.e. a and b) introduced in [13], hence (23) can be rewritten as (24). From (22) and (24) the G_R can be calculated by (25).

$$R_{bias} = \frac{R_{on}}{f \cdot \alpha \cdot a \cdot b}; \qquad (24)$$

$$\frac{1}{G_R} = \frac{R_L}{R_{DC}} = \frac{f \cdot \alpha \cdot a \cdot b}{\frac{R_{on}}{R_L}} + \frac{8}{\pi^2}\frac{1}{\left(1 + \frac{R_{on}}{R_L}\right)}. \quad (25)$$

From (14), (21) and (22), the η can be calculated by:

$$\frac{1}{\eta} = \frac{\pi^2}{8} \cdot \frac{1}{G_R} \cdot \left(1 + \frac{R_{on}}{R_L}\right)^2. \qquad (26)$$

Using (25), the equation (26) can be rewritten as:

$$\eta = f_{D_{CMOS}}\left(\mathbf{G}_R\right) = \frac{1}{\left(1 + l(\mathbf{G}_R)\right)\left(1 + k\left(1 + \frac{1}{l(\mathbf{G}_R)}\right)\right)}; \quad (27)$$

where $k = \frac{\pi^2}{8}\alpha \cdot a \cdot b \cdot f$ and the function $l(x)$ is given by:

$$l(x) = \frac{1}{2}\left((A + B)x - 1 \pm \sqrt{1 + (2A - 2B)x + (A + B)^2 x^2}\right); \quad (28)$$

where, $A = f \cdot \alpha \cdot a \cdot b$ and $B = \frac{8}{\pi^2}$. This efficiency is maximum when (29) is forced by design. In this case, the maximum value of η is given by (30).

$$l\left(\mathbf{G}_{R_{opt}}\right) = \sqrt{\frac{k}{k + 1}} \qquad (29)$$

$$f_{D_{CMOS}}\left(\mathbf{G}_{R_{opt}}\right) = \frac{\sqrt{k^2 + k}}{\left(k + \sqrt{k^2 + k}\right)\left(k + 1 + \sqrt{k^2 + k}\right)}. \tag{30}$$

V. PA DESIGN METHODOLOGY

In the application described in Fig. 2, the design goal is to maximize the power delivered to the load (R_{IL}). This methodology achieves this goal by the maximization of both the power supplied by the harvester and the PA efficiency. Therefore, the methodology finds the optimum R_{DC} ($R_{DC_{opt}}$) that dissipated the P_{avs} of the power EPS at the highest possible voltage for a particular implementation. Following, the methodology finds the optimum R_L ($R_{L_{opt}}$) based in the

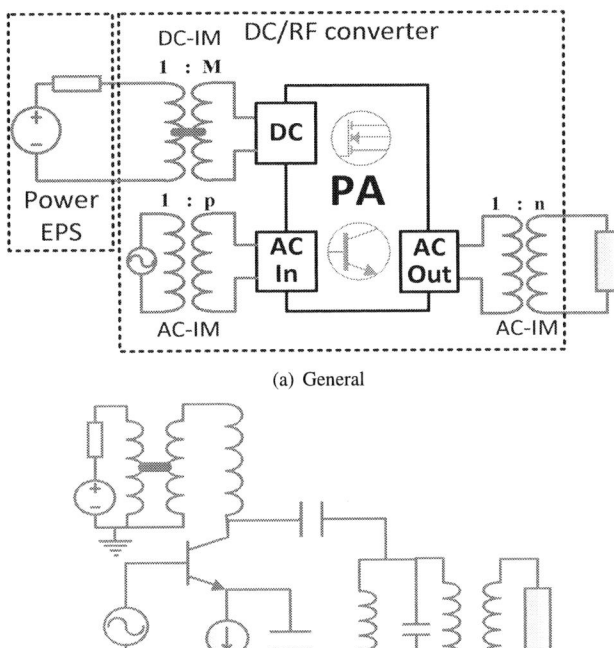

(a) General

(b) Class-A

Fig. 6. DC/RF converter topologies with impedance matching networks

TABLE III. DESIGN METHODOLOGY STEPS

Step	Step Description	Equation
1	Find the P_{avs} of the power EPS and its related variables: optimum load impedance (R_{pavs}), load voltage (V_{pavs}) and current (I_{pavs}).	e.g. for a thermoelectric generator, the internal series resistor of the EPS (R_s) is constant, therefore: $R_{pavs} = R_s$, $V_{pavs} = \frac{V_{DC}}{2}$ and $I_{pavs} = \frac{V_{DC}}{2R_s}$.
2	Fix the voltage in the PA DC-port as the highest for a particular implementation restriction, e.g. the technology voltage of the CMOS process.	$V_{DC_{opt}} = V_{\max}$
3	Find the DC current that extracts the P_{avs} of the EPS	$I_{DC_{opt}} = \frac{P_{avs}}{V_{\max}}$
4	Find the impedance of the PA DC-port that maximizes the power extracted from the harvester.	$R_{DC_{opt}} = \frac{V_{DC}}{I_{DC}} = \frac{V^2_{\max}}{P_{avs}}$
5	Find the optimum load value for maximizing PA efficiency	$R_{L_{opt}} = G_{R_{opt}} R_{DC_{opt}}$
6	Find the specifications of the DC and AC impedance matching networks (DC-IM and AC-IM).	$M = \sqrt{\dfrac{R_{DC_{opt}}}{R_{pavs}}} = \sqrt{\dfrac{V^2_{\max}}{P_{avs} \cdot R_{pavs}}}$ $n = \sqrt{\dfrac{R_L}{R_{L_{opt}}}} = \sqrt{\dfrac{P_{avs} \cdot R_L}{G_{R_{opt}} V^2_{\max}}}$

amplifier design specifications are summarized in Table IV. The simulation setup uses the harmonic balance simulation technique in the Advanced Design System (ADS®) software. Furthermore, two parametric sweeps, the input signal (V_{AC}) sweep and the R_{PA} sweep, were implemented. The R_{PA} sweep was implemented by a fixed current (I_{PA}=1mA) and a V_{DC} sweep. Additionally, the circuit was simulated with and without the output LC tank filter. Further, the assumption that P_{AC} is negligible compared to P_{DC} was verified (i.e. $P_{AC} < 8\mu W$ for all simulations). The simulation results are summarized in Table V.

The PA was implemented without the output LC tank and the fundamental components were obtained by FFT processing on the time-domain signals in order to avoid the influence of the parasitic elements of the LC tank. In the experimental setup (Fig. 7b), the I_{PA} was fixed to 1 mA and the R_{DC} was set with the V_{DC}. In this setup, V_{AC} was incremented until the PA operates at the limit of the class-A operation ($I_m = I_{PA}$ or $V_m = V_C$). At this point the efficiency was calculated through the measured values (i.e. I_{PA}, V_{DC}, and V_m). The voltage operation limit ($V_m = V_C$) was measured with a digital oscilloscope with FFT option. On the other hand, the voltage

calculated $R_{DC_{opt}}$ and the $G_{R_{opt}}$ of the used PA topology. In particular the $G_{R_{opt}}$ (calculated using (19) and (27)) of the analyzed PAs in Section IV is summarized in Table II. Finally, the $R_{DC_{opt}}$ is matching to the power EPS impedance by a DC IM network (DC-IM) and the $R_{L_{opt}}$ is matching to the impedance of the real load by a AC IM network (AC-IM), as is illustrated in Fig 6(a). These networks was modeled as an ideal transformers (i.e. assuming IM networks with negligible losses). Further, the DC-IM implementation could be an DC/DC converter, and the AC-IM implementation could be a L or π network. The design methodology steps were proposed and summarized in Table III. The application of this methodology is illustrated in Section VI.

TABLE II. G_{opt} FOR THE ANALYZED PAs

BJT ClassA PA	CMOS ClassD PA
1	$\dfrac{\sqrt{(k+1)k}+k}{(A+B)\sqrt{(k+1)k}+A(k+1)}$ $k = \frac{\pi^2}{8}\alpha \cdot a \cdot b \cdot f$ $A = f \cdot \alpha \cdot a \cdot b$ and $B = \frac{8}{\pi^2}$

VI. PA STUDY CASE

In order to verify (19) experimentally, a class A PA (Fig. 6(b)) was designed, simulated and implemented. The

TABLE IV. PA SPECIFICATIONS

V_{DC}	I_{PA}	f	$R_{L_{opt}}$
1V	1mA	100 kHz	1kΩ

TABLE V. PA SIMULATION RESULTS

R_{PA} (Ω)	Limit type	$\|v_{ac}\|_{\phi_t}$	P_{DC} w/o_LC (mW)	P_{DC} w_LC (mW)	P_L w/o_LC (mW)	P_L w_LC (mW)	η w/o_LC (%)	η w_LC (%)
0.5	$V_m=V_C$	0.530	0.564	0.564	0.138	0.138	24.5	24.5
1.0	$V_m=V_C$	1.181	1.127	1.127	0.545	0.545	48.3	48.2
1.5	$I_m=I_{PA}$	1.180	1.691	1.691	0.545	0.545	32.2	32.1
2.0	$I_m=I_{PA}$	1.180	2.254	2.254	0.545	0.545	24.2	24.1

978-1-5090-2737-8/16 $31.00 © 2016 IEEE

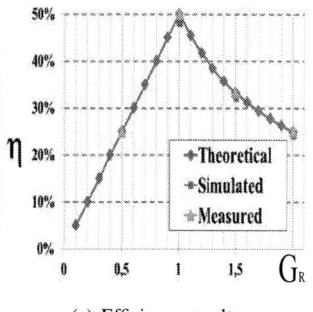
(a) Efficiency results

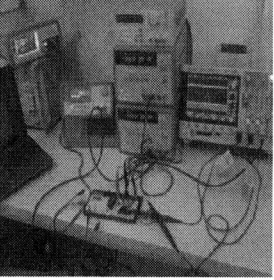
(b) Experimental setup

Fig. 7. Implemented class-A PA

operation limit ($I_m = I_{PA}$) was estimated from the V_m and the measured load value (R_L= 997.74 Ω). The experimental results are summarized in Table VI. The predicted efficiency by the model and the results (simulated and experimental) are plotted in the Fig. 7a, considering this figure, the equation (19) is an appropriate expression for model η.

In order to verify experimentally the proposed methodology without the practical limitations of the commercial harvesters and the impedance matching networks (AC and DC), we choose a scenario with the following specifications: a power EPS with P_{avs} = 1mW and R_{pavs} = 1kΩ, a resistive load of R_L = 1kΩ, and an ideal sinusoid input signal with the required amplitude and frequency of 100kHz. The specifications of the PA and the IM networks were calculated following the methodology steps listed in the Table III. The results are summarized in Table VII. In the experimental setup, the AC input was incremented until the PA operates at its maximum efficiency. The DC/RF converter efficiency was calculated through the measured values, and the EPS efficiency was calculated based on its DC current and the output open voltage. The results are summarized in Table VIII. They reflect that the designed converter extracts the P_{avs} of the power EPS with the maximum PA efficiency (i.e. 50%).

VII. CONCLUSIONS

A design methodology for a generic PA fed by a power EPS was proposed. Using this methodology, a class-A PA was designed, implemented and tested. The results reflect that the designed PA extracts the maximum available power of the source with its maximum efficiency. Further, the PA modeling based on its impedance ports (DC and AC) was proposed. Furthermore, the models of class-A PA and class-D PA were developed as examples of the proposed modeling technique. Also, this paper presents an open challenge on EPSs, Energy converters (i.e. PAs) and circuits that could take advantage of the use of power specifications instead of predefined voltage or current condition between them.

TABLE VI. PA EXPERIMENTAL RESULTS

R_{PA} (Ω)	Limit type	$|v_{ac}|_{\phi_t}$	LC tank	P_{DC} (mW)	P_L (mW)	η (%)
0.50	$V_m = V_C$	0.679	w/o	0.5002	0.1242	24.8
1.00	$V_m = V_C$	1.516	w/o	10.003	0.5004	50
1.50	$I_m = I_{PA}$	1.516	w/o	15.002	0.5010	33.3
2.00	$I_m = I_{PA}$	1.516	w/o	2.002	0.5010	25

TABLE VII. DC/RF CONVERTER SPECIFICATIONS

$V_{DC_{opt}}$	$I_{DC_{opt}}$	$R_{DC_{opt}}$	$R_{L_{opt}}$	M	n
1V	1mA	1kΩ	1kΩ	1	1

TABLE VIII. DC/RF CONVERTER EXPERIMENTAL RESULTS

R_{DC}	P_{EPS}	P_{DC}	P_{RF}	η_{EPS}	$\eta_{DC/RF}$
1.001kΩ	2.004mW	1000.3 μW	500.4μW	50%	50 %

VIII. ACKNOWLEDGMENT

The first author would like to thank COLCIENCIAS and the Pontificia Universidad Javeriana for the financial support. Also, the authors would like to thank the CNPq and the INCT/NAMITEC for their partial financial support and all the students of the Radio Frequency integrated circuits Group (GRF-UFSC) for the important discussions.

REFERENCES

[1] A. Whitmore, A. Agarwal, and L. Xu, "The internet of things, a survey of topics and trends," *Information Systems Frontiers*, vol. 17, no. 2, pp. 261–274, 2014.

[2] A. Fajardo and F. Rangel de Sousa, "A taxonomy for learning, teaching, and assessing wireless body area networks," in *Proc. IEEE 7th Latin American Symp. on Circuits and Syst. (LASCAS2016)*, 2016.

[3] S. Movassaghi, M. Abolhasan, and J. Lipman, "Wireless Body Area Networks: A Survey," *IEEE Commun. Surveys and Tutorials*, pp. 1–29, 2013.

[4] A. Fajardo and F. Rangel de Sousa, "Ideal energy power source model and its implications on battery modeling," in *Proc. 22th IBERCHIP Workshop*, February Florianopolis, Brasil. 2016, pp. 19–25.

[5] S. Bandyopadhyay and A. P. Chandrakasan, "Platform architecture for solar, thermal, and vibration energy combining with mppt and single inductor," *IEEE Journal of Solid-State Circuits*, vol. 47, no. 9, pp. 2199–2215, Sep. 2012.

[6] K. Niotaki, A. Collado, A. Georgiadis, S. Kim, and M. M. Tentzeris, "Solar/electromagnetic energy harvesting and wireless power transmission," *Proceedings of the IEEE*, vol. 102, no. 11, pp. 1712–1722, Nov 2014.

[7] A. Georgiadis and A. Collado, "Solar powered class-e active antenna oscillator for wireless power transmission," in *2013 IEEE Radio and Wireless Symp. (RWS2013)*, Jan 2013, pp. 40–42.

[8] G. A. Rincon-Mora, "Powering microsystems with ambient energy," in *Energy Harvesting with Functional Materials and Microsystems*. New York: CRC Press, 2013, pp. 1–30.

[9] X. Liu and E. Sanchez-Sinencio, "An 86 % efficiency 12 uw self-sustaining pv energy harvesting system with hysteresis regulation and time-domain mppt for iot smart nodes," *IEEE Journal of Solid-State Circuits*, vol. 50, no. 6, pp. 1424–1437, June 2015.

[10] J. C. Rudell, V. Bhagavatula, and W. C. Wesson, "Future integrated sensor radios for long-haul communication," *IEEE Commun. Mag.*, vol. 52, no. 4, pp. 101–109, 2014.

[11] R. W. Erickson and D. Maksimovic, *Fundamentals of power electronics*. New York: Springer, 2001.

[12] K. K. Clarke and D. T. Hess, *Communication circuits: analysis and design*. New York: Addison-Wesley Reading, 1971.

[13] B. R. W. Stratakos, Anthony J. and S. R. Sanders, "High-efficiency low-voltage dc-dc conversion for portable applications." in *Proc. Int. Workshop on Low-Power Design*, April Napa, CA. 1994, pp. 21–27.

Integrated CMOS Class-E Power Amplifier for Self-Sustaining Wireless Power Transfer system

Arturo Fajardo Jaimes*†
* Department of Electronics Engineering
Pontifical Xavierian University, Bogota, Colombia
Email: fajardoa@javeriana.edu.co

Fernando Rangel de Sousa †
† Radiofrequency Laboratory
Department of Electrical and Electronics Engineering
Federal University of Santa Catarina (UFSC), Florianópolis, Brazil
Email: rangel@ieee.org

Abstract—In this paper is proposed a methodology for designing a CMOS class-E PA used to drive an inductive link. In order to satisfy the operating conditions imposed by the PA specifications and the available technology, a differential class-E PA with split slab inductor and high level of integration was both designed and simulated. The proposed methodology uses the analytical solution of the ideal class-E PA equations as the first point of an iterative procedure for solving the optimization of the PA. Further, the proposed design set solve the trade-off between ON-resistance and gate capacitance of the switches, resulting in the optimal choice of the power transistors width for a class-E PA with finite DC-feed inductance. In the post-layout simulation, the PAE of the PA was 45.7% when 20.7 dBm.

Keywords—*Class-E, power amplifier, power efficiency, wireless power transfer.*

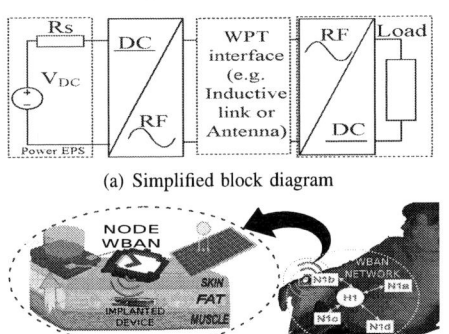

(a) Simplified block diagram

(b) for powering implanted device

Fig. 1. Self-sustaining WPT system

I. INTRODUCTION

Currently, there is an increasing interest in providing energy autonomy to electric devices in order to implement concepts such as the Internet of Things (IoT), and Wireless Body Area Networks (WBAN) [1]–[3]. Energy-harvesting technologies allow this autonomy, collecting energy from primary energy-sources (e.g. solar, thermal, kinetic, or electromagnetic). When energy in the environment are insufficient, secondary energy-sources are used to power the self-sustaining device. For instance, a satellite collects solar energy and then radiates this energy to other satellites where it can be used [4]. On a smaller scale, the non-electromagnetic energy sources and wireless power transfer (WPT) have been used to synthesize artificial energy sources [5]. This kind of scheme, illustrated in the Fig. 1(a), is referred as self-sustaining WPT systems. The integrated power amplifier (PA) proposed in this paper is part of a self-sustaining WPT system for powering an implanted device, this concept is shown in the Fig. 1(b).

Integrating a PA in a CMOS system-on-chip is challenging, mainly due to the low breakdown voltage of CMOS devices and low quality factor (Q) of the integrated passive components (i.e. inductors with low Q). Several approaches have been used in order to increase the power added efficiency (PAE). For instance in [6], it was explored the analytical solution of the trade-off between the ON-resistance and gate capacitance for finding the transistor size that maximizes the PAE of a integrated class-D PA. As another example, in [7], it was implemented a PA with high PAE using an Class-E power oscillator with injection-locking and only the transistor integrated (i.e. passive components off-chip). This system incurs in a significant penalty of area, cost, and hardware complexity. As

an intermediate solution between area and efficiency (η), the use of the bond-wire connections as inductances was reported in [8]. But, the inductance values are limited and not well controlled. On the other hand, in [9], it was proposed a fully-integrated class-E PA with high PAE using an on-chip slab inductor combined with an adaptive class-E PA.

The Class-E PA with a finite dc-feed inductance instead of an RF-choke has been explored in several works [10]–[13]. This topology is shown in the Fig. 2(a). For the same supply voltage, output power and load, using finite dc-feed inductance has significant benefits [12]: more efficient output matching network, implementation in low-voltage technologies and higher frequency of operation. The published papers design the PA based on analytical equations or based on iterative procedures [13]. Further, when the switch on-resistance and the inductor resistance are taken into account, the class-E PA solution (i.e. the optimum operation of the non-ideal PA for maximum η) results in nonlinear analytic equations that must be solved numerically [11] or in iterative design procedures even more lengthy and complex [14]. Furthermore, this solution occurs outside the nominal operation of the class-E PA (i.e. ZVS and DZVS) [14]. As an alternative option, the methodology proposed in this paper uses an analytical design set for calculating the start point of an optimization process, hence its complexity decrease.

In this work a methodology for designing a CMOS class-E PA that drives an inductive link is proposed. This methodology uses an expanded version of the Class-E design set proposed in [10]. This set solves the trade-off between ON-resistance and gate capacitance of the switches, resulting in a suboptimal

978-1-5090-2737-8/16 $31.00 © 2016 IEEE

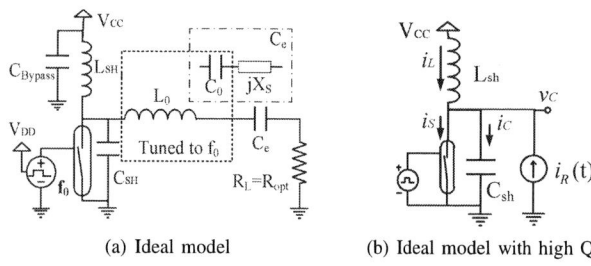

(a) Ideal model   (b) Ideal model with high Q

Fig. 2. Class-E PA with a finite dc-feed inductance

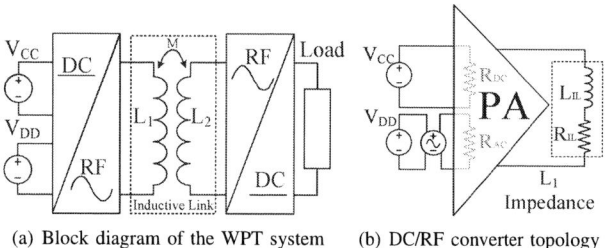

(a) Block diagram of the WPT system   (b) DC/RF converter topology

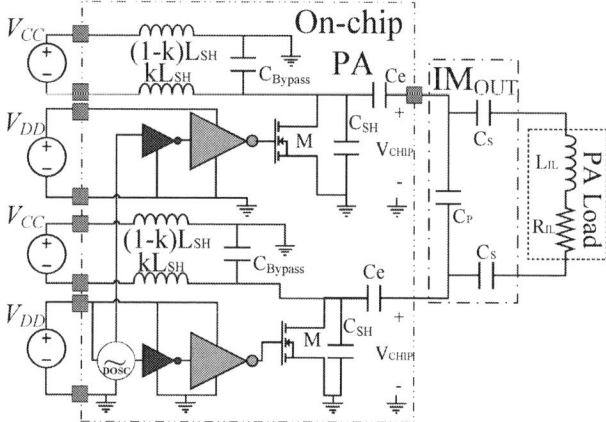

(c) Topology of the diferencial Class E PA with split slab inductor

Fig. 3. System description a top-down approach

choice of the power transistors width, near to its optimum value. In order to satisfy the operating conditions imposed by the PA specifications (e.g. load and source impedance are not equal to 50 Ω) and the available technology (i.e. 180nm), a differential class-E PA with split slab inductor was designed and simulated. In the post-layout simulation, the PAE of the PA was 45.7% when 20.7 dBm was delivered to its load.

II. PA DESIGN

The power amplifier (PA) proposed in this paper is part of a self-sustaining WPT system for powered an implanted device. This system transfers energy from a wireless body area network (WBAN) node to an implanted device, as illustrated in Fig. 1(b). To achieve energy autonomy, the WBAN node harvest energy from the body environment (i.e. solar and thermal energy) for powering two DC/DC converters which in turn powered the DC/RF converter. The WPT interface is implemented with an inductive link, where the harvested energy is transferred through magnetic coupling between two

inductors: the primary inductor is connected to the source and the secondary is embedded into an implanted device. The inductive link needs an RF signal which frequency must be chosen to optimize the link efficiency [15]. The system can be summarized as shown in Fig. 3(a). The implementation of the DC to RF converter is based on an oscillator that drives a switched PA as shown in the Fig. 3(b).

A. PA Specifications

PA Load: The PA proposed was designed for driving an inductive link whose primary inductor is described in [6]. The secondary side is a fully-integrated wireless power receiver explained in [16]. The power delivered to the primary inductor was specified to be at least 20 dBm at a frequency of 990 MHz (f_0). The inductive link is expected to operate under the weak coupling regime, for this reason the input impedance of the link is approximately the impedance of the primary inductor. This impedance can be represented as an inductor ($L_{IL} = 9.3nH$) in series with a resistor ($R_{IL} = 1.8\Omega$), as presented in Fig. 3(c).

PA Energy Source: The power supply will be a multi-input energy-harvesting system with solar and thermal energy inputs, intermediate storage, shared inductor, and regulated outputs ($V_{DD} = 1.8V, V_{CC} \leq 1.8V$). The topology will be similar to the proposed in [17].

General: The PA must has high efficiency, narrow band operation and high level of integration. Additionally, in order to avoid extensive electromagnetic simulations, only the available components of the design kit were used. The principal limitations of these technology are: $V_{nom} = 1.8$ V, $V_{I/O} = 5$ V, $L_{min} = 180$ nm, $IL_{DC_{max}} = 150$ mA, $IL_{RMS_{max}} = 228$ mA, $L_{SH} < 2$ nH, $C_{SH} > 10$ pF.

B. Proposed PA Topology

The diagram of the proposed differential class-E PA is shown in Fig. 3(c). The capacitors C_S and C_P form an impedance transformation network (IM_{OUT}). The capacitor C_S must be off-chip because the voltage at load-terminals exceeds the maximum value of the chip technology. The equivalent capacitance C_e combines the impedance jX_s and C_0 at the operation frequency (Fig. 2(a)). On the other hand, the choice of the C_e integration (Fig. 3(c)) has at least two advantages [6]: 1) low harmonic level at the chip output, which is translated into lower power losses. 2) higher output voltage, which means a reduction in the output current for the same power level and hence lower losses in the wire-bond parasitic resistance. In order to provide flexibility to the design, the C_P will be an external device that allows to adjust the resonance of the load and the unmodeled parasitic capacitances. This capacitive value is limited by the output PAD and wire-bond capacitances. The DC energy sources of the amplifiers do not share the same reference, hence the inductor L_{SH} may be split ($k \leq 1$). The advantages of this approach are: 1) simplify the layout when slab inductors are implemented, 2) decrease both gain reduction and source degeneration. The transistors (M in Fig. 3(c)) act as switches, its widths (W) are in the order of mm, therefore its gate capacitance is high, given that the use of drivers are necessary. In the chip a differential oscillator ($DOSC$) and a driver (i.e. Two cascaded inverters) are integrated.

C. Ideal PA model

The PA was analyzed by its half-circuit shown in Fig.2(a). The transistor is represented as an ideal switch. The ideal class-E can be modeled as the circuit shown in Fig. 2(b) when the control signals have zero time transitions at a frequency f (near to f_0) and the series resonant circuit L_0, C_e and R_L has a high loaded quality factor. Therefore, the output current is given by (1). As a result of development presented in [10], for ideal class-E nominal operation (i.e. with ZVS and DZVS), the PA currents and PA voltages, are given by the equations (2) to (7).

$$i_R(t) = I_P \sin(\omega t + \varphi) = \sqrt{\frac{2P_{OUT}}{R_L}} \sin(2\pi f t + \varphi); \quad (1)$$

$$v_{C_{SH_{on}}}(t) = i_{C_{SH_{on}}}(t) = i_{S_{off}}(t) = 0; \quad (2)$$

$$i_{L_{SH_{on}}}(t) = \frac{V_{CC}}{L_{SH}}t - I_P \sin(\varphi); \quad (3)$$

$$i_{S_{on}}(t) = \frac{V_{CC}}{L_{SH}}t + I_P(\sin(\omega t + \varphi) - \sin(\varphi)); (4)$$

$$i_{L_{SH_{off}}}(t) = \frac{V_{CC}}{L_{SH}}t - \int_{\frac{2\pi D}{\omega}}^{t} \frac{v_{C_{SH}}(\tau)}{L_{SH}}d\tau - I_P \sin(\varphi); \quad (5)$$

$$v_{C_{SH_{off}}}(t) = V_{CC} + \frac{C_1 \cos(q\omega t) + C_2 \sin(q\omega t)}{-\frac{q^2}{1-q^2}pV_{CC}\cos(\omega t + \varphi)}; \quad (6)$$

$$i_{C_{SH_{off}}}(t) = \frac{\frac{V_{CC}}{L_{SH}}t - \frac{1}{L_{SH}}\int_{\frac{2\pi D}{\omega}}^{t} v_{C_{SH}}(\tau)d\tau}{+I_P(\sin(\omega t + \varphi) - \sin(\varphi))}; \quad (7)$$

where, X_{on} means that the expression X is valid when the switch is in the ON state ($0 < t < \frac{2\pi D}{\omega}$). On the other hand, X_{off} means that the expression X is valid when the switch is in the OFF state ($\frac{2\pi D}{\omega} < t < \frac{2\pi}{\omega}$). The constants C1 and C2 are analytic functions of p,q,φ and V_{DD}, and they were found in [10]. The variables p and q were introduced by [10], in order to simplify the math analysis and are defined as:

$$q = \frac{1}{\omega\sqrt{L_{SH}C_{SH}}} = \frac{\omega_{SH}}{\omega}; p = \frac{\omega L_{SH}I_P}{V_{CC}} = \frac{Z_{LSH}}{R_\omega}; \quad (8)$$

where, ω_{SH} is the natural frequency of the LC_{SH} network , Z_{LSH} is the impedance of L_{SH}, and R_ω is:

$$R_\omega = \frac{V_{CC}}{I_P} = \sqrt{\frac{P_{in}R_{DC}}{2P_{out}/R_L}} = \frac{\sqrt{R_L R_{DC}}}{\sqrt{2}}; \quad (9)$$

where, R_L is the PA load , P_{in} is the power delivered by V_{CC}, P_{out} is the power dissipated by R_L, and R_{DC} is the equivalent resistance that the amplifier imposes to V_{CC}, introduced by [18]. It can be calculated as:

$$R_{DC} = \frac{V_{DC}}{I_{DC}} = V_{CC} \bigg/ \frac{\omega}{2\pi}\int_0^{\frac{2\pi}{\omega}} i_s(t)dt = \frac{R_\omega}{g}; \quad (10)$$

from (10) and (9), we find $R_{DC} = R_L/2g^2$, where

$$g(q, D) = \left\{ \begin{array}{l} \left(\frac{1-\cos(2\pi D)}{2\pi}\right)\cos(\varphi) \\ + \left(\frac{\sin(2\pi D)}{2\pi} - D\right)\sin(\varphi) + \frac{D^2\pi}{p} \end{array} \right\}. \quad (11)$$

TABLE I. LOSSES PARAMETERS

Parameter h
$h(D, q) = \begin{array}{l} h_0(D, q) + \frac{2}{p}\{h_1(D, q)\cos(\varphi) + h_2(D, q)\sin(\varphi)\} \\ + h_3(D, q)\cos(2\varphi) + h_4(D, q)\sin(2\varphi) \end{array}$
Where,
$h_0(D, q) = \frac{8}{3}\frac{D^3\pi^3}{p^2} + 2\pi D - \sin(2\pi D)$
$h_1(D, q) = \sin(2\pi D) - 2\pi D\cos(2\pi D)$
$h_2(D, q) = \cos(2\pi D) + 2\pi D\sin(2\pi D) - \left(1 + 2D^2\pi^2\right)$
$h_3(D, q) = \sin(2\pi D) - \frac{1}{4}\sin(4\pi D) - \pi D$
$h_4(D, q) = -2(\sin(\pi D))^4$
Parameters P_{SS}
$P_{ss_1}(q, D) = \frac{g(q,D)}{p(q,D)}\left((2p(q, D)g(q, D))^2 + (\pi D)^2\right)$
$P_{ss_2}(q, D) = \frac{2}{\pi}g(q, D)\sqrt{h(q, D)}$

It is important to emphasize that the expressions from (1) to (11) can be calculated in terms of V_{CC}, ω, R_L, and P_{OUT} only if p,q,φ and D are known, but in [10] was demonstrated that both φ and p could be solved as an analytic function of both q and D. Additionally, the peak value of the v_C must be less than the breakdown voltage of M. In [13] is proposed a simple analytic relationship between the DC input voltage and the peak value of the v_C. This expression was based in the numerical solution of the analytical equations of the ideal class-E proposed by [10], [12]. This relationship is:

$$V_{c_{max}} \approx V_{CC}\left(\frac{1.83}{1 - D}\right). \quad (12)$$

D. Modeling losses and sizing the transistors

Assuming the same waveforms of the ideal PA and modeling M as an ideal switch and a series resistance (R_{on}), and L_{SH} as an ideal inductor and a series resistor (R_{SH}), the power dissipated (P_{Loss}) can be approximated by:

$$P_{Loss} = P_{L_{SH}} + P_{R_{on}} + P_{driver}; \quad (13)$$

$$P_{L_{SH}} = i_{L_{rms}}{}^2 R_{SH}; \quad (14)$$

$$P_{R_{on}} = i_{S_{rms}}{}^2 R_{on} = i_{S_{rms}}{}^2 \frac{b}{W}; \quad (15)$$

$$P_{driver} = V_{DD}{}^2 f\alpha C_G = V_{DD}{}^2 f\alpha aW; \quad (16)$$

where, P_{SH} is the power dissipated by R_{SH}, $P_{R_{on}}$ is the power dissipated by R_{on}, a and b are technology parameters introduced in [19], P_{driver} is the power used to charge and discharge the NMOS gate capacitance C_G, and the factor α represents the capacitance excess due to the driver implementation. Hence, solving the trade-off between the ON-resistance and the gate capacitance, the W that minimize (13) is:

$$W = \frac{\sqrt{a/b}}{V_{DD}}\sqrt{\int_0^{\frac{1}{f}} i_s(t)^2 dt} = \frac{g(D,q)}{1/\sqrt{h(D,q)}}\frac{2V_{CC}\sqrt{\frac{b}{a\omega}}}{R_L V_{DD}}; \quad (17)$$

therefore, the minimum losses are given by:

$$P_{LOSS} = \frac{V_{CC}{}^2}{R_L}\left(P_{SS1}\frac{1}{Q_{L_{SH}}} + P_{SS2}\frac{\sqrt{\alpha ab}\sqrt{\omega}}{V_{CC}/V_{DD}}\right); \quad (18)$$

where $Q_{L_{SH}}$ is the quality factor of the L_{SH}, and all the loss parameters (i.e. h, P_{SS_1} and P_{SS_2}) are listed in the Table I. For $D = 50$ %, the P_{SS} functions are plotted in the Fig.4(c),

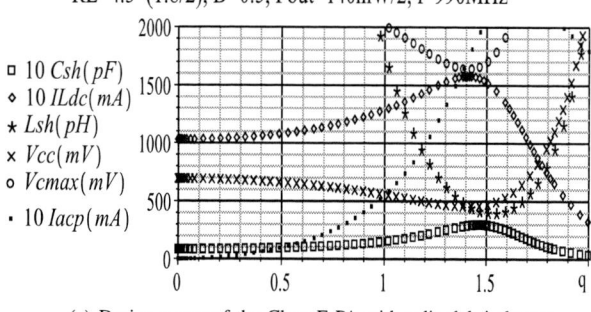

RL=4.3*(1.8/2), D=0.5, Pout=140mW/2, f=990MHz

□ 10 Csh(pF)
◇ 10 ILdc(mA)
★ Lsh(pH)
× Vcc(mV)
○ Vcmax(mV)
· 10 Iacp(mA)

(a) Design space of the Class E PA with split slab inductor

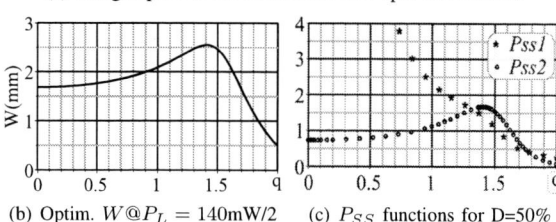

(b) Optim. $W @ P_L = 140mW/2$ (c) P_{SS} functions for D=50%

Fig. 4. Parametric sweep of q

TABLE II.   DESIGN SET GAINS

Circuit element gains
$K_P(q, D, V_{CC}, R_L) = Pout/R_L = 2g(q,D)^2 \left(\frac{V_{CC}}{R_L} \right)^2$
$K_{C^{-1}}(q, D, \omega) = (C_{SH})^{-1}/R_L = \frac{q^2 p(q,D)}{2g(q,D)} (\omega)$
$K_L(q, D, \omega) = L_{SH}/R_L = \frac{p(q,D)}{2 \cdot g(q,D)} \left(\frac{1}{\omega} \right)$
$K_{X_S}(q, D) = \frac{X_s}{R_L} = \frac{v_{CI}/i_{RQ}}{v_{CQ}/i_{RQ}} = \frac{\left(\frac{-2}{I_P T} \int_0^T v_C(t) \cos(\omega t + \varphi) dt \right)}{\left(\frac{-2}{I_P T} \int_0^T v_C(t) \sin(\omega t + \varphi) dt \right)}$
$K_{L_0}^{-1}(\omega, Q_L) = L_0/R_L = \frac{Q_L}{\omega}$
$K_{C_0}^{-1}(\omega, Q_L) = (C_0)^{-1}/R_L = Q_L \omega$
$K_{C_e}^{-1}(\omega, Q_L, q, D) = (C_e)^{-1}/R_L = \omega \left(Q_L - K_{X_s}(q, D) \right)$
Specifications gains
$K_{R_{DC}} = R_L/R_{DC} = 1/2g(q,D)^2$
$K_P(q, D, V_{CC}, R_L) = Pout/R_L = 2g(q,D)^2 \left(\frac{V_{CC}}{R_L} \right)^2$
$K_{IL_{DC}}(q, D, V_{CC}, R_L) = i_{LDC}/R_L = 2g(q,D)^2 \frac{V_{CC}}{R_L^2}$
$K_{IL_{RMS}}(q, D, V_{CC}, R_L) = \frac{i_{LDC}}{R_L} = \left(\begin{array}{c} 4g(q,D)^4 \\ +2\pi^2 \left(\frac{g(q,D) \cdot D}{p(q,D)} \right)^2 \end{array} \right) \frac{V_{CC}^2}{R_L^3}$
$G_P(q, D, V_{CC}, R_L) = P_{out}/V_{CC} = 2g(q,D)^2 \left(\frac{V_{CC}}{R_L} \right)$
$G_V(q, D) = v_{Cmax}/V_{CC} \approx \left(\frac{1.83}{1-D} \right)$

it is clear that for high q values the driver losses are more significant than the inductor losses. Contrary, for low q values the inductor losses are significant. Further, both losses are quasi equals for the q near to 1.412, this q value maximize the output power [10].

E. Proposed design set

The relations between the input parameters (i.e. q, D, Q_L, and ω), the circuit element values (i.e L_{SH}, C_{SH}, C_e, L_0 and R_L), and the circuit specification (e.g. $I_{L_{DC}}, I_{L_{RMS}}, R_{DC}, P_{OUT}, V_{CC}$) should be known for calculating the PA design space. These relations are referred as the design set. in order to limit the calculated design space by technology constraints (e.g. RMS current in the inductor), we expanded the set proposed in [10]. This proposed set is summarized in the Table II. Further, their relations are illustrated in Fig.5. It is important to emphasize that the expressions summarized can be calculated only if p and φ are known, but both can be solved as analytic functions of both q and D. This set was implemented in the software Maple® in order to analyze all the involved trade-offs.

F. Design Methodology Steps

1) Design class-E PA for nominal operation: Calculate the design space of the PA assuming an ideal class E PA, a drain efficiency value (i.e. $\eta = 70\%$), the proposed design set, the operating conditions and a given technology. In this phase, a value for q, D, and R_L that satisfies all the constraints must be identified. Parametric sweeps of the variables R_L and q were made for a fixed P_{out}. In the designed PA the value of D was chosen in order to simplify the design of the differential oscillator ($D = 50\%$). The possible R_L range was found (from 1.8 to 5). In Fig. 4(a) is shown the final sweep of the variable q with R_L fixed in its optimum value ($R_L = 3.87 \ \Omega$), and a $P_{out} = 140/2$ mW (i.e $P_{out} = P_{in}\eta$). In the range $1 < q < 1.1$, all the goals of the analyzed PA are simultaneously achieved; in the superior limit of q, the DC current achieves its maximum possible value (i.e. 150 mA); in the inferior limit of q, the inductance and V_C values (2 nH and 2 V) are both limited by technology constraints. The q chosen was 1.05.

2) Estimate the Sizing of the transistors: Simulate or measure the transistors to obtain the parameters b and a of the technology. The parameters a and b were estimated by simulation using the MOSFET model ($b = 4.3$m $\Omega \cdot$ m, $a = 7.8$ nF/m). Assuming a driver overhead (i.e. $\alpha = 1.5$ overhead of 50%) estimate the optimal value for the transistor width using (17). In the Fig.4(b) is shown this optimum value for the PA as a q function, for $D = 50\%$, $R_L = 3.87 \ \Omega$ and $P_{out} = 140$ mW.

3) Calculate impedance transformation network: Calculate the transformation network that transforms the load to the optimum load. Considering the Fig. 3, the capacitors C_S and C_P form an impedance transformation network (IM_{OUT}) that transforms R_{IL} and L_{IL} in R_{opt} and L_0 respectively. We calculate the capacitances (C_S, C_P and C_e) using the equations listed in the Table III. These equations were developed using a procedure similar to [6]. Further, the maximum voltage parameter (V_{max}) was set as 4 V because in the technology the allowed voltage for I/O is 5 V.

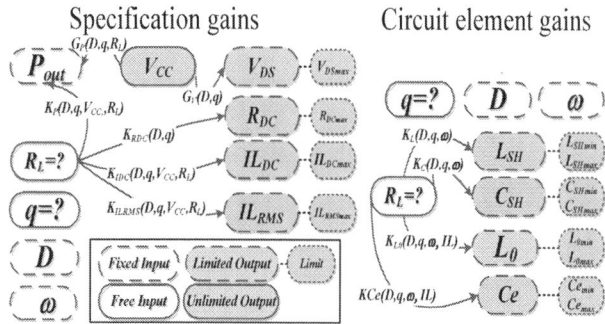

Fig. 5. Proposed design set

TABLE III. CAPACITANCE EXPRESSIONS

$$\frac{1}{C_S} = \omega^2 L_{LI} - \omega R_{LI}\sqrt{\left(\frac{V_{max}^2}{2P_{out}R_{LI}} - 1\right)}$$

$$\frac{2}{C_P} = \frac{\omega\frac{Rs}{R_{opt}}\left(\left(\omega L_{LI} - \frac{1}{\omega C_S}\right)^2 + R_{LI}^2\right)}{\left(\omega L_{LI} - \frac{1}{\omega C_S}\right) - \sqrt{\left(\omega L_{LI} - \frac{1}{\omega C_S}\right)^2\frac{R_{LL}}{R_{opt}} - R_{LI}^2 + \frac{R_{LI}^3}{R_{opt}}}}$$

$$\frac{1}{C_e} = \omega\left(\sqrt{\frac{R_{opt}}{4R_{LI}}\left(\left(\omega L_{LI} - \frac{1}{\omega C_S}\right)^2 - R_{opt}R_{LI} + R_{LI}^2\right)} - K_{X_S}R_{opt}\right)$$

4) Optimization process: From an efficiency point of view, the nominal waveform (ZVS and DZVS) is optimum only if the parasitic components are negligible. For each circuit, the optimum occurs out of the nominal class-E operation. Therefore, the values computed in the previous steps are the initial values of the optimization process assisted by simulation. In the optimization process there are many interactions that are being adjusted. In the available design kit the models of the passive elements do not allow automation, therefore a suboptimal optimization was made, where parametric sweeps of some circuit values were used to find the local maximums of the goal function. These sweeps were performed in a predefined order with the available variables, e.g. in post-layout optimization only V_{CC} and the off-chip component values are available. The optimization process is summarized in the tables IV and V. In these tables the identifier *Opt* means that the set values presented was optimized, as well as *Vsin* or *Vpul* means that the PA was driven by a single-tone or a pulse-signal in the input of the driver for the pre-layout and post-layout simulations, or a single-tone or a pulse-signal in the gate of the transistors for the others simulations.

III. CMOS CLASS-E IMPLEMENTATION

As is shown in the Fig. 6(a), in addition to the PA, the chip includes a PLL for generating the differential input signal for the PA and an envelope detector (ED) to sense the backscattered response at the inductive link (their implementation is out of the scope of this paper). The implemented PA circuit is based on the diagram of Fig. 3(c). The PA pinout is shown in the Fig. 6(a). The PADs were organized in such way to facilitate the wirebonding process to the printed board. The detailed layout of the PA is shown in Fig. 6(b). Each transistor (M) and its driver were divided in sixteen cells. The channel length was 180 nm for all transistors, and their widths are summarized in the Table VI. The driver transistors were sized

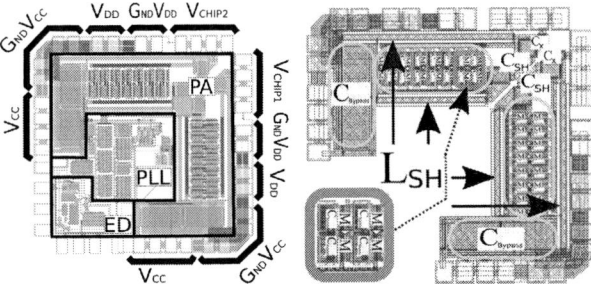

| (a) Blocks and pinout diagram | (b) Detailed layout of the PA |

Fig. 6. CHIP Layout

to drive the switches at the specified frequency considering that transitions must be slow while turning the switches on and fast while turning them off as is shown in the waveform plotted on the Fig. 7(a). Dual Metal-Insulator-Metal (MIM) bypass capacitors were included for the digital voltage source (V_{DD}) for filtering the supply voltage. They were divided and positioned near of each M cell in a routing star-like approach, in that way the AC current is well distributed.

The capacitors C_e and C_{SH} were integrated using the MIM option which provides a higher quality factor than dual MIM option. The inductor L_{SH} was splitted using a $k = 0.612$ in order to simplify the layout. Each slab inductor was implemented in the last metal layer with the higher available width for the slab inductor model (i.e. $W = 25$ um for the used design kit) for high quality factor. The chip floorplanning was very important since the current through the V_{DD} supply nodes can be up to 220 mA rms, and a DC current of 150mA. Hence, a large number of PADs were used for source connections in order to reduce the parasitic impedance of the bondwire, and for increasing its current carrying capacity. For the bypass capacitors the Dual-MIM option was used for high density capacitance value in order to reduce the AC current on the voltage source.

TABLE IV. TABLE OF THE CHANGED CIRCUIT VALUES IN THE OPTIMIZATION

Name	q	V_{CC} (mV)	L_{SH} (nH)	Q	C_{SH} (pF)	Q	W (mm)	C_e (pF)	Q	C_P (pF)
Ideal	1,05	535	1,5	-	15,7	-	-	7,7	-	-
Integrated Lsh	1,05	535	1,44	17	15,7	-	-	7,7	-	-
Integrated Csh	1,05	535	1,50	-	15,9	98	-	7,7	-	-
Int. MOS W	1,05	535	1,50	-	15,7	-	2,63	7,7	-	-
Int. L,C and M	1,05	535	1,44	17	15,9	98	2,63	6,4	-	6,2
Int. L,C and M Opt.	1,10	600	1,40	17	14,6	105	2,27	6,3	185	6,2
Int. (L,C, M, Cx). Opt.	1,10	600	1,40	17	14,6	105	2,36	6,2	210	6,2
Int. (L,C, M, Cx). Vsin Opt.	1,10	600	1,40	17	14,6	105	3,87	6,2	210	6,2
PreLayout	1,10	600	1,22	15	14,8	57	4,56	6,2	210	6,2
PostLayout	1,10	600	1,22	15	14,8	57	4,56	6,2	210	6,2
PostLayout Opt.	1,10	660	1,22	15	14,8	57	4,56	6,2	210	5,8
PostLayout Vsin	1,10	660	1,22	15	14,8	57	4,56	6,2	210	5,8

TABLE V. TABLE OF THE MERIT FIGURES USED AS GOALS IN THE OPTIMIZATION

Name	P_L (mW)	η (%)	PAE (%)	R_{DC} W	V_{max} (mV)	$I_{L_{DC}}$ (mA)	R_L type
Ideal	148	99,9	99,9	3,89	2017	137	Ropt
Integrated Lsh	120	86,4	86,4	4,13	1833	129	Ropt
Integrated Csh	138	97,0	97,0	4,03	2063	133	Ropt
Int. MOS W	100	84,2	74,1	4,82	1662	111	Ropt
Int. L,C and M	80	72,1	55,3	5,13	1683	104	LI+IM
Int. L,C and M Opt.	106	73,0	61,5	4,98	1886	121	LI+IM
Int. (L,C, M, Cx). Opt.	100	66,1	55,3	4,75	1629	126	LI+IM
Int. (L,C, M, Cx). Vsin Opt.	114	62,8	62,3	3,99	1476	150	LI+IM
PreLayout	100,1	63	47,3	4,55	1854	132	LI+IM
PostLayout	83,77	59,8	42,2	5,14	1706	117	LI+IM
PostLayout Opt.	116,3	60,3	46,3	4,52	1659	146	LI+IM
PostLayout Vsin	117,5	59,4	45,7	4,41	1682	150	LI+IM

TABLE VI. THE WIDTHS OF TRANSISTORS OF THE PA

MOS Type	1th driver stage		2th driver stage		Switch (M)
	N	P	N	P	N
W (um)	2	4	2,48	2,48	2,85
Fingers	1	1	5	5	100
We (um)	32	64	198	198	4560

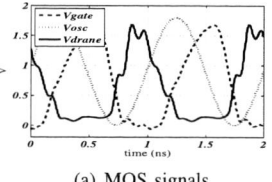
(a) MOS signals

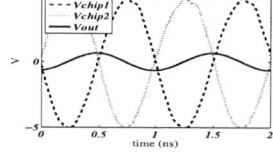
(b) Output voltages

Fig. 7. PA simulated waveforms

A. Simulated Results and State of Art

The designed PA was simulated using Cadence® Virtuoso® and Advanced Design System (ADS®). The circuit and the parasitic resistances and capacitances were extracted from the layout. The layout view of the PA is shown in Fig. 6. As shown in the Table IV, the calculated values of the circuit elements (Int. L,C and M) and the final values of the proposed PA (PostLayout Vsin) had good agreement, i.e. percentage change less than 20% in all circuit values except the V_{cc} and (W). The performance of the designed PA is compared with the PAs found in literature in Table VII. The efficiency achieved is the best between the references with integrated inductor. The simulated waveform of both the differential chip outputs and the load voltage is plotted in the Fig. 7(b). The voltage excursion of the PAD is between the designed specifications.

TABLE VII. COMPARISON TABLE OF THE PERFORMANCE

Ref	f (MHz)	P_L (dBm)	PAE %	Area (mm²)	Tech. (nm)	R_L oms	$L_{DC_{feed}}$ type	PA class	Tested type
[7]	820	29	70,7	0,5	180	50	External	E	Exp.
[6]	990	25,1	58	1,5	180	1,8	w/o L	D	Sim.
This Work	990	20,7	45,7	1,5	180	1,8	On-chip Split L	E	Sim.
[8]	900	29,5	41	4	250	50	Bondwire	E	Exp.
[9]	800	28	40	1,5	180	50	On-chip Transformer	E	Exp.

IV. CONCLUSIONS

An analytic design methodology to find the optimum transistor width of a class-E PA was presented. An integrated differential class-E PA with split slab inductor was designed and simulated with good agreement between the initial values of the circuit elements (calculated), and the final values after the optimization process. Hence, the complexity of this process decreases thanks to the suboptimal first point found by the proposed analytical solution. In the post-layout simulations, the PAE of the PA was 45.7 % when 20.7 dBm was delivered to its load and a 2.3 dBm was supplied by the AC source. This good result was obtained thanks to the proposed design methodology. The PA designed in this work is highly integrated, only two capacitors are left outside the chip. The silicon area is kept small because the PA uses slab inductors and splitted inductor class E topology.

ACKNOWLEDGMENT

The first author would like to thank COLCIENCIAS and the Pontificia Universidad Javeriana for the financial support. Also, the authors would like to thank the CNPq and the INCT/NAMITEC for their partial financial support and all the students of the Radio Frequency integrated circuits Group (GRF-UFSC) for the important discussions.

REFERENCES

[1] A. Whitmore, A. Agarwal, and L. Xu, "The internet of things, a survey of topics and trends," *Information Systems Frontiers*, vol. 17, no. 2, pp. 261–274, 2014.

[2] A. Fajardo and F. Rangel de Sousa, "A taxonomy for learning, teaching, and assessing wireless body area networks," in *Proc. IEEE 7th Latin American Symp. on Circuits and Syst. (LASCAS2016)*, 2016.

[3] S. Movassaghi, M. Abolhasan, and J. Lipman, "Wireless Body Area Networks: A Survey," *IEEE Commun. Surveys and Tutorials*, pp. 1–29, 2013.

[4] K. Niotaki, A. Collado, A. Georgiadis, S. Kim, and M. M. Tentzeris, "Solar/electromagnetic energy harvesting and wireless power transmission," *Proceedings of the IEEE*, vol. 102, no. 11, pp. 1712–1722, Nov 2014.

[5] A. Georgiadis and A. Collado, "Solar powered class-e active antenna oscillator for wireless power transmission," in *2013 IEEE Radio and Wireless Symp. (RWS2013)*, Jan 2013, pp. 40–42.

[6] F. Cabrera and F. Rangel de Sousa, "A 25-dbm 1-ghz power amplifier integrated in cmos 180nm for wireless power transferring," in *Integrated Circuits and Systems Design, 2016.Proceedings*, 2016.

[7] J.-S. Paek and S. Hong, "A 29 dbm 70.7amplifier for pwm digitized polar transmitter," *Microwave and Wireless Components Letters, IEEE*, vol. 20, no. 11, pp. 637–639, Nov 2010.

[8] C. Yoo and Q. Huang, "A common-gate switched 0.9-w class-e power amplifier with 41vol. 36, no. 5, pp. 823–830, May 2001.

[9] W.-Y. Kim, H. S. Son, J. H. Kim, J. Y. Jang, I. Y. Oh, and C. S. Park, "A fully integrated triple-band cmos class-e power amplifier with a power cell resizing technique and a multi-tap transformer," *Microwave and Wireless Components Letters, IEEE*, vol. 23, no. 12, pp. 659–661, Dec 2013.

[10] M. Acar, A. Annema, and B. Nauta, "Analytical design equations for class-e power amplifiers," *Circuits and Systems I: Regular Papers, IEEE Transactions on*, vol. 54, no. 12, pp. 2706–2717, Dec 2007.

[11] R. Sadeghpour and A. Nabavi, "Design procedure of quasi-class-e power amplifier for low-breakdown-voltage devices," *Circuits and Systems I: Regular Papers, IEEE Transactions on*, vol. 61, no. 5, pp. 1416–1428, May 2014.

[12] M. Acar, A. Annema, and B. Nauta, "Generalized design equations for class-e power amplifiers with finite dc feed inductance," in *Microwave Conference, 2006. 36th European*, Sept 2006, pp. 1308–1311.

[13] A. Fajardo and F. Rangel de Sousa, "Simple expression for estimating the switch peak voltage on the class-e amplifier with finite dc-feed inductance," in *Proc. IEEE 7th Latin American Symp. on Circuits and Syst. (LASCAS2016)*, 2016.

[14] N. Sokal and A. Mediano, "Redefining the optimum rf class-e switch-voltage waveform, to correct a long-used incorrect waveform," in *Microwave Symposium Digest (IMS), 2013 IEEE MTT-S International*, June 2013, pp. 1–3.

[15] F. Cabrera, Fabian L. Rangel de Sousa, "Optimal design of energy efficient inductive links for powering implanted devices," in *Biomedical Wireless Technologies, Networks, and Sensing Systems (BioWireleSS), 2014 IEEE Topical Conference on*, Jan. 2014, pp. 37–39.

[16] F. Cabrera and F. Rangel de Sousa, "A CMOS fully-integrated wireless power receiver for autonomous implanted devices," in *Circuits and Systems (ISCAS), 2014 IEEE International Symposium on*, Jun. 2014, pp. 1408–1411.

[17] S. Bandyopadhyay and A. P. Chandrakasan, "Platform architecture for solar, thermal, and vibration energy combining with mppt and single inductor," *IEEE Journal of Solid-State Circuits*, vol. 47, no. 9, pp. 2199–2215, Sep. 2012.

[18] A. Fajardo and F. Rangel de Sousa, "Modeling and design of high-efficiency power amplifiers fed by limited power sources," in *29th Symp. on Integrated Circuits and Syst. Design (SBCCI 2016)*, 2016.

[19] B. R. W. Stratakos, Anthony J. and S. R. Sanders, "High-efficiency low-voltage dc-dc conversion for portable applications." in *Proc. Int. Workshop on Low-Power Design*, April Napa, CA. 1994, pp. 21–27.

978-1-5090-2737-8/16 $31.00 © 2016 IEEE

Low-Power Hardware Design for the HEVC Binary Arithmetic Encoder Targeting 8K Videos

Fábio Luís Livi Ramos[1,3], Jones Goebel[2], Bruno Zatt[2], Marcelo Porto[2], Sergio Bampi[1]

[1]PPGC – UFRGS – Federal University of Rio Grande do Sul
Porto Alegre, RS, Brazil

[2]UFPel – Federal University of Pelotas
Pelotas, RS, Brazil

[3] Unipampa – Federal University of Pampa
Bagé, RS, Brazil

fabioramos@unipampa.edu.br, {jwgoebel, zatt, porto}@inf.ufpel.edu.br, bampi@inf.ufrgs.br

Abstract—The HEVC (High Efficiency Video Coding) has risen as the state-of-the-art video coding standard, targeting the processing of high quality video. As one of its main components, the entropy encoding is based on the CABAC (Context-adaptive binary arithmetic coding) algorithm, which allows parallelization at higher video data representation (i.e. slices, tiles, and wavefront parallel processing) whereas posing serious challenges for low-level parallelization. Low-power dissipation is a sough-after goal for hardware video encoder designs, along with high performance. This work presents a hardware architecture for the sub-module BAE (Binary Arithmetic Encoder) of the CABAC, featuring low power techniques, such as clock gating and operand isolation. The techniques were inserted after a careful analysis, based on the statistical behavior of CABAC inputs for real video sequences. A 4-cores BAE core structure was implemented in order to provide high performance. As results, power reduction ranging 25-40% was achieved by applying the aforementioned techniques in specific parts of the architecture, while still accomplishing the requirements for UHD 8K real-time processing.

Keywords—HEVC; CABAC; BAE; Low-power; Clock Gating; Operand Isolation.

I. INTRODUCTION

Video applications have been of increasing interest for both academia and industry for the past years. This situation led to an increasing search for higher quality videos along smaller output bitstream representation. As consequence, improvements and updates in older video coding standards begat the newest standard, called High Efficiency Video Coding (HEVC) [1], created in collaboration between ITU-T (VCEG) and ISO/IEC (MPEG). Compared to the previous standard, H.264/AVC [2], HEVC achieves twice the compression while maintaining the same objective video quality of its antecessor [3].

The HEVC standard allows only one type of entropy algorithm, which is CABAC (Context Adaptive Binary Arithmetic Coding), due to its higher compression gains at the cost of increased computational complexity.

The new parallelization tools introduced in HEVC [1], such as tiles and wavefront parallel processing, allow the parallelization of CABAC cores for higher processing rate, which is a desirable goal, depending on the application. Nevertheless, the challenge to achieve high throughput and low-level parallelization at a single CABAC core fashion is still a desirable goal.

CABAC module has been subject of several studies in the past few years, targeting hardware solutions [4-10], where many important contributions for architectural solutions were presented, and we shall highlight some of the most prominent works. One-round renormalization approach is proposed in [4]. The default pipeline alternative to speed-up the design, while still respecting the data dependencies between bins (i.e. syntactic elements transformed to binary values at the beginning of CABAC flow) is shown in [5]. The solution presented in [6] aggregates both previous mentioned works together with remarkable improvements on the critical paths of the design, such as pre-renormalization and hybrid path-coverage, to achieve higher clock frequency and thus higher bins throughput.

Although the aforementioned works contribute for a high performance improvement, achieving in some cases the Ultra High Definition (UHD) requirements for real-time processing, few of them focus on reducing/analyzing the power consumption of the design [4,7]. The study shown in [11] presents a parallel approach at the algorithm level of CABAC to increase frequency and throughput. Nevertheless, at the algorithm level, it is very difficult to assess the implementation of fine-grain techniques at the architectural level of the design.

Our work focuses on CABAC critical sub-block (BAE – Binary Arithmetic Encoder), which represents the higher complexity of the whole component [6]. The main contributions of this paper follow:

- **Analysis focusing low-power design**: A careful assessment of the state-of-the-art CABAC architecture (targeting BAE sub-module) was done, considering also the statistical behavior of its inputs, to analyze possible low-power technique insertions on a fine-grain level.

978-1-5090-2737-8/16 $31.00 © 2016 IEEE

- **Development of the low-power BAE architecture**: an architecture of the BAE sub-module was done and the chosen low-power techniques (clock gating and operand isolation) were implemented, achieving improved power consumption at a minor implementation effort, while keeping the requirements for real-time 8K UHD video processing.

The paper is divided as follows: Section II gives an overview of CABAC algorithm. Section III presents the analysis made for low-power opportunities on the BAE core. The proposed architecture is described on Section IV. Section V presents the results and comparisons, whereas Section VI concludes the work.

II. CABAC ALGORITHM

The CABAC is the only algorithm used on HEVC standard for entropy encoding and decoding. The module function is based on recursive interval subdivision [12], where, at each round of the algorithm, the interval used is updated using the values of the interval relative to the current symbol.

The inputs of the entropy encoder in the context of HEVC are the syntactic elements coming from all other modules of the encoder (i.e. data processed at the other stages of encoding), such as transform residues, motion vectors, etc [1]. The CABAC process flow is presented in Fig. 1, which consists of three main parts: Binarizer (Binarization), Context Modelling and Binary Arithmetic Encoding.

Since CABAC can only process binary elements, all syntactic elements must be *binarized* at Binarizer, which will be called after that process simply as *bins*.

The bins are classified as regular or bypass and this classification will affect their processing after the binarization. By bypass, we consider a bin whose probability to occur is assumed to be uniformly distributed and which the Context Modeling will be bypassed (i.e. the occurrence probability is 50% for '0' or '1'). The regular bins, on the other hand, must undergo the Context Modeling, which will insert data dependencies between each bin processed as regular [1]. The Context Modeling sets the probability of the regular bins, based on its previous values.

Regular bins have yet another division: as already mentioned, CABAC processes only binary values, thus bins could be '0' or '1'. Hence, the current bin being processed could be the Most Probable Symbol (MPS) or the Less Probable Symbol (LPS). This classification is defined on the fly, according to the statistics and characteristics of the symbols, taken from the standard [1].

The last CABAC stage corresponds to the Binary Arithmetic Encoder (BAE), which will effectively calculate the new interval for the current bin, based on two main variables: *Range* and *Low* (respectively the overall range of probabilities value, and the lower bound value of this range of probabilities). This is the block responsible for the most complex process of the whole CABAC, since Binarizer and Context Modeling have no or little data dependencies, and thus are easier to be pipelined/parallelized than BAE [6].

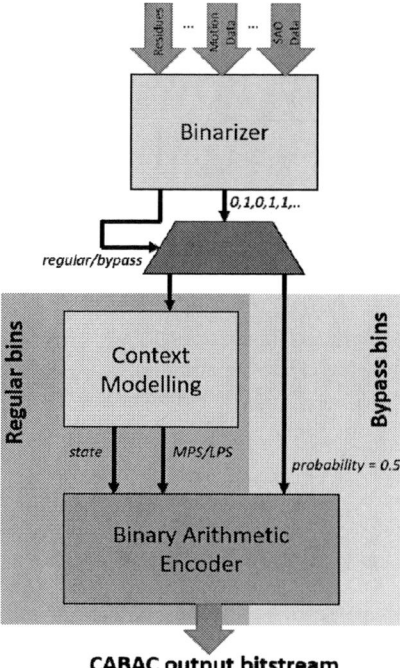

Fig. 1. CABAC coding flow

A. Binary Arithmetic Encoder

The pseudo-code presented in Fig. 2 shows the behavior of BAE function, where rLPS and rMPS represent the range probability for the Less Probable Symbol and Most Probable Symbol, respectively. The value of rLPS is calculated on a LUT-based fashion (i.e. the values of rLPS are pre-calculated by the standard [1]), while rMPS is the Range variable minus rLPS. State is a Context Modelling variable, which is updated depending if the current bin is a LPS or MPS, following the transitions defined on the standard [1]. The gray-shaded lines represent the parts of the algorithm that are processed at Context Modelling stage.

The *Range* (and thus *Low*) variable may fall out of the minimum value expected after the update process (i.e. when the updated *Range* has a value less than 256. *Range* is a nine-bit variable and *Low* a ten-bit variable). Hence, a renormalization may be needed after the process represented by Fig.2. This may lead to the generation of *Outstanding Bits* (OB), which will have to be calculated for the correct CABAC output generation [1].

As depicted in Fig. 3, an example is shown where only two possible symbols can occur, the same way as in CABAC ('0' or '1'). Each of these symbols has a probability to occur at each round. For instance, the first column in Fig. 3 shows that the symbol '1' has an approximate probability of 80% to occur (i.e. the MPS of the example), while the probability of symbol '0' is approximately 20% (i.e. the LPS of the example). These values are quantized to an overall range varying from 0 to 510 (which corresponds to 100% of the probability), while from 0 to 100 corresponds to the occurrence probability of 20% from symbol '0', and 100 to 510 to the occurrence probability of 80% from symbol '1'.

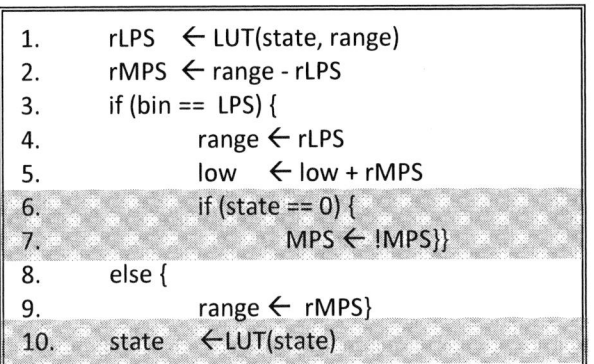

```
1.      rLPS   ← LUT(state, range)
2.      rMPS   ← range - rLPS
3.      if (bin ==  LPS) {
4.              range ← rLPS
5.              low   ← low + rMPS
6.              if (state == 0) {
7.                      MPS ← !MPS}}
8.      else {
9.              range ← rMPS}
10.     state   ←LUT(state)
```

Fig. 2.   BAE algorithm pseudo-code

After each symbol occurs, the probability overall range is updated, and the new range will be contained within the value that represents the amount of probability of the previous processed symbol. According to the example in Fig. 3, at the first round the first symbol was a '1' (the MPS), thus, the new updated range for the next processing symbol will be modified to 410. The probabilities for the occurrences of '0' and '1' are updated, considering this new range value (the symbol that occurred will have its occurrence probability increased for the next round).

At the third round of the example on Fig. 3, a '0' occurs (the LPS). Again, the overall range probability for the next round will have the value of the current processed symbol. Since it was a LPS, the new low boundary of the overall range will consist of the current low boundary summed with the range probability of the MPS (as seen in Fig. 2). In the example, the low bound for the fourth round is updated to 282.

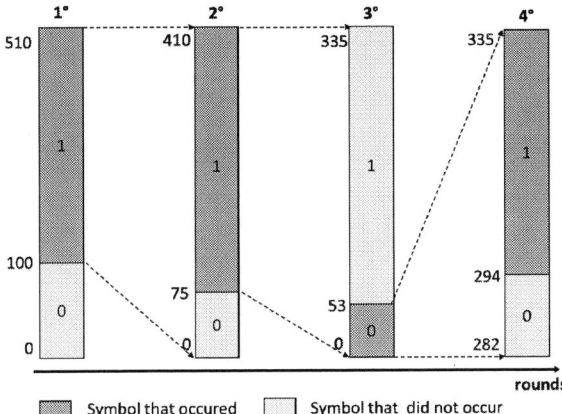

Fig. 3.   Example of recursive interval subdivision

III. BAE ANALYSIS TARGETING LOW-POWER DESIGN

The proposed BAE sub-module hardware architecture is shown in Fig. 4. This architecture is based on the work presented in [6] that implements a hardware architecture for the CABAC module, which presents the highest throughput found in the literature.

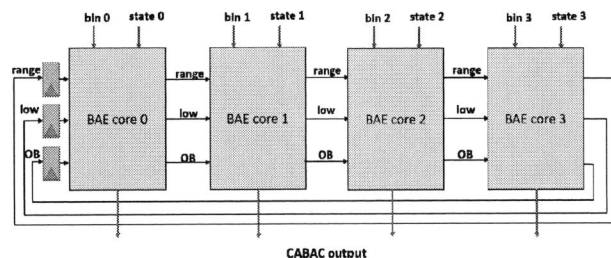

Fig. 4.   Four BAE-core architecture

The design is composed of four identical BAE cores appended one to the other. The choice to use four cores comes from the fact that by using more cores than four would certainly lead to more bins being processed per cycle, but also at the cost of lower clock frequency. This happens because the critical path would increase (the interconnections between each core inputs-outputs go through a purely combinational path, and thus the operations among all cores are done in a single clock cycle). As presented in [6], the throughput results for more cores appended (e.g. five or six) are not considerably higher than with four cores, at the cost of more area/resources.

The outputs to one core to another are the updated *Range*, *Low* and OB, which are the values that the next BAE core needs to calculate the recursive interval algorithm for the bin it will process, in order to generate the output bitstream. That variables run through an entire combinational path between one BAE to the next, until they are registered after the fourth core. This way, it is easy to see why more cores would lead to smaller clock frequency, as mentioned before.

Before we take a further look at the architecture, we shall analyze how the statistical behavior of the inputs is, according its three basic possible situations: regular MPS bin, regular LPS bin, or bypass bin.

Table I presents the statistical distribution of the bins according their types, as reported in [6] for some common used video test sequences with QP values of 22 and 37. As can be seen, regular MPS bins correspond to an average of 55% of all bins distribution, while regular LPS bins happen on an average of 20%, and bypass on an average of 25%.

TABLE I.   PROPORTION OF BINS BY TYPE FOR HEVC LOW-DELAY AND RANDOM-ACESS AND QPS 22 AND 37

Sequence	P_{MPS}	P_{LPS}	P_{Bypass}
BasketeballDrive	54.8%	18.1%	27.1%
Traffic	56.4%	21.1%	22.5%
PeopleOnStreet	50.9%	19.9%	29.2%
BQTerrace	61.6%	18.4%	20.0%
Kimono	52.7%	18.2%	29.1%
Average	55.3%	19.2%	25.6%

A bin could follow one of three possible flows inside BAE, and these flows do not necessarily share all the same components and circuitry. Since the probabilities of bins type occurrence vary along time and depending on video characteristics, this may lead to unnecessary switching components at certain moments.

IV. PROPOSED LOW-POWER BAE ARCHITECTURE

The requirement for UHD 8K videos (7,680x4,320 pixels) real-time processing at level 6.2 high tier is 800 Mbits/s [1] and is our goal for performance. At the scope of BAE, we need to convert that information to a bin-ratio per second. According [13], the bin-to-bit ratio of 800 Mbit/s will correspond to approximately 1 Gbin/s, at the worst case across multiple frames. Thus, as state on the previous section, since the proposed work has a static throughput of 4-bin per clock cycle, we will need a clock frequency of approximately 250-260 MHz to achieve the requirements.

Fig. 5 shows the proposed BAE architecture for a single BAE core, which will be appended to other three to compose the whole BAE architecture (as seen on Fig. 4). The proposed BAE architecture consists in a four-stage pipeline [6]. The first stage corresponds to the pre-normalization step for LPS bins and the first part of *Range* updating process. The second stage calculates and renormalizes rMPS and thus concludes the *Range* updating process by choosing between rMPS or rLPS. The third stage processes the *Low* update for either regular or bypass bins; and the fourth stage is responsible for the output bitstream logic.

As presented in Fig. 5, there are different datapaths followed by bins whether the current symbol is a regular MPS bin or a regular LPS bin or a bypass bin. This will lead to different registers into the pipeline barriers to be actually set.

The renormalization process, which either *Range* or *Low* variables may undergo, is a shit-left logic by some amount of bits, until a minimum value required is reached (already

mentioned on Section II). This is done using Leading Zero Detector (LZD) [14] and shift-left circuits, which do the renormalization within one clock cycle (i.e. using combinational logic). The LZD circuit indicates the position of the most significant '1' value into the vector. Since *Range* is a nine-bit variable and the smaller rLPS value is two, three bits are necessary to represent the shift-left amount for each of the four candidates.

The shift-left amount is the same for *Range* and *Low* variables. That is the reason why the rLPS shift-amount travels to the second pipeline stage (i.e. twelve registers named as rLPS shift amount on Fig. 5) and to the third pipeline stage (named as shift amount on Fig. 5).

One might notice that the pre-renormalization of rLPS is done at the first pipeline stage, and the decision of which of the four rLPS value candidates will be used to calculate rMPS is decided at the second stage (because it depends on *Range* in, that is fed at the second stage). Hence, thirty-two registers are needed at the pipeline barrier between these stages (named as rLPS candidates on Fig. 5).

The *Low* update (third pipeline stage) of a LPS bin consists of the sum between *Low* input and rMPS (second pipeline stage) as seen on Fig. 2. This situation leads to nine registers (i.e. nine bits are required for that variable) to transmit the rMPS (named the same way on Fig.5) through the pipeline stages when a LPS bin occurs.

When a bypass bin with value '1' is the current symbol, the *Low* update process (third stage) also requires the input *Range* value (which is supplied at the second stage) before update (i.e. 2 * Low + Range is the calculation applied [1]). Again, more nine registers are necessary for that transmission (named as range on Fig. 5).

Before *Low* update process (at the third stage of pipeline), regular and bypass bitstreams shall merge [6] (bypass bins do not have to update the *Range* variable). Moreover, the logic for

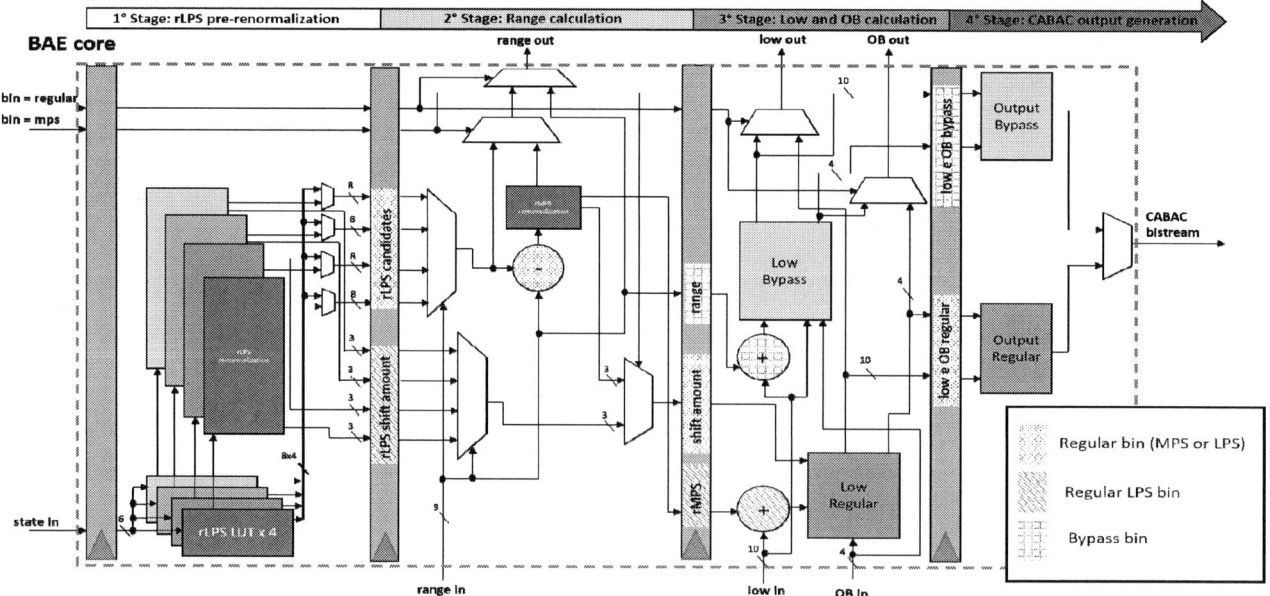

Fig. 5. BAE core architecture

978-1-5090-2737-8/16 $31.00 © 2016 IEEE   46

calculating *Low* variable is different for regular and bypass bins, as depicted in [4]. Nevertheless, the output logic (fourth stage) is also different for regular and bypass bins, also shown in [4].

The output logics of a either a regular or a bypass bin need the updated *Low* and OB values (as reported in [4]). These values require respectively twenty registers and eight registers (ten and four for each type of bin) to go through the third to the fourth stage (name as low and OB regular and low and OB bypass on Fig. 5).

The combinational logic within the pipeline barriers is composed of operators that are not needed in all situations, such as: the nine-bit input subtractor for rMPS calculation (second stage); the ten-bits adder for LPS Low calculation and the eleven-bit input adder for *Low* update logic for bypass bins (both at the third stage). All of these components are shown in the architecture datapaths on Fig. 5 with the respective colors according the type of bins that will require them.

Considering the behavior mentioned above, we decided to apply the following low-power techniques on the BAE architecture presented before:

A. Clock Gating

This technique aims the internal dynamic power consumption of flip-flops at the instants they do not have to be updated [15] (i.e. when the inputs are enabled to be fed into the register). By gating the clock, not only the internal dynamic power of the flip-flops is decreased, but also no update into the registers values outputs is accomplished and thus the combinational logics connected to them also is stalled at those situations.

Clock gating technique is worthy to be used if the registers that will have its input clock blocked shall not be update all the time during the component's operation, and if at least one clock gating cell blocks the clock of more than one register (since the cell consists of a latch/flip-flop and an AND gate).

The flip-flops elected for clock gating are the following, considering the components presented in the last section (refer to Fig. 5 for information):

- **From first stage to second stage**: the *twelve registers* of rLPS shift amount related to pre-renormalization of a regular LPS bin, and the *thirty-two registers* related as rLPS candidates, which are used to effectively discover the actual value of rLPS and thus to calculate *Range* and rMPS for regular bins of both types.

- **From second stage to third stage**: the *nine registers* related to the *Low* update in case the bin is a regular LPS (to be used in the sum *Low* + rMPS) In addition, the *nine flip-flops* that feed the *Range* input value to be used when the current bin is a bypass with value '1'. Finally, the *three flip-flops* used to renormalize *Low* variable (i.e. Shift Amount), in case a regular bin of either type has occurred.

- **From third stage to fourth stage**: the *ten* and *four registers* related respectively to *Low* and OB variables

that are needed for bypass bin output logic, as long as the *ten* and *four registers* related to the same variables, but for regular bins output. The choice to separate these registers in two groups was made because we can completely disable one of the groups, according the type of current bin. Hence, the combinational logic for either regular or bypass bins output at fourth state is stalled (at cost of twice the number of registers originally required, and using two clock gating cells – one for each group).

B. Operand Isolation

This technique targets the switching/capacitive dynamic power of considerable big combinational logics (e.g. adder, multipliers, etc.). No switch into the aimed logic will occur if the inputs are kept with the same previous values for the time they are not required [16].

This approach is only worthy if the combinational logic is big enough and it is only needed to be active during some parts of the operation (since AND gates or latches are used to block the inputs of the logic).

A latch-based approach is applied in the proposed design to avoid any input changes coming from *Range* input to the subtractor at the second stage when a bypass bin is the current one (i.e. the *Range* update is not necessary). In addition, both adders of third stage will have the *Low* input blocked, according to the conditions they are used (i.e. one adder for regular LPS bins only, while the other for bypass bins only – refer to Fig. 5).

V. SYNTEHSIS RESULTS AND COMPARISONS

The architecture was described in VHDL and synthesized for 45 nm Nangate FreePDK cells library with and without the low-power techniques mentioned at the previous section.

A power consumption analysis was made for the pre-layout design considering that four bins will be fed to the four identical BAE cores simultaneously, and with an input bitstream for the whole BAE design that respects the percentage of bins type as shown in Table I. Total power results are presented in Table II (at 0.95v and -40°C). Even that each BAE core is identical, the input bin stream may vary between them, which leads to the different power dissipations presented. Static power is negligible (around 1-2% of the total consumption).

As shown, the insertion of clock gating resulted in power savings ranging from 10% to 16%, while operand isolation resulted in power gains ranging from 12% to 28%, both compared with the baseline architecture, without any power saving technique. When both techniques are used simultaneously, the power savings range from 25% to 40%.

The overall results and comparisons are shown in Table III. Notice that the power consumption for our work consists of the sum of power values from the four cores presented in Table II when using both clock gating and operand isolation. The power dissipation of [7] and the gates count of [10] are related to the whole CABAC design. All other results consider only BAE architecture.

978-1-5090-2737-8/16 $31.00 © 2016 IEEE

TABLE II.   POWER RESULTS FOR THE PROPOSED BAE ARCHITECTURE

	Baseline	Clock Gating	Operand Isolation	Clock Gating + Operand Isolation
BAE 0	1.076 mW	0.971 mW	0.839 mW	0.675 mW
BAE 1	1.002 mW	0.887 mW	0.728 mW	0.604 mW
BAE 2	0.897 mw	0.756 mW	0.667 mW	0.545 mW
BAE 3	0.893 mW	0.793 mW	0.792 mW	0.670 mW
Power Saving from Baseline (Range)	-	10%-16%	12%-28%	25%-40%

TABLE III.   COMPARISONS WITH RELATED WORKS

Design	Liu [4] '11	Chen [5] '10	Zhou [6] '15	Kuo [7] '06	Fei [8] '11	Peng [9] '13	Vizzotto [10] '15	Our Work
Clock Frequency (MHz)	238	222	420	-	279	357	380	280
# of bins/cycle	2	1~8	4.37	-	4	1.43	2.37	4
Mbins/s	634	1776	1836	200	1116	439	900	1120
Technology (nm)	90	130	90	180	90	130	130	45
Gates count (k)	3.9	14.7	7.98	-	8.22	24.9	31.1	9.95
Power Dissipation (mW)	4.9	-	-	20.7	-	-	-	2.49
mW/Gbin	7.72	-	-	103	-	-	-	2.24
Suports 8K videos	No	Yes	Yes	No	Yes	No	No	Yes

The proposed work together with [5], [6] and [8] are the only able to process 8K videos. Our work has a higher gate count than [6] and [8], but this is expected, since the insertion of clock gating and operand isolation would lead to increasing gate count. Only [4] and [7] present power results (for pre-layout design also), but compared with the BAE in [4] (at 0.9v), even with the scaling of the technology (considering it would scale linearly [17]), our architecture still has lower power dissipation cost by Gbin being processed.

VI. CONCLUSION

The present work proposes a low-power architecture for the BAE core of CABAC algorithm. After an analysis, based on the statistical inputs of BAE core and state-of-the-art designs, clock gating and operand isolation were chosen to be inserted into our architecture. The application of these techniques leads to a reduction of power consumption ranging from 25% to 40%, while keeping the requirements for UHD 8K videos real-time processing.

ACKNOWLEDGMENT

We have a special acknowledgement to CNPq, CAPES and FAPERGS to support this work.

REFERENCES

[1] High Efficiency Video Coding document ITU-T H.265/ISO/IEC 23008-2 HEVC, 2013.

[2] International Telecommunication Union "ITU-T Recommendation H.264 (03/05): Advanced Video Coding for Generic Audiovisual Services", 2005.

[3] G. J. Sullivan, J. R. Ohm, W.J. Han and T. Wiegand, "Overview of the High Effciency Video Coding (HEVC) Standard," IEEE Transactions on Circuits and Systems for Video Technology (TCSVT), vol. 22, no. 12, pp. 1649-1668, Dec. 2012.

[4] Z. Liu and D. Wang, "One-round renormalization based 2-bin/cycle H.264/AVC CABAC encoder," in Proc. 18th IEEE International Conference on Image processing (ICIP), Sep. 2011, pp.369-372.

[5] J. W. Chen, L.C.Wu, P.S. Liu and Y.L. Lin, "A high-thoughput fully hardwired CABAC encoder for QFHD H.264/AVC main profile video," IEEE Transactions on Consumers Electronics (TCE), vol. 56, no. 4, pp. 2529-2536, Nov. 2010.

[6] D. Zhou, J. Zhou, W. Fei and S. Goto, "Ultra-high-throughput VLSI architecture of H.265/HEVC CABAC encoder for UHDTV applications," IEEE Transactions on Circuits and Systems for Video Technology (TCSVT), vol. 25, no. 3pp. 497-507, Mar. 2015.

[7] C.C. Kuo and S.F. Lei, "Design of low power architecture for CABAC encoder in H.264," in Proc. IEEE Asia Pacific Conference on Circuits and Systems (APCCAS), Dec. 2006, pp. 243-246.

[8] W. Fei, D. Zhou and S. Goto, "1 Gbin/s CABAC encoder for H.264/AVC," in Proc. 19th European Signal Processing Conference (EUSIPCO), Sep. 2011, pp. 1524-1528.

[9] B. Peng, D. Ding, X. Zhu and L. Yu, "A hardware CABAC encoder for HEVC, " in Proc. IEEE Internacional Symposium on Circuits and Systems (ISCAS), May 2013, pp. 1372-1375.

[10] B. Vizzotto, V. Mazui and S. Bampi, "Area efficient and high throughput CABAC encoder architecture for HEVC, " in Proc. 15th IEEE International Conference on Electronics, Circuits, and Systems (ICECS), Dec. 2015, pp. 572-575.

[11] V. Sze, A.P. Chandrakasan, M. Budagavi and M. Zhou, "Parallel CABAC for low power video coding," in Proc. 15th IEEE International Conference on Image Processing (ICIP), Oct. 2008, pp. 2096-2099.

[12] D. Marple, H. Schwarz and T. Wiegand, "Context-based binary arithmetic coding in the H.264/AVC video compression standard," IEEE Transactions on Circuits and Systems for Video Technology (TCSVT), vol. 13, no. 7, pp. 620-636, Jul. 2003.

[13] Y. H. Chen and V. Sze, "A deeply pipelined CABAC decoder for HEVC supporting level 6.2 high-tier applications, " IEEE Transactions on Circuits and Systems for Video Technology (TCSVT), vol. 25, no. 5, pp. 856-868, May 2015.

[14] G. Dimitrakopoulos, K. Galanpoulos, C. Mavrokefalidis and D. Nikolos, "Low-power leading-zero counting and anticipation logic for high-speed floating point units," IEEE Transactions on Very Large Scale Integration Systems (TVLSI), vol. 16, no. 7, pp. 837-850, Jul. 2008.

[15] Q. Wu, M. Pedram and X. Wu, " Clock-gating and its applications to low-power design of sequential circuits," IEEE Transactions on Circuits and Systems I: Fundamental Theory and Applications, vol. 47, no. 3, pp. 415-420, Mar. 2000.

[16] A. Correale, "Overview of the power minimization techniques in the IBM powerPC 4xx embedded controllers," in Proc. International Symposium on Low Power Electronics Design (ISLPED), 1995, pp. 75-80.

[17] L. Greggain and B. White, "Predicting and Scaling Power Consumption in CMOS ASICs," in Proc. Second Annual IEEE ASIC Seminar and Exhibit, Sep. 1989, pp. P8-6.1-P8-6.4.

A 0.3 V, High-PSRR, Picowatt NMOS-Only Voltage Reference using zero-V_T Active Loads

David Cordova*†, Arthur C. de Oliveira†, Pedro Toledo*†, Hamilton Klimach†, Sergio Bampi† and Eric Fabris*†

*NSCAD Microeletronica †Graduate Program on Microelectronics - UFRGS - Porto Alegre, RS, Brazil

{david, toledo}@nscad.org.br, oliveira.arthurc@gmail.com, hamilton.klimach@ufrgs.br, {bampi, fabris}@inf.ufrgs.br

Abstract—A low-voltage high-PSRR CMOS voltage reference operating with picowatt power consumption is presented. The voltage reference is generated from the threshold voltage (V_T) difference of two transistors biased in weak inversion. The V_T difference is achieved through its dependence with the transistor dimensions. The high-PSRR is obtained using zero-V_T transistors as active loads. The final circuit was designed in a 130 nm CMOS process and occupies around 0.0007 mm^2 of silicon area while consuming just 18.5 pW at 27°C. Post-layout simulations present a 62 mV reference voltage with a temperature coefficient of 15 ppm/°C, for a temperature range from -25 to 125 °C and a Power Supply Rejection Ratio (PSRR) of -68.7 dB at 0.3 V of supply voltage.

Index Terms—Low-voltage, low-power, voltage reference, zero-V_T Transistor.

I. INTRODUCTION

The aggressive device and power scaling in recent CMOS technologies has entailed circuits operating with supply voltages below 1 V, that is the case of analog circuits that are able to take advantage of it [1]. Although, lower supply voltages require biasing at lower voltage levels which results in worse transistor properties, and hence yield circuits with lower performance [2].

One of the most important building block in analog design is the voltage reference, which is employed in several more complex structures. In newer technologies nodes, for which the power supply voltage is being progressively reduced, the need of low-voltage high-precision references with good rejection to supply variation is still a major concern.

Low-voltage low-power voltage references have been developed using two MOSFETs with different threshold voltages operating in weak inversion (WI) to obtain a temperature-compensated voltage reference [3]–[5]. In [6] the same idea was applied, but instead of using different MOSFETs, it was exploit the dependence of the threshold voltage on the transistor size.

In this paper we present a low-voltage low-power voltage reference with enhanced PSRR designed in 130 nm CMOS technology. The circuit is based on the topology of [3] to generate the reference voltage, but using the same idea of [6], where the V_T dependence with the transistor dimensions is exploited. The paper is organized as follows: Section II, presents the circuit analysis; in Section III, the PSRR analysis and enhancement; the design and simulation of the voltage reference are shown in Section IV; Section V presents the conclusions.

II. CIRCUIT ANALYSIS

The voltage reference schematic is presented in Fig. 1(a). The proposed circuit is formed by the self-cascode MOSFET (SCM) M_1-M_2 biased by a current source I_{LOAD}, the reference voltage V_{REF} is obtained from the difference between the gate-to-source voltages of the SCM structure. In order to begin with

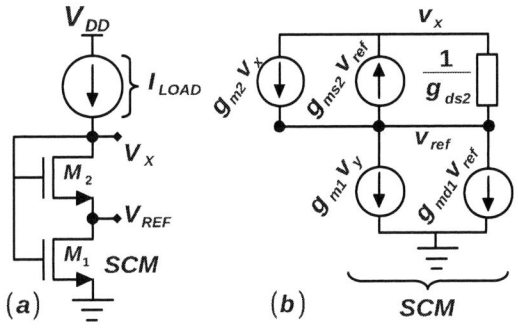

Fig. 1. (a) Simplified circuit of the voltage reference; (b) Equivalent circuit of the self-cascode MOSFET (SCM) M_1-M_2.

the circuit analysis we need a appropriate transistor model for weak inversion (WI), which is the most appropriate inversion for low-power operation. In WI, the drain current is determined by diffusion [7] and is given by

$$I_D = 2eI_{SQ}Se^{\frac{V_G-V_T}{n\phi_t}}\left(e^{-\frac{V_S}{\phi_t}} - e^{-\frac{V_D}{\phi_t}}\right) \quad (1)$$

I_{SQ} represents the specific current and is given by

$$I_{SQ} = \mu_n C'_{ox} n \frac{\phi_t^2}{2} \quad (2)$$

V_G, V_S and V_D are the gate, source and drain voltages referred to the bulk, respectively; μ_n is the mobility; C'_{ox} is the oxide capacitance per unit area; n is the slope factor; $S = $ W/L is the aspect ratio; and $\phi_t = kT/q$ is the thermal voltage, where k is the Boltzmann constant, q is the elementary charge, and T is the absolute temperature.

Knowing that for the SCM transistor M_1 can be in triode, while M_2 must be saturated, by applying (1) for both transistors (M_1 and M_2) we have

$$I_{D1} = 2eI_{SQ1}S_1e^{\frac{V_X-V_{T1}}{n_1\phi_t}}\left(1 - e^{-\frac{V_{REF}}{\phi_t}}\right) \quad (3)$$

$$I_{D2} = I_{SQ2}S_2e^{\frac{V_X-V_{T2}}{n_2\phi_t}} \cdot e^{-\frac{V_{REF}}{\phi_t}} \quad (4)$$

978-1-5090-2737-8/16 $31.00 © 2016 IEEE

Since $I_{D1} = I_{D2}$, a general expression for V_{REF} can be obtained as

$$V_{\text{REF}} = \phi_t \ln\left(1 + \frac{I_{SQ2}S_2}{I_{SQ1}S_1} e^{\frac{V_{T1}n_2 - V_{T2}n_1}{n_1 n_2 \phi_t}} e^{V_x \frac{n_1 - n_2}{n_1 n_2 \phi_t}}\right) \quad (5)$$

Considering that $\ln(1+x) \to \ln(x)$ for $x \gg 1$ and $n_1 = n_2$, a simplified expression for V_{REF} is obtained as

$$V_{\text{REF}} = \frac{V_{T1} - V_{T2}}{n} + \phi_t \ln\left(\frac{I_{SQ2}S_2}{I_{SQ1}S_1}\right) \quad (6)$$

It is usually assumed [8] that V_T depends on temperature as

$$V_T(T) = V_T(T_0) + \alpha_{V_T}(T - T_0) \quad (7)$$

where T_0 is the reference temperature, and $\partial V_T/\partial T = \alpha_{V_T}$ is a negative constant. Replacing (7) into (6)

$$V_{\text{REF}} = \frac{V_{T1}(T_0) - V_{T2}(T_0)}{n} + \frac{(|\alpha_{V_{T2}}| - |\alpha_{V_{T1}}|)(T - T_0)}{n} + \frac{kT}{q}\ln\left(\frac{I_{SQ2}S_2}{I_{SQ1}S_1}\right) \quad (8)$$

By setting $\partial V_{\text{REF}}/\partial T|_{T=T_0} = 0$, a value of S_2/S_1 for optimal temperature compensation can be expressed as

$$\left(\frac{S_2}{S_1}\right)_{\text{OPT}} = \frac{I_{SQ1}}{I_{SQ2}}\exp\left[\frac{q}{k}\left(\frac{|\alpha_{V_{T1}}| - |\alpha_{V_{T2}}|}{n}\right)\right] \quad (9)$$

Replacing (9) in (8) the temperature compensated V_{REF} is given by

$$V_{\text{REF}} = \frac{V_{T1}(T_0) - V_{T2}(T_0)}{n} + \frac{(\alpha_{V_{T1}} - \alpha_{V_{T2}})(T_0)}{n} \quad (10)$$

The first term of (10) gives us a reference voltage equal to the difference of V_T ($V_{T1} > V_{T2}$) from the transistors of the SCM (M_1-M_2). In order to guarantee a low temperature coefficient (TC), the variation of α_{V_T} for different V_Ts should be minimum, and S_2/S_1 is the ratio of transistors (M_1-M_2) which are limited by the maximum transistor area. This constraints will be used during the design procedure.

III. PSRR ANALYSIS AND ENHANCEMENT OF V_{REF}

A. PSRR Analysis

Consider the simplified circuit of the voltage reference depicted in Fig. 1(a), the PSRR of the V_{REF} is given by

$$\text{PSRR} = \frac{v_{\text{ref}}}{v_{dd}} = \frac{v_x}{v_{dd}} \cdot \frac{v_{\text{ref}}}{v_x} \quad (11)$$

Where (11) represents the voltage transfer function between the supply voltage (v_{dd}) and the reference voltage (v_{ref}) is separated in two transfer functions: v_x/v_{dd} and v_{ref}/v_x, which are the supply voltage contributions of the current load (I_{LOAD}) and the self-cascode MOSFET to the reference voltage.

The voltage reference becomes insensitive to supply variations when v_x/v_{dd} or in turn v_x/v_{dd} and v_{ref}/v_x are minimized. The value v_{ref}/v_x is calculated using the equivalent of Fig. 1(b), yielding

$$\frac{v_{\text{ref}}}{v_x} = \frac{g_{m2} - g_{m1} + g_{ds2}}{g_{ms2} + g_{ds2} + g_{md1}} \approx \frac{g_{m2} - g_{m1}}{g_{ms2}} \quad (12)$$

where g_m, g_{ms} and g_{ds} are the gate, source and drain transconductance, respectively. This expression is defined for a specific design value of the voltage reference.

Fig. 2 shows the circuit of I_{LOAD}, it can be implemented using a PMOS or NMOS transistor acting as a current source, in this case we are considering both transistors saturated, the circuit design will be discussed later. The value of v_x/v_{dd} is calculated from its small-signal analysis and yields

$$\left(\frac{v_x}{v_{dd}}\right)_{\text{PMOS}} \approx \frac{g_{ms}}{g_{ds}} \vee \left(\frac{v_x}{v_{dd}}\right)_{\text{NMOS}} \approx \frac{g_{ds}}{g_{ms}} \quad (13)$$

where g_{ms} and g_{ds} are the source transconductance and drain conductance, respectively. Since, $g_{ms} > g_{ds}$, v_x/v_{dd} is higher than one for a PMOS, worsening the PSRR, leading us to discard this current load. On the other hand, for a NMOS is lower than one, which is desired to improve the PSRR.

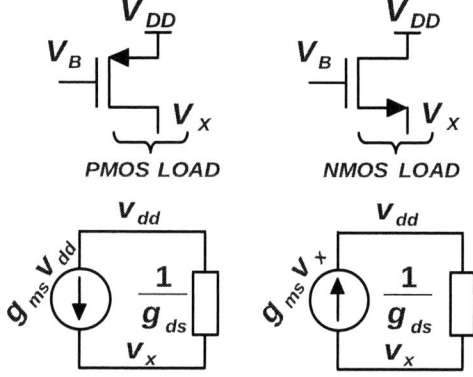

Fig. 2. PMOS and NMOS implementation of I_{LOAD} of Fig. 1(a). V_B is constant bias voltage.

Taken into account that the voltage reference circuit is working in weak inversion and the gate voltage of the NMOS of Fig. 2 is ($V_B < V_{DD}$), we have two options for V_B:

- For the case $V_B = V_X$, v_x/v_{dd} is increased from g_{ds}/g_{ms} to g_{ds}/g_{mb}, Fig. 3(a).
- For the case $V_B = 0$ V, v_x/v_{dd} remains the same as (13), Fig. 3(b).

From the two previous cases, $V_B = 0$ V is chosen for presenting the lowest v_x/v_{dd} to guarantee the highest PSRR. The SCM is biased with the leakage current of the I_{LOAD}, resulting in a NMOS with a large or small area depending on its current density (I_D/W). Since, I_{LOAD} only acts as a current source, its design is constrained to obtain the bias current of the SCM for the lowest achievable area. Fig. 4 shows the different current densities of the available transistors for $V_{GS} \leq 0$ V, the lower I_D/W of the standard-V_T or low

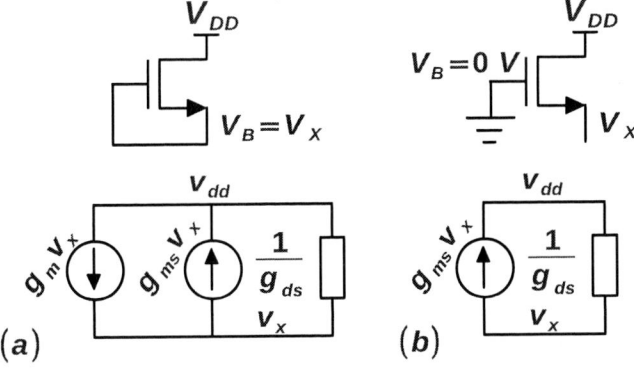

Fig. 3. Implementations of I_{LOAD} using self-biased NMOS transistors. (a) gate-to-source load; (b) gate-to-ground load.

power-V_T transistors will yield larger areas in contrast to the zero-V_T or low-V_T transistors. Due this fact, the last ones will be used to design I_{LOAD} and will be covered in the following subsection.

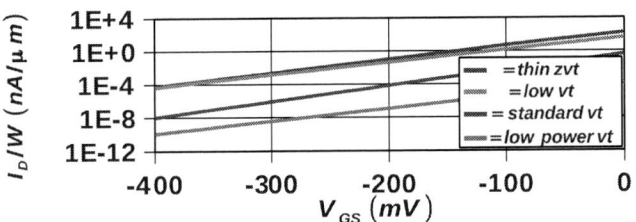

Fig. 4. Current densities (I_D/W) for zero-V_T, low -V_T, standard-V_T and low power-V_T transistors. $V_{DS} = 0.1$ V and W/L= 4 μm/1 μm for all transistors.

B. PSRR Enhancement with zero-V_T MOSFETs

Special attention has been given to the properties of the zero-V_T transistor due to its high drive capability at low voltages [7]. Since $V_T \approx 0$ V, a zero-V_T transistor with a gate voltage equal or lower than to 0 V can be saturated with a small value of drain-to-source voltage, $V_{DS} \approx 100$ mV. Fig. 5 shows the simulated drain current for three different transistors (thin zero-V_T, thick zero-V_T and low-V_T), among them the thin zero-V_T and thick zero-V_T present the highest and lowest current density, respectively.

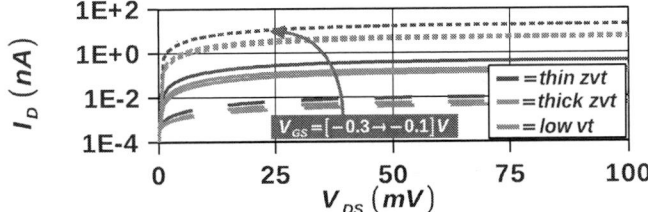

Fig. 5. $I_D \times V_{DS}(V_B = 0$ V) characteristics of a thin zero-V_T, thick zero-V_T and low-V_T transistors. W/L= 4 μm/1 μm for all transistors.

I_{LOAD} will be designed using one of the previously mentioned transistors. Fig. 6 shows the I_{LOAD} implementation using one and two zero-V_T transistors.

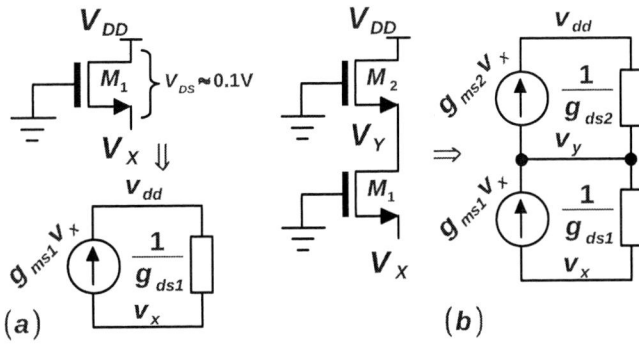

Fig. 6. I_{LOAD} implementation with (a) One zero-V_T transistor; (b) Two zero-V_T transistors.

Using the above equivalent circuit, v_x/v_{dd} is obtained for one and two zero-V_T transistors

$$
\begin{aligned}
\left(\frac{v_x}{v_{dd}}\right)_{\text{1-ZVT}} &\approx \frac{g_{ds1}}{g_{ms1}} \\
\left(\frac{v_x}{v_{dd}}\right)_{\text{2-ZVT}} &\approx \frac{g_{ds1}}{g_{ms1}} \cdot \frac{g_{ds2}}{g_{ms2}}
\end{aligned}
\tag{14}
$$

Through (14) it can be shown that by adding another transistor in series v_x/v_{dd} is reduced by a factor of g_{ms}/g_{ds}, consequently improving the overall PSRR of the voltage reference. Fig. 7 presents v_x/v_{dd} as a function of V_{DS} for three different simulated transistors. Using just one transistor, we can obtain an attenuation of above -30 dB using only 0.1 V of V_{DS}, for the two transistor implementation can go low as -60 dB approximately.

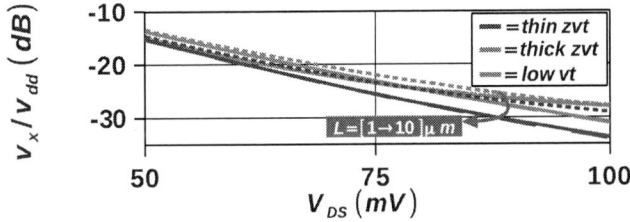

Fig. 7. v_x/v_{dd} as function of V_{DS} for the simulated transistors. $W = 4\mu$ for all transistors.

IV. LV CMOS Voltage Reference

A. Design of the SCM

The first step to design our voltage reference is choose the transistors of the SCM. Since our goal is low-power operation, we will choose from the process the two available transistors with high-V_T. The voltage reference is designed in IBM 130 nm CMOS process. Fig. 8 shows the normalized α_{VT} variation, where the low power V_T transistor presents the lowest variation of α_{VT} which is desired, recalling that the second term of (10) should be minimized.

Fig. 9 shows the threshold voltage (V_T) variation against channel length (L) for different transistors ratios W/L, having L as the major contributor to the variation of V_T. Transistor M_1 is sized small (W/L between 1 to 2) to satisfy the first

978-1-5090-2737-8/16 $31.00 © 2016 IEEE

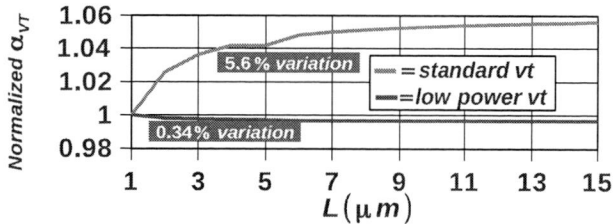

Fig. 8. Variation of normalized α_{VT} for different transistor channel length.

term of (10), $V_{T1} > V_{T2}$, while M_2 is sized taking in to consideration the following:

- $L_2 > L_1$, condition of the first term in (10).
- To prevent charging damage in the polysilicon, the maximum thin oxide area for any single gate in the process is 230 μm².
- In order to guarantee an efficient layout area and compliant with $(W \times L)_2 \leq 230$ μm², a range of $[1-2]$ for $(W/L)_2$ and $L_2 = [10-15]$ μm is defined, as illustrated in Fig. 10.

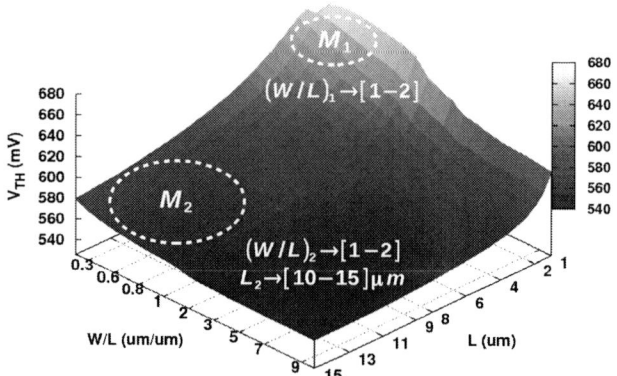

Fig. 9. (a) V_{TH} as function of W/L and L

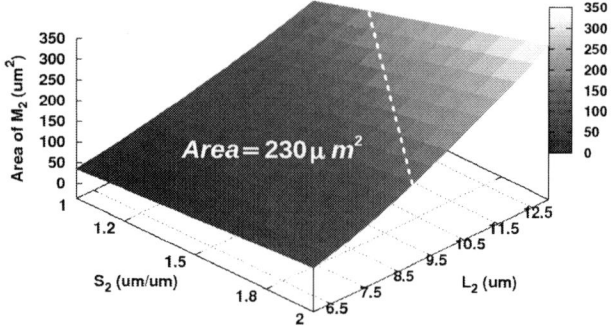

Fig. 10. W/L vs L of a transistor with a maximum thin oxide area of 230 μm²

From the previous considerations, the maximum value for V_{REF} is calculated using the first expression of (10), as presented in Fig. 11, resulting in a maximum value of 70 mV. This value of V_{REF} is used to defined the minimum voltage for the SCM, which is $V_X = V_{REF} + V_{DS2}$; with a minimum value of $V_{DS2} = 4\phi_t$ to guarantee saturation [7].

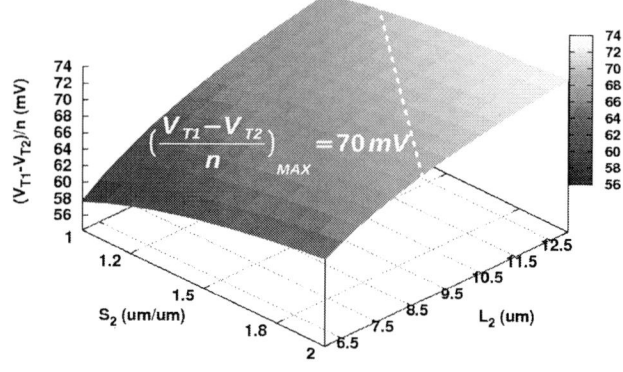

Fig. 11. $\frac{V_{T1}-V_{T2}}{n}$ as function of M_2 ratio (S_2) and channel length. $(W/L)_1 = 1$ and $n = 1.35$

Using the range defined for $(W/L)_2$ and L_2 as starting point, the voltage references are designed in order to achieve the lowest achievable temperature coefficient (TC) for the minimum V_X, as illustrated in Fig. 12. The same procedure can be implemented to obtain the lowest current consumption, Fig. 13. As previously stated, the area o M_2 is verified to satisfy the constraint of $(W \times L)_2 \leq 230$ μm².

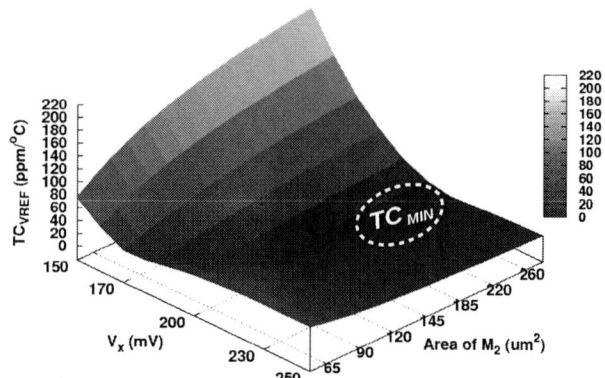

Fig. 12. TC as function of V_x and M_2 area.

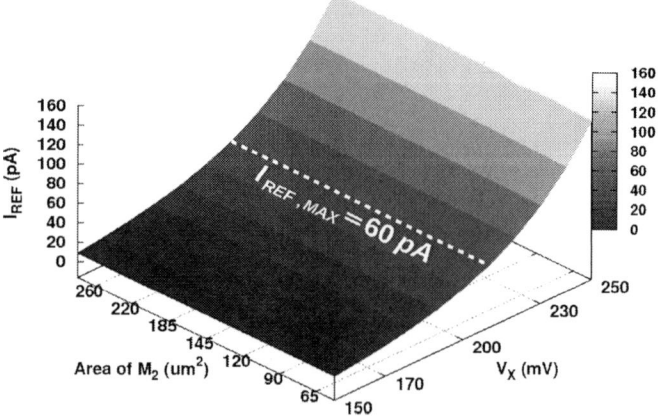

Fig. 13. I_{REF} as function of V_X and M_2 area

The current reference (I_{LOAD}) for V_{REF} was designed with one and two series transistors for three different transistors (thin zero-V_T, thick zero-V_T and low-V_T) as depicted in Fig. 14. Table I shows the transistor sizing for the designed circuits. The ratio S_2/S_1 for optimal temperature compensation predicted by (9) is around 0.9, while the chosen after simulations is $S_2/S_1 = 0.826$. The proximity of the predicted and simulated values for temperature compensation validates the theoretical approach.

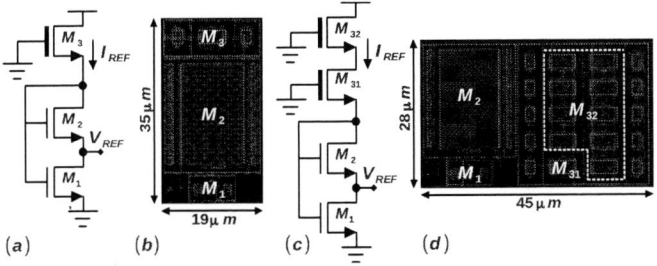

Fig. 14. Designed LV Voltage Reference. (a) schematic (1 zero-V_T); (b) layout of (a); (c) schematic (2 zero-V_T); (d) layout of (c).

TABLE I
SIZING OF THE LV VOLTAGE REFERENCE CIRCUIT

(W/L)	IBM 130nm
M_1	$\frac{2\ \mu m}{1\ \mu m}$ (LP)
M_2	$\frac{19\ \mu m}{11.5\ \mu m}$ (LP)
$M_{3,31} - M_{32}$	$\left[\frac{3\ \mu m}{3.9\ \mu m} - 9\left(\frac{3\ \mu m}{3.9\ \mu m}\right)\right]$ (ZVT); $\left[\frac{7\ \mu m}{4\ \mu m} - 9\left(\frac{7\ \mu m}{4\ \mu m}\right)\right]$ (LVT)

B. Simulation Results

The proposed voltage reference circuits is validated with post-layout simulations using the models provided by the foundry. Fig. 15(a) shows the voltage against temperature in the range of -25 to 125 ^{o}C for the designed voltage references, presenting an average voltage value of 62 mV for the different implementations. Regarding the temperature coefficient (TC), it was obtain a minimum and maximum of 15 and 17.3 ppm/^{o}C, respectively. Fig. 15(b) presents the PSRR of V_{REF}, for the circuits with a 300 mV of supply voltage an outstanding average value of -66 dB was achieved, in contrast to the -93 dB for the ones with 400 mV of supply voltage. Also, it can be seen that the predicted 30 dB improvement from using two instead of one transistor for I_{LOAD} at the cost of just 100 mV of supply voltage.

Additionally the PSRR was simulated in the temperature range of -25 to 125 ^{o}C (steps of 15°C) performing 100 Monte Carlo simulations per temperature point for the circuits with 300 mV and 400 mV of supply voltage, respectively, as illustrated in Fig. 16, our voltage reference attains for the worst case of temperature a PSRR below -45 and -59 dB, for the 300 mV and 400 mV V_{DD} circuits, showing a very robust design. All relevant performance figures, like line sensitivity, PSRR, temperature coefficient and other important metrics are presented in Table II.

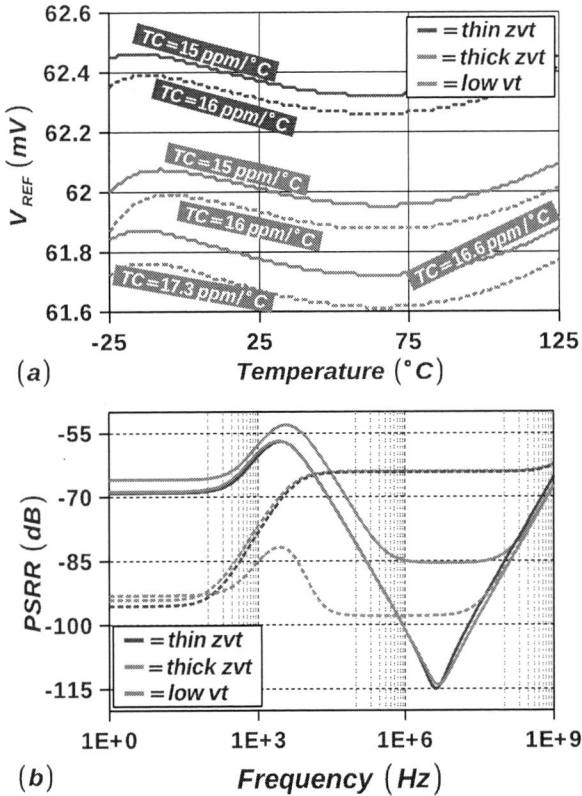

Fig. 15. (a) V_{REF} vs Temperature. (b) PSRR of V_{REF}, for one transistor (solid) and two transistors (dashed).

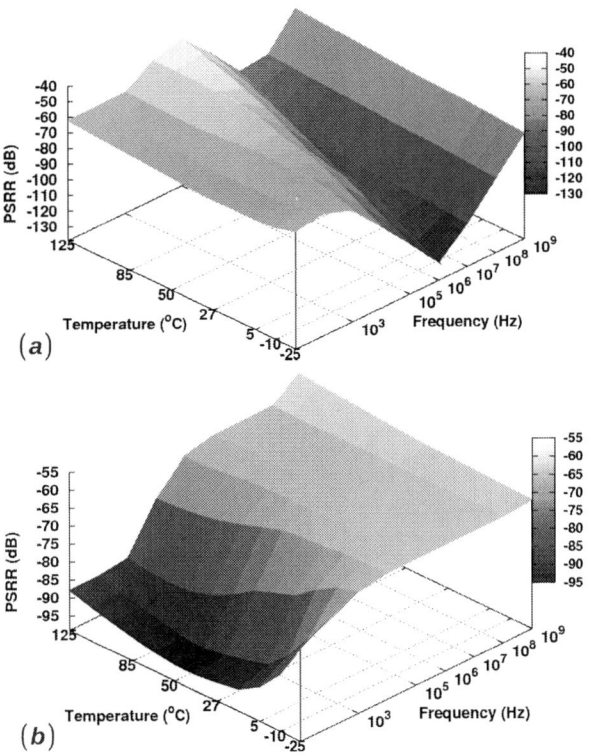

Fig. 16. PSRR vs Temperature for ZVT$_{thick}$ loads, (a) $V_{DD} = 0.3$ V. (b) $V_{DD} = 0.4$ V.

978-1-5090-2737-8/16 $31.00 © 2016 IEEE

TABLE II
SIMULATED PERFORMANCE OF THE VOLTAGE REFERENCE

CMOS Process				130nm			Unit
V_{DD}		0.3			0.4		V
LOAD	LVT	ZVT$_{thin}$	ZVT$_{thick}$	2-LVT	2-ZVT$_{thin}$	2-ZVT$_{thick}$	
V_{REF}†	61.8	62.4	62	61.7	62.3	61.9	mV
I_{REF}	89.2	73.2	61.7	98	73.4	62.5	pA
LR (V_{REF})	512	361	374	22	17	20	μV/V
$(\sigma/\mu)_{V_{REF}}$†	3.4	3.4	3.4	3.5	3.3	3.5	%
TC (V_{REF})	16.6	15	15	17.3	16	16	ppm/^{o}C
PSRR$_{DC}$†	-66	-69	-68.7	-93	-95.5	-94	dB
Power	26.8	22	18.5	68.6	29.4	25	pW
Area	0.0007	0.0007	0.0007	0.0014	0.0011	0.0013	mm^2

†Process and Mismatch (1000 runs)

TABLE III
COMPARISON WITH LOW-VOLTAGE LOW-POWER CMOS VOLTAGE REFERENCES

	This Work$^\theta$	[6]	[4]	[5]	Unit
CMOS Process	130	180	130	180	nm
V_{DD}	0.3-1.2	0.15-1.8	0.5-3	0.45-2	V
V_{REF}	62	17.69	176.1	263.5	mV
LS	0.06	2.03	0.033	0.44	%/V
$(\sigma/\mu)_{V_{REF}}$	3.4	1.6	0.72	3.9	%
TC (V_{REF})	31.5†	1462.4	62	142.1	ppm/^{o}C
Temp range	-25:125	0:120	-20:80	0:100	^{o}C
PSRR$_{@100\ Hz}$	-68.3*	-64	-53	-49.4	dB
Power	18.5*	26.1	2.2	3150	pW
Area	0.0007	0.0012	0.0014	0.0430	mm^2

$^\theta$Simulations Results, †Process and Mismatch (1000 runs), *$(27^oC, V_{DD_{MIN}})$

Using the thick zero-V_T implementation as reference, it is compared to the other works in the literature, as shown in Table III. Our voltage reference presents the lowest supply voltage compared to [4], [5] and best trade-off between LS, $(\sigma/\mu)_{V_{REF}}$, TC, PSRR, power and area compared to other implementations.

V. CONCLUSIONS

A low-voltage low-power CMOS voltage reference with high-PSRR was presented. Different than the standard approach, that uses different V_T transistors, in this work we have exploited the dependence of V_T with transistor dimensions in order to obtained a voltage reference through its difference. Post-layout results shows that the designed references work at picowatt power consumption range and achieve a high-PSRR while keeping the remaining performance figures competitive, showing the great advantage of using zero-V_T and low-V_T transistors as active loads.

VI. ACKNOWLEDGMENTS

This work was partially supported by CNPq, CAPES and by IC-Brazil Program and MOSIS for access to chip fabrication services.

REFERENCES

[1] P. Kinget, "Designing analog and RF circuits for ultra-low supply voltages," in *Solid State Circuits Conference, 2007. ESSCIRC 2007. 33rd European*, Sept 2007, pp. 58–67.

[2] A.-J. Annema, B. Nauta, R. van Langevelde, and H. Tuinhout, "Analog circuits in ultra-deep-submicron CMOS," *Solid-State Circuits, IEEE Journal of*, vol. 40, no. 1, pp. 132–143, Jan 2005.

[3] W. Yan, W. Li, and R. Liu, "Nanopower CMOS sub-bandgap reference with 11 ppm/c temperature coefficient," *Electronics Letters*, vol. 45, no. 12, pp. 627–629, June 2009.

[4] M. Seok, G. Kim, D. Blaauw, and D. Sylvester, "A Portable 2-Transistor Picowatt Temperature-Compensated Voltage Reference Operating at 0.5 V," *Solid-State Circuits, IEEE Journal of*, vol. 47, no. 10, pp. 2534–2545, Oct 2012.

[5] L. Magnelli, F. Crupi, P. Corsonello, C. Pace, and G. Iannaccone, "A 2.6 nW, 0.45 V Temperature-Compensated Subthreshold CMOS Voltage Reference," *Solid-State Circuits, IEEE Journal of*, vol. 46, no. 2, pp. 465–474, Feb 2011.

[6] D. Albano, F. Crupi, F. Cucchi, and G. Iannaccone, "A Sub-kT/q Voltage Reference Operating at 150 mV," *Very Large Scale Integration (VLSI) Systems, IEEE Transactions on*, vol. 23, no. 8, pp. 1547–1551, Aug 2015.

[7] C. Galup-Montoro, M. Schneider, and M. Machado, "Ultra-Low-Voltage Operation of CMOS Analog Circuits: Amplifiers, Oscillators, and Rectifiers," *Circuits and Systems II: Express Briefs, IEEE Transactions on*, vol. 59, no. 12, pp. 932–936, Dec 2012.

[8] Y. Tsividis, *Operation and Modeling of the MOS Transistor*, 2nd ed. McGraw-Hill, 1998.

A Standard Cell Characterization Flow for Non-Standard Voltage Supplies

Matheus Gibiluka, Matheus Trevisan Moreira, Walter Lau Neto, Ney Laert Vilar Calazans

GAPH – Faculty of Computer Science – PUCRS – Porto Alegre – RS – Brazil

{matheus.gibiluka, matheus.moreira, walter.lau}@acad.pucrs.br, ney.calazans@pucrs.br

Abstract—Static timing analysis (STA) is a widely used technique to perform timing verification of digital circuits analytically. These models are available in standard cell libraries, and are usually generated based on data acquired from timing characterization performed by foundries. Even though advanced node libraries are characterized for some corners, they currently do not cover voltage levels that will likely drive ultra-low power applications as those present in domains like IoT and wearable devices, a trend for upcoming years. This work contributes to solve this issue, by proposing a characterization flow that can be applied to any standard cell set. The flow is foundry-independent and relies solely on information that can be obtained from any cell library. It employs commercial tools incremented by a set of in-house scripts. As a case study, the article explores the characterization of a commercial standard cell library for a 28nm FDSOI technology. The analysis shows that this library can be characterized for voltages as low as 250mV and still guarantee that its cells, or a subset of these, work as intended by the technology rules. A set of experiments show that the flow obtains characterization results that match those of the associated commercial library within 5% of error, for those voltages where the latter provides information.

I. INTRODUCTION

The cell-based approach to design integrated circuits (ICs), and particularly the method based on the use of standard cell libraries are often referred as the key success factor for the rapid growth of the VLSI market [1]. The wide availability of commercial versions of these libraries allows a much faster route to design an IC, by providing huge amounts of pre-characterized data that can be used to compute timing, power, variability and reliability data on specific designs using automated commercial tools.

Static timing analysis (STA) [2] is a widely used technique to perform timing verification of digital circuits analytically. Different from dynamic (i.e. simulation-based) timing analysis, which is often time consuming and input-dependent, this method computes timing information based on pre-existing delay models. These models are included in the standard cell libraries and are usually generated based on data acquired from timing characterizations performed by foundries. Even though advanced node libraries are characterized for several corners, they usually do not cover the voltage levels that will likely drive ultra-low power IoT and wearable applications in the upcoming years [3], [4], [5].

For instance, the libraries available with the ST-Microelectronics 28nm FDSOI design kit, which is the target technology in this work, are only characterized for supply voltages of 1.0V (the nominal supply voltage), 0.9V, and 0.8V at the typical process corner. Therefore, to leverage STA to analyze circuit behaviour under other supply voltage levels (e.g. at near-threshold or even sub-threshold values), additional research and tools are required to obtain methods able to produce characterization data for the voltages of interest.

The goal of the paper is to contribute with a new method and associated tools. The authors could not find similar works in the literature with the same approach. The remaining of this paper comprises four Sections. Section II provides basic definitions regarding the characterization of standard cells. Section III then presents the proposed Multi-voltage Characterization flow, while Section IV explores a case study of the application of the flow. Finally, Section V advances a set of conclusions of the work.

II. CHARACTERIZATION OF STANDARD CELLS

The electrical characterization of standard cells is a well established process that can be performed automatically by library characterization tools [6], [7]. These tools usually take as input characterization settings and a SPICE-level post-layout netlist containing references to specific transistor models, resistances, and capacitances for each library cell. The main output is a database that often contains the logic function, timing, power, and noise models for each cell. Figure 1 illustrates a high-level view of a typical cell characterization flow. Initially, the tool analyzes the transistor circuits in the cell SPICE netlist to identify the logic function and type – such as combinatorial logic, sequential logic, pass transistor logic, among others – of each cell.

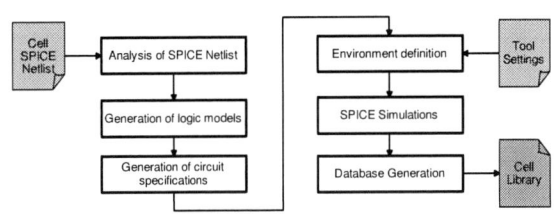

Fig. 1.   High-level view of a typical standard cell characterization flow.

Based on the previous analysis, a logic model for each cell is created, followed by the generation of circuit specifications (i.e. pin direction, pin-to-pin delay, etc). Next, parameters, such as supply voltage, temperature, process corner, output load and input slew vectors are read from a settings file and used to set the characterization environment. Finally, SPICE simulations are performed and the electrical characteristics of a cell are measured during these simulations. Results are then extracted,

978-1-5090-2737-8/16 $31.00 © 2016 IEEE

processed, and stored on a database that can be exported to a library file.

III. THE MULTI-VOLTAGE CHARACTERIZATION FLOW

The multi-voltage characterization (MVC) flow proposed in this work is the implementation of a method designed to extend the voltage corners supported by a given standard cell library. Figure 2 depicts an overview of the flow. MVC provides a systematic way to characterize libraries that ensures the proper scaling of voltage-dependent parameters, such as signal slew and voltage levels used to determine delays. At the circuit level, MVC provides the means to enable STA of circuits operating at virtually any supply voltage within the limits of the library silicon technology.

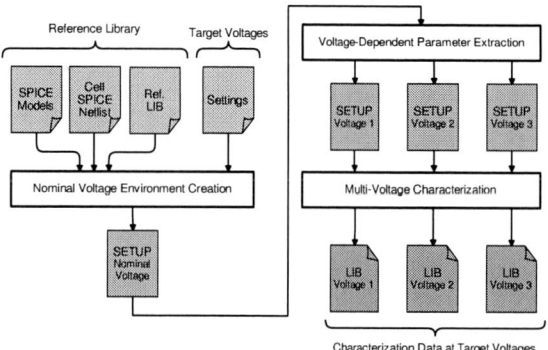

Fig. 2. Overview of the MVC Flow. White rectangles identify the main tasks comprising MVC. Yellow boxes denote the main input files of MVC, green boxes designate outputs, and blue boxes indicate intermediate files.

The MVC flow extracts voltage-dependent parameters from a reference library characterized at nominal voltage and scales them to create characterization environments compatible with the target voltages. These environments are loaded into an commercial characterization tool such as the Encounter Library Characterizer (ELC) by EDA vendor Cadence, to characterize the standard cell libraries and export the data in a library format supported by the tool, e.g. the Liberty format. The MVC flow also includes a tool proposed here to validate characterization data, by comparing the generated libraries with the reference. As Figure 2 illustrates, the flow takes as input a reference library – composed by a set of Liberty files originated from cells characterized at nominal voltage (Ref. LIB), the technology models (SPICE Models) and cell netlists (Cell SPICE Netlist) in SPICE format – and a settings file (Settings), containing the target supply voltages for the characterization. The outputs are libraries characterized at each target voltage (LIB Voltage 1, 2,3 in Figure 2). The flow consists of three sequentially applied tasks: Nominal Voltage Environment Creation, Voltage-Dependent Parameter Extraction, and the Multi-Voltage Characterization in itself. The next sections detail these tasks, all of which appear in Figure 3.

A. Nominal Voltage Environment Creation

Refer to Figure 3(a). This Section presents the task that has as goal to create a characterization environment at the technology nominal voltage. This environment can generate libraries with data equivalent to the reference library. Since it will be the basis to create the characterization environments for each target supply voltage, it is vital to ensure that the task can produce correct results. The input to this task is the reference library (Ref. LIB) and the output is a characterization environment file (SETUP Nominal Voltage) tuned to the nominal voltage. Figure 3(a) shows the four steps of the task and their interaction.

The initial step is the extraction of parameters from the reference library characterized at the nominal voltage. The goals are to create a Cell Property File, containing information about the cells that are not used by the characterization tool (i.e. cell area and footprint), and a reference characterization environment file (Ref. SETUP). The creation of the former uses text manipulation tools, while the latter is automatically extracted using characterization tools. The Ref. SETUP file contains the input slew and output capacitance vectors used in the reference library, along with simulation and corner settings. The next step consists in creating a characterization environment file (SETUP Nominal Voltage) based on the Ref. SETUP file, with parameters grouped by drive strength of the cells. This eases the process of scaling voltage dependent parameters that is executed in the second task of the MVC flow. Next, the cell library is characterized using SETUP Nominal Voltage as the characterization environment. The resulting library is compared with the reference library using a custom Library Comparator (detailed in Section III-B), and the delay differences are plotted as an error histogram. If the difference is not within the acceptable bounds, SETUP Nominal Voltage must be tuned to reduce the error. Automated characterization tools offer several approaches for environment tuning. For instance, the delay values extracted from SPICE simulations can be scaled or offset. Also, the input slew and capacitance vectors and the driver cell can be adjusted to reduce the error. Once the environment is tuned, the output SETUP Nominal Voltage file becomes the input of the next step of the flow.

B. The Library Comparator

The Library Comparator (LC) is an automated environment for comparing timing models of Liberty libraries. The creation of a custom in-house environment for this purpose was necessary, as the library comparator tool bundled with ELC (libdiff), the characterization tool used in this work, does not support the Non-Linear Delay Models (NLDM) used in the reference library. As Figure 4 shows, LC works by extracting logic path delays from a set of netlists using the provided libraries and comparing the obtained results. It relies on Synopsys PrimeTime to extract path delays and on an in-house tool called Timing Report Comparator to compare the results. The Timing Report Comparator encompasses two Python scripts of around 500 lines each.

To ensure that each netlist is thoroughly analyzed, Prime-Time generates unique timing reports for each path that connects a register output (or primary input) to a register input (or primary output) without any intervening registers – thus, guaranteeing that data from all timing paths are acquired. In addition, PrimeTime evaluates up to eight types of delay for each path, considering the combinations of: i) minimum and maximum delays; ii) falling and rising input signals; iii) falling

978-1-5090-2737-8/16 $31.00 © 2016 IEEE

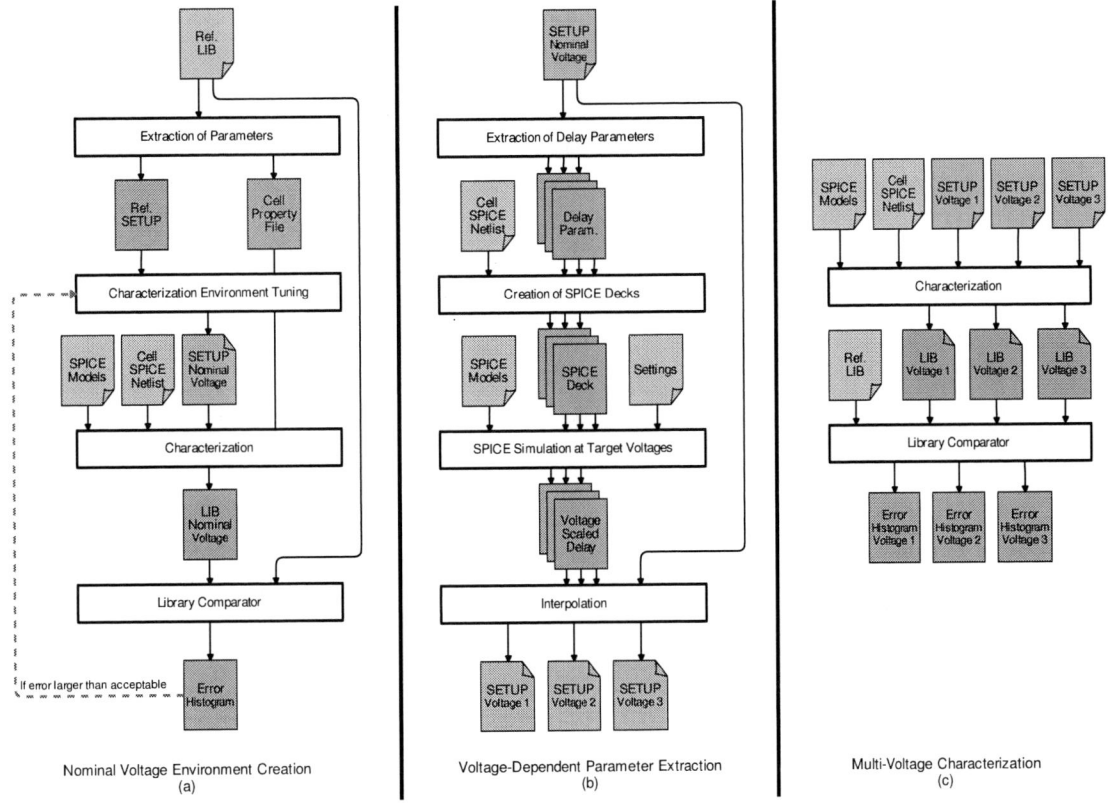

Fig. 3. Detailed view of the three steps in the MVC flow: (a) Nominal Voltage Environment Creation; (b) Voltage-Scaled Parameter Extraction; (c) Multi-Voltage Characterization. Yellow boxes denote the main input files of the flow, green boxes designate outputs, and blue and violet boxes indicate intermediate files of MVC and of each task, respectively. White rectangles correspond to tasks steps.

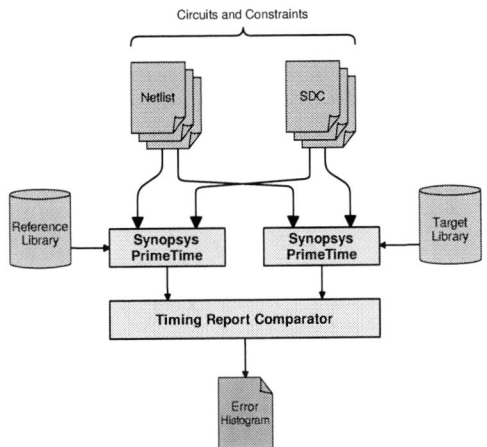

Fig. 4. Automated environment used to compare data models from Liberty libraries. STA is performed on a set of circuits using Synopsys PrimeTime. The Timing Report Comparator parses the STA reports, compares the data between reference and target libraries, and generates an error histogram with the results.

and rising output signals. The Timing Report Comparator parses each timing report and, using Equation 1, computes cell and path delay errors between the reference and target libraries. LC combines the error data obtained from all netlists as error histograms. From the error histograms one can determine if both libraries have equivalent timing models – that is, if both libraries generate similar STA reports.

$$error = (d_{target} - d_{reference})/d_{reference} \qquad (1)$$

The reason behind the development of LC is the need to validate the characterization environments created by the MVC flow. In other words, it is necessary guarantee that libraries generated with this flow are accurately characterized. To assess this, the MVC-generated libraries are compared with the ones supplied in the design kit (i.e. the reference library) using LC. If the error distribution shown in the histogram presents a small standard deviation and average error close to zero, it is assumed that the characterization is good. This validation approach is based on two assumptions: *i)* reference libraries are accurate; *ii)* libraries that present near-zero error when compared to the reference are also viewed as accurate. Based on observations of the delay differences seen between SPICE simulations of logic paths and STA reports, which can be in the order of 5-10%, a target error and standard deviation of up to 0.05 (5% error) is considered acceptable in this work. In this way, library differences which are within this target are considered accurate.

C. Voltage-Dependent Parameter Extraction

Refer to Figure 3(b). Characterization environments are collections of settings that establish how characterization tasks should take place. Some of these settings, such as temperature of operation and output load vectors, do not depend on supply voltage and, consequently, do not need to be modified to support other voltage corners. On the other hand, voltage-dependent parameters need to be carefully scaled when the target supply voltage of a characterization changes. Therefore, to enable accurate characterizations, each target supply voltage of the MVC flow requires a unique characterization environment with voltage-dependent parameters appropriately scaled. The goal of this second task of the MVC Flow is to create such environments.

The first step in this task is to extract delay parameters from the *SETUP Nominal Voltage* file. The voltage-dependent settings contained in the environments produced by the MVC flow can be categorized in two classes: *i)* signal parameters and *ii)* simulation parameters. The first class determines how signal voltage levels are interpreted by the characterization tool. These parameters are defined relatively to the voltage corner and are, consequently, scaled automatically by commercial characterization tools. Class *ii* specifies the input slews that each standard cell will be subjected to during characterization. These values are expressed as absolute delays and need to be scaled to reflect the slew that a circuit would be exposed to when operating under a specific target supply voltage. Timing analysis tools use the transition delay of one gate to determine the input slew of the next one. If the input slew computed by the tool is outside the range defined in the delay model (i.e. values specified in class *ii*), it cannot reliably estimate cell and transition delays. Therefore, scaling these parameters in a realistic manner is imperative to avoid inaccurate delay computations.

This work proposes a technique based on SPICE simulations to scale input slew vectors in a way that reflects the slew range found when operating on a given target supply voltage. First the technique determines the minimum and maximum signal slews that library cells produce when operating in a certain voltage corner. To do so, a SPICE deck was designed to reproduce these extreme values using the standard cells that exhibit the smallest and largest transition delay. As Figure 5 illustrates, this deck comprises a step function generator, a pair of input inverters, the test cell, and load capacitances.

When analyzing minimum slew, the test cell is replaced by the gate that presents the smallest transition delay. Similarly, the standard cell with the largest transition delay is used as test cell to analyze maximum slew. This way, when the supply voltage scales, the minimum/maximum slew characteristics are maintained. The simulation stimulus produced by the step generator goes through a pair of inverters to produce a realistic non-linear input signal to the test cell. Once the load capacitances are calibrated to generate the minimum and maximum slew values employed in the reference environment,

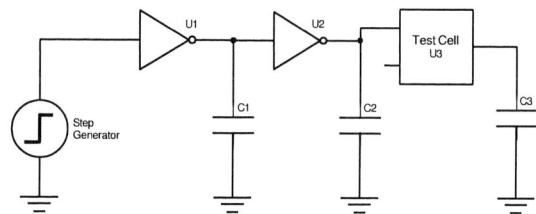

Fig. 5. Generic SPICE deck employed in parameter scaling. The test cell is selected according to the characteristics of the parameter to scale. Capacitances are tuned to reflect the expected slews.

SPICE simulations are performed in all target supply voltages, to obtain the scaled values. These results determine the two extreme slew values for each voltage. With the extreme values determined, intermediate values can be interpolated using cross-multiplication. This is exemplified in Table I, where min and max values were determined using the SPICE-based technique. Reference values represent the input slew vector at nominal voltage obtained from the reference library. The example assumes a slew vector of 5 values. Larger input vectors can be supported by adding additional columns.

TABLE I.   EXAMPLE OF INPUT SLEW SCALING THROUGH
CROSS-MULTIPLICATION BETWEEN REFERENCE AND SCALED VALUES.

	Input Slew (ns)				
Reference	a	b	c	d	e
Scaled	min	$f = g * b/c$	$g = h * c/d$	$h = max * d/e$	max

The second task of the MVC flow uses the scaling technique previously described to create the environments required to enable library characterization on all target supply voltages. This task, as Figure 3(b) details, takes as input the characterization environment created in the previous task and, after four steps, generates a collection of new environments, each tuned to a specific target supply voltage. The first step within this task consists in extracting the voltage-dependent parameters from the input environment. As previously mentioned, only the slew indexes used to create the NLDMs need to be explicitly scaled and, therefore, only these are extracted. Next, SPICE decks representing each of these parameters are created and simulated in all target supply voltages to obtain scaled values of minimum and maximum slews. These values are interpolated using the cross-multiplication technique exemplified in Table I to generate scaled input slew vectors. Finally, the original slew vectors are replaced by the interpolated ones, thus creating unique simulation setup files scaled for each target supply voltage.

D. The Multi-Voltage Characterization Validation

The final step of the MVC flow consists on the cell library characterization at each target supply voltage. As Figure 3(c) illustrates, this step takes as input the characterization environments created in the previous task and outputs a set of libraries, one for each target supply voltage. If the design kit includes libraries characterized at any of the target supply voltages of the MVC flow, these can be employed on an additional validation step using the LC environment.

978-1-5090-2737-8/16 $31.00 © 2016 IEEE   58

IV. A CASE STUDY OF THE MVC FLOW APPLICATION

Using the MVC flow, a subset of cells from the ST-Microelectronics 28nm FDSOI CORE library was characterized in the supply voltage range from 1V (nominal) to 250mV in steps of 50mV. According to [8], libraries restricted to a maximum fan-in of three avoid excessive transistor stacking, which reduces noise margin-related issues when operating in low supply voltages. For this reason, the subset of cells used in the case study was restricted to cells with, at most, 3 input pins. In addition, according to our experiments, gates from the target library cannot operate correctly when subjected to supply voltages under 250mV[1]. When operating below this voltage level, some standard cells are not able to output signals with voltage levels above the lower limit for a logic high signal. The characterizations presented in this case study were performed using the Encounter Library Characterizer tool by EDA vendor Cadence.

As previously described in Section III, the first step of the MVC flow is the creation of a characterization environment tuned for the library nominal voltage. A modified version of ST-Microelectronics 28nm FDSOI CORE library that contains only the set of gates that are part of the reduced library (all gates with 3 or less inputs) was used as reference library. From this library, the input slew and output capacitance vectors used in the characterization setup file were extracted. To increase characterization accuracy, a midsized inverter is used as the driver cell. With this environment, the reduced cell library was characterized at nominal voltage (1V), and the results compared with the reference. A set of 58 benchmark circuits composed of 4 cryptographic circuits, 15 combinational circuits from the ISCAS'85 benchmark [9], 38 sequential circuits from the ISCAS'89 benchmark [10], and a network-on-chip router [11] were used by LC to validate the library.

Figure 6(a) shows a histogram of path delay error grouped in bins of 0.01 (i.e. 1% error). The histogram, which includes errors computed for each path of each circuit analyzed by LC, points to a mean path error delay of -0.025, with a standard deviation of 0.015. A similar approach was taken to evaluate the delay error of individual cells. Figure 6(b)shows this analysis. Here, delay errors were grouped by logic cell function and binned using the same criteria as before. In other words, each value accounted in the histogram is the average error for a given logic function. The results indicate a mean cell delay error of -0.026, with a standard deviation of 0.043. When compared to the reference library, the mean error measured for both path and cell delay are under the 5% error limit established for this work. The statistical data, therefore, attests the accuracy of the characterization environment tuned for nominal voltage.

The second step of the MVC flow takes as input the characterization environment validated for the nominal voltage. In this step, voltage-dependent parameters are scaled according to the method defined in Section III-C. An inverter with the maximum drive strength was used as test cell to determine the

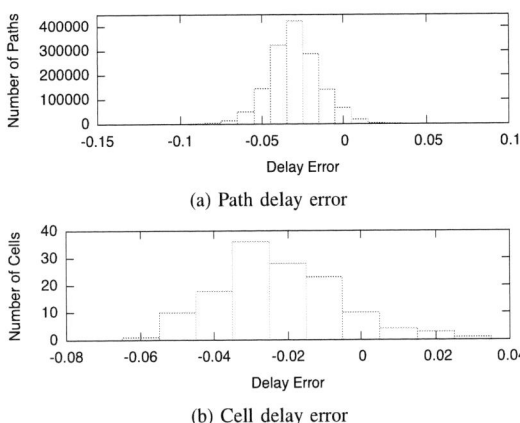

(a) Path delay error

(b) Cell delay error

Fig. 6. Error histograms comparing the reduced cell library characterization at nominal voltage to the reference library. Histogram (a) shows a mean path delay error of -0.025, with a standard deviation of 0.015. Histogram (b) indicates a mean cell delay error of -0.026, with a standard deviation of 0.043.

minimum input slew for each voltage level. Maximum input slews were scaled using a 4-input NOR gate. Even though this gate is not present in the reduced cell library, it was employed in this analysis because it presents the slowest transition delay of the original library. This step resulted in 15 unique characterization environments, each targeting a supply voltage in the range from 950mV to 250mV. These files are based on the characterization environment tuned for nominal voltage, with the input slew vectors for nominal voltage replaced by the scaled ones.

The last step of the MVC flow consists in performing library characterizations employing the simulation setup files created in the previous step. Figure 7 gives results that demonstrate the accuracy of the achieved results. The outcome of the flow is a set of Liberty libraries that extend the voltage corners supported by the reduced cell library to the range of 1V to 250mV. In addition to providing characterization data at the nominal voltage (1V), the ST-Microelectronics 28nm FDSOI CORE library includes timing models characterized at 900mV (nominal supply - 10%) and 800mV (nominal supply - 20%). These models were used as additional verification points for the reduced cell library characterization. The approach taken is similar to what was previously done to validate the characterization environment at the nominal voltage. Modified versions of the 900mV and 800mV ST-Microelectronics 28nm FDSOI CORE libraries were created and used as the reference library for LC. These contain only the 3-or-less-input gates that are part of the reduced library.

The 900mV library presented a mean path delay error of -0.010, with a standard deviation of 0.019, and a mean cell delay error of -0.010, with a standard deviation of 0.045. At 800mV, LC points to a mean path error delay of 0.022, with a standard deviation of 0.025, and a mean cell delay error of 0.025, with a standard deviation of 0.053. Since the statistical data displays mean path and cell delay errors within the 5% error limit established for this work, we can attest the accuracy of the timing characterizations performed by the MVC flow.

---

[1]In the context of this work, the notion of *correct operation* refers to the standard cell being able to generate output signals within the bounds defined in the cell library assumptions. If a set of less strict assumptions is provided, *correct operation* could encompass voltages lower than 250mV, but this complex change is outside the scope of the work. The analysis here also disregards other effects, such as noise and crosstalk.

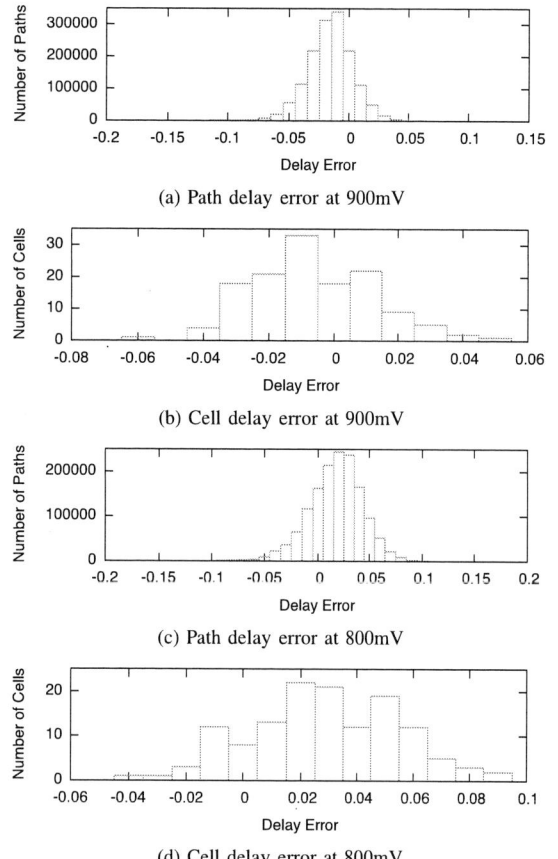

(a) Path delay error at 900mV

(b) Cell delay error at 900mV

(c) Path delay error at 800mV

(d) Cell delay error at 800mV

Fig. 7. Error histograms comparing the 800mV and 900mV characterizations of the reduced cell library to the reference. Histograms (a) and (b) show, respectively, a mean path delay error of -0.010, with a standard deviation of 0.019, and a mean cell delay error of -0.010, with a standard deviation of 0.045 for the library characterized at 900mV. Histograms (c) and (d) furnish, respectively, a mean path error delay of 0.022, with a standard deviation of 0.025, and a mean cell delay error is 0.025, with a standard deviation of 0.053 for the library characterized at 800mV.

V. Conclusions

This work proposes a method to extend the voltage corners supported by standard cell libraries. The Multi-voltage Characterization (MVC) flow incorporates and implements the method, which provides a systematic way to characterize cell libraries for a wide range of target supply voltages, ensuring the proper scaling of voltage-dependent parameters. The voltages can be super-, near- and sub-threshold values for virtually any technology. This clearly advances the state of the art in the use of cell-based design in modern and future ultra-low power applications, such as IoT and others. Also, the MVC flow can support designs intended to trade-off speed and power figures based on the exploration of alternative supply voltages not directly considered by commercial standard cell libraries. The MVC flow was used to successfully characterize a reduced cell library from 1V down to 250mV in a state of the art commercial technology. In addition to the flow, the article proposes and validates LC, a tool designed to validate cell characterizations by the automated comparison of STA reports.

References

[1] H. Eriksson, P. Larsson-Edefors, T. Henriksson, and C. Svensson, "Full-custom vs. standard-cell design flow - an adder case study," in *Asia and South Pacific Design Automation Conference (ASP-DAC)*, Jan. 2003, pp. 507–510.

[2] J. M. Rabaey, A. Chandrakasan, and B. Nikolic, *Digital Integrated Circuits*, 3rd ed. Upper Saddle River, NJ, USA: Prentice Hall Press, 2008.

[3] T. D. Burd and R. W. Brodersen, "Design issues for dynamic voltage scaling," in *International Symposium on Lower Power Electronics and Design (ISLPED)*, 2000, pp. 9–14.

[4] L. Yuan and G. Qu, "Analysis of energy reduction on dynamic voltage scaling-enabled systems," *IEEE Transactions on Computer-Aided Design of Integrated Circuits and Systems*, vol. 24, no. 12, pp. 1827–1837, Dec 2005.

[5] B. Zhai, D. Blaauw, D. Sylvester, and K. Flautner, "Theoretical and practical limits of dynamic voltage scaling," in *Design Automation Conference (DAC)*, 2004, pp. 868–873.

[6] *Encounter Library Characterizer User Guide, Version 13.1*, Cadence Design Systems, Inc, Apr. 2013.

[7] M. T. Moreira, C. M. Oliveira, N. L. V. Calazans, and L. C. Ost, "LiChEn: Automated Electrical Characterization of Asynchronous Standard Cell Libraries," in *Euromicro Conference on Digital System Design (DSD)*, Sep. 2013, pp. 933–940.

[8] J. Kwong, Y. Ramadass, N. Verma, M. Koesler, K. Huber, H. Moormann, and A. Chandrakasan, "A 65nm Sub-Vt Microcontroller with Integrated SRAM and Switched-Capacitor DC-DC Converter," in *IEEE International Solid-State Circuits Conference (ISSCC)*, Feb. 2008, pp. 318–319, 616.

[9] F. Brglez and H. Fujiwara, "A Neutral Netlist of 10 Combinational Benchmark Circuits and a Target Translator in FORTRAN," in *IEEE International Symposium on Circuits and Systems (ISCAS), Special Session on ATPG and Fault Simulation*, Jun. 1985, pp. 695–698.

[10] F. Brglez, D. Bryan, and K. Kozminski, "Combinational Profiles of Sequential Benchmark Circuits," in *IEEE International Symposium on Circuits and Systems (ISCAS)*, May 1989, pp. 1929–1934 vol.3.

[11] M. T. Moreira, L. H. Heck, G. Heck, M. Gibiluka, N. L. V. Calazans, and F. G. Moraes, "The YeAH! NoC Router," Faculty of Informatics, PUCRS, Tech. Rep. 083, Dec. 2014.

Design and Analysis of the HF-RISC Processor Targeting Voltage Scaling Applications

Felipe Todeschini Bortolon*, Sergio Johann Filho[†], Matheus Gibiluka[†],
Sergio Bampi*, Ney Laert Vilar Calazans[†], Fabiano Passuelo Hessel[†], Matheus Trevisan Moreira[†]

* Universidade Federal do Rio Grande do Sul – Porto Alegre – RS – Brazil

† Pontifícia Universidade Católica do Rio Grande do Sul – Porto Alegre – RS – Brazil

{ftbortolon, bampi}@inf.ufrgs.br

{sergio.filho, ney.calazans, fabiano.hessel}@pucrs.br {matheus.gibiluka, matheus.moreira}@acad.pucrs.br

Abstract—**This paper presents the design and analysis of HF-RISC, a 32-bit RISC processor, targeting voltage scaling applications. We start proposing a design flow that enables the processor to operate at multiple voltage levels and explore how this flow enables designers to leverage the advantages of low voltage designs. Next, we present a set of case study designs of HF-RISC in a 28nm FD-SOI technology assessing their area, performance and power figures. Using the collected data we discuss how our flow can enable better design space exploration for voltage scaling applications and define guidelines for achieving lower power and better power efficiency. Accordingly, the obtained results indicate that the proposed flow allows 9.5% lower power overall and 25.5% better energy efficiency in HF-RISC design.**

Keywords—*Voltage Scaling; Low Voltage; Energy Efficiency; MIPS; FD-SOI; Design Space Exploration; Embedded systems;*

I. INTRODUCTION

Internet-of-Things (IoT) applications typically require relatively high processing power combined to low energy consumption. In such applications, several architectural characteristics have to be evaluated and processors must be designed for specific operating conditions and scenarios. Due to the the need for integration of IoT devices with other computer systems through network protocols such as IPv6 [1], 32-bit processor cores provide better performance and energy trade-offs, compared to 8 or 16-bit devices [2]. In RISC designs it is advantageous to keep the number of pipeline stages low to reduce the penalties originated from hazards. A simpler pipeline improves the number of executed instructions per cycle (IPC), simplifies the design [3] and reduces energy consumption [4].

With this shallow pipeline strategy, HF-RISC, a 32-bit RISC processor was designed, as presented in [5]. HF-RISC greatly simplifies the instruction set architecture (ISA) implementation, as no interlocks or forwarding units are needed to fix hazards. This approach is useful for low power design, where applications aim at better energy/MHz ratios, rather than high clock frequencies at a high energy cost. The industry currently employs the same principle, using 32-bit processors like the ARM Cortex-M family [6] of processors with only 2 or 3 pipeline stages [2], in place of 8- and 16-bit microcontrollers. In this way, the design choices in HF-RISC aim to improve both performance and energy efficiency as detailed in [5].

At the circuit design level, an interesting approach to decrease energy consumption is to reduce the supply voltage while the device does not need to deliver symbolic performance. This is commonly known as supply voltage scaling or simply voltage scaling (VS). This technique is very effective because the power dissipation in an integrated circuit is quadratically proportional to the supply voltage [7]. Therefore, small voltage reductions lead to significant energy savings. For this reason several publications explore the design of VS systems, however a thorough discussion about the complete design flow is still missing. More specifically, there is no guidelines available considering how Static Timing Analysis (STA) tools can assist the design of a VS system, even though they are largely employed for circuit optimization.

This paper aims to fulfill this gap by analyzing the trade-offs offered by the synthesis tool when using a cell library characterized from sub to super-threshold. After a review of the state-of-art in Section II-A to demonstrate the novelty of this investigation, the target processor for the case-study, the HF-RISC, is presented in Section II-B . Next, Section III explains the synthesis flow and how it is used to implement several case studies. Section IV presents the results and delves into the details of the trade-offs that might be achieved using this flow. Accordingly, Section V uses the obtained results to present the final conclusions and to define guidelines for better exploring the synthesis flow using STA tools.

II. STATE-OF-THE-ART AND BACKGROUND

A. Processors for Voltage Scaling Applications

According to [8], VS leverages the computational bursti-ness of processors used in portable electronics devices, where typically only a fraction of the computation utilizes its full performance. Several works available in contemporary literature explore this profile to demonstrate the benefits of VS in integrated circuits. From a mathematical perspective, different authors analyzed at distinct angles how to stress VS and achieve its limits. In [9], for example, Yuan and Qu define three different types, called models, of VS systems based on how voltage can be scaled. Their purpose is to define under different constraints, e.g. voltage scaling range and delay, guidelines for the designer of how much energy savings might be achieved. Their analysis provides solid equations and simulation results to answer this question; however their results do not advance to the sub-threshold region.

This topic is of particular importance as stated in [10], where authors discover that the minimum operating voltage point for maximum energy-efficiency lies on that region. Although the optimal point depends on the workload characteristics of a specific processor, they determine that under typical load, operating voltage should scale to approximately

30% (V_{min}) of its maximum value. A more recent work [11] shows an interesting investigation of VS benefits for future technologies using LEON3 processor as their case-study. Their analysis compare two different FinFET nodes against two traditional CMOS bulk nodes and conclude that the latter takes more advantage of scaling over the first due to their layout differences.

While some authors provide a theoretical perspective of this technique, there are other contemporary reports that demonstrate experimental applications of VS [12]. Accordingly, the work published by Craig et al. in [13] presents a 32-bits, 90 nm data flow processor that employs dynamic voltage scaling from sub-threshold to high performance. The design can scale at three different V_{DD} settings (0.7V, 0.8V and 1.2V) which can switch in a single clock cycle through selectively controlling a set of PMOS transistors. Each transistor has its source terminal connected to different V_{DD}s; therefore if it is in the on state, it lets the selected V_{DD} voltage to supply the data path. The authors present an interesting discussion about the special considerations taken for the sub-threshold PMOS switch design and the architectural modifications to allow blocks to operate at different voltages, i.e. memory on 1.2V and data path on 0.7V. Nevertheless, the synthesis process, e.g. cell pruning and selection, is not explored to optimize the design.

Authors in [14] fabricated a 65 nm, 32-bits sub-threshold RISC processor using a custom standard cell library which was designed to optimize noise margins, switching energy and propagation delay simultaneously. For comparison, an additional core based on conventional standard cells is added to the design. The chip implementation uses a hierarchical multi-mode multi-corner (MMMC) synthesis to account for different timing conditions under varying supply voltage. Even though conventional standard cell core instance achieves 260MHz at 1.2V it fails to operate bellow 700mV, while the custom can operate over a supply voltage range from 200mV to 1.2V with clock frequencies from 10 kHz to 94 MHz for the best samples. Another work that explores custom standard cell library is presented in [15] for a 32nm, 32-bits processor that operates from 280mV to 1.2V with clock frequency from 3MHz to 915MHz. To optimize their design for robust and reliable ultra-low voltage operation, they prune the cell library to eliminate circuits which exhibit DC failures or extreme delay degradation due to reduced transistor on/off current ratios and increased sensitivity to process variation. In contrast to [14], they argue that in the absence of MMMC tools that it is important to identify the optimal design such that the results achieved in one corner do not compromise operation at other corners. Therefore, their resulting library is characterized for three different supply voltages: 0.5V, 0.75V and 1.05V and it was identified that using 0.5V yield the best trade-offs for their case.

This increasing concern to achieve better performance and energy trade-offs with VS is crucial to new power-aware embedded systems, since stricter power dissipation constraints are imposed to devices that have higher transistor integration density. The works analyzed here explore different techniques to that extent. However only [14] and [15] delve into the synthesis process. Regarding library pruning, both of them confirm that considering noise margins and delay as cell

selecting criterion provides more robust and reliable circuits. Nonetheless, [14] uses MMMC while [15] uses standard cell library characterization on different voltages. Even though, MMMC guarantees that constraints for all voltages are meet, this approach does not actually ensure that the tool will explore design trade-offs for performance and energy. On the other hand, [15] motivates this exploration by pointing to the importance of standard cell library characterization for different voltages to achieve those better trade-offs, however their approach is restricted to only three corners.

In conclusion, none provided a thorough exploration to determine how to deal with different voltage levels in synthesis flows or how to guide synthesis tools to achieve the best trade-offs between sub and super-threshold voltage operation, i.e. the synthesis optimal point. Because the majority of designs rely on STA tools to synthesize and optimize integrated circuits (ICs), this type of analysis can provide better insight to allow the designer to better explore VS design space. Therefore, this work stands off by analyzing the different performance, power and area outcomes from the synthesis tool with standard cell libraries characterization for different supply voltages considered in the design phase.

B. The HF-RISC Processor

The HF-RISC Instruction Set Architecture (ISA) comprises a small subset of the MIPS I ISA introduced in [16] targeting compatibility with existing tools and optimizing compilers. Additionally, its core has specific features for low-power design such as fewer pipeline stages and a more compact organization of its components. The most relevant differences between this implementation and the classic 5-stage MIPS are:

- Short, 3-stage pipeline, to simplify core design and to reduce chip area and energy consumption;

- No hazard or forward units, due to the short pipeline;

- Shared instruction and data memories, i.e. a Von Neumann organization.

- Fully synchronous, single clock edge design: registered memories are interfaced directly;

- 3-cycle branch delay when taken, with 2 branch delay slots;

- No unaligned loads/stores; no MMU; no exceptions;

- No co-processor; only memory-mapped peripherals (EPC, MASK, STATUS, VECTOR, and CAUSE registers);

- Configurable HW multiply unit; no HW division unit;

- A set of MCU-like peripherals: an optional UART, an interrupt controller, a running counter, two programmable counters, compare registers and a debug interface.

Most HF-RISC instructions take just one clock cycle, but load and store instructions take three cycles each, due to the memory bus multiplexing and pipeline refill. Also, multiply instructions take several cycles, depending on the chosen hardware configuration - 4 cycles for a parallel multiplier and from 11 to 35 for a serial multiplier. A side effect of

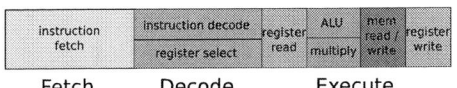

Fetch　　　　Decode　　　　Execute

Fig. 1: The HF-RISC 3-stage pipeline and the stage tasks [5].

the simple pipeline is the absence of explicit load delay slots of conventional MIPS organizations. Two branch delay slots arise due to pipeline design, as the outcome of branches is discovered on the third pipeline stage. This simplifies the datapath, as the same ALU is used for both arithmetic and branch target calculation. Other advantages are the lack of a hazard detection unit and forwarding paths. The compiler can schedule instructions in the first branch slot, reducing branch penalty to 2 cycles. Although there are several different configuration possibilities for the processor, we chose one which offers the best performance/power ratio [5]. In this paper, we have conducted our experiments with the serial multiplier version, as it offers good performance (1.61 Coremark/MHz) and low area overhead.

Fig. 1 depicts the stages of the HF-RISC pipeline and the tasks executed in each of these. In the *fetch* stage, memory is accessed and an instruction becomes available in one cycle. In this same cycle the PC is updated. In the *decode* stage an instruction is fed into the decoding and control logic, so values are registered for the next stage. Pipeline bubble insertion is performed in this stage for memory and branch operations. In the *execute* stage the register file is accessed and the ALU calculates the result of the operation. Address and data are put on the data bus (on store operations) or data are copied to the register file (on load operations). On logic/arithmetic operations, the ALU result is written to the register file. Branch outcomes are computed in this stage. Multiply operations write the result to HI and LO registers. The register file is accessed only in the *execute* stage in order to simplify the design, as it allows avoiding the addition of extra hardware to cope with data hazards. To evaluate the performance of the architecture and build an adequate programming environment for measurements, we created a hardware abstraction layer (HAL), small C and runtime libraries, and used the GNU tools based on GCC 4.9.3 and Binutils 2.24. The compiler backend was modified to support all different processor configurations, including the absence of multiply and divide instructions and other microarchitecture features.

III. Voltage Scaling Aware Design and Analysis Flow

Electronic Design Automation (EDA) tools usually rely on Register-Transfer Level (RTL) descriptions to capture an IC specification, and on cell libraries models to synthesize and map this specification to a specific technology. During this process, the design goes through optimization steps that are driven by STA tools, which provide delay and power estimations for the synthesized circuit. These results are calculated using data available on the cell library models that describe cell's logic behavior and electrical characteristics for a set of operating conditions, e.g. temperature and voltage. In this way, an efficient synthesis and analysis of a circuit for VS applications requires cell libraries characterized for a specific set of voltage levels. However, these models are not conventional or easy to find, which impairs the capability of a

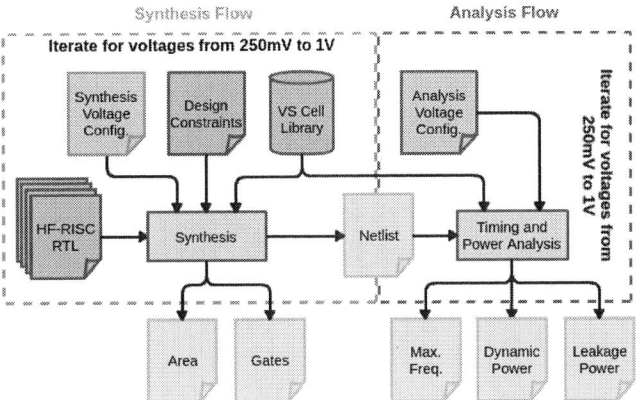

Fig. 2: Voltage scaling aware design and analysis flow.

designer to optimize a circuit for a given voltage level, or for a range. Hence, to explore the design space for VS applications this work characterized and pruned a library of cells targeting the STMicroelectronics 28nm FD-SOI technology, which will be referenced as VS Cell Library. Note that the selected node is FD-SOI because it allows operation in a wide range of voltages, suitable for VS applications, and its migration from bulk is relatively straightforward [17]. In summary, first the available library provided for nominal V_{DD} is reduced to a subset of cells that have a maximum of 3 inputs as advised by Kwong et al. in [18]. Then, the selected cells to compose the VS library were characterized using Cadence ELC^{TM} for voltages from 250 mV to 1 V in steps of 50 mV. For a complete description of this process refer to [19], [20].

With the availability of these different models, it was devised a voltage scaling aware design and analysis flow, as showed in Fig. 2, to evaluate the HF-RISC for different voltage levels. As the figure shows, the flow was divided in two main steps: synthesis and analysis. The synthesis has as its inputs the RTL description, a set of design constraints, the VS cell library and the voltage configuration. Note that the design constraints were defined to allow a realistic synthesis scenario, where the outputs of the design had realistic intrachip capacitance values and the inputs had realistic transition and insertion delays. Such figures are all equivalent to those of a slice register, commonly employed in microprocessor SoCs. This step was iterated 16 times, sweeping the defined voltage configuration from 250 mV to 1 V in steps of 50 mV. At each iteration the HF-RISC is synthesized to its maximum frequency generating a netlist that is optimized for maximum performance at each voltage. This process is undertake for every library model previously generated, thus all 16 library models are combined to the 16 different operating voltages. To achieve maximum frequency it is necessary to perform a binary search in the defined clock and resynthesize the circuit multiple times. By adopting this approach is possible to stress the STA tool and explore its capability to optimize circuits for different voltages levels.

As Fig. 2 shows, after synthesis, each of the generated netlists are considered in the analysis step. Additionally to synthesis output, this step generates maximum operating frequency, dynamic power and leakage power reports. Because the flow can be configured for different voltage levels, it is possible to analyze how each synthesized version of HF-

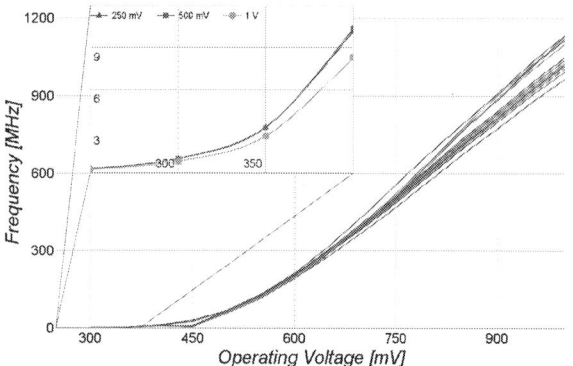

Fig. 3: Frequency vs. Operating Voltage for different Synthesis Voltage

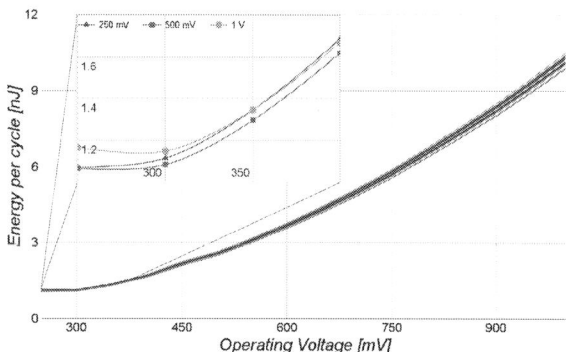

Fig. 4: Energy / Cycle vs. Operating Voltage for different Synthesis Voltage

RISC behaves at each voltage. Furthermore, results help to understand how frequency and power metrics scale for each synthesized netlist with the variation in the voltage configuration. Maximum frequency was measured by analyzing the maximum clock period that allowed non-negative slack in the design as the voltage was scaled in the analysis step. Power figures were measured by simulating the CoreMark benchmark in the synthesized circuit, exporting the internal activity of the nets and performing static power analysis. After all iterations of the complete flows, the output collected a total of 16 designs (one for each voltage level) and the respective metrics of each for 16 distinct voltage levels.

IV. RESULTS ANALYSIS

The first step to understand the trade-offs between the different generated netlists is to evaluate how their maximum frequency scales with voltage. This analysis, which is depicted in Fig. 3, shows how the frequency of a circuit synthesized at 250 mV varies as we sweep its operating voltage and how the same metric varies for the circuit synthesized at all other voltages. Therefore, the designer will have better insights on what is the best voltage to characterize a library for VS applications. Interestingly, the depicted results for the 250 mV synthesis exhibit an average of 24% higher frequency than 1 V synthesis while operating on the sub-threshold region, and a decrease of 11% on the super-threshold. An even better trade-off is obtained at the 500 mV synthesis, where the sub and super-threshold frequencies increase and decrease, respectively, by 22% and 5% compared against 1V synthesis.

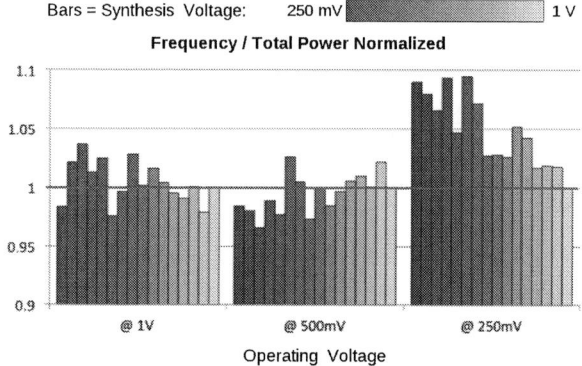

(a)

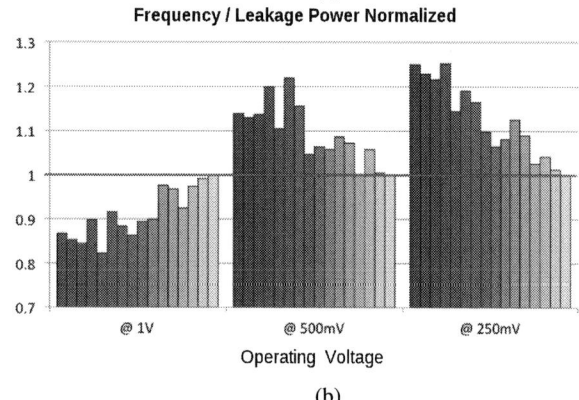

(b)

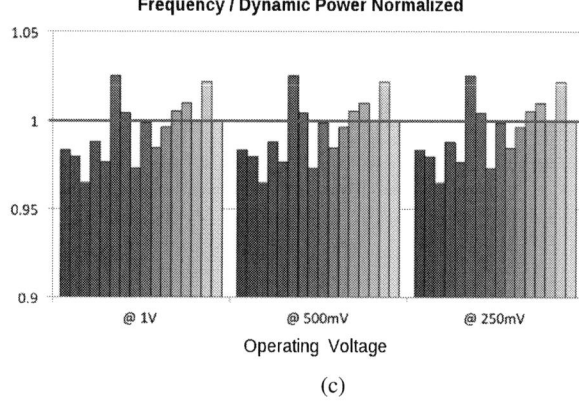

(c)

Fig. 5: Frequency over (a) Total Power, (b) Leakage Power and (c) Dynamic Power, all normalized to the 1V synthesis. In each is depicted three different operating voltages: @ 1V, @ 500mV and @ 250mV, and synthesis voltages varies from 250mV to 1V (left to right) for each.

Extending this investigation to power consumption, the results depicted in Fig. 4 demonstrate that exploring voltage during synthesis can yield extra energy savings at the sub-threshold region. For example, as the charts show, the netlists synthesized at 250 mV and 500 mV present a better minimum energy point when compared to the netlist synthesized at 1 V. Note that these 3 voltage levels were isolated because they exhibit superior behavior for the 3 operating modes: sub-threshold, near-threshold and nominal. Diving more deeply into this scenario, Fig. 5 details this analysis for dynamic and leakage power and normalizes the results to the netlist synthesized at 1 V to determine the energy gains for each different synthesis voltage. Fig. 5 (a) demonstrates that the netlist synthesized at 250 mV allows a 9% improvement on energy efficiency for an operating voltage of 250 mV, while presenting less than 2% degradation at both 500 mV and 1 V supply voltages. The 500 mV synthesis once again exhibits better improvements than 250 mV, with an increase of 9.5% and 2.5%, and a decrease of 2.5% at respectively 250 mV, 500 mV and 1 V. Interestingly, those energy savings derive from the leakage power improvements (Fig. 5 (b)) that significantly supersedes dynamic power (Fig. 5 (c)) deterioration. Since the off-current does not reduces as strongly as the on-current when the voltage scales below the threshold [10] optimizing leakage results in the depicted energy saving enhancement.

To further scrutinize energy optimization opportunities on the sub-threshold region, Fig. 6 plots the power efficiency measured as maximum coremarks divided by total power for three different synthesis: 250 mV, 500 mV and 1 V. As explored in [5], the maximum coremarks figure represents the processor efficiency in terms of its frequency since it correlates architectural and technology metrics of a design. Because all case studies share a common architecture, this analysis permits to explore the impact of technology specific figures on the core performance, in this case the voltage level used during synthesis. Combining this metric with total power, it is possible to understand the impact of such choice in a processor's design, more specifically, its effects on overall power efficiency. Results in Fig. 6 demonstrate that lower synthesis voltage achieves interesting improvements. Here the gains provided by 500 mV synthesis over 250 mV are more noticeable, which confirms the same results derived from previous analysis.

Another interesting metric is the relationship between design area and performance through synthesis variation. To that extent, Fig. 7 shows the Maximum Coremark achieved over design area (μm^2) for three supply voltages and four different synthesis normalized to the 1 V synthesis. The collected results are summarized in Table I. At each operating voltage the maximum performance is achieved at their respective synthesis voltages, i.e. 750 mV synthesis has maximum gain at 750 mV supply voltage. In the other cases the graph exhibits different improvements considering how close is the synthesis voltage to the supply voltage, i.e. staircase behavior at 1 V and 250 mV supply voltages. Nonetheless, similar to the previous graphs, overheads at super-threshold region for 500 mV and 250 mV are not so pronounced as the advantages that they offer for the sub-threshold. Therefore, the designer should carefully consider synthesizing at smaller voltages if the circuit targets voltage scaling.

The benefits demonstrated thus far rely on the synthesis tool capabilities to select gates that are more suited to operate at different voltage levels. Cell libraries characterized at lower voltage have precise information of the high delays that gates present when operating at lower voltage levels, thus allowing the tool to chose higher strength gates to minimize overall timing, i.e. maximize frequency. Fig. 8 supports this statement demonstrating the difference on cell strength selection for three distinct netlists using the 1 V as their reference. In the 500 mV synthesis, for example, 34.65% cells with strength higher than 30x is added comparing to the number of similar cells on the 1 V synthesis. Therefore, it is possible to conclude that in order

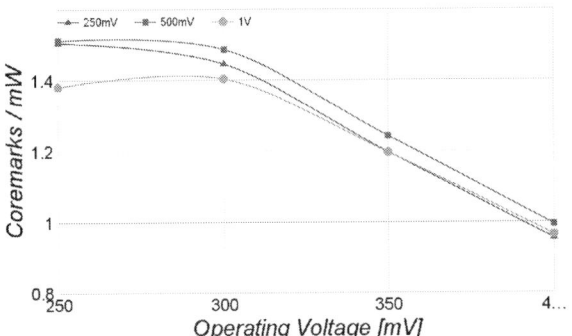

Fig. 6: Maximum Coremarks over Total Power vs. Operating Voltage for three different synthesis voltages: 250mV, 500mV and 1V

TABLE I: Maximum Coremark over Area performance comparison against 1V synthesis.

	Synthesis Voltage		
Supply Voltage	@250mV	@500mV	@750mV
@1V	↓ 13.2%	↓ 8.5%	↓ 1.3%
@750mV	↓ 7.2%	↓ 1.3%	↑ 6.9%
@500mV	↑ 13.3%	↑ 21.2%	↑ 8.7%
@250mV	↑ 25.5%	↑ 20.2%	↑ 12.5%

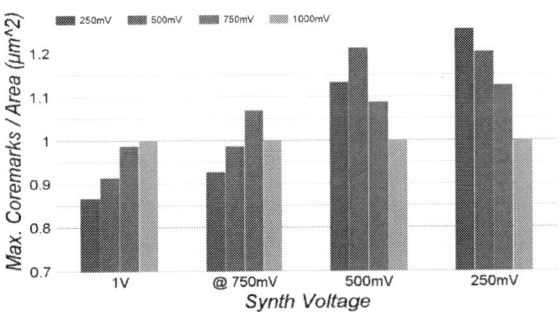

Fig. 7: Maximum Coremarks over Area vs. Synthesis for three different synthesis voltages: 250mV, 500mV and 1V

to meet the timing constraints the tool uses cells with higher strengths. This analysis is in agreement with previous works such as [21], which demonstrates that optimizing transistor sizing for the minimum energy operating point gives symbolic performance and power improvements. Moreover, presented results demonstrate the importance of having cell libraries characterized at different voltage levels for VS aware design and the capability of synthesis tools to leverage this data.

V. DISCUSSION AND CONCLUSIONS

Embedded designs often have to trade performance improvements for overall power and area. Thus, performance, power and area ratios, are more relevant than the traditional area, speed and power measurements isolated. To that extent, this work proposed a design flow that explores HF-RISC processor, i.e. target circuit, outcomes from synthesis using libraries characterized at distinct voltage levels. Apart from Jain et al. [15] and Lutkemeier et al. [14], which do not delve into the same details, the state-of-the-art does not explore the synthesis process to balance defined metrics. Results obtained

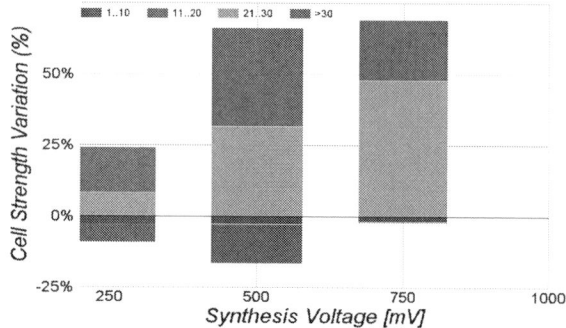

Fig. 8: Cell strength variation for 250mV, 500mV and 750mV synthesis compared to 1V.

from our VS aware design flow indicate that it is possible to achieve 20% higher clock frequencies in the sub-threshold with an overhead of only 5% at nominal voltage 1 V. Additionally, it is interesting to note that waiving 5% of performance at nominal supply voltage can offer further 9.5% and 2% more energy savings at the minimum and maximum voltage values, respectively. As depicted in Fig. 5, the dominant savings on energy derives from the leakage current. Considering that the target technology of this work, i.e. 28 nm FD-SOI, already offers reduced device off-current, we consider a fair assumption that traditional CMOS bulk technologies could achieve higher order of magnitude savings. Therefore, future work may explore other technology nodes or stress energy savings through body-biasing, technique of which is very suitable for FD-SOI.

To conclude, the results obtained in our analysis indicate that for synthesizing our processor for VS applications, extra cell library models are required in order to better explore the design space. Specifically for the HF-RISC processor, the 500 mV synthesis achieves the best trade-off between presented metrics, slightly superseding 250 mV. Note that the latter offers, in some cases, better improvements at a higher degradation cost, however. Examining Fig. 8 the probable reason for this outcome is the higher strength cells used for 500 mV, which ensure superior behavior at near and sub-threshold region. Altogether, this work demonstrates that proposed VS design flow can significantly enhance overall design efficiency regarding power and frequency. Moreover, results offer important insights for the IC designer on the behavior of the synthesis tool when considering different cell libraries. For example, exploring two extra voltage level models, one for sub- and another for near-threshold, can provide improvements for VS applications, without the overhead implied from finer grain voltage steps library characterization. This is an important observation, because this process is an extensive and laborious task.

REFERENCES

[1] J. Gubbi, R. Buyya, S. Marusic, and M. Palaniswami, "Internet of Things (IoT): A Vision, Architectural Elements, and Future Directions," *Fut. Gen. Comp. Syst.*, vol. 29, no. 7, pp. 1645–1660, Sep. 2013.

[2] J. Yiu, *The Definitive Guide to the ARM Cortex-M0.* Newnes, 2011.

[3] M. Labrecque, P. Yiannacouras, and J. G. Steffan, "Custom Code Generation for Soft Processors," *SIGARCH Comput. Archit. News*, vol. 35, no. 3, pp. 9–19, Jun. 2007.

[4] O. Azizi, A. Mahesri, B. Lee, S. Patel, and M. Horowitz, "Energy-performance Tradeoffs in Processor Architecture and Circuit Design: A Marginal Cost Analysis," in *37th Annual International Symposium on Computer Architecture*, 2010, pp. 26–36.

[5] S. J. Filho, M. T. Moreira, N. L. V. Calazans, and F. P. Hessel, "The HF-RISC Processor: Performance Assessment," in *LASCAS*, 2016.

[6] ARM, "Cortex-m0+ technical reference manual," Tech. Rep., 2012. [Online]. Available: http://infocenter.arm.com/help/topic/com.arm.doc.ddi0484b/DDI0484B_cortex_m0p_r0p0_trm.pdf

[7] J. Kwong and A. Chandrakasan, "Variation-driven device sizing for minimum energy sub-threshold circuits," in *International Symposium on Low Power Electronics and Design*, 2006, pp. 8–13.

[8] T. D. Burd and R. W. Brodersen, "Design issues for dynamic voltage scaling," in *International Symposium on Lower Power Electronics and Design (ISLPED)*, 2000, pp. 9–14.

[9] L. Yuan and G. Qu, "Analysis of energy reduction on dynamic voltage scaling-enabled systems," *IEEE Transactions on Computer-Aided Design of Integrated Circuits and Systems*, vol. 24, no. 12, pp. 1827–1837, Dec 2005.

[10] B. Zhai, D. Blaauw, D. Sylvester, and K. Flautner, "Theoretical and practical limits of dynamic voltage scaling," in *Design Automation Conference (DAC)*, 2004, pp. 868–873.

[11] X. Lin, A. Shafaei, S. Chen, T. Cui, and M. Pedram, "Impact of technology and voltage scaling on LEON3 processor performance and energy," in *Microelectronics Technology Unified Conference (S3S)*, Oct 2015, pp. 1–2.

[12] M. Ashouei, J. Hulzink, M. Konijnenburg, J. Zhou, F. Duarte, A. Breeschoten, J. Huisken, J. Stuyt, H. d. Groot, F. Barat, J. David, and J. V. Ginderdeuren, "A voltage-scalable biomedical signal processor running ECG using 13pj/cycle at 1mhz and 0.4v," in *Solid-State Circuits Conference Digest of Technical Papers (ISSCC), 2011 IEEE International*, Feb. 2011, pp. 332–334.

[13] K. Craig, Y. Shakhsheer, S. Arrabi, S. Khanna, J. Lach, and B. H. Calhoun, "A 32 b 90 nm Processor Implementing Panoptic DVS Achieving Energy Efficient Operation From Sub-Threshold to High Performance," *IEEE Journal of Solid-State Circuits*, vol. 49, no. 2, pp. 545–552, 2014.

[14] S. Lutkemeier, T. Jungeblut, H. K. O. Berge, S. Aunet, M. Porrmann, and U. Ruckert, "A 65 nm 32 b Subthreshold Processor With 9t Multi-Vt SRAM and Adaptive Supply Voltage Control," *IEEE Journal of Solid-State Circuits*, vol. 48, no. 1, pp. 8–19, Jan. 2013.

[15] S. Jain, S. Khare, S. Yada, V. Ambili, P. Salihundam, S. Ramani, S. Muthukumar, M. Srinivasan, A. Kumar, S. K. Gb, R. Ramanarayanan, V. Erraguntla, J. Howard, S. Vangal, S. Dighe, G. Ruhl, P. Aseron, H. Wilson, N. Borkar, V. De, and S. Borkar, "A 280mv-to-1.2v wide-operating-range IA-32 processor in 32nm CMOS," in *Solid-State Circuits Conference Digest of Technical Papers (ISSCC), 2012 IEEE International*, Feb. 2012, pp. 66–68.

[16] J. Hennessy, N. Jouppi, F. Baskett, and J. Gill, "MIPS: A VLSI Processor Architecture," in *CMU Conference on VLSI Systems and Computations*, 1981, pp. 337–346.

[17] B. Nikoli, M. Blagojevi, O. Thomas, P. Flatresse, and A. Vladimirescu, "Circuit design in nanoscale fdsoi technologies," in *Proc. 29th International Conference on Microelectronics*, may 2014, pp. 3–6.

[18] J. Kwong, Y. Ramadass, N. Verma, M. Koesler, K. Huber, H. Moormann, and A. Chandrakasan, "A 65nm Sub-Vt Microcontroller with Integrated SRAM and Switched-Capacitor DC-DC Converter," in *IEEE International Solid-State Circuits Conference - Digest of Technical Papers*, Feb. 2008, pp. 318–616.

[19] "A standard cell characterization flow for non-standard voltage supplies," in *29th Symposium on Integrated Circuits and Systems Design (SBCCI)*, 2016, p. 16.

[20] M. Gibiluka, "Analysis of Voltage Scaling Effects in the Design of Resilient Circuits," Master's thesis, Faculdade de Informatica, Pontificia Universidade Catolica do Rio Grande do Sul, 2016.

[21] A. L. R. Rosa, L. B. Soares, K. H. Stangherlin, and S. Bampi, "Designing cmos for near-threshold minimum-energy operation and extremely wide v-f scaling," in *Proceedings of the 28th Symposium on Integrated Circuits and Systems Design*, 2015, p. 1:11:6.

A Placement and Routing Algorithm for Quantum-dot Cellular Automata

Alyson Trindade
and Ricardo Ferreira
Departamento de Informática
Universidade Federal de Viçosa
Viçosa, Brazil
Email: ricardo@ufv.br

José Augusto M. Nacif
Instituto de Ciências Exatas e Tecnológicas
Universidade Federal de Viçosa
Florestal, Brazil
Email: jnacif@ufv.br

Douglas Sales
and Omar P. Vilela Neto
Departamento de Ciência da Computação
Universidade Federal de Minas Gerais
Belo Horizonte, Brazil
Email: omar@dcc.ufmg.br

Abstract—The beyond silicon nanotechnologies create the possibility to develop new logic devices. In special, Quantum-dot Cellular Automata (QCA) technology opens new opportunities for saving power and achieving high performance clock rates. QCA technology circuit design imposes several challenges to be adressed by design tools. This paper proposes a new placement and routing algorithm to automatically design QCA logic circuits. The proposed solution ensures the correct operation by using an universal clock scheme that creates a layer to decouple the low level synchronization task from logic level. The technology layer is regular and scalable. The automatic generated layouts are competitive in terms of area when compared to handmade layouts. In comparison to previous automatic generators, the proposed approach saves area and provides more flexibility.

I. INTRODUCTION

QCA is a promising nanotechnology that can replace CMOS with advantages such as less energy consumption and higher clock rates. However, the nature of QCA cells imposes more complex synchronization techniques and, as a result, most part of QCA circuits are not designed using a well defined methodology. An usual approach to synchronize QCA signals is the 4-phase clock scheme using uni-directional layers [1], [2]. However, this approach normally increases feedback connections costs occupying additional area to balance reconvergent paths and to avoid wire crossing. Recently, USE (Universal, Scalable, Efficient) QCA clock scheme has been proposed [3]. Although the USE scheme simplifies the QCA wiring to allow more compact layouts, there are no automatic placement and routing tools available.

Placement and routing (P&R) are well-know NP-complete problems [4], [5] in FPGA and VLSI target technologies. However QCA P&R has different constraints imposing new challenges to create a robust and scalable solution. FPGA P&R algorithms explore the device layout performing routing using regular channels and switches between the Configurable Logic Blocks (CLBs). Switches are placed on all vertical and horizontal channel intersections. Most common approaches use simulated-annealing based placement and path finder routing algorithms. Although an FPGA CLB can be implemented on the QCA technology as shown in [6], [7], designing an FPGA switch and the bus channel sections is a complex task on QCA circuits. VLSI approaches use multiple wire layers to

avoid wire crossing. Moreover, VLSI as well as FPGA have a simpler clock scheme in comparison to QCA technology. QCA P&R algorithms were proposed in [1], [2]. However, these works adopt clock schemes based on irregular-sized zones and uni-directional propagation, which can seriously reduce scalability. Moreover, previous P&R [1], [2] are executed at a high abstraction level and no simulation is performed at QCA layout level.

This work proposes a novel P&R algorithm at QCA layout level. Our tool automates the realization of a QCA circuit, and also verifies wire crossing and clock synchronization at lower level. We route reconvergent paths on two dimensions by employing a regular bi-directional clock scheme. Moreover, we automatically generate QCA layouts that are competitive in terms of area when compared with handmade layouts. In comparison to previous automatic generator as QCA-LG tool [8], our approach reduces the total square area as well as the wire and the non-used area.

This work is organized as follows. Section II introduces the QCA technolgy and the clock scheme. Section III presents the previous QCA P&R algorithms. We describe our P&R algoritm in section IV. Finally, sections V and VI present experimental results, conclusion and future work.

II. BACKGROUND

This section describes the QCA technology, QCA cells and clock schemes. The basic units of QCA circuit are QCA cells, which are typically composed of four quantum dots located at the corners of a square. A dot, in this context, is just a region where an electric charge can be located or not. Each cell has two free and mobile electrons which are able to tunnel between adjacent dots. Also, the cell occupancy is controlled by a back plane voltage. Tunneling to the outside of the cell is not allowed due to a high potential barrier. The Coulomb interaction between the electrons tends to locate them at opposing diagonals, as shown in Fig. 1. An isolated cell may be in one of two equivalent energy states. These states are called cell polarizations P = +1 and P = -1. So, it is possible to codify binary information by considering that P = +1 represents the value 1 and that P = -1 represents the value 0.

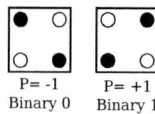

Fig. 1. Possible polarizations of QCA cells with four quantum dots. Black dots represent the electrons positions.

When two cells are placed near each other, the polarization of one cell will influence the polarization of the other cell. This feature is shown in Fig. 2(a). Following the same rule, a wire can be built by placing several QCA cells in a row, as shown in Fig. 2(b). QCA logic devices are designed by selecting the placement of QCA cells in a way that leverages the interaction between them. Fig. 3 depicts two fundamental QCA gates, inverter and majority gates.

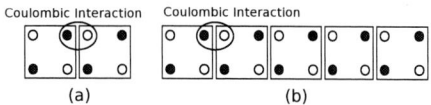

Fig. 2. (a) Coulomb interaction between two QCA cells and (b) A QCA wire.

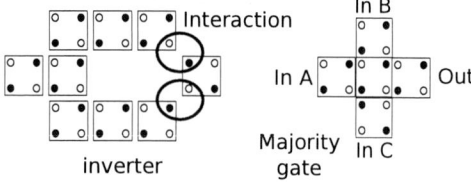

Fig. 3. QCA Inverter and Majority gates.

One of the main issues concerning QCA circuits is the switching of QCA arrays. QCA takes advantage of the physical ground state. If the input to a QCA array (eg. a wire) is switched suddenly the array will be momentarily in some combination of exited states (eg. cells with polarizations -1 and 1 in the same wire). As has been pointed by Landauer [9], the presence of metastable states could cause a significant delay in the system reaching its new ground state or even remain in a metastable state, not getting the necessary logic. The solution for this problem is to apply an adiabatic switch.

This being said, the QCA clock has four different phases: Switch, Hold, Release, and Relax. In the switch phase, the rising inter-dot barriers allow polarization level changes. In the hold phase, the cell is insusceptible to external influences and keeps its polarization level. In the release phase, the falling inter-dot barriers are responsible lead to the depolarization of the cell. In the final relax phase, the cell remains depolarized until a new cycle restarts with the following switch phase.

A QCA circuit is usually divided into subsections. Thus, the same clock signal is applied to more than one cell at a time. A set of cells under the same clock signal is named clock zone. The zones arrangement enables the transport and processing

of information in a pipeline fashion. More information about QCA and QCA clocking can be found in [10].

Recently a new clocking scheme, namely USE, has been proposed for clock distribution in QCA circuits [3]. It solves one of the most limiting factors of existing clock schemes, the implementation of feedback paths and easy routing of QCA-based circuits. Consequently, USE considerably facilitates the development of standard cell libraries and design tools for this technology, besides avoiding thermodynamics problems. It is based on the principle exposed earlier that the clock zones must be sequentially arranged in order to establish the information flow. A 4x4 grid ensures the right arrangement for the clock zones, by associating the cells located within each of its division boundaries to one of the four clock zones. Each grid division has square dimensions that are defined by the designer.

In this clocking scheme, two adjacent rows or columns always pass on the information by opposite directions, which enables several possibilities for routing and an easy establishment of feedback paths. Moreover, USE allows the implementation of coplanar and multilayer wire crossings and can be easily extended by positioning multiple grids side by side in both directions (horizontal and vertical) as depicted in Fig. 4. More information regarding USE, including possibilities for its physical implementation, can be found in [3].

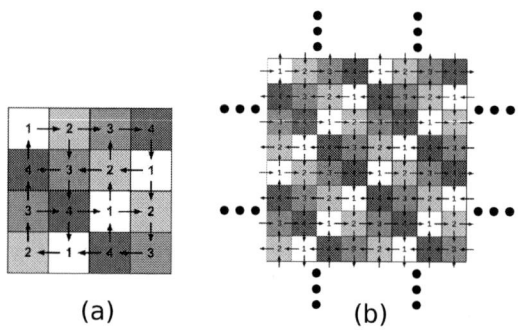

Fig. 4. (a) A single 4x4 USE grid. Each number indicates the clocking zone of the particular square. Note that information flow manner allows several possibilities for QCA structures design and routing. (b) An extended USE grid, which can be achieved by positioning the individual grids side by side.

III. RELATED WORK

Regarding QCA P&R, the approaches proposed in [1], [2], [11] implement 4-phase clocking by using timing zones with layers as depicted in Fig 5(a). This scheme is uni-directional, and therefore more "square-like" aspect ratio layouts and feedback are complex due the one direction flow. In addition, since all reconvergent paths starting at the same zone should have the same length, the QCA P&R should implement a strategy that will ensure balanced paths. The P&R proposed in [1] introduces dummy nodes as depicted in Fig 5(b). Therefore, the row with the largest number of gates defines the width of the entire zone. To reduce the area, the rows with a large number of gates are folded into two or more rows, and then dummy nodes are inserted. However, as shown in Fig 5(b),

978-1-5090-2737-8/16 $31.00 © 2016 IEEE

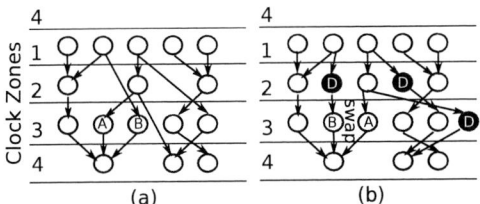

Fig. 5. (a) Each layer represents one timing zone in a 4-phase sequence: $1, 2, 3, 4, 1, 2 \dots$ (b) Dummy nodes (labeled by D) are inserted to balance paths, and nodes are swapped to reduce wire crossing between the layers as shown for nodes a and b in layer 3.

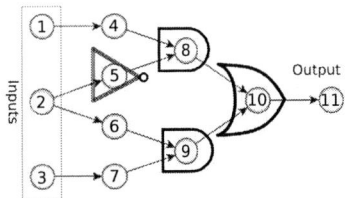

Fig. 6. A direct acyclic graph (DAG) for 2:1 multiplexer.

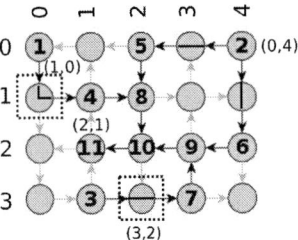

Fig. 7. P&R for the 2:1 multiplexer at USE level.

the P&R generates several wire crossing between layers. The authors of [1] implement two approaches to swap nodes inside the row to reduce wire crossing: the first is based on an analytical solution (the k-layer bipartite graph algorithm) and the second uses the simulated annealing algorithm. However, all cost measures in [1] are estimated at graph level where every node is a gate or a dummy node, which is not a realist assumption for QCA technology because the wires consume more area than gate cells.

In the P&R approach presented in [2], the QCA gates are duplicated to avoid wire crossing. In addition, heuristics have been implemented to minimize the number of duplicated gates. The authors mention a custom QCA P&R tool to convert the QCA graph to a QCA layout, however there is neither references nor descriptions of how this important step is conducted. Even at the graph level, the authors state that duplication essentially antecipates all wire crossing to the primary inputs, which results in more inputs or complex wire crossing at first level. Bubna *et al.* [11] propose a P&R that is also implemented by an abstract model that does not evaluate the final layout logical correctness and feasibility. The main contribution of this work is a more elaborate optimization approach at QCA graph level to reduce wire crossing by swapping nodes in the same level (clock zone).

The QCA-LG layout generator tool has been presented in [8]. In comparison to previous P&R approaches, the design is detailed at QCA dot level and it can be exported to be simulated and validated by using QCAdesigner [12]. QCA-LG approach is also based on gate duplication and, even more, a sub-graph duplication at QCA graph level to eliminate the internal cross wiring. Similar to the approach proposed in [2], the inputs are duplicated and a complex wiring should solve at first level. The tree structure simplifies the planar layout with higher cost due to the ternary structure and the duplicated node overhead.

In comparison to previous work, our approach: a) is based on a scalable bi-directional regular clock scheme; b) avoids gate duplication; c) validates the routed circuit in QCAdesigner.

IV. PLACEMENT AND ROUTING ALGORITHM

This section presents the proposed QCA P&R algorithm. Two examples are employed to drive the explanation: a 2:1 multiplexer and a 2-input XOR gate. A direct acyclic graph

(DAG) at gate level is the algorithm start point as shown in Fig. 6. In the presence of cycles, one or more edges will be removed to perform the initial P&R and these feedback edges are added at final routing steps. Since the P&R maps the circuit onto an USE clock scheme as shown in Fig. 4, the feedback edges have flexibility to route in all directions in the two-dimension space in comparison to most other QCA approaches [1], [2] based on uni-directional clock scheme.

The first step is the technology mapping at gate level. The DAG nodes are remapped to majority and inverter gates. The maximum gate fan-in is three and the maximum gate fan-out is two. For ease of explanation, let us consider a 2-input multiplexer as our first example. The technology mapping is straightforward, and all gates are replaced by majority and inverter gates. The output of technology mapping is the graph shown in Fig. 6. Then, the second step will balance all paths by inserting dummy nodes. These nodes are symbolic, and it will be replaced by wires. For this example, nodes 4, 6, and 7 are dummy nodes.

The P&R starts from the DAG outputs through the DAG inputs in breadth-first order. For each level, all nodes are simultaneously placed and routed. Once there is a valid placement for all nodes in the current level, the routing is executed. The placement will be feasible if there is a free node at the minimum distance (measured in clock zones) for all nodes in the current level. In addition, the routing prioritizes to avoid wire crossing. If it is not possible, the routing can allow few wire crossings. The maximum number of wire crossings is a P&R parameter. If there is no feasible routing for the current minimum distance, the algorithm increases the minimum value in one unit, and the P&R step is performed again.

The first P&R example is the 2:1 multiplex DAG depicted in Fig. 6. The output cell is initially placed on a generic cell i, j, where i is the grid row and j is the grid column. We will use numbers instead of generic indexes for ease of explanation. Fig. 7 presents the QCA layout at USE grid level for the DAG

978-1-5090-2737-8/16 $31.00 © 2016 IEEE   69

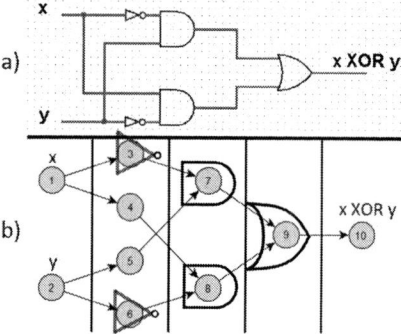

Fig. 8. (a) XOR function; (b) XOR DAG.

from Fig. 6. First, suppose the output node 11 placed at cell (2,1), then the output gate node 10 is placed at cell (2,2) as shown in Fig. 7. The P&R traverses the DAG in breadth-first order. The next level has two internal gate nodes: 8 and 9. Starting from the minimum distance of one cell, both nodes can be positioned at grid cells (1,2) and (2,3) and the routing is straightforward. The next level has three input nodes. The minimum distance (included the dummy nodes) is one for nodes 1 and 3, as shown by using dash line boxes at cells (1,0) and (3,2) in Fig. 7. However node 2 has fanout two, and it should be placed with equal distance from nodes 5 and 6. In order to satisfy this requirement, the minimum distance must be at least two. One solution is to place nodes 1, 2, and 3 at cells (0,0), (0,4), and (3,1), respectively. Finally, the routing is feasible, and the P&R is successful finished.

The next P&R example is the 2-input XOR shown in Fig. 8. Fig. 9 depicts the P&R main steps level by level. First, the primary output node 10 is placed. There are two possible placements for the output gate node 9 as shown in Fig. 9(a). We assume that node 9 is placed at the right side of node 10. There are also two placements at minimum distance of one cell for the nodes 7 and 8 in the next level as depicted in Fig. 9(b). We assume that node 7 is placed on the right side and node 8 on the bottom of node 9. The next level has four nodes, where the nodes 4 and 5 are dummy nodes inserted to balance the paths. Fig. 9(c) shows the placement options. For this example, all nodes until now are placed and routed with minimum distance of one. Suppose the placement depicted in Fig. 9(d). Node 1 has one placement option at distance of three cells to reach nodes 3 and 4. Node 2 has one option at distance of two cells from nodes 5 and 6. However, nodes 1 and 2 are primary inputs, and both nodes should start at same timing zone. Therefore, node 2 should also be placed at distance of three cells as shown in the final P&R depicted in Fig. 9(e).

Although the previous two examples depicts the P&R results in a two-dimensional abstract layer, the generated layouts are validated on QCAdesigner [12]. Fig. 10 shows the layout overlapped by grid graph for the multiplexer 2:1. All cells have same size QCA dots (5x5) and the clock zones area is also displayed. The clock scheme is scalable and regular as introduced in section II. Moreover, it is the first P&R tool for two-dimensional and bi-directional QCA clock scheme.

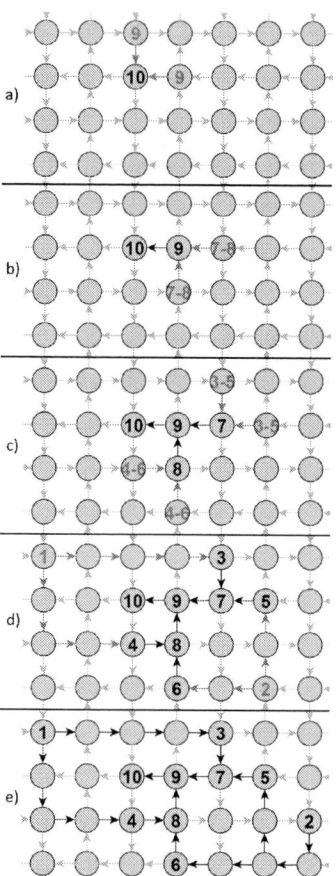

Fig. 9. P&R for the XOR graph.

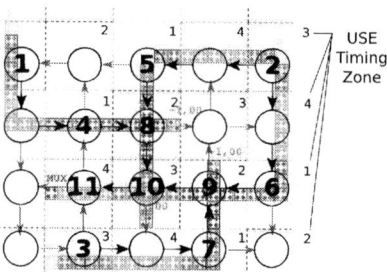

Fig. 10. Final QCA layout with 5x5 cells for the Multiplexer 2:1 overlapped by the grid graph P&R with bi-directional and two dimension timing zones.

Fig. 11 depicts another view of the final layout at QCA level.

V. EXPERIMENTAL RESULTS

Fig. 12 presents a 2:1 multiplexer generated by QCA-LG [8], where there is no abstract layer and the timing zone size is *ad-hoc*. We propose a P&R with regular timing zone, requiring 4 grid rows and 5 grid columns as depicted in Fig. 11. We add squares in the layout from [8] to provide an easy comparison to our approach. Regarding the size, the QCA-LG layout requires more cells, at least 5 grid rows and 6 grid columns. Moreover, the QCA-LG generates long wires, and more non-used area. In addition, the signal flow is uni-directional.

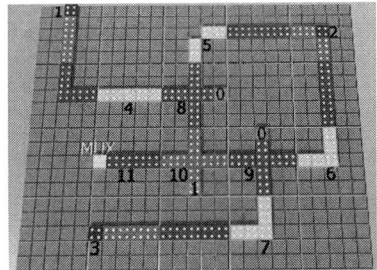

Fig. 11. Final layout for the multiplexer 2:1 by using 5x5 USE QCA.

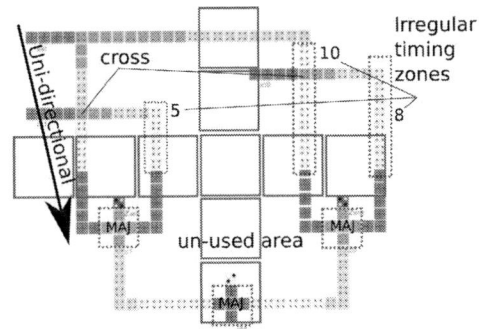

Fig. 12. Automatic QCA layout for the Multiplexer 2:1 generated by QCA-LG [8].

Fig. 13(a) shows an adder generated by QCA-LG [8], and Fig. 13(b) depicts the adder generated by our P&R tool. Our adder layout uses 30% more area due the basic cell size of 5x5 QCA dots. The QCA-LG adder has several timing zones with only 3 QCA dots. In addition, QCA-LG reduces the layout by using 5 wire crossings. Our design has only three, which increases the routing complexity. Moreover, QCA-LG adder has long parallel wires separated by only two QCA dots. Our layout has only five local parallel wire separated by short distance.

Since there is no other tool based on regular abstract bi-directional QCA layer to perform P&R steps, we propose to compare the results of our approach with handmade and optimized layouts from [13]–[19].

Table I shows minimum layout of handmade layouts from

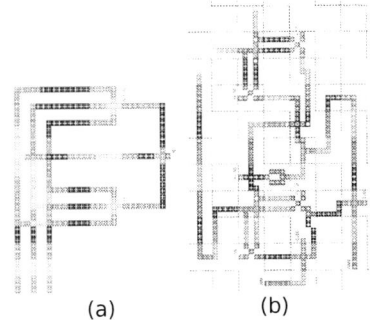

Fig. 13. (a) Adder generated by QCA-LG [8]; (b) Our adder mapped on USE layout with 5x5 cells.

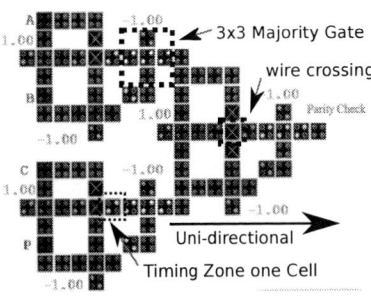

Fig. 14. Handmade layout for a parity checker from [14].

the literature to compare to our layout which is automatically generated. The area and the critical path are computed in QCA dots. Table I also depicts the number of cross wiring and majority gates. In both approaches, the area is dominated by wires. Finally, the total of occupied dots is depicted in column cell "*Occ. dots*". Small size circuits are used for the preliminaries results. The circuits are: multiplexer 2:1, a 2-input XOR, a full adder 1-bit, a 2-input XNOR, a parity generator, and a parity checker. First, the handmade layout approaches do not use any regular clock scheme. Therefore, the area can be significantly reduced by irregular *ad-hoc* timing zones. Fig. 14 shows the layout of the parity checker [14], where there is a timing zone of one QCA dot. Our approach uses a regular grid of 5x5 dots per cell. In addition, each majority gate occupies 3x3 dots in [14] as shown in Fig. 14. The signal propagation is uni-directional and wire crossing is extensively used to generated a compact layout. Therefore, the scalability is reduced turning the routing of complex designs unfeasible because of *ad-hoc* strategies used on handmade circuit design.

Regarding medium and large size circuits, previous P&R work only estimates cross wiring and wire length at the planar graph level. However, these assumptions are not accurate, and the designs must be validated at layout level. Our proposed P&R is based on a virtual layer which give us the abstraction, and at the same time, the layouts are validated on QCAdesigner [12]. Moreover, the signal propagation can be performed in all directions of the two-dimension space.

Regarding the unused QCA area due clock scheme constrains, Fig. 12 depicts a large number of unused cells in the automatic layout generated by QCA-LG tool [8], as well as irregular timing zone sizes. Our clock approach saves area by using the bi-directional approach. Fig. 15(a) shows one example of a long wire with no unused area thanks to the bi-directional propagation provided by USE layer [3]. Moreover, feedback and bi-directional wires provide communication between sub-circuits, and it can be also efficiently implemented in USE layer as shown in Fig. 15(b).

VI. CONCLUSIONS

QCA is a promising technology to replace CMOS, however, it is crucial to implement design automation tools to build arbitrary size and feasible QCA circuits. Scalability thanks to a regular design, portability by using an abstract layer and low

TABLE I
HANDMADE VERSUS P&R LAYOUTS.

QCA Structures	Previous Structures					Proposed Structures				
	Area	Critical path	Cross	Maj	Occ. dots	Area	Critical path	Cross	Maj	Occ. dots
MUX 2x1 [3]	13x15	23	0	3	55	18x21	30	0	3	86
XOR [3]	20x20	35	1	3	79	16x31	34	0	3	108
Full-adder 1-bit [3]	40x41	75	2	7	226	51x35	98	3	7	339
XNOR [13]	13x12	24	0	4	65	26x26	38	0	3	130
Parity Generator [14]	27x8	58	2	6	99	41x46	66	1	6	228
Parity Checker [14]	18x15	40	3	9	145	46x66	66	2	9	284

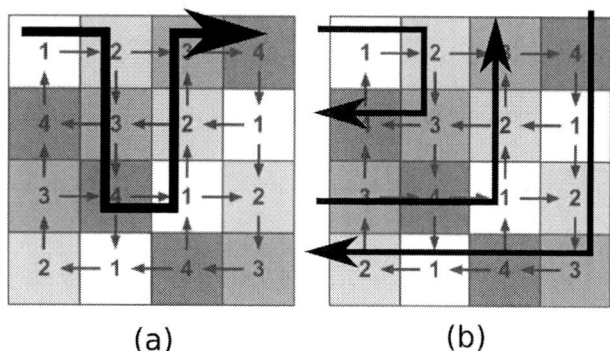

Fig. 15. (a) Long wire without wasted area; (b) A feedback and bi-directional wires.

level layout verification are key features on the QCA design flow. This work presented a novel placement and routing (P&R) that automatically generates a QCA floorplan layout from a direct acyclic graph, while obeying the design rules and main QCA concerns. The layout is generated as an abstract layer onto the USE clock scheme [3]. Since QCA features and constraints can change due to technological developments, the design rules are specified in unit length by using grid squares.

Experimental results show small size circuits, where the correctness of all generated layouts have been verified in QCADesigner [12]. The USE layer is scalable, and therefore, various graph partition heuristics can be applied to manage large circuits. Previous work on QCA P&R are restrict to planar graph level estimations [1], [2], [11], and the circuit area and functionality are not verified at layout level. The QCA wires process and forward information within cells, and can be only effiently evaluated at the final layout level.

Further, the USE clock scheme is bi-diretional in comparison to other QCA approaches based on uni-directional signal propagation. Therefore, both the intracell and the intercell interconnections be performed to save area. These features also introduce new challenges to be addressed during the partition step, in addition to balance path and wire crossing concerns. Future work includes partitions in the 2-D grid and the intercell interconnections with the help of USE layers.

ACKNOWLEDGMENT

The authors would like to thank FAPEMIG, CNPq, and CAPES.

REFERENCES

[1] R. Ravichandran, N. Ladiwala, J. Nguyen, M. Niemier, and S. K. Lim, "Automatic cell placement for quantum-dot cellular automata," in *Proc. of ACM Great Lakes symposium on VLSI*. ACM, 2004, pp. 332–337.

[2] A. Chaudhary, D. Z. Chen, K. Whitton, M. Niemier, and R. Ravichandran, "Eliminating wire crossings for molecular quantum-dot cellular automata implementation," in *Proc. of IEEE/ACM International conference on Computer-aided design*, 2005, pp. 565–571.

[3] C. Campos, A. Marciano, O. P. Vilela Neto, and F. Sill Torres, "Use: A universal, scalable, and efficient clocking scheme for qca," *Computer-Aided Design of Integrated Circuits and Systems, IEEE Transactions on*, vol. 35, no. 3, pp. 513–517, 2016.

[4] T. H. Cormen, *Introduction to algorithms*. MIT press, 2009.

[5] R. Ferreira, L. Rocha, A. Santos, J. Nacif, S. Wong, and L. Carro, "A run-time graph-based polynomial placement and routing algorithm for virtual fpgas," in *IEEE Int. Conf. on Field programmable Logic and Applications (FPL)*, 2013.

[6] M. Kianpour and R. Sabbaghi-Nadooshan, "A conventional design for clb implementation of a fpga in quantum-dot cellular automata (qca)," in *Proc. of IEEE/ACM International Symposium on Nanoscale Architectures*, 2012, pp. 36–42.

[7] P. Marciano, A. Luis, A. B. Oliveira, J. Nacif, and O. P. V. Neto, "An efficient fpga implementation in quantum-dot cellular automata," in *Integrated Circuits and Systems Design (SBCCI)*, 2013.

[8] T. Teodósio and L. Sousa, "Qca-lg: A tool for the automatic layout generation of qca combinational circuits," in *Norchip*. IEEE, 2007.

[9] M. E. Welland and J. K. Gimzewski, *Ultimate limits of fabrication and measurement*. Springer Science & Business Media, 2012, vol. 292.

[10] C. S. Lent and P. D. Tougaw, "A device architecture for computing with quantum dots," *Proceedings of the IEEE*, vol. 85, no. 4, pp. 541–557, 1997.

[11] M. Bubna, S. Roy, N. Shenoy, and S. Mazumdar, "A layout-aware physical design method for constructing feasible qca circuits," in *Proceedings of the 18th ACM Great Lakes symposium on VLSI*. ACM, 2008.

[12] K. Walus, T. J. Dysart, G. A. Jullien, and R. A. Budiman, "Qcadesigner: A rapid design and simulation tool for quantum-dot cellular automata," *Nanotechnology, IEEE Transactions on*, vol. 3, no. 1, pp. 26–31, 2004.

[13] H. S. Jagarlamudi, M. Saha, and P. K. Jagarlamudi, "Quantum dot cellular automata based effective design of combinational and sequential logical structures," *World Academy of Science, Engineering and Technology*, vol. 60, 2011.

[14] F. Ahmad and G. Bhat, "Novel code converters based on quantum-dot cellular automata (qca)," *International Journal of Science and Research*, vol. 3, no. 5, pp. 364–371, 2012.

[15] B. Sen, M. Goswami, S. Mazumdar, and B. K. Sikdar, "Towards modular design of reliable quantum-dot cellular automata logic circuit using multiplexers," *Computers & Electrical Engineering*, vol. 45, 2015.

[16] S. K. Lakshmi, "Efficient design of logical structures and functions using nanotechnology based quantum dot cellular automata design," *International Journal of Computer Applications*, vol. 3, no. 5, 2010.

[17] A. Sarker, A. N. Bahar, P. K. Biswas, and M. Morshed, "A novel presentation of peres gate (pg) in quantum-dot cellular automata (qca)," *European Scientific Journal*, vol. 10, no. 21, 2014.

[18] M. R. Beigh, M. Mustafa, and F. Ahmad, "Performance evaluation of efficient xor structures in quantum-dot cellular automata (qca)," *Circuits and Systems*, vol. 4, no. 2, 2013.

[19] I. Hänninen and J. Takala, "Binary adders on quantum-dot cellular automata," *Journal of Signal Processing Systems*, vol. 58, no. 1, pp. 87–103, 2010.

A Novel Pruned-Based Algorithm for Energy-Efficient SATD Operation in the HEVC Coding

Leonardo Bandeira Soares[1], Cláudio Machado Diniz[2], Eduardo Antonio César da Costa[2], Sergio Bampi[1]

[1]Graduate Program on Microelectronics (PGMicro)
Federal University of Rio Grande do Sul (UFRGS), Porto Alegre, Brazil
[2]Graduate Program on Electronic Engineering and Computing (PPGEEC)
Catholic University of Pelotas (UCPEL), Pelotas, Brazil
{lbsoares, bampi}@inf.ufrgs.br; {claudio.diniz, eduardo.costa}@ucpel.edu.br

Abstract— This paper proposes a novel approximate computing algorithm for the Sum of Absolute Transformed Differences (SATD) to meet energy efficiency in CMOS accelerator circuits. It is based on the pruning of least significant coefficients in the 2-D Hadamard Transform (HT) which is the most compute intensive kernel in the SATD. The SATD is a metric for block matching that is used in video coding standards like the new High Efficiency Video Coding (HEVC). This metric is used to provide better results in mode decision when compared to the Sum of Absolute Differences (SAD) at the expense of larger amount of arithmetic operations as well as higher energy consumption. We present 6 different approximate SATD 4x4 architectures that were synthesized for a 45 nm PDK. Results for the approximate architecture with 10 discarded HT coefficients show energy per operation reduction of 70.7% and BD-PSNR reduction of just -0.008 dB, for a 1080p video sequence.

Keywords—Approximate Computing; CMOS Energy Efficiency; Video Coding; Low Power CMOS design.

I. INTRODUCTION

The semiconductor industry faces challenges at each new CMOS technology node. One of them is the power density increase, which is related to the physical limitations to scale the nominal supply voltage by the same factor used to scale down the device channel length. The main consequence is that the power per silicon area has been substantially increased, so that operational blocks of a chip need to be turned off frequently to allow for heat dissipation [1]. Another CMOS power concern is associated with portable and battery-powered devices which demand intensive computation and power-hungry applications. Nowadays, it is mandatory to shift from performance-driven to energy-efficient oriented CMOS digital circuit design techniques. Energy efficiency is defined in [2] as being the maximum number of operations per energy budget or the minimum consumed energy per unit operation (*i.e.* instruction fetching/decoding, addition, etc).

Among the vast and different proposed energy-efficient techniques, the approximate computing emerged to increase performance and to reduce power dissipation [3]. The key approach in approximate hardware is to reduce the accuracy in the computation by designing simpler circuits to speed up the critical path timing, and to consume less power. Approximate computing techniques take advantage of error-tolerant applications which do not need high accuracy all the time but only good enough results for output perceptual quality. For example, multimedia applications (e.g. video coding, audio filtering, image processing, and so on), highly demanded by current portable devices, are intrinsically related with human senses. Since human sensors process analog information and have difficulty to realize digital approximations [4], then multimedia signals are in fact error-tolerant applications. In other words, it is possible to adopt approximate computing techniques to improve energy efficiency in multimedia applications by properly exploiting the user experience at different levels of quality.

Video coding is one application which demands high computational effort to provide high video compression ratio. HEVC is the new video coding standard [5], which improves compression by 50% compared to the previous H.264/AVC standard [6], while increasing computational effort by up to 3.2x [7]. Hence, energy-efficient techniques for HEVC are mandatory to accomplish the severe power/performance requirements. In terms of block matching for Fractional Motion Estimation (FME), the HEVC adopts the use of SATD. According to [7], for random access configuration in the HEVC Test Model (HM) reference software [8], FME and SATD represent 55% and 16% of the computational effort demanded by the entire encoding process, respectively.

Since the SATD is a compute intensive kernel in HEVC, we can leverage the condition of good enough block matching in order to improve the energy-efficiency. In [9] the use of approximate adders is proposed in the less complex and less accurate SAD metric for block matching technique used in MPEG encoders. In that work, maximum power savings of 42% for very low video resolutions (CIF resolution) are shown. Our approximate algorithm copes with new challenges imposed by the more complex HEVC standard and the use of higher video resolutions (e.g. 1080p and 1600p). To the best of our knowledge, there is no previous work that addresses approximate computing techniques for SATD. In our work the key point is to perform approximation in the 2-D HT, since this is the most compute intensive kernel in the SATD. Our approximate algorithm is driven by pruning the least significant coefficients from the 2-D HT. We show 6 different approximate architectures for the case study on SATD 4x4. The approximate architecture which discards 10 coefficients presents 70.7% of energy per operation reduction while the BD-PSNR decreases 0.008 dB in comparison with the precise SATD 4x4 for a 1080p video sequence.

The remainder of this paper is organized as follows: Section II presents a basic overview in SATD metric for HEVC coding. Our approximate algorithm is presented in Section III followed by the case study on SATD 4x4 results in Section IV. Finally, the conclusions are drawn in Section V.

II. SUM OF ABSOLUTE TRANSFORMED DIFFERENCES OVERVIEW

The SATD is a metric that measures distortion between two blocks in video coding. This distortion is evaluated by block matching methods for motion estimation. For the SATD computation the differences between two blocks are transformed by the 2-D forward HT as shown in (1).

$$W = H \cdot Y \cdot H^T \qquad (1)$$

In (1), W and Y denote the transformed coefficients and the input matrix of differences, respectively. The H and H^T refer to the transform matrix and its transposed form, respectively. For instance, the H can be seen in (2) considering a block size of 4x4 pixels.

$$H = \frac{1}{2}\begin{bmatrix} 1 & 1 & 1 & 1 \\ 1 & -1 & 1 & -1 \\ 1 & 1 & -1 & -1 \\ 1 & -1 & -1 & 1 \end{bmatrix} \qquad (2)$$

Based on that, the SATD is calculated with the aforementioned HT as shown in (3), where $w_{i,j}$ denotes the HT coefficient from the ith row and j^{th} column.

$$SATD = \sum_{i,j} |w_{ij}| \qquad (3)$$

III. OUR COEFFICIENT PRUNING-BASED APPROXIMATION ALGORITHM

In the following subsections we show the computational cost of the baseline SATD architecture. Then we detail next our coefficient pruning-based algorithm to perform

approximation in the HT, and its resulting approximate architectures with an analysis of computational cost reduction.

A. Computational Cost of the Baseline SATD Architecture

In this work we assume as baseline a parallel and precise SATD hardware architecture. For example, the parallel precise SATD 4x4 architecture is shown in Fig. 1. The input x_{ij} denotes the difference between two blocks in the ith row and j^{th} column. The first step in the architecture is the horizontal transform. The transformed samples y_{ij} are registered and are delivered to the vertical transform. Then, the absolute operator is applied in the HT coefficients w_{ij} followed by the adder tree sum. One can observe that these last two steps of absolute operators and sum implement the SAD 4x4 architecture.

Both the horizontal and vertical transforms (*i.e.* the first two steps in Fig. 1) are implemented by the internal structure presented in Fig. 2. One can observe that inside the vertical transform we use the intermediate term v_{ij} that is important for our coefficient pruning-based algorithm as explained further on Section II-B. For the horizontal transform we define the intermediate term as being h_{ij}. The horizontal and vertical transforms refer to the HT presented in (1) and represent an additional computational cost when compared to the SAD computation. Based on that, Table I summarizes the computational cost in terms of operators (*i.e.* adders and absolute operator) for the SATD and SAD with three different block sizes of 2x2, 4x4, and 8x8, while considering architectures of 2-D Hadamard Transform as shown in Fig. 1.

TABLE I.   SATD AND SAD COMPUTATIONAL COSTS

	SATD			SAD		
block size	2x2	4x4	8x8	2x2	4x4	8x8
adders	11	79	447	3	15	63
absolute operators	4	16	64	4	16	64
total	15	95	511	7	31	127

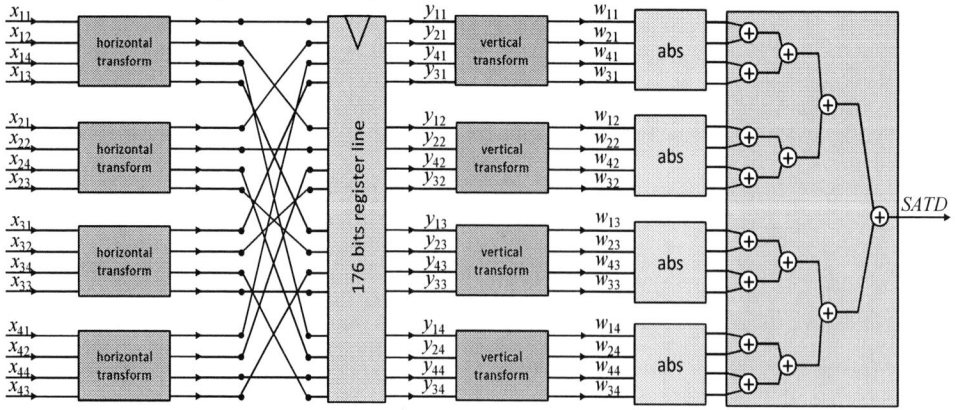

Fig. 1.   Precise SATD 4x4 hardware accelerator parallel architecture.

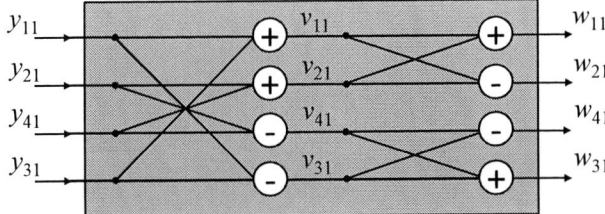

Fig. 2. Internal structure of horizontal and vertical transforms.

Table I clearly shows the additional hardware cost of 3.1X when performing fully parallel the SATD 4x4 instead of the SAD 4x4. This cost is 4X higher when considered the SATD 8x8 in comparison with the SAD 8x8. Based on that, this is our motivation to propose the pruning-based algorithm in order to approximate the HT by reducing its hardware cost in terms of adders count, as well as power consumption.

B. The Proposed Pruning-based Algorithm

The proposed pruning-based algorithm shown in Algorithm 1 is independent from the block size of the SATD. Our algorithm uses two data structures: i) a stack S containing the HT absolute coefficients w_{ij} organized from the top to the bottom with the least to the most significant coefficients, ii) a tree R of dependencies among the prior terms used in the HT to compute each coefficient w_{ij}. The stack is important to decide the order of the coefficients being discarded from the matrix W shown in (1). The tree is used by the proposed algorithm to solve dependencies among the discarded coefficients w_{ij} and prior terms that can be pruned from the architecture because they are not used by the remaining HT coefficients. The proposed algorithm also takes as input the number of coefficients n to be discarded. The output is the approximate SATD RTL VHDL netlist generated according to the remaining terms in the pruned tree R.

Before explaining the algorithm, an example of the tree of dependencies is shown in Fig. 3, since this is the central structure for the algorithm. This example refers to the sub-tree related to the vertical transform presented in Fig. 2. All the coefficients w_{ij} are leaves in the tree. They are computed by the prior terms denoted as father nodes. For instance, the prior terms v_{31} and v_{41} are used only for the computation of the w_{31} and w_{41} coefficients. Hence, if those coefficients are discarded, the proposed algorithm must prune from the tree v_{31}, v_{41}, and $y_{11} - y_{31} + y_{41} - y_{21}$ terms. It is worth to mention that all the remaining terms from the tree must be preserved, since the coefficients w_{11} and w_{21} were not discarded.

The proposed algorithm has an external loop that is performed n times to remove at each iteration the next least significant coefficient and its exclusive associated prior terms. The n parameter can range from 1 up to 16 (*i.e.* no use of HT). The *pruning* Boolean variable is used to control the pruning iteration in line 5. Each coefficient is extracted from the top of the stack S with the *pop* function. Then, the extracted coefficient is searched in the leaves of the tree R. Once the coefficient is found, then its location is stored in the variable l and the pruning process starts. The next step is to visit its father node and store the location in variable k. Since the location of

---

Algorithm 1 The pruning-based algorithm for SATD

Input: S – a stack of coefficients sorted by significance
$\quad\quad\quad$ R – a tree of dependencies among HT terms
$\quad\quad\quad$ n – the number of coefficients to be discarded
Output: Q – the approximate SATD RTL netlist

```
1: while n > 0 do
2:      pruning := TRUE
3:      s := pop(S)
4:      l := search_leaf(s,R)
5:      while pruning do
6:              k := father(l,R)
7:              remove(l,R)
8:              if has_child(k,R) then
9:                      pruning := FALSE
10:             else
11:                     if is_root(k,R) then
12:                             remove(k,R)
13:                             pruning := FALSE
14:                     else
15:                             l := father(k,R)
16:                             remove(k,R)
17:                             if has_child(l,R) then
18:                                     pruning := FALSE
19:                             end if
20:                     end if
21:             end if
22:     end while
23:     n := n -1
24: end while
25: Q := generate_netlist(R)
```

---

the coefficient to be discarded is stored in variable l, then we use this information to effectively remove this coefficient from the tree. The next step is to evaluate if the current father node has a remaining child. If this verification is true, the pruning must stop, since there are coefficients or terms that still depend on the current father node. On the other hand, the current father node can also be removed. Before removing the current father node, it is important to verify if this node is also the root node.

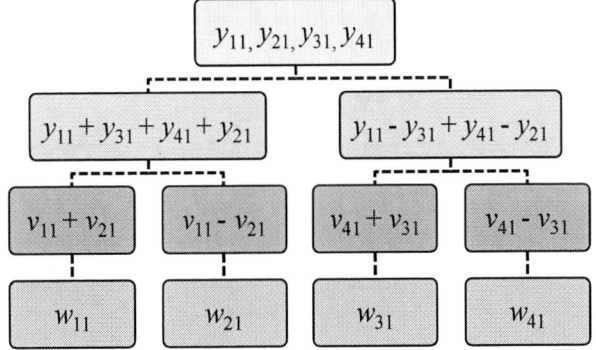

Fig. 3. Example of a tree of dependencies for the vertical transform.

If true, then this node is removed and the pruning must be stopped. This is because there are no more coefficients or terms to be pruned. Otherwise, is important to store the location of the grandfather in *l* before removing the current father. The next step is to verify if the grandfather node has not remaining child. When false, the pruning iteration (*i.e.* line 5) must continue to remove the grandfather node and possibly more prior terms. On the other hand, this pruning has to stop. This is because the grandfather still has child and cannot be pruned from the tree.

Until now, this paper has not discussed how the significance can be measured among the HT coefficients. In this work we define significance as being the average magnitude of each Hadamard transformed coefficient considering a set of video sequences from different classes. Once the stack of most significant coefficients is determined, then we perform the proposed algorithm to prune the SATD architecture based on prior average information. For example, in Fig. 4 we show the average magnitude for each coefficient in 4x4 HT considering 4 video sequences (*i.e.* from class A, B, C and D, as presented in Table II).

For instance, if we want to discard the ten least significant coefficients from the precise example in Fig.1, then the following coefficients will be removed: w_{44}, w_{43}, w_{24}, w_{42}, w_{23}, w_{34}, w_{33}, w_{22}, w_{14}, and w_{41}. One can observe that, based on our proposed algorithm, for each discarded coefficient the tree of dependencies is pruned to reduce the number of cells in the circuit. The resulting approximate SATD architecture can be verified in Fig. 5, where the approximate architecture does not have a complete horizontal or vertical transform. The pruned horizontal and vertical transform blocks have fewer adders than the complete ones. The reduction in the number of adders for the approximate SATD architecture in comparison with the

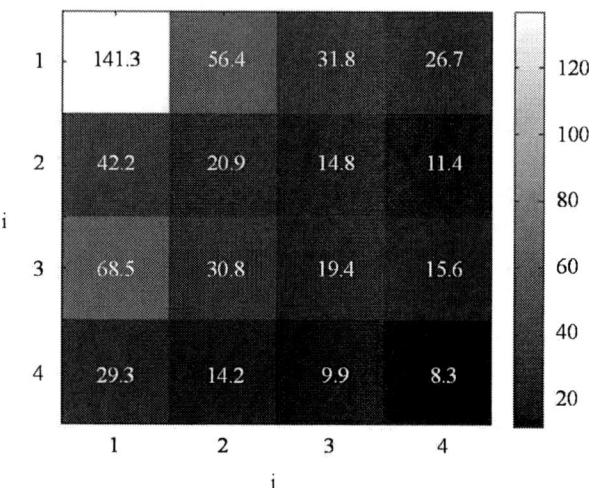

Fig. 4. Average magnitude of each coefficient in 4x4 HT (4 videos).

precise one is 37.9%. In conclusion, the additional cost of the approximate SATD architecture shown in Fig. 5 is 1.7X when compared to SAD 4x4 architecture. In Table I, we show that the precise SATD 4x4 has an additional cost of 3.1X which has 1.8X more cost than the approximate SATD architecture with 10 least significant coefficients being discarded.

IV. RESULTS

In this section we show results for the case study on SATD with block size of 4x4. First, we present the experimental setup followed by the results in terms of video quality, compression and energy efficiency.

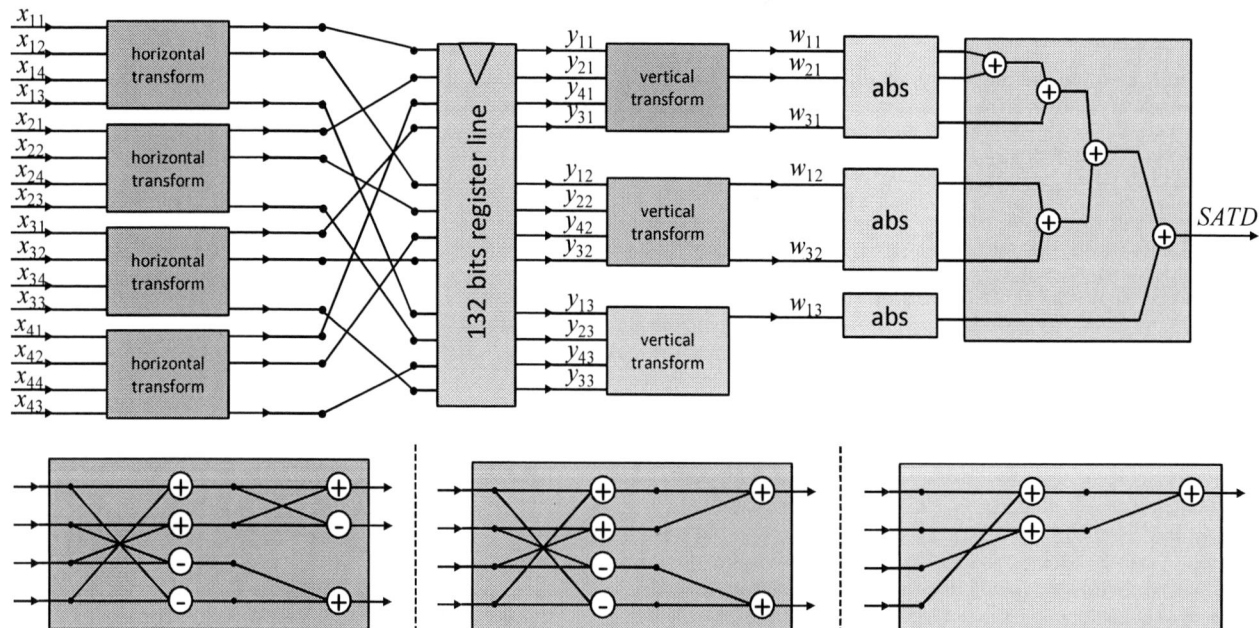

Fig. 5. Approximate SATD architecture with 10 discarded coefficients.

A. Experimental Setup

In order to analyze the video quality and compression, we use the HM 16.0 [8]. In the HM scope we encode 64 frames of 4 videos from the common test conditions [10] as shown in Table II.

TABLE II.   VIDEO SEQUENCES SPECIFICATION

Class	Video sequence name	Resolution
A	Traffic	2560x1600
B	BQTerrace	1920x1080
C	RaceHorses	832x480
D	BlowingBubbles	416x240

We have selected the random access configuration (instead of all intra or low delay) because it is the most complex and most generic configuration for picture ordering. In order to evaluate the quality loss and the bit-rate increase, we run each video sequence for the following quantization parameters: 22, 27, 32, and 37. This is because we integrate the resulting PSNR and bit-rate points to evaluate the Bjøntegaard-Delta bit-rate (BD-BR) and PSNR (BD-PSNR). For BD-BR and BD-PSNR metrics we use the precise SATD 4x4 results as reference.

The encoding process is performed for the precise SATD 4x4 and for the range of 1 to 16 discarded coefficients. Since we consider the average significance to prune the SATD operation, the discarded coefficients for all the video sequences are performed in the same order. For the approximation where the 16 coefficients are discarded we define as being the SAD 4x4.

For the synthesis results we perform a multiple timing synthesis using the Cadence RTL Compiler™ based on the bisection method to find the maximum frequency of operation for each design under evaluation. All the circuits are mapped to a 45nm Nangate FreePDK [11]. For the iso-performance analysis we selected the maximum frequency achieved by the precise SATD 4x4 architecture since this design present the

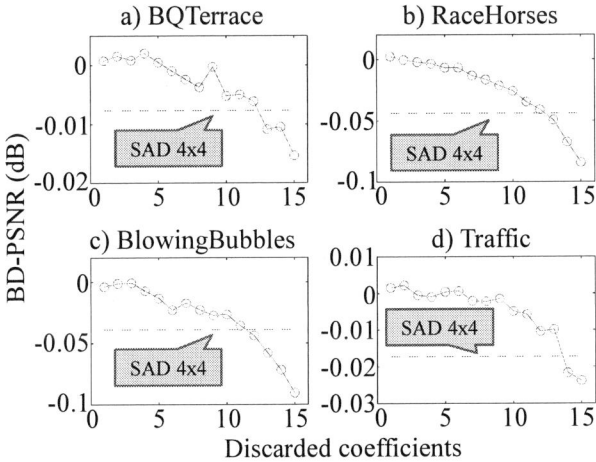

Fig. 6. BD-PSNR results for approximate SATD architectures.

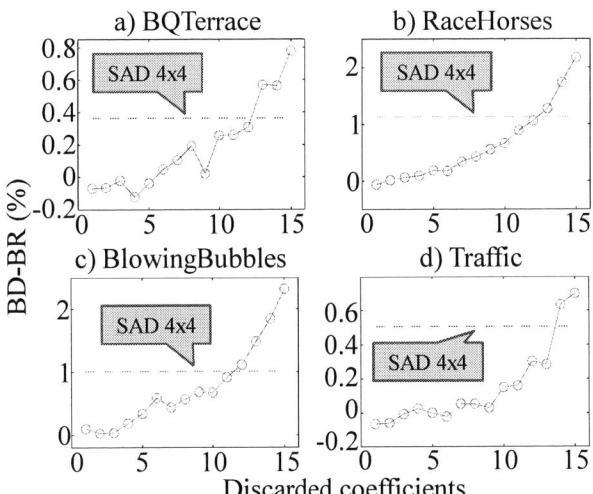

Fig. 7. BD-BR results for approximate SATD architectures.

lowest maximum achievable frequency when compared to the other approximate designs. The power results were obtained through switching activity extraction from 10,000 real 4x4 blocks with residues from 4 video sequences.

B. Video Quality and Compression Results

Video quality (BD-PSNR) and video compression (BD-BR) results can be seen in Fig. 6 and Fig. 7, respectively. In the graphs we show discrete points for each configuration of discarded coefficients and the dotted red line represents the SAD 4x4 (*i.e.* when 16 coefficients are discarded in the 4x4 HT). In fact, this line represents a boundary to evaluate if there are configurations ranging from 1 to 15 discarded coefficients that has worse results than using the SAD 4x4. Based on that, for all the cases the SAD 4x4 presents better results than some pruned configurations. For instance, for the BQTerrace sequence both BD-PSNR and BD-BR SAD 4x4 results has lower degradation and higher compression than configurations from 13 up to 15 discarded HT coefficients. Therefore, is preferable to use the SAD 4x4 instead of those configurations because SAD 4x4 will be a more energy-efficient solution with better results regarding BD-BR and BD-PSNR.

In terms of BD-PSNR, higher resolution videos BQTerrace and Traffic has lower degradation in quality. For this video sequences, when ranging from 1 to 10 discarded coefficients BD-PSNR results range from slightly above zero to -0.01 dB. One can realize that some configurations present better BD-PSNR results than the reference precise SATD 4x4. Our hypothesis is that some pruned HT (*e.g.* the 2 discarded coefficient HT) could present slightly better properties than the complete HT in terms of energy concentration. For lower resolution videos the BD-PSNR ranges from slightly above zero to -0.05 dB when considered the 1 discarded coefficient HT to the SAD 4x4.

The BD-BR results are just a complementary evaluation in relation to the BD-PSNR. We show that for higher resolution videos (*i.e.* a) and d)) and considering the range from 1 discarded coefficient up to the SAD 4x4, the BD-BR configuration increases up to 0.4% and 0.5%, respectively. This is an interesting result, since those videos has huge amount of data and do not suffer a large bit-rate overhead when using approximation for the SATD 4x4. For lower resolution videos in b) and c), the increase in BD-BR are not higher than 1%.

C. Energy Efficiency Results

According to the findings in subsection IV-B, we decided to evaluate the energy efficiency for the following approximate architectures configurations: 2, 4, 6, 8, 10 discarded coefficients HT, the SAD 4x4 (*i.e.* 16 discarded coefficients), and the precise SATD 4x4. These pruning choices were evaluated since higher number of discarded coefficients showed worse results in relation to the totally pruned HT configuration of SAD 4x4.

The maximum frequency achieved for each design was 911.3 MHz, 757.2MHz, 741.1 MHz, 675.1MHz, 676.6 MHz, 658.4 MHZ, and 650 MHZ, respectively. The SAD 4x4 and the 10 discarded coefficients SATD 4x4 architectures present reductions in the critical path timing of 28.7% and 14.2% when compared to the SATD 4x4, respectively. In other words, this represents approximate circuits 1.4X and 1.2X faster than the precise one. For the iso-performance analysis, in which all versions clock at the same frequency, we chose the maximum frequency achieved by the precise SATD 4x4, since all the approximate designs can reach this frequency of operation.

Table III shows the iso-performance analysis at the frequency of 650 MHz. Area is presented as the number of NAND2 X1 equivalent gate count (EG). When using approximate designs, the area reduction ranges from 75.9% down to 6.6%. For the 10 discarded coefficients approximate architecture this area reduction reaches 52.75%. The area overhead of this configuration when compared to the SAD 4x4 is of 1.9X, while comparing to the precise SATD 4x4 the area reduction is 2.1X. In terms of energy efficiency, the mean energy per SATD operation (MEOp) shows reductions of 89.5 % for the SAD 4x4 when compared to the precise SATD 4x4 architecture. The approximate architecture with 10 least significant coefficients being discarded presents a significant reduction of 70.7% in energy per operation. When considering the less pruned architectures with 6 and 2 discarded coefficients, the energy per operation has a reduction of 38.5 % and 12.9%.

TABLE III.   AREA AND ENERGY EFFICIENCY RESULTS @ 650 MHz

# of discarded coefficients	Area (EG)	# cells	Leakage Power (µW)	Dynamic Power (µW)	MEOp (pJ)
SAD	3847	2171	74.9	1556.9	2.5
10	7553	3305	146.5	4389.1	7
8	9426	4299	187.2	6214.4	9.9
6	11863	5632	249.8	9305.4	14.7
4	12813	6177	272.5	10216.2	16.2
2	14930	7339	333.2	13160.7	20.8
Precise SATD	15986	8082	354.3	15170.7	23.9

Based on that, one can conclude that our proposed pruning-based algorithm brings substantial energy reduction for the case study on SATD 4x4 and video sequences under evaluation in Table II. Furthermore, for those video sequences we show that good enough video quality results are achieved since the higher BD-PSNR degradation is just -0.05 dB without affecting the compression (i.e. maximum BD-BR of 1%).

V. CONCLUSION

This work proposed a novel pruned-based algorithm for SATD architectures regarding HEVC coding in order to obtain energy-efficient 45nm CMOS accelerator circuits. Our algorithm performs coefficient pruning according to the number of least significant HT coefficients to be discarded. We also show that the SAD architecture can be defined as the case where the entire HT is discarded from the SATD. For the case study on SATD 4x4, results showed good enough video quality with the highest BD-PSNR degradation of just -0.05 dB regarding the lower-bound SAD 4x4 approximate solution. In terms of compression, the increase in BD-BR is minimal, of up to 1%. Area reduction is provided by our resulting approximate architectures with a maximum of 75.9% for SAD 4x4 and 52.75% for the 10 discarded coefficients SATD 4x4 approach. Energy efficiency is high for the latter, with an average of 7 pJ per SATD 4x4 operation while the precise one consumes 23.9 pJ - an energy per operation reduction of 70.7%.

REFERENCES

[1] H. Esmaeilzadeh et al. Dark silicon and the end of multicore scaling. In Computer Architecture (ISCA), 2011 38th Annual International Symposium on, pp 365 –376.

[2] D. Markovic, V. Stojanovic, B. Nikolic, M. A. Horowitz, and R. W. Brodersen, "Methods for true energy-performance optimization," *IEEE Journal of Solid-State Circuits*, vol. 39, no. 8, pp. 1282–1293, Aug. 2004.

[3] J. Han and M. Orshansky, "Approximate computing: An emerging paradigm for energy-efficient design," in Test Symposium (ETS), 2013 18th IEEE European, Avignon, 2013, pp. 1–6.

[4] N. Zhu, W. L. Goh, W. Zhang, K. S. Yeo, and K. S. Kong, "Design of Low-Power High-Speed Truncation-Error-Tolerant Adder and Its Application in Digital Signal Processing," *IEEE Transactions on Very Large Scale Integration (VLSI) Systems*, vol. 18, no. 8, pp. 1225–1229, Aug. 2010.

[5] ITU-T and ISO/IEC, High Efficiency Video Coding, ITU-T Recommendation H.265 and ISO/IEC 23008-2, 2013.

[6] ITU-T and ISO/IEC JTC 1, Advanced video coding, ITU-T Recommendation H.264 and ISO/IEC 14496-10 (MPEG-4 AVC), 2011.

[7] J. Vanne et. al. "Comparative Rate-Distortion-Complexity Analysis of HEVC and AVC Video Codecs". *IEEE Transactions on Circuits and Systems for Video Technology*, v. 22, n. 12, Dec 2012

[8] HEVC Test Model (HM) version 16.0. http://hevc.hhi.fraunhofer.de/

[9] V. Gupta, D. Mohapatra, A. Raghunathan, and K. Roy, "Low-Power Digital Signal Processing Using Approximate Adders," *IEEE Transactions on Computer-Aided Design of Integrated Circuits and Systems*, vol. 32, no. 1, pp. 124–137, Jan. 2013.

[10] F. Bossen, "Common test conditions and software configurations", JCT-VC-L1100, JCT-VC, Geneva, Jan. 2013

[11] NanGate 45 nm Open Cell Library, www.nangate.com/?page_id=22.

Architectural Exploration of Last-Level Caches targeting Homogeneous Multicore Systems

[1]Rodrigo Cataldo, [1]Guilherme Korol, [1]Ramon Fernandes, [2]Debora Matos, [1]César Marcon

[1]Pontifícia Universidade Católica do Rio Grande do Sul (PUCRS), Porto Alegre, Brazil
[2]Universidade Estadual do Rio Grande do Sul (UERGS), Porto Alegre, Brazil
{rodrigo.cataldo, guilherme.korol, ramon.fernandes}@acad.pucrs.br; debora.motta@gmail.com, cesar.marcon@pucrs.br

Abstract— **The Last-Level Cache (LLC) influences the overall system performance and power dissipation in multicore systems significantly. This paper evaluates five LLC architectures targeting execution time, dynamic and static power dissipation, and area consumption. They are measured using the widely adopted PARSEC benchmark suite for parallel shared-memory systems. Employing Gem5 full-system simulator and 32 nm technology characterization of the McPAT framework, this work had two interesting findings: (i) the shared LLC has the overall best performance under the PARSEC parallel workload, even for applications with less than 20% of shared data. (ii) A privately accessed cache can reduce up to 20 times the dynamic power dissipation on 32 nm technology and 25 times the area consumption when compared to shared-accessed caches.**

Keywords—multicore system; last-level cache; Gem5

I. INTRODUCTION

Shrinking process technologies and the power wall phenomenon have resulted in the development of multicore architectures to achieve the increasing demand of processing and bandwidth requirements [1]. Many interesting challenges arise in these architectures such as the appropriate sharing of on-chip resources across multiple cores and threads, as well as operating under a tight thermal design point [2]. In addition, highly complex heatsink are not suitable for such designs [3]. Caches are essential to hide main memory latency and, thus, to give support to highly efficient cores.

While the use of privately-accessed first cache level (L1) is predominant on multicore designs, for LLC, there is not a dominant architecture paradigm [1][4]. The trend of the recent commercial multicore (IBM POWER8, SPARC 64X and M7, Samsung Exynos 5) is employing shared LLC due to its efficiency of sharing data while many cores (Intel Xeon Phi, Tilera-Gx) focused on private designs due to its multi-hop interconnect nature.

Software-defined cache partitioning was proposed to adapt the LLC architecture for a given application set dynamically. However, only a set of cache parameter is available at the software level, which limits the potential of this approach and can even lead to an increase of execution time [5]. Therefore, trade-offs analysis of LLC on-chip architectures is a significant topic for the computer architecture community.

The main contributions of this paper are:

- Evaluation of the effects for committing to a given LLC architecture targeting homogeneous multicore systems. The

first metric analyzed is the application time achieved through the execution of eight application from the PARSEC benchmark [6].

- An LLC architecture analysis related to dynamic and static power dissipation and area consumption. These two metrics are achieved through the 32 nm technology characterization of the McPAT framework [7]. Although recently made 28nm chips have been able to reduce its static power dissipation using newer fabrication processes [8], LLC still can produce significant static power due to its memory footprint [9]. Besides, advances on minimizing power dissipation are reaching the believed theoretical minimal possible [10].

The experimental results enable to get several interesting conclusions. Applications that share even less than 20% of data can still benefit significantly from a shared LLC [11]. However, shared designs have a single large footprint that requires more power dissipation for every access to its content. Therefore, we observed the opposite scenario for both area consumption and power dissipation. Privately accessed designs were up to twenty-two times and twenty-five times more efficient than a traditional shared LLC design, respectively.

This paper is organized as follows. Section II discusses some related works. Section III describes the system and LLC architecture. Section IV shows our methodology for evaluating architecture designs. Section V presents and Section VI discusses the experimental results. Finally, Section VII presents our conclusions and future works.

II. RELATED WORK

There have been some studies of the impact of LLC on multiple core systems over the years. There is not a single LLC architecture that is perfect for all workloads. Therefore, it is imperative to state the intended group of requirements that the LLC must attain.

Asaduzzaman et al. [1] analyze the impact of sharing LLC space on the performance and power of homogeneous multicore embedded systems. They compare shared and distributed cache organizations over a single and an 8-core system. Experimental results have shown that the FFT application mean delay is reduced by 64% for distributed LLC and 18% for shared LLC when an 8-core is compared to a single-core system. For the three applications analyzed (FFT, MI, and DFT), the distributed LLC always outperforms the shared LLC. Furthermore, the system consumed less power

978-1-5090-2737-8/16 $31.00 © 2016 IEEE

when employing the distributed organization. The authors do not consider the impact of the system interconnect (it is assumed to be negligible) and the small set of applications is restricted to fit in the L1 cache. Therefore, they do not benefit instruction-wise from sharing LLC space.

Sabry et al. [4] investigated the best L2 cache architecture for embedded MPSoCs. They propose a framework to study four L2 cache architectures: private, multi-port shared L2, single-port multiplexed L2 and a hybrid architecture that uses a small cache for shared data and private caches for private data. Their work uses a fixed shared L3 cache for the LLC architecture. The shared designs of L2 cache dissipated the most power (as they consume more power per access) and had the worst execution time. The multi-port shared L2 had a better performance since it provides multiple entries and avoids congestion. This cache organization occupied a large area due to its wiring scheme. For applications with shared data, the hybrid cache had the best performance and the lowest energy consumption of all designs analyzed. Unfortunately, the framework proposed and employed in this work is not readily available for other researchers.

Yun et al. [5] evaluate the effectiveness of employing a page-coloring technique for cache partitioning on multicore platforms. This technique is intended to isolate LLC space according to the system's need – in other words, the LLC architecture is controlled by the operating system and can be shared, private or a hybrid between the two. Unfortunately, as the cache hardware is not entirely available for software parametrization, negative side effects can occur. The authors observed up to 14 times slowdown in out-of-order cores due to interference in accessing the LLC – precisely the opposite of the desired effect. The cause is attributed to the contention in the Miss Status Holding Registers (MSHR) of the LLC.

Cheng et al. [9] conduct a study for energy-efficient LLCs in Chip Multiprocessors. They propose a runtime strategy to disable portions of a shared LLC for energy reduction through hardware monitors. Using similar tools (Gem5, PARSEC, and McPAT) the authors analyze the power-saving and performance impact with different power management policies. They observed power reduction with less than 2% of performance degradation and less than 3% of area overhead.

III. THE PROPOSED ARCHITECTURE

This section defines the baseline architecture for evaluating LLC designs. Next, we propose four additional LLC designs to evaluate against the baseline.

A. Baseline Architecture

The baseline architecture comprises eight homogeneous 1GHz ARM-v7a cores capable of executing out-of-order instructions on a seven-stage pipeline. All cores have private access to L1 cache of data and instructions, each supporting 32KiB of data. Both L1 caches use a 2-way associativity and an MSHR queue of six slots. The L1 instruction cache and data cache have a 1ns latency (one core cycle) and a 2ns latency (two core cycles), respectively. The baseline LLC is an L2 cache shared by all cores. Coherence is maintained by a cache-coherent interconnect based on the ARM CCI-400 [12] and the MOESI coherence protocol.

The Wide I/O version 2 is employed as the main memory here. Wide I/O is a standard memory interface that increases the memory bandwidth at a lowest possible power dissipation. The key is to stack multiple memory channels on top of the system and interconnect them through TSV (Through Silicon Via). Recently, JEDEC[1] published the second standard of Wide I/O that presents significant improvements. We use eight memory channels clocking at 266MHz (real clock).

Figure 1 depicts a schematic representation of the baseline architecture. In this figure, only four cores are shown instead of the eight actually present.

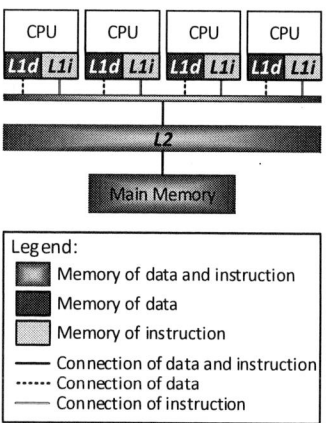

Figure 1. Schematic representation of the baseline architecture.

B. LLC Design Exploration

The LLC design exploration will be achieved using five cache designs, which three of them are shared by the cores, and two of them are privately accessed. Figure 2 represents the four additional LLC designs to the baseline. Their definition is as follows.

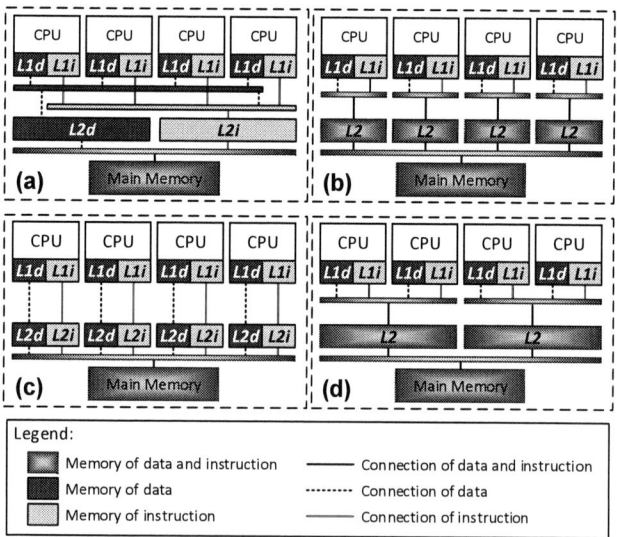

Figure 2. The LLC design exploration.

---

[1] JEDEC is a global leader in developing open standards for the microelectronics industry, with more than 3,000 volunteers representing nearly 300 member companies (source: www.jedec.org/about-jedec).

(i) **Shared LLC** (Figure 1) is the baseline cache design. There is no distinction between instructions and data on L2 cache; (ii) **Shared LLC I+D** (Figure 2(a)) has the same shared design as the previous cache, yet it separates L2 in instructions and data cache; (iii) **Paired Shared LLC** (Figure 2(d)) is based on the LLC proposed by SPARC M7 [13]. It shares an LLC by a subset of the total core count. In our evaluation, there is a shared LLC for every two cores; (iv) **Private LLC** (Figure 2(b)) has one cache for each core in the system. For our evaluation, this results in eight LLCs. Besides, it shares data and instructions indiscriminately; and (v) **Private LLC I+D** (Figure 2(c)) has two caches for each core in the system.

Table 1 presents the configuration parameters for all LLC designs. We use the following principle for size configuration: all caches must have the same overall size. Thus, a single LLC (Shared L2) have a total of 1MiB of unified data. The Shared L2 I+D has two LLCs for the system – consequently, each one of them has 512KiB of data, and so on.

Table 1. Cache parameters for five LLC designs.

	Shared L2	Shared L2 I+D	Private L2	Private L2 I+D	Paired shared L2
Size	1 MiB	512 KiB + 512 KiB	128 KiB	64 KiB + 64 KiB	256 KiB
Hit latency	12 ns	10 ns + 10 ns	8 ns	4 ns + 4 ns	10 ns
Associativity	16-way	16-way + 16-way	16-way	16-way + 16-way	16-way
MSHR queue size	16	14 + 14	12	8 + 8	14

The hit latency and MSHR queue size are adjusted according to their respective cache size. We use conservative values based on previous work using McPAT [14]. Bigger data space has higher access latency because we employ uniform memory access. Associativity is fixed at 16-way as this is a value used by the ARM family on L2 caches [15].

IV. METHODOLOGY

We used two frameworks to achieve the experimental results. Firstly, we employed the Gem5 full system simulator, a widely applied framework for architecture exploration [16]. Secondly, we used the McPAT framework, as it is one of the most accurate tool for this purpose [4], to evaluate the power dissipation and area consumption of every LLC design.

We used the out-of-order CPU model, which is the most detailed mode of execution on Gem5, to provide an accurate characterization of the application set.

Our application set is the PARSEC benchmark suite due to its broad range of application domains, parallel techniques and emerging workloads [17]. Table 2 highlights the subset of PARSEC used here. This subset gives us a diverse application domain and all combinations of parallelization model, working set size and data usage. We simulated the entire application time using the largest input intended for simulators available (*simlarge*). All applications were compiled with the GNU Compiler Collection using the O3 optimization flag. Moreover, every application was executed three times, and their average results were normalized.

The Linux Kernel 3.03-rc3 was compiled, and a suitable filesystem was built to execute PARSEC. The effects of

mapping techniques are outside of this work. They are complementary for our evaluation. As such, the standard Linux Kernel scheduler, Completely Fair Scheduler, was used. Finally, we targeted our system for 32 nm processing technology and SRAM technology for LLCs.

Table 2. The eight applications of PARSEC benchmark suite employed in this work and its characteristics [17].

Program	Application domain	Parallelization		Working set	Data usage	
		Model	Granularity		Sharing	Exchange
Blackscholes	Financial analysis	Data-parallel	Coarse	Small	Low	Low
Bodytrack	Computer vision	Data-parallel	Medium	Medium	High	Medium
Canneal	Engineering	Unstructured	Fine	Unbounded	High	High
Dedup	Enterprise storage	Pipeline	Medium	Unbounded	High	High
Fluidanimate	Animation	Data-parallel	Fine	Large	Low	Medium
Swaptions	Financial analysis	Data-parallel	Coarse	Medium	Low	Low
Vips	Media processing	Data-parallel	Coarse	Medium	Low	Medium
X264	Media processing	Pipeline	Coarse	Medium	High	High

V. EXPERIMENTAL RESULTS

We first analyze the performance of the five cache designs using the execution time of the *simlarge* input set for eight applications of PARSEC. Then, we analyze the area and power consumption for the same cache designs.

A. Execution time for eight PARSEC benchmarks

Figure 3 summarizes the results obtained. All values are relatively normalized according to the baseline LLC; i.e., for all applications, the execution time of the baseline cache is 0% and the remaining values are perceptual deviations of this reference. The first significant observation from these results is the fact that no other cache organization executes substantially faster than the baseline. Blackscholes, Fluidanimate and Swaptions are the cases where some cache organization executes faster; however, for a limited amount (less than 3%). For the remaining five applications, the new organizations are up to 90% slower, as in the case of Bodytrack.

To clarify how these cache organizations are affecting the performance of the applications, we discuss the LLC miss rate for three applications: Blackscholes, Bodytrack, and x264, as illustrated in Table 3.

Table 3. LLC miss rate for Blackscholes, Bodytrack, and x264.

	Core 0	Core 1	Core 2	Core 3	Core 4	Core 5	Core 6	Core 7		
Blackscholes	4.0%									Baseline
	0.2%								I	Shared L2
	10.9%								D	
	16.2%	14.8%	16.5%	14.0%	3.1%	12.1%	11.1%	12.4%		Private L2
	50.8%	45.1%	47.7%	47.3%	0.3%	34.4%	33.2%	50.5%	I	Private L2
	20.1%	19.5%	20.5%	18.1%	20.5%	17.6%	16.7%	18.0%	D	
	12.5%		11.5%		3.27%		8.78%		Paired Shared L2
Bodytrack	1.0%									Baseline
	0.2%								I	Shared L2
	1.4%								D	
	31.0%	30.9%	30.7%	31.6%	31.5%	31.0%	30.7%	30.9%		Private L2
	2.2%	1.5%	2.2%	1.9%	3.5%	2.0%	1.4%	1.1%	I	Private L2
	61.1%	61.0%	61.0%	61.0%	59.6%	60.1%	60.9%	61.2%	D	
	3.9%		3.3%		6.0%		3.2%		Paired Shared L2
x264	12.8%									Baseline
	0.1%								I	Shared L2
	25.5%								D	
	33.4%	34.1%	31.6%	33.2%	33.1%	31.7%	33.8%	33.2%		Private L2
	25.2%	24.9%	25.1%	26.0%	25.3%	23.8%	25.0%	25.1%	I	Private L2
	37.5%	38.6%	38.0%	39.4%	36.2%	37.1%	37.2%	37.7%	D	
	22.2%		21.7%		22.2%		21.2%		Paired Shared L2

978-1-5090-2737-8/16 $31.00 © 2016 IEEE

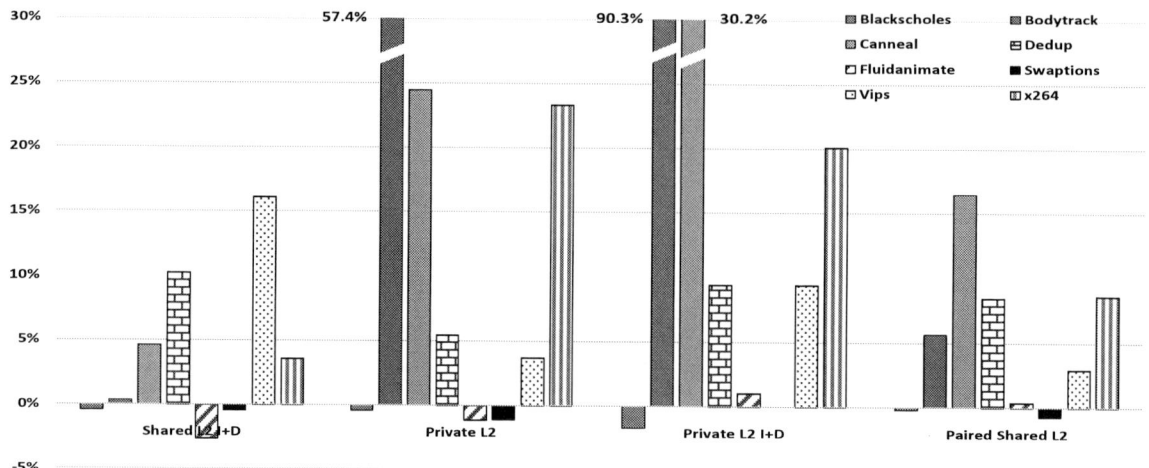

Figure 3. Experimental results for the execution time of Parsec applications targeting four cache architectures.

Using these three applications, we have a representative set to show the effects of cache organization. They represent, respectively: a simple application with scarce communication; an application with huge synchronization barriers that limits its achievable speedup and; high-parallel application.

For the Blackscholes application, the baseline has a 4.04% miss rate while the miss rate of the shared LLC I+D is 0.18% and 10.87% for instructions and data, respectively. These miss rates imply that the instruction cache size is overestimated for this scenario. Both private LLC and paired shared LLC have an increase of miss rates for the majority of cores.

From the configuration values and the parallel nature of the application set, it is expected that LLCs that share data will have a distinct benefit. Bienia et al. [17] have shown that PARSEC applications have at least 10% of share data and can reach up to 50% on a 4MiB LLC (including false sharing).

The private LLC I+D cache organization shows a remarkable phenomenon that affects all applications analyzed in this work, since instructions miss surpasses the 40% rate. Even though we set PARSEC parameters the intent of using an equal number of threads and cores, it internally uses additional threads [17]. In this case, Blackscholes employ one additional thread, which cannot be pinned to a single core, and must hop around the available cores thrashing the cache data with its working set. This thrashing affects the performance of private LLC for parallel applications negatively. In the particular case of Blackscholes, the execution time is not affected even with exorbitant L2 miss rates because Blackscholes is the simplest applications of the PARSEC suite.

Bodytrack LLC miss rates are similar to Blackscholes. Once again, the baseline cache has a low LLC miss rate (i.e., 1.03%). The shared L2 data cache has approximately the same miss rate as the baseline while the instruction cache has near zero miss rate (0.15%). Conversely, the private LLC I+D has thrashing of data information instead of instructions, as was the case of Blackscholes.

Bodytrack has a significant amount of data sharing. Luo et al. [11] showed that Bodytrack can share up to 10% of the cache size across all cores – in the case studied by them, this means that 10% of data is shared by eight cores, which is the highest sharing encountered when seven applications of PARSEC were compared. For privately accessed caches, this phenomenon can degrade performance considerably. The use of an MSI-like coherence protocol (as it is done in our system) can result in two effects: copy & invalidate operation on a store instruction (i.e., one cache has its line invalidated and another gains the exclusive/owned access to that line) and a "ping-pong" effect of two caches writing to the same cache line. Both events increase latency to exchange cache lines. An additional execution with increased data size (64KiB to 128KiB) and decreased instruction size (64KiB to 16KiB) was done to assert the effect of data sharing in this application. The average data miss was impacted significantly – falling from 61% to 28%.

The Bodytrack results from Figure 3 depict a different scenario than the one presented in the Blackscholes results. The L2 miss rate increases and the data sharing discussed earlier rises the execution time significantly. The private LLC organization showed 58% and 90% raise for the unified and I+D organizations, respectively. The shared LLC I+D shows approximately the same execution time as the baseline cache – corroborated by similar miss rates. The paired shared LLC organization increases 0.5% the execution time and 2 percent points for miss rate.

The x264 application has an interesting characteristic: it creates a vast number of threads – for the *simlarge* input, 257 threads are created during the entire execution. The baseline cache has 12.75% of miss rate. The Shared LLC I+D has the lowest instruction miss rate of all applications (0.03%) but increases the data miss rate to 25.54%. As in the case of the Blackscholes, yielding up the instruction size for data size would benefit the execution time. The private LLC I+D organization again shows high miss rate for both instruction and data, which are in average 25.04% and 37.69%, respectively. The private LLC organization presents a miss rate between the instruction and data caches of the private LLC I+D organization. Finally, the paired shared LLC shows a better miss rate than the shared L2 D (i.e., 21.83%, in average).

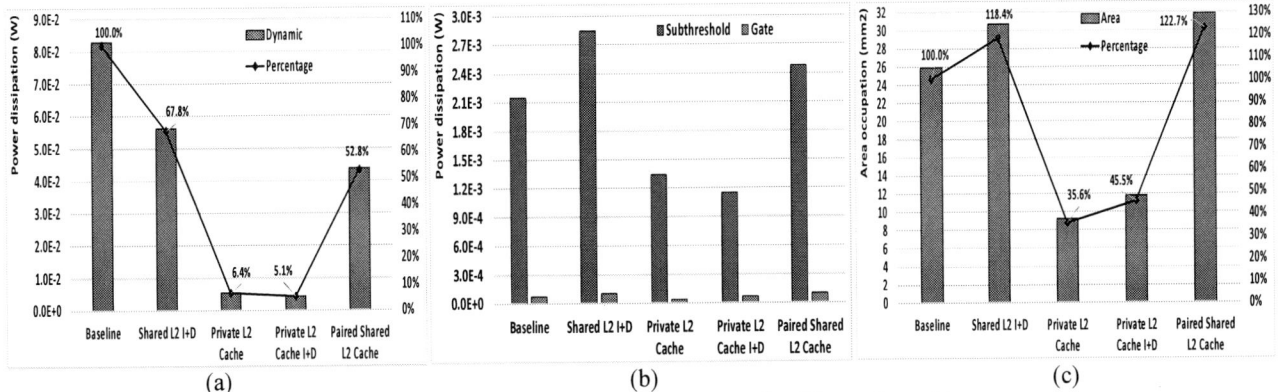

Figure 4. Power dissipation and area consumption results of five cache architectures.

The shared LLC I+D and paired shared LLC have an increase of 4% and 9% of execution time from the baseline cache, respectively. Private LLC I+D and private LLC have an increase of 23% and 20% of execution time from the baseline cache, respectively. These results match the findings of increased LLC miss rate.

The remaining applications of Figure 3 had the following results. Canneal and Vips had more than 10% of performance variance. For Canneal, private LLC organization had the worst execution time; i.e., 24% and 30% of increase when comparing the unified and I+D schemes with the baseline, respectively. The paired shared LLC organization also surpass the 10% mark, increasing performance by 16% compared to the baseline. For the Vips application, only the shared LLC I+D exceeded the 10% mark – also increasing performance by 16% compared to the baseline. The variance of all other cache organizations is under 10%. Three cache organizations are approximately at the 10% mark for Dedup application – shared LLC I+D (10%), LLC private I+D (9%), and paired shared LLC (8%). The L2 private designs showed the best execution time of the new cache designs, albeit it is still worse compared to the baseline.

B. Power Dissipation and Area Consumption

This section shows the dynamic and static power dissipation, and area consumption results regarding a 32 nm CMOS technology, which is characterized by the McPAT framework. The dynamic power dissipation covers the power dissipated for readings and writings accesses; whereas the static power dissipation comprises two types of static power dissipated due to circuit leakage – subthreshold and gate.

Figure 4(a) and (b) illustrate the dynamic and static power dissipation of five cache organizations using the parameters established in Table 1, respectively. The comparison is made summarizing all caches available. For instance, for the private LLC, the subthreshold and gate leakage of eight caches are summed up. Note that all caches have the same 1MiB data size restriction when summed up.

Figure 4(a) depicts that the baseline dissipates more dynamic power than all other cache organizations; e.g., the privative cache organizations dissipate almost 20 times less dynamic power than the baseline. This low power dissipation is an exciting aspect in contraposition of the notorious reduction of performance when using the baseline cache organization. Besides, Figure 4(b) shows that only the private cache organization dissipates less static power than the baseline cache architecture. The baseline cache dissipates 60% and 100% more subthreshold and gate leakage power than the private LLC organization, respectively. However, this changes to 87% more and 5% less subthreshold and gate leakage power than the private LLC I+D organization, respectively. Additionally, the remaining two cache organizations shared LLC I+D and paired shared LLC, dissipates more static power than the baseline for both cases.

Figure 4(c) depicts the area consumption of the entire LLC system. The attained values for individual caches are 25.92 mm² for the baseline cache design; 15.35 mm² for one shared LLC I+D; 1.15 mm² for one private LLC; 0.73 mm² for one private LLC I+D; and 7.95 mm² for one paired shared cache. These values are within the expected range because, considering a rough assumption of halving the cache size results in halving the area consumption, the results have the following disparity: 2.38 mm² (+18%), -2.08mm² (-65%), -0.88 mm² (-55%), 1.47mm² (22%). Clearly, many extra factors are influencing the attained area, such as floor planning design, MSHR queue size, write-back buffer size, and so on.

The experiments show that only the private cache organization diminish the area consumption when compared to the baseline. Both private caches consume under 50% of the baseline area. The fact that shared LLC I+D consumes more area is not surprising since it essentially doubles the amount of logic required to control the same cache size. The paired shared design needs four caches, and each one has additional area requirements described earlier. The result is that this cache organization consumes 22% more area than the baseline.

VI. EXPERIMENTAL FINDINGS

In general, the baseline LLC architecture produces the smaller execution times for the evaluated benchmark, as the workload comprises parallel applications intended to share data across tasks. The private cache organizations reach higher execution time due to the small cache sizes and the increase of coherence overhead of multiple LLCs. The additional shared designs is an intermediate solution increasing the execution time up to 16%.

978-1-5090-2737-8/16 $31.00 © 2016 IEEE

Although the private cache designs degrade the performance of the system, they dissipate less static power than all shared designs. Hence, private caches are much smaller and dissipate less static power. Of all shared designs, the traditional shared LLC showed the minimum static power dissipation.

The baseline cache organization is the architecture that far more dissipates dynamic power which was fifteen and nineteen times the dynamic power dissipation of the set of private LLC and private LLC I+D organizations, respectively. A similar trend occurs with the area occupation where they have decreased twenty-two and thirty-five times compared to the baseline for the same cache set. When the shared organizations were compared to the baseline, they showed a drawback (an increase of area occupation) and a benefit (decrease of power dissipation). The set of shared LLC I+D dissipates two-thirds of the dynamic power and occupy six-fifths of the area compared to the baseline, approximately. The set of paired shared L2 has half of the power dissipation and six-fifths of the area occupation compared to the baseline, approximately.

The paired shared LLC did not achieve the middle ground we expected between performance and power dissipation. Although it had lower dynamic power dissipation, its performance was on par with the shared LLC I+D, and we expected better than this. However, it is important to note that refined mapping techniques can change this scenario.

VII. CONCLUSIONS

This work presents an analysis of execution time, power dissipation and area consumption of five cache organizations on real scenarios. We have shown that the cache organizations provide compelling tradeoffs for these requirements allowing the designer to employ a design that suits his needs. For the lowest energy consumption, we recommend employing privately small LLC. In the exact opposite of this scenario; i.e., highest performance, we recommend applying large shared LLCs. The additional shared L2 organization explored in this work are the middle ground of these scenarios and can be refined further by cache-aware mapping techniques. The cache organizations employed in this work are available as Gem5 patches on this link: *http://reviews.gem5.org/r/3506/*.

For future work, we intend to expand this study for Non-Uniform Cache Architectures (NUCA) LLC and some cache-coherence protocol schemes.

ACKNOWLEDGMENT

This work is partially funded by CNPq-Brasil (process number 132778/2014-9).

REFERENCES

[1] A. Asaduzzaman; F. Sibai; M. Rani. **Impact of Level-2 Cache Sharing on the Performance and Power Requirements of Homogeneous Multicore Embedded Systems**. *Microprocessors and Microsystems,* vol. 33, Issues 5-6, pp. 388-397, Aug. 2009.

[2] H. Esmaeilzadeh et al. **Dark Silicon and the End of Multicore Scaling**. *International Symposium on Computer Architecture (ISCA)*, pp. 365-376, 2011.

[3] M. Hajkazemi; M. Tavana; H. Homayoun. **Wide I/O or LPDDR? Exploration and Analysis of Performance, Power and Temperature Trade-offs of Emerging DRAM Technologies in Embedded MPSoCs**. *International Conference on Computer Design (ICCD)*, pp. 62-69, 2015.

[4] M. Sabry; M. Ruggiero; P. Valle. **Performance and Energy Trade-Offs Analysis of L2 on-chip Cache Architectures for Embedded MPSoCs**. *Great Lake Symposium on VLSI (GLVLSI)*, pp. 305-310, 2010,

[5] H. Yun; P. Valsan. **Evaluating the Isolation Effect of Cache Partitioning on COTS Multicore Platforms**. *Workshop on Operating Systems Platforms for Embedded Real-time Applications (OSPERT)*, pp. 45-50, 2015.

[6] PARSEC team. **The PARSEC Benchmark suite**. parsec.cs.princeton.edu/.

[7] McPAT research team. **McPAT**. www.hpl.hp.com/research/mcpat/.

[8] Altera Corporation. **Meeting the Low Power Imperative at 28 nm**. *Whitepaper*, pp. 1-12, Sep. 2012.

[9] H.-Y. Cheng et al. **EECache: A Comprehensive Study on the Architectural Design for Energy-Efficient Last-Level Caches in Chip Multiprocessors**. *ACM Transactions on Architecture and Code Optimization (TACO)*, vol. 12, Issue 2, pp. 1-22, Jul. 2015.

[10] SureCore Technology. **Technology Whitepaper**. *White Papers Repository*, pp 1-6, 2013.

[11] H. Luo; P. Li; C. Ding. **Parallel Data Sharing in Cache: Theory, Measurement, and Analysis**. *Technical Report TR-994*, pp. 1-25, Mar. 2015.

[12] U. Wiener. **Modeling and Analysis of a Cache Coherent Interconnect**. *Thesis Report, Eindhoven University of Technology*, pp. 1-83, Aug. 2012.

[13] R. Sivaramakrishnan; S. Jairath. **Next Generation SPARC Processor Cache Hierarchy**. *Presentation at Hot Chips (HC)*, pp. 1-28, 2014.

[14] D. Woo; N. Seong; D. Lewis; H.-H. Lee. **An Optimized 3D-Stacked Memory Architecture by Exploiting Excessive, High-Density TSV Bandwidth**. *International Symposium on High Performance Computer Architecture (HPCA)*, pp 1-12, 2010.

[15] ARM. **Cortex A-15**. *Technical Reference Manual*, pp. 1-364, 2011.

[16] N. Binkert et al. **The gem5 Simulator**. *ACM SIGARCH Computer Architecture News*, vol. 39, Issue 2, pp. 1-7, May 2011.

[17] C. Bienia et al. **The PARSEC Benchmark Suite: Characterization and Architectural Implications**. *International Conference on Parallel Architectures and Compilation Techniques (PACT)*, pp. 72-81, 2008.

[18] G. Southern; J. Renau. **Deconstructing PARSEC Scalability**. *Annual Workshop on Duplicating, Deconstructing and Debunking of International Symposium on Computer Architecture (ISCAWDDD)*, pp. 1-10, 2015.

Side Channel Attack on NoC-based MPSoCs are practical: NoC Prime+Probe Attack

Cezar Reinbrecht, Altamiro Susin
Instituto de Informática
PPGC - UFRGS
Porto Alegre, Brazil
cezar.reinbrecht@inf.ufrgs.br

Lilian Bossuet
Lab. Hubert Curien
UMR CNRS 5516 - Univ. of Lyon
Saint-Etienne, France
lilian.bossuet@univ-st-etienne.fr

Georg Sigl, Johanna Sepúlveda
Inst. for Security in Information Technology
Technical University of Munich
Munich, Germany
johanna.sepulveda@tum.de

Abstract—**Many authors have shown how to break the AES cryptographic algorithm with side channel attacks, specially the timing attacks oriented to caches, like Prime+Probe. In this paper, we present a practical timing attack on NoC that improves Prime+Probe technique. Our attack targets the communication between an ARM Cortex-A9 core and a shared cache memory. Furthermore, we evaluate a secure enhanced NoC applied as a countermeasure of the timing attack. Finally, we demonstrate that attacks on MPSoCs through the NoC are a real threat and need to be further explored.**

Index Terms—**Network-on-Chip, Security NoC, Timing Attack, Timing Side-channel Attack.**

I. INTRODUCTION

In the past few years, Systems-on-Chip (SoC) became a reference platform for a high variety of products, being present in the life of the people, dealing with bank account information, passwords, etc. As a consequence, such systems have been attacked since then. Side channel attacks (SCA) [1] [2] [3] are one of the most dangerous treat that targets hardware components. In complex hardware architectures, such as SoCs, the side channel based on timing behavior became very popular. The typical timing leakage used by attackers are the cache memory, taking advantage that a cache miss and cache hit responses results in a huge difference on delay, where even the software can observe it. Previous Advanced Encryption Standard (AES) timing attacks focused on cache access monitoring are shown in [4][5][6][7][8][9][10][11][12]. One of the most efficient timing attack technique is the Prime+Probe, proposed by Osvik et al. [8]. Prime+Probe attack uses a spy process in the same CPU of the victim, in order to monitor the cache accesses during cryptographic operations. The cache accesses in algorithms as AES are key-dependent. Thus, by forcing cache misses, the spy process can identify the memory positions accessed by the AES in the cache. As a result, the secret key is revealed.

New market demands, like high performance and low energy consumption, resulted in flexible platforms known as Multi-processors Systems-on-Chip (MPSoCs). MPSoCs are architectures composed of several processing elements, custom IP cores and memories interconnected by a Network-on-Chip (NoC). MPSoCs are the key enabler technology for the new computation paradigm Internet-of-Things (IoT). In such scenario, the added connectivity to these devices, source of huge benefits, creates new and remote attack channels. MPSoCs are target of remote attacks. Security implementation

at MPSoCs that execute critical tasks or store and process sensitive information is mandatory.

MPSoC security presents new challenges and opportunities for the attacker perspective. Regarding the challenges, current MPSoC protection techniques wraps sensitive IP cores in secure zones as in [13]. Inside the security zone the IP cores run trusted applications, which allow the processing of sensitive information in a highly controlled environment. Such features make impossible to execute the Prime+Probe attack technique, since a spy process cannot run on the same victim's IP core. Regarding the opportunities, shared cache memories MPSoC architectures and NoC shared resources are the new leakage sources of the MPSoC. In this work, we explore these two components in order to implement a novel Prime+Probe attack.

The NoC as a leakage source for a timing attack was already addressed by [14][15][16]. They propose the integration of random NoC behavior as a way of protection. Despite their good results, a practical attack on NoC-based MPSoC has never been shown. Our work presents the first practical timing attack on MPSoC based on the Prime+Probe technique. In our work the spy process and the victim process run in separate IP cores. By monitoring the NoC traffic throughput the sensitive information is leaked. In addition, in this work we verify the timing security of a low area overhead timing protected NoC *Gossip NoC* [17]. The main contributions of our work are:

- Perform a practical timing attack on MPSoCs;
- Verify the timing protection of the *Gossip NoC*;
- Evaluate the performance and cost of the timing protection mechanism.

This paper is divided in seven sections. Section II presents the related works, regarding cache timing attacks. In section III, a background information regarding the proposed platform and cipher algorithm are explained. The proposed timing attack are presented in Section IV. The countermeasure, defined as Gossip NoC, can be observed in Section V. Section VI presents all experiments and results. Finally, we conclude the paper in Section VII.

II. RELATED WORKS

Timing attacks on caches were first mentioned by Kocher [4] and Kelsey et al. [5]. According to Kocher, cryptographic algorithms running on a platform always presents timing leakages that can be used for performing side channel attacks

(SCA). Tsunoo et al. [6] presented for the first time a practical timing attack on caches, breaking the DES algorithm. This work opened a new front on SCA. Bernstein [7] adapted Tsunoo attack to break AES cryptography. At the same time, Osvik et al. [8] proposed three techniques to perform very efficient timing attacks on cache: i) Evict+Time; ii) Prime+Probe; and iii) Asynchronous. The Prime+Probe was further applied in [9][10][11][12].

Xinjie et al. [9] propose a novel analysis strategy to implement Prime+Probe. Instead of using the accessed lines of the cache, it targets the non accessed lines. The objective was to reveal the key with a reduced number of traces. Liu et al. [11] present a Prime+Probe attack targeting the last level cache of a bus-based MPSoC. It employs two techniques to break an ElGamal decryption: i) Probe cache sets without knowing the virtual-address mapping; and ii) identify victims security-critical accesses using temporal access patterns.

Prime+Probe technique is very suitable to attack MPSoCs platforms, since its components can access directly the hardware information, such as physical addresses and communication infrastructure. However, a practical timing attack on NoC-based MPSoC has not been reported yet. The exploration of NoC channels as a leakage source was addressed by Yao et al. [14], Wassel [15] and Sepúlveda et al. [16]. The works of [14] [15] propose the integration of hard QoS mechanism to isolate the sensitive information. They include temporal network partitioning, based on high [14] and bounded [15] priorities arbitration schemes. Furthermore, the work of Sepúlveda [16] proposes random arbitration and adaptive routing as protection techniques. Despite the different levels of protection, these works does not show the practical execution of the timing attack. Our work aims to contribute by showing for the first time a timing attack in a NoC-based MPSoC and to verify the timing protection of Gossip NoC. Table I presents the state-of-the-art of the attacks in SoCs and MPSoCs.

III. BACKGROUND

This section presents the concepts required to understand the Prime+Probe attack. First subsection presents the MPSoC environment. Second subsection presents the AES algorithm and the cache accesses.

A. MPSoC Architecture

The target MPSoC architecture is composed of homogeneous processing IP cores (ARM Cortex-A9), external interfaces, memories and a NoC. The memory hierarchy follows the proposal of [18], where the processing IP cores have access to a local and a shared caches. This memory strategy increases the performance and reduces the area overhead when compared to full local cache strategy. The NoC is characterized by a mesh topology and deterministic routing algorithm. Moreover, the MPSoC implements a security zone, as in [13], defined in design time. The IPs inside this zone are considered trusty among them and are responsible for performing sensitive operations, such as the AES cryptography. Hence, typical Prime+Probe attack, which requires a spy process running in the same cryptography IP core is impossible. In order to overcome this protection, new Prime+Probe mechanisms

Fig. 1: Homogeneous MPSoC platform arranged in a 4x4 mesh topology.

are required in MPSoC environments. Figure 1 presents an example of our target MPSoC, composed of 12 IP cores, one shared cache (IP 0), one shared memory (IP 15) and two external interfaces (IP 3 and IP 11) in a 4x4 mesh-based NoC.

B. Memory Access in AES

The timing attack targets to reveal the secret key used in the AES cryptography. AES is the preferred cryptography in many (specially commercial) applications. It is based on a substitution-permutation strategy, composed of a key expansion step and several rounds that depends on the key size (10 rounds for a key of 128 bits). It can be implemented using just logical and arithmetic operations. However, in order to obtain better performance, this cipher is implemented with the main operations already pre-computed and stored in huge tables, as presented in [19]. Figure 2 presents the AES dataflow.

The performance-oriented AES has two main phases. The first one is to calculate the expanded key ($K \rightarrow k_0, .., k_{10}$), that will be used for all round computations. The second phase is to compute the intermediate states, given by x^{round}. The same operations are repeated according to the number of rounds. Each round is composed of algebraic the operations SubBytes, ShiftRows, and MixColumns (Figure 2). The result is XORed with the round key x^{round}. These three operations

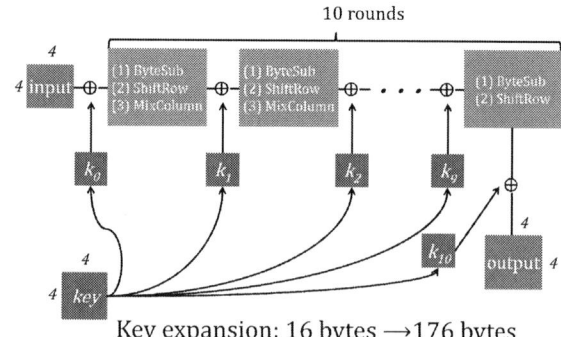

Key expansion: 16 bytes →176 bytes

Fig. 2: AES Dataflow.

TABLE I: Summary of Prime+Probe Attacks.

Work	Platform	Timing Leakage Source	Attacker Method	Traces Used
Osvik et al. [8]	SoC (single core)	L1 Cache	Spy process	16000
Xinjie et al. [9]	SoC (single core)	L1 Cache	Spy process	350
Liu et al. [11]	Bus-based MPSoC	LLC - L3 Cache	Spy process	33600
Oren et al. [12]	Bus-based MPSoC	LLC - L3 Cache	Browser process	5000
Yao et al. [14]	NoC-based MPSoC	NoC	Spy process	Not mentioned
Wassel [15]	NoC-based MPSoC	NoC	Spy process	Not mentioned
Sepúlveda et al. [16]	NoC-based MPSoC	NoC	Spy process	Not mentioned
This work	NoC-based MPSoC	NoC (Shared Cache)	Spy process	80

are pre-computed and stored in four tables (T_0, T_1, T_2, T_3). Therefore, given a 16-byte plaintext $p = (p_0, .., p_{15})$, encryption proceeds by computing a 16-byte intermediate state at each round r as presented by equation 1:

$$x^r = (x_0^r, .., x_{15}^r) \tag{1}$$

The initial state x^0 is computed by equation 2:

$$x_i^0 = p_i \oplus k_i \text{ para } i = 0, .., 15 \tag{2}$$

The initial calculation is very important for the attack analysis, as presented in the section IV-D. Then the first 9 rounds are computed updating the intermediate state as follows in the equation 3, for $r = 0, .., 8$:

$$
\begin{aligned}
(x_0^{r+1}, x_1^{r+1}, x_2^{r+1}, x_3^{r+1}) &\leftarrow T_0[x_0^r] \oplus T_1[x_5^r] \oplus T_2[x_{10}^r] \oplus T_3[x_{15}^r] \oplus K_0^{r+1} \\
(x_4^{r+1}, x_5^{r+1}, x_6^{r+1}, x_7^{r+1}) &\leftarrow T_0[x_4^r] \oplus T_1[x_9^r] \oplus T_2[x_{14}^r] \oplus T_3[x_3^r] \oplus K_1^{r+1} \\
(x_8^{r+1}, x_9^{r+1}, x_{10}^{r+1}, x_{11}^{r+1}) &\leftarrow T_0[x_8^r] \oplus T_1[x_{13}^r] \oplus T_2[x_2^r] \oplus T_3[x_7^r] \oplus K_2^{r+1} \\
(x_{12}^{r+1}, x_{13}^{r+1}, x_{14}^{r+1}, x_{15}^{r+1}) &\leftarrow T_0[x_{12}^r] \oplus T_1[x_7^r] \oplus T_2[x_6^r] \oplus T_3[x_{11}^r] \oplus K_3^{r+1}
\end{aligned} \tag{3}
$$

The last round is computed by repeating the equation given in 3 with $r = 9$, except that $T_0, ..., T_3$ is replaced by $T_0^{10}, ..., T_3^{10}$. The resulting x^{10} is the ciphertext. These lookup tables combines all algebraic operations, which are ShiftRows, MixColumns and SubBytes. The last table is different, because it does not include the MixColumns computation as observed in Figure 2.

IV. NoC PRIME+PROBE

The objective of the NoC Prime+Probe attack is to use the leakage channel of the NoC to monitor the accessed sets of the shared cache during the cryptographic tasks. The route between the cipher and the cache is known as sensitive path. By using the analysis presented by [9], the secret key can be revealed. Identifying the cipher accesses to the cache through the NoC is possible by exploring the traffic throughput. When the victim sends or receives packets, the throughput of the nodes that intersects the sensitive path is decreased. In order to perform the attack, the spy process must be executed in an IP core that intersects the sensitive path. The preconditions to perform the NoC Prime+Probe attack are:

- Attacker knows IPs' mapping on MPSoC
- Attacker knows the routing algorithm used in the NoC
- Attacker knows the cache configuration
- Attacker generates the encryption plaintext
- Attacker knows the location of the AES lookup tables in memory
- Attacker can access the shared cache
- Attacker can control an IP core inside the MPSoC

Based on these knowledge, an attacker performs the NoC Prime+Probe attack by means of four stages: i) Infection; ii) Prime; iii) Probe; and iv) Analysis. The attack model is represented by the flow chart in Figure 3.

A. Infection

The *Infection* starts when the attacker stores a malware into the MPSoC. Such malicious software may be spread into several IPs. The infected IPs must intersect the sensitive path. Several infected IPs may aid in the malicious monitoring task, thus improving the efficiency of the attack.

B. Prime

The *Prime* consists in the preparation of the cache by the infected IP. The goal is to guarantee that there are no AES lookup tables in the cache before the attack. By spreading the random vector AT_DATA data in the cache, the attacker overwrites several cache memory locations. After the cache is ready, the infected IP sends a random plaintext for the AES crypto-processor in order to force the execution of the cryptographic tasks.

C. Probe

The *Probe* aims to monitor the accessed cache locations during the AES execution. This stage takes place into two phases:

1) Identification: The infected IP core throughput is monitored to detect the AES access to the cache. The

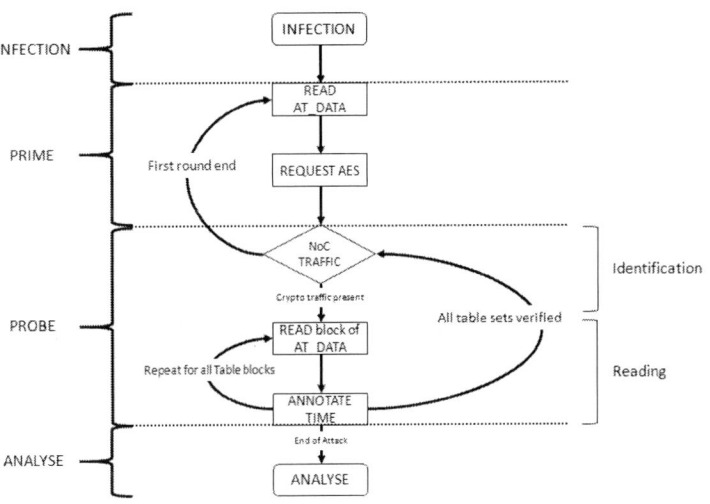

Fig. 3: Flow chart of the attack stages.

978-1-5090-2737-8/16 $31.00 © 2016 IEEE

identification phase is based on the technique used by [16]. By continuously requesting NoC communication, the infected IP core can monitor the collisions with the sensitive packets. The throughput of the infected IP reveals the access pattern and the volume of communication over the sensitive path. Considering the mesh-based and XY routing NoC presented in Figure [18], the infected IP core can be placed at IP 1. This position guarantees that cache responses to the victim will collide with the attacker's packets. Moreover, a second infected IP can be placed at IP 4, in order to identify the cache requests from the crypto-processor. The cooperation of the infected IPs reduce the probability of a false-positive, which can be caused by other processes accessing the cache.

2) Reading: The infected IP core fetches the the random vector AT_DATA from the cache memory. Longer fetching times reveal cache misses, thus the accessed memory locations by the AES.

These two phases are repeated until AES finishes the first round of the algorithm, which causes about 16 accesses. The attacker awaits the end of the cryptography and a new *Prime* stage is performed. This workflow is repeated until the sufficient number of samples required to break the cipher has been collected.

D. Analysis

The *Analysis* stage follows the algorithm presented in [9]. We perform the first round analysis, where only the accesses during the first AES round are used for the mathematical test. The values generated after the first round follow the expression $x_i^0 = p_i \oplus k_i (i = 0, ..., 15)$. Therefore, by testing the data acquired in the *Probe* stage, it is possible to identify which sets have not been used by the crypto-processor, which means the non accessed indexes. By assuming that $x_i^0 \neq p_i \oplus k_i (i = 0, ..., 15)$ and knowing the plaintext byte p_i, is possible to prove that $k_i \neq p_i \oplus x_i^0 (i = 0, ..., 15)$. Hence, possible key candidates can be removed. This strategy reduces the key search space within the brute force process.

V. NoC Protection

The *Gossip NoC* is a security enhanced architecture able to protect the MPSoC against timing attacks. Gossip NoC is based in two protection strategies: i) Detection, which includes the bandwidth monitoring and an alert message (gossip) generation in the presence of an abnormal behavior; and ii) protection, triggered when any gossip message is received and which is able to modify the route of the packet from XY to YX. The alert messages reinforce the suspect of an attack, avoiding false-positives. If an attack is detected, the router changes the routing algorithm for the packets that follows the sensitive path. The usage of XY and YX routing algorithms together is guaranteed as deadlock and livelock free [20] [21].

The *Gossip* router microarchitecture is based on a typical NoC router composed by routing scheme (XY and YX), Round-Robin arbiter and FIFO memory. Additional three main components are integrated:

- Gossip In Block: It controls the internal state of the gossip router according the values of the input signals. When the number of gossip messages received from neighbor routers overcomes the threshold *gossip confidence*, an attack is confirmed. As a result the routing the *gossip switching* is modified.
- Gossip switching: It commutes the incoming packets fron an input to an output. Under attack, the traffic is commuted according to the YX algorithm. Otherwise the XY route is implemented.
- Gossip Generator: It monitors the traffic bandwidth. When it exceeds a protection bandwidth threshold, a signal indicating a possible attack is activated and transmitted to the Gossip In Block of all the other routers.

Due to page limitations we refer to [17] for detailed description.

VI. Experiments and Results

In this section we present the experimental setup, the evaluation of the NoC Prime+Probe attack and the Gossip NoC security efficacy, efficiency and cost evaluation.

A. Experimental Setup

The MPSoC was implemented in an [22] FPGA. An ARM hard IP executes the AES cryptography. Other processing elements were modeled by means of synthesized traffic generators. The AES access the lookup tables from the shared cache of the MPSoC. It is a 16-way set associative cache with 32KB. The cache response time is 16 cycles and the main memory access, in a case of a cache miss, adds 100 additional cycles. However, the total latency of a cache access must consider also the latency of the network interfaces and the NoC routers. Each network interface adds 5 cycles in a congestion-free scenario. The routers of the NoC adds 2 cycles. The proposed environment follows the concept presented in Figure 1.

We have tested an attack with 100 cryptography requests, with random plaintexts generated by the infected IP. We performed the attack as described in section IV. We collected the cache accesses latencies for the *analysis*. The same attack was repeated for the *Gossip NoC* in order to evaluate the security efficacy.

B. Identification Step Evaluation

The objective of this step during the attack is to identify the moment that the victim CPU access the shared cache. First, we evaluate the successful rate and accuracy of the attacker measurements. The successful rate is the quantity of percentage of the identified sensitive accesses. The accuracy represents the quality of these results. In order to extract these metrics, we evaluated 5000 samples. Each sample corresponds to the latency of sending a message to the NoC during a cryptography task.

Figure 4 presents the latency results. It can be observed that the typical latency of simple packets transmitted by the network interface is 5 cycles. Then, any variation on this latency can be interpreted as a big packet passing through attackers router, probably a cache response. The Table II presents the relation of true cache responses and the estimated by the attack. We tested two scenarios:

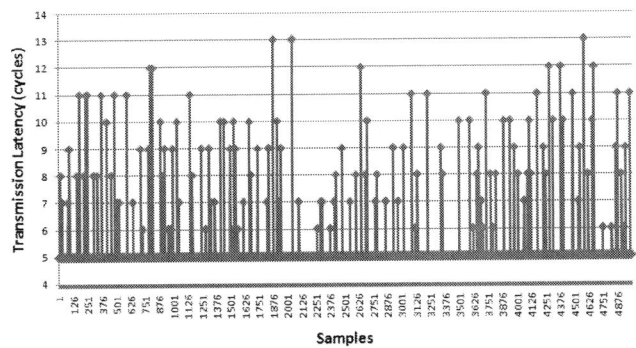

Fig. 4: Transmission latency of attacker. Latencies higher than five cycles represents collisions with sensitive data in the NoC.

1) Scenario 1 - Communication noise do not access shared cache
2) Scenario 2 - Communication noise can access shared cache

TABLE II: Relation of true and estimated cache responses by the attack.

Timing Attack	Estimated	Real	Successful Rate	Accuracy
Scenario 1	6647	16000	41.54%	100%
Scenario 2	8663	16000	41.64%	76.72%

Despite our measurements can identify only 41% of all accesses occurred, the accuracy is 100% in a case where no other IP accesses the shared cache during the attack. This accuracy decreases when extra IPs use the cache, because it is not possible for the attacker to identify the destination of the traffic. However, if a second infected IP is included in the system, the accuracy can be increased again, since it is possible to determine when the victim's access takes place.

The identification step can be performed correctly with this high accuracy and sampling of 41% of all accesses. This success rate is not a significant disadvantage of our work, because we use such information to trigger the reading of the cache. Since AES algorithm does not make sequential accesses to the same table, it is possible to perform the correct cache probe even with some triggers missing. For example, after table T_0 access, the next three access (regarding T_1, T_2 and T_3) can also perform a proper trigger. The only problem is that it is not guaranteed that we can observe the end of the first round, which corresponds to 16 cache accesses. Based on that, our attack uses the information of the first three accesses in the target table during the first round. As a consequence, further samples are required to reveal the key.

C. Reading Step Evaluation

At this step, the attacker reads specific values, annotating the access time of each data aligned with the target table. Note that the time required to perform the cache probe should not be superior to the next AES access. Analysing the results presented by the identification step evaluation allow to elaborate a strategy to overcome such difficulty. Figure 4 shows the latency measured by the attacker. We took the

minimum time measured between two requests to define a time execution limit of reading step, which is 1150 cycles. Then, we developed an experiment, where our attacker performed several cache requests to measure the accumulated latency. Our results considered several cache hits and only one miss, as expected in our attack scenario due to Prime phase. Table III presents the results of latency and the target limit.

TABLE III: Shared cache access latencies in cycles and possible reads count between AES accesses.

Cache Hit	Cache Miss	Limit	Max. Possible	Required
50 cycles	150 cycles	1150 cycles	20 reads	64 reads

As observed in Table III, it is possible to make 20 access between AES accesses, representing 31.25% of the reading required for the attack. As a result, our attack must perform the cryptography with the same plaintext at least 4 times. Therefore, our attack requires 4 times more execution time.

D. Analysis

After all practical considerations regarding identification and reading steps, our attack were configured to collect the data. In summary, the main modifications in NoC Prime+Probe attack from the theory presented in section IV to the practical implementation were:

- Analyze only the three first cache accesses to the target table instead four, due to NoC timing attack limitation; and
- Divide the reading step, performing the same cryptography 4 times.

The Prime+Probe attack was performed analysing 100 cryptography tasks, where all data collected was stored in attacker local memory, and sent to the host PC for analysis. In Figure 5 the number of encryption required versus the number of the revealed key bytes through the analysis process is presented.

Our result shows the high efficiency of our NoC Prime+Probe attack. Since our attack can perform the cache probe with high accuracy and coordinated with AES algorithm, only 80 encryption tasks (20 different plaintexts) are needed to recover the bytes of the key. It is not possible to recover all bytes in our technique because we avoid to evaluate the

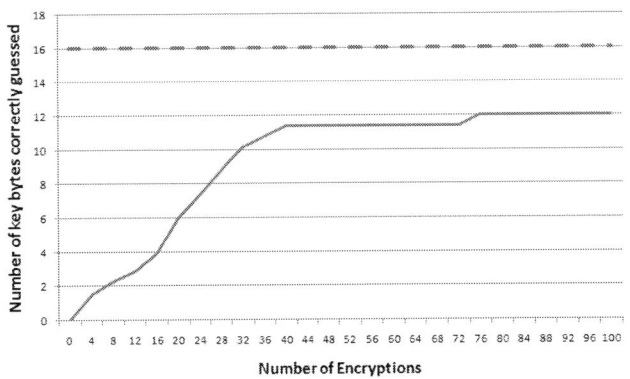

Fig. 5: Number of key bytes correctly recovered vs number of encryptions needed.

978-1-5090-2737-8/16 $31.00 © 2016 IEEE

last four accesses during the first round. Hence, our limit is 12 bytes of the 16 bytes. Therefore, a brute force technique is used to reveal the entire key, exploring the 2^{32} possibilities.

E. Countermeasure

The final experiment was to evaluate the timing security efficacy of the *Gossip NoC* under the proposed attack. Results show that *Gossip NoC* is able to protect the MPSoC due to the traffic deviation. As a consequence, the identification step was not performed, even for the first cache access, becoming impossible to perform the reading step. A second attacker could perform a timing attack in the NoC at the same time to discover the new sensitive path. However, since the main attacker triggers the reading step, a synchronization between infected IPs should be implemented.

F. Hardware Costs

Table IV presents the synthesis results, regarding logic utilization, block memories and power.

TABLE IV: FPGA Cyclone V SoC synthesis results of the NoC, the processing elements, caches and the network interfaces (NI).

	Logic (in ALMs)	Registers	Power (mW)
HPS (ARM core)	n/a	n/a	n/a
Cache	7131.9	12969	198.75
Attacker	47.5	96	0.27
Core NI	198.6	280	2.70
Cache NI	2560	2725	16.02
Router	499.8	738	4.44
NoC	6254.8	9273	63.17
MPSoC Platform	17,826	27323	885.61
Gossip Router	605,6	770	5,16

As observed in Table IV, *Gossip router* increases the logic area of the NoC about 21%. However, the unprotected NoC that represented 35% of the current MPSoC, became 42.5%, meaning that the true logic overhead in the sytem was 7.5%. When calculating the same impact for power, we obtain only 1.18% of power overhead in the system.

VII. CONCLUSION

Our work presented the first detailed timing attack on NoC-based MPSoCs, the NoC Prime+Probe technique. The practical execution has limitations regarding synchronization and accuracy, but with 80 encryptions was possible to reduce the search space of the key to 2^{32}. Besides, we evaluate the efficiency of the *Gossip NoC*, and propose strategies to improve the attack against such countermeasure. Future works aims to study the effects of different mapping of the attackers in MPSoC platforms.

REFERENCES

[1] R. Karri, K. Wu, P. Mishra, and Y. Kim, "Concurrent error detection of fault-based side-channel cryptanalysis of 128-bit symmetric block ciphers," in *Design Automation Conference, 2001. Proceedings*, 2001, pp. 579–584.

[2] P. Bayon, L. Bossuet, A. Aubert, V. Fischer, F. Poucheret, B. Robisson, and P. Maurine, "Contactless electromagnetic active attack on ring oscillator based true random number generator," May 2012, pp. 151–166.

[3] A. Moradi, O. Mischke, and C. Paar, "One attack to rule them all: Collision timing attack versus 42 aes asic cores," *IEEE Transactions on Computers*, vol. 62, no. 9, pp. 1786–1798, Sept 2013.

[4] P. C. Kocher, "Timing attacks on implementations of diffie-hellman, rsa, dss, and other systems," in *Proceedings of the 16th Annual International Cryptology Conference on Advances in Cryptology*, ser. CRYPTO '96, 1996, pp. 104–113.

[5] J. Kelsey, B. Schneier, D. Wagner, and C. Hall, *Side channel cryptanalysis of product ciphers*. Berlin, Heidelberg: Springer Berlin Heidelberg, 1998, pp. 97–110. [Online]. Available: http://dx.doi.org/10.1007/BFb0055858

[6] Y. Tsunoo, T. Saito, T. Suzaki, M. Shigeri, and H. Miyauchi, *Cryptographic Hardware and Embedded Systems - CHES 2003: 5th International Workshop, Cologne, Germany, September 8–10, 2003. Proceedings*. Berlin, Heidelberg: Springer Berlin Heidelberg, 2003, ch. Cryptanalysis of DES Implemented on Computers with Cache, pp. 62–76.

[7] D. J. Bernstein, "Cache timing attacks on aes," April 2005.

[8] D. A. Osvik, A. Shamir, and E. Tromer, *Cache Attacks and Countermeasures: The Case of AES*. Berlin, Heidelberg: Springer Berlin Heidelberg, 2006, pp. 1–20.

[9] Z. Xinjie, W. Tao, M. Dong, Z. Yuanyuan, and L. Zhaoyang, "Robust first two rounds access driven cache timing attack on aes," in *Computer Science and Software Engineering, 2008 International Conference on*, vol. 3, Dec 2008, pp. 785–788.

[10] Y. A. Younis, K. Kifayat, Q. Shi, and B. Askwith, "A new prime and probe cache side-channel attack for cloud computing," in *Computer and Information Technology; Ubiquitous Computing and Communications; Dependable, Autonomic and Secure Computing; Pervasive Intelligence and Computing (CIT/IUCC/DASC/PICOM), 2015 IEEE International Conference on*, Oct 2015, pp. 1718–1724.

[11] F. Liu, Y. Yarom, Q. Ge, G. Heiser, and R. B. Lee, "Last-level cache side-channel attacks are practical," in *Security and Privacy (SP), 2015 IEEE Symposium on*, May 2015, pp. 605–622.

[12] Y. Oren, V. P. Kemerlis, S. Sethumadhavan, and A. D. Keromytis, "The spy in the sandbox: Practical cache attacks in javascript and their implications," in *Proceedings of the 22Nd ACM SIGSAC Conference on Computer and Communications Security*, ser. CCS '15. New York, NY, USA: ACM, 2015, pp. 1406–1418. [Online]. Available: http://doi.acm.org/10.1145/2810103.2813708

[13] J. Sepulveda, D. Florez, and G. Gogniat, "Efficient and flexible noc-based group communication for secure mpsocs," in *2015 International Conference on ReConFigurable Computing and FPGAs (ReConFig)*, Dec 2015, pp. 1–6.

[14] W. Yao and E. Suh, "Efficient timing channel protection for on-chip networks," in *Networks on Chip (NoCS), 2012 Sixth IEEE/ACM International Symposium on*, May 2012, pp. 142–151.

[15] H. Wassel, G. Ying, J. Oberg, T. Huffmire, R. Kastner, F. Chong, and T. Sherwood, "Networks on chip with provable security properties," *Micro, IEEE*, vol. 34, no. 3, pp. 57–68, May 2014.

[16] M. Sepulveda, J.-P. Diguet, M. Strum, and G. Gogniat, "Noc-based protection for soc time-driven attacks," *Embedded Systems Letters, IEEE*, vol. 7, no. 1, pp. 7–10, March 2015.

[17] C. Reinbrecht, A. Susin, L. Bossuet, and J. Sepulveda, "Gossip noc - avoiding timing side-channel attacks through traffic management," in *IEEE Computer Society Annual Symposium on VLSI (ISVLSI '16)*, July 2016.

[18] G. Sievers, J. Daberkow, J. Ax, M. Flasskamp, W. Kelly, T. Jungeblut, M. Porrmann, and U. Rckert, "Comparison of shared and private l1 data memories for an embedded mpsoc in 28nm fd-soi," in *Embedded Multicore/Many-core Systems-on-Chip (MCSoC), 2015 IEEE 9th International Symposium on*, Sept 2015, pp. 175–181.

[19] J. Daemen and V. Rijmen, *The Design of Rijndael*. Secaucus, NJ, USA: Springer-Verlag New York, Inc., 2002.

[20] A. Borhani, A. Movaghar, and R. Cole, "A new deterministic fault tolerant wormhole routing strategy for k-ary 2-cubes," in *Computational Intelligence and Computing Research (ICCIC), 2010 IEEE International Conference on*, Dec 2010, pp. 1–7.

[21] K. Tatas, S. Sawa, and C. Kyriacou, "Low-cost fault-tolerant routing for regular topology nocs," in *Electronics, Circuits and Systems (ICECS), 2014 21st IEEE International Conference on*, Dec 2014, pp. 566–569.

[22] Altera, "Cyclone v soc - overview," (Date last accessed 17-April-2016). [Online]. Available: https://www.altera.com/products/soc/portfolio/cyclone-v-soc/overview.tablet.html

978-1-5090-2737-8/16 $31.00 © 2016 IEEE

Cache Sizing for Low-Energy Elliptic Curve Cryptography

Felipe Piovezan, Tarcísio E. M. Crocomo, Luiz C. V. dos Santos
Computer Science Department
Federal University of Santa Catarina, Florianopolis, SC, Brazil

Abstract—Public-key (asymmetric) cryptography (PKC) is one the pillars of secure communication, since it includes user authentication, key exchange, and digital signatures. PKC dominates energy consumption for transactions shorter than 100KB. Elliptic Curve Cryptography (ECC) allows for shorter key and signature sizes while maintaining the same security level as other PKC algorithms. Since ECC algorithms become more computationally expensive with growing key sizes and since many communicating artifacts are battery-powered devices (e.g. smartphones, smart sensors, and implantable medical devices), secure communication requires energy-efficient ECC as the security level raises and/or as transaction length shrinks. That is why this paper breaks down the energy spent in the memory system for full-software ECC implementations running in a 32-bit RISC processor under growing cache sizes (from 1KB to 16KB) and for the different levels of security provided by distinct elliptic curves (P224, P256/P25519, P384, and P521). We report the ideal cache sizes for each elliptic curve when targeting a 32nm technology node. We show that energy-efficient ECDH requires an instruction cache of around 8KB (or larger) and a data cache or scratchpad at least 4 times as small. We also show that the adoption of P25519, instead of P256, leads to 15 times less energy consumption in the memory system and half the required cache capacity, while keeping the same level of security.

I. Introduction

Security processing is crucial to devices communicating sensitive data across wireless networks. To reach security objectives, distinct classes of cryptographic algorithms are used. Primarily, symmetric algorithms are used for confidentiality, asymmetric algorithms for authentication and non-repudiation, and hash algorithms for verifying the integrity of the exchanged message. At least 40% of the energy spent in security processing is due to non-cryptographic protocol tasks regardless of transaction length [1]. The remaining of the energy consumption is either dominated by asymmetric algorithms for short transactions or by symmetric and hash algorithms for long transactions. Therefore, the use of a hardware accelerator (for either an asymmetric or a symmetric algorithm) leads to the highest energy efficiency only at one of the extremes of the transaction-length spectrum.

Public-key (asymmetric) cryptography (PKC) dominates energy consumption for transactions shorter than 100KB. Even for a 100KB transaction, PKC typically requires 4 times more energy than symmetric and hash algorithms altogether and as much energy as non-cryptographic protocol tasks [1]. Elliptic Curve Cryptography (ECC) [2] is a subclass of PKC that allows for shorter key and signature sizes for a given level of security.

Since most artifacts communicating through a wireless network are battery-powered devices, cryptography is constrained by low-energy (e.g. personal mobile devices) or even ultra-low energy budgets (e.g. smart sensors and implantable medical devices). As ECC algorithms become more computationally expensive with growing key sizes but are constrained by low-energy budgets, secure communication requires energy-efficient ECC as the security level raises and/or as transaction length shrinks.

The energy efficiency obtained with an ECC hardware accelerator can be 6 times higher as compared to a full-software baseline built upon a 32-bit RISC processor without cache [3]. Although the use of a hardware accelerator for ECC would lead to the highest energy efficiency, its impact is reduced as transaction length grows. Besides, its use may be precluded in applications where reconfigurability is desirable or mandatory (e.g. wireless sensor networks or implantable medical devices). Finally, since no cryptographic accelerator can ever reduce the energy spent in non-cryptographic protocol tasks (40% of the energy budget), its impact on the overall energy efficiency is limited.

On the other hand, in designs where the use of a hardware accelerator might be less effective or inappropriate, the inclusion of an instruction cache has the potential of making the energy efficiency 1.5 times higher as compared to the baseline [4]. In other words, such an improvement is obtained by simply reducing the cost of fetching instructions directly from non-volatile storage. Note that, as opposed to hardware accelerators, an instruction cache has also the potential to improve the energy efficiency of the non-cryptographic protocol tasks.

Therefore, the judicious use of an instruction cache becomes a promising key to energy efficiency for architectures that cannot rely on hardware accelerators or need to provide reconfigurability. In such a scenario, however, the improvement on energy efficiency becomes limited by the cache consumption itself. This motivated us to break down ECC energy consumption, so as to investigate how the cache energy per access could be reduced without

978-1-5090-2737-8/16 $31.00 © 2016 IEEE

increasing the consumption in main memory.

This paper reports the memory energy breakdown for full-software ECC implementations, when used by the ECDH [5] and ECDSA [6] protocols. To assess the impact of the level of security on energy, we report the consumption for standard elliptic curves of increasing key size (P224, P256, P384, and P521). We also report the energy breakdown for a curve recently made available (P25519 [7]), which provides the same level of security as P256. To the best of our knowledge, no previous reports on memory energy consumption are available in the literature for either ECDH (whatever the curve) or P25519 (whatever the protocol). Finally, we determine the cache sizes leading to the overall minimum consumption in the memory system.

The main contributions of this paper are:

- The first evaluation of memory energy profiles ever reported for curve P25519.
- The first analysis of memory energy consumption ever reported for the ECDH protocol.
- A cache sizing that paves the way towards energy-efficient solutions (for 32-bit RISC full-software implementations of elliptic curves).

The remaining of this paper is organized as follows. Section II reviews previous work on ECC energy consumption. Section III describes our experimental setup. Sections IV and V report the impact of cache sizing on memory energy consumption for ECDH and ECDSA. Finally, Section VI shows our conclusions and plans for further work.

II. RELATED WORK

Table I gives a panorama of the main reports on ECC energy efficiency. For each related work, the table identifies target applications such as wireless sensor networks (WSN), implantable medical devices (IMD), and RFID. Besides, it summarizes the target instruction-set architecture (ISA) and accounts for whether or not an instruction cache (I-cache) or a hardware accelerator (HWA) are exploited in the microarchitecture. The table's last column summarizes the ranges of energy efficiency obtained for distinct key sizes. Observe that they are reported for largely different processors and for operations of different granularity (protocol-level or primitive operations).

The first two works adopt a compound operation (Sign + Verify) to evaluate the energy efficiency of an ECDSA protocol for a baseline 32-bit RISC architecture (without cache), which is extended under distinct hardware configurations. For instance, they report that the energy efficiency becomes 1.5 or 6 times higher when either an instruction cache or a hardware accelerator is added to the baseline, respectively. Similarly, the work by Wander et al. [8] compared the energy cost of RSA and ECC-based protocols, showing the feasibility of the latter on extremely resource-constrained environments. Furthermore, that work reports that ECC-based public key cryptography accounts for more than 70% of the total energy

spent on small Transport Layer Security (TLS) transactions (1KB). Unfortunately, the adoption of a compound (Sign + Verify) operation for the evaluation of energy efficiency of the ECDSA protocol is questionable. In a typical application, the server node is responsible for the signature and the peer node for its verification, as opposed to the works by Targhetta et al. [3], [4], which assume that signature and verification are both performed at the same node. Besides, since those works report only the energy consumption of the compound operation, it is not possible to infer the actual consumption in the peer node, since sign and verify operations are largely different in terms of computational costs [6].

Instead of evaluating the energy spent by protocol-level operations, the works by Wenger et al. [9] and Fan et al. [10] focus on one primitive operation, namely point multiplication, which is exploited by both ECDH and ECDSA protocols. While point multiplication largely dominates energy consumption for ECDH, it contributes much less for the energy consumed by ECDSA. Therefore, the use of point multiplication as a single typical ECC operation would lead to imprecise estimations of energy efficiency for the latter protocol.

For these reasons, this work neither focuses on point multiplication (as in [9][10]) nor uses unrealistic compound operations (as in [3][4]). Instead, we prefer to analyze the actual energy consumption for both asymmetric protocols (ECDH and ECDSA) so as to identify lower-bound cache sizes as starting points towards energy-efficient solutions.

The next sections apply the devised approach to the memory system for energy-efficient cache sizing.

III. EXPERIMENTAL SET UP

We relied on the OpenSSL open-source library [11] to provide the implementations for ECC operations. Only the relevant subset of the code was compiled (without multi-thread support). We employed only C code (avoiding optimizations written in assembly) and relied on Clang (version 3.9) as a cross compiler.

We adopted ARMv7 as the target instruction-set architecture. Since security processing is one among Cortex-R's target applications, we selected memory parameters compatible with those recommended for that microarchitecture. The memory hierarchy contains separate instruction and data caches at the first level and main memory (512 KB) at the second. We assumed both caches with the same size and the same 2-way associativity. Unless stated otherwise, the blocksize was set to 32 bytes for all caches. To evaluate the impact of cache size on energy consumption, we repeated the evaluation for growing cache sizes (1 KB, 2 KB, 4 KB, 8 KB, 16 KB).

To run the cryptographic algorithms, we adopted the gem5 infrastructure [12] as a design representation for the microarchitecture (under the setup CPU model = "arm-detailed", memory model = "simple"). To estimate memory energy consumption, we relied on gem5 for keeping

TABLE I
MAIN REPORTS OF ENERGY EFFICIENCY FOR ECC

Work	Application domain	Processor		I-cache	HWA	Main Results
		ISA	Clock			
[3][4]	IMD, WSN, RFID	32-bit RISC	100MHz	no	no	ECDSA Sign+Verify baseline: 300 to 1550 μJ/op (key size: 192 to 384 bits)
				yes	no	2/3 of baseline values
				no	yes	1/6 of baseline values
[8]	WSN	8-bit μC	4MHz	no	no	ECDSA Sign+Verify: 68 to 183 mJ/op (key size: 160 to 224 bits)
[9]	IMD, WSN, RFID	8-bit and 16-bit μC, 32-bit RISC	10MHz	no	no	Point multiplication implementations: 202 to 1914 μJ/op (key size: 160 to 256 bits)
[10]	IMD	8-bit μC	847MHz	no	yes	Point multiplication implementations: 5.1 μJ/op (key size: 160 to 256 bits)

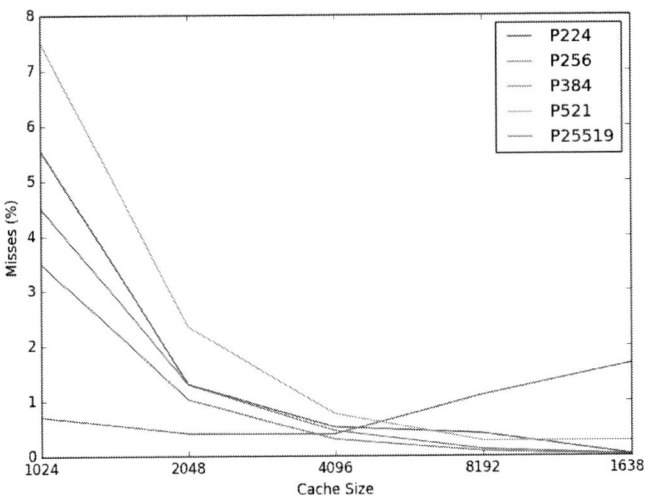

Fig. 1. Data cache miss rates (ECDH)

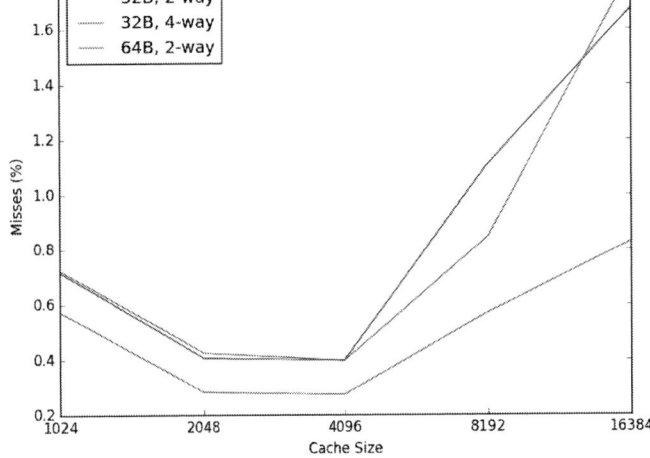

Fig. 2. Data miss rates for P25519 (ECDH)

track of the number of read and write accesses. Then we used the CACTI tool [13] for energy estimation under uniform cached access when targeting a 32nm low standby-power (itrs-lstp) technology node.

IV. EVALUATION FOR ECDH

Figure 1 shows miss rates in the data cache for increasing cache sizes and distinct security levels. Note that for a 4KB cache, the miss rate is inferior to 1% regardless of security level. Notice that the level of security barely affects miss rate for caches larger than 4KB for all curves but P25519. This means that, in general, the working set for ECC data is largely independent from the required security level, it largely fits in a 2-way 4KB cache with 32-byte blocks, and more items could be accommodated in cache simply by increasing the number of blocks. On the other hand, the behavior of P25519 indicates that either a larger block size or a higher associativity would be required for benefiting from a higher cache size.

Such conjecture is supported by the extra results in Figure 2. Note that, for a cache larger than 4KB, the miss rate largely decreases when the block size is increased, but it is almost unaffected when the associativity is in-

creased. Therefore, a block size larger than 32 bytes would be required for capturing more *spatial* locality so as to accommodate extra working set items into a larger cache.

Figure 3 shows instruction miss rates for growing cache sizes and distinct security levels. Note that, for a given cache size, the miss rates do not exhibit any correlation with security levels. This indicates that the sizes of the working sets do not vary monotonically with the required level of security. This might be seen as an evidence that conventional on-demand fetching might be unable to uniformly exploit the instruction cache for improving energy efficiency regardless of security level. This indicates that prefetching is a promising optimization for further exploiting the instruction cache.

Since the miss rates in Figures 1 and 3 differ from one order of magnitude, they provide early evidence that the data cache could be smaller or might be replaced by a small scratchpad mapped to a distinct address space.

Figure 4 shows the components of the energy consumption in the memory system when running P256: the dynamic energy spent when accessing data and instruction in caches (D-cache and I-cache) or in main memory (D-main and I-main), as well as the energy due to leakage in

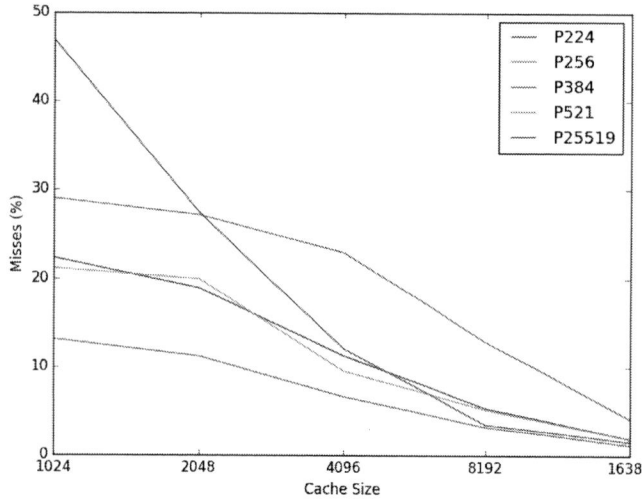

Fig. 3. Instruction cache miss rates (ECDH)

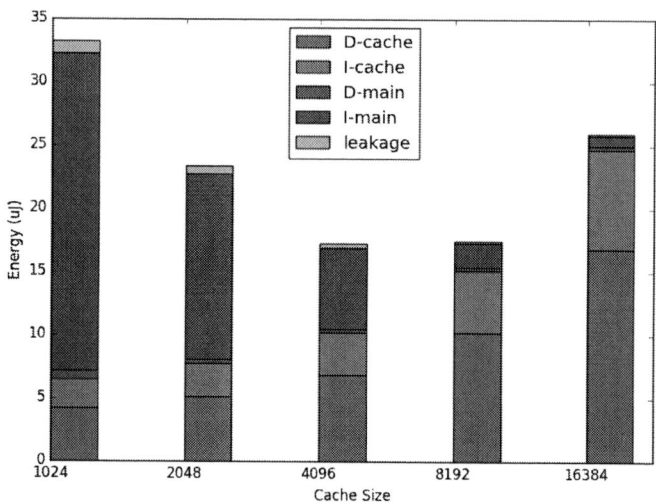

Fig. 5. Energy breakdown for P25519 (ECDH)

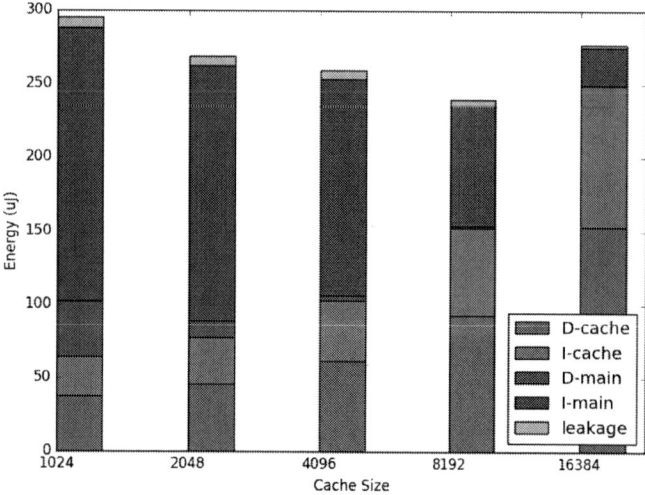

Fig. 4. Energy breakdown for P256 (ECDH)

the main memory. Observe that the energy spent with leakage in main memory decreases with growing cache sizes as a result of shrinking execution times, although the impact of leakage can be neglected in face of the much higher dynamic energy components. Notice that the minimum energy consumption when accessing instructions (I-cache + I-main) corresponds to an instruction cache capacity around 8KB. Finally, note that the energy spent when accessing data in main memory can be neglected (less than 2% of the overall energy) for a 4KB or larger data cache.

Figure 5 shows the energy breakdown when running P25519 (which provides the same level of security as P256). Despite a behavior somewhat similar to the one observed in Figure 4, energy consumption was reduced in one order of magnitude for the same level of security! This results from two new features of P25519 that are absent

from P256: 1) The adoption of a Montgomery algorithm [7] for point multiplication, 2) the fact that no key validation is ever required by the new curve. Those features largely reduce the number of required accesses to memory. We observed that P25519 requires the execution of 10 times less instructions than P256 (which largely affects the I-cache and I-main components). Besides, we observed that P25519 requires the execution of 8 times less loads and 10 times less stores than P256 (which largely affects the D-cache and D-main components). Notice that the overall minimum energy consumption corresponds to 4KB caches, although the minimum energy consumption when accessing instructions only corresponds to an instruction cache of around 8KB. Note also that the energy spent when accessing data in main memory can be neglected (less than 2% of the overall energy) for a 1KB or larger data cache, i.e. for a capacity four times as small as compared to P256.

Since it would be cumbersome to exhaustively report energy breakdowns for all curves, Table II summarizes our results by showing the minimum *dynamic* consumption under two distinct scenarios for all evaluated curves. The first scenario (A) assumes equally sized instruction and data caches; the second (B), relaxes such assumption. From the energy breakdown of each curve, we first identified the (I/D) cache size in scenario A leading to the minimum total dynamic energy when accessing instructions or data, either in cache or in main memory, i.e. min E_T = min $(E_{Icache} + E_{Imain} + E_{Dcache} + E_{Dmain})$. Then we identified the instruction cache size in scenario B leading to the minimum energy when accessing *instructions* only, either in cache or in main memory, i.e. min E_I = min $(E_{Icache} + E_{Imain})$. Next, we identified the data cache size in scenario B leading to the minimum energy when accessing *data* only, either in cache or in main memory, i.e. min E_D = min $(E_{Dcache} + E_{Dmain})$. Finally, we computed the *combined* dynamic energy that would be obtained if the instruction and cache sizes obtained separately for scenario B were adopted, i.e. $E_C = \min E_I + \min E_D$.

On the one hand, in scenario A, the cache size minimizing dynamic consumption is 4KB for all curves but P256 (minimized by a cache size of 8KB). On the other hand, in scenario B, as expected from the largely distinct instruction and data miss rates reported in Figures 1 and 3, the minimum overall dynamic energy is obtained by making the data cache smaller than the instruction cache. All curves but one (P256) require an 8KB instruction cache and all but one (P25519) require a 2KB data cache. Note that scenario B leads to less dynamic consumption than scenario A for all curves. This was achieved at the expense of a larger overall cache capacity for all curves. On average, a 19% reduction in dynamic consumption was obtained with a 15% increase in cache capacity.

Note that the highest energy reduction (32%) from scenario A to B was observed for curve P25519, which was also the one requiring the least amount of data cache. Therefore, it could be argued that, if instructions and data were mapped to disjoint address spaces, the data cache could even be replaced by a scratchpad leading to even smaller energy consumption. This conjecture is supported by the evidence in Figure 2: what prevented curve P25519 from further reducing the miss rate was the bound in block size, not the lack of associativity. Therefore, the data working set is likely to fit into a non-associative memory slightly larger than the required data cache size.

V. EVALUATION FOR ECDSA

To report the results obtained for the ECDSA protocol, we measured miss rates and evaluated energy breakdowns for every curve and for both operations, verify and sign. The verify operation is the one consuming more energy. In spite of that, it is typically performed in the peer node, which is often a battery-operated device. For these reasons, we first focus on the analysis of the results for the verify operation (Figures 6 and 7) and then we summarize the results for both sign and verify (Table III).

Since data and instruction miss rates for ECDSA are quite similar to ECDH's and therefore do not add value to the analysis, we omit them for simplicity.

Figures 6 and 7 show the energy breakdown for curves P256 and P25519. Observe that the values for ECDSA are always smaller than those for ECDH (see Figures 1 and 3). This stems from the fact that the ECDH protocol requires the generation of the key, whereas the ECDSA only verifies with a predefined key. This results in a smaller number of accesses to memory. For instance, when using P25519, we observed that ECDSA requires 18.7% less instructions than ECDH, 18.5% less loads, and 21.7% less stores.

Table III summarizes all ECDSA results for both operations, verify and sign. Note that, for a peer node, the cache sizing for ECDSA (verify) leads to *exactly* the same cache capacity as for ECDH (for all levels of security), except for curve P25519, which would require larger caches

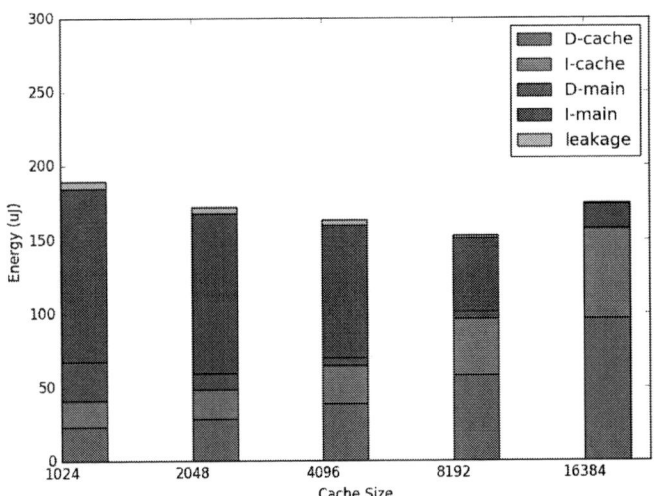

Fig. 6. Energy breakdown for P256 (ECDSA verify)

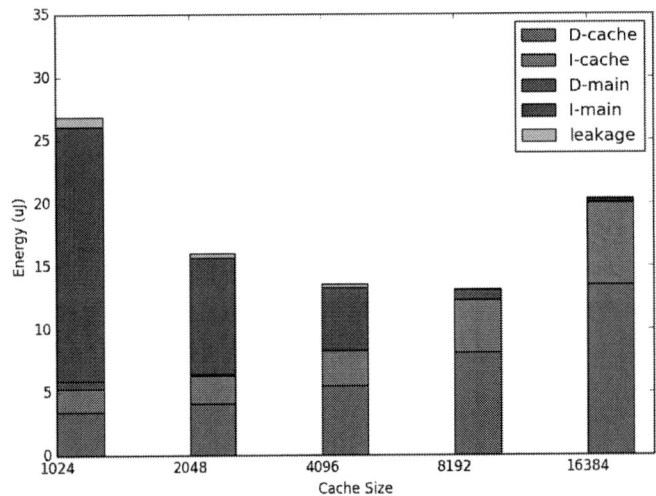

Fig. 7. Energy breakdown for P25519 (ECDSA verify)

in scenario A. However, since B is the best scenario, the proposed sizing leads to the minimum memory energy for *both* ECDH and ECDSA.

Note that, when standard curves are used in scenario B, the verify operation spends on average 1.13 times more energy in the memory than the sign operation. When P25519 is used instead, the verify operation spends 1.68 times more energy than the sign operation. Notice that, as expected, ECDH always requires more energy than ECDSA. When standard curves are used in scenario B, ECDH spends, on average, 1.57 times more energy than ECDSA (verify). When P25519 is used instead, ECDH spends 1.30 times more energy. Thus, when P25519 replaces P256, the energy decreases by one order of magnitude, but the dominance of ECDH is reduced, which increases the share of the verify operation in the overall consumption, making it a promising optimization target.

TABLE II
SCENARIOS LEADING TO MINIMUM DYNAMIC ENERGY (ECDH)

Curve	A		B				E_C (uJ)
	min E_{Total}		min E_{Instr}		min E_{Data}		
	E_{Total} (uJ)	I/D-cache	E_{Instr} (uJ)	I-cache	E_{Data} (uJ)	D-cache	
P224	148.8	4KB	79.7	8KB	43.7	2KB	123.4
P256	235.5	8KB	121.3	16KB	55.9	2KB	177.2
P384	339.4	4KB	185.6	8KB	114.7	2KB	300.3
P521	1179.1	4KB	625.6	8KB	424.9	2KB	1050.5
P25519	17.1	4KB	6.7	8KB	4.9	1KB	11.6

TABLE III
SCENARIOS LEADING TO MINIMUM DYNAMIC ENERGY (ECDSA)

Curve	A		B				E_C (uJ)
	min E_{Total}		min E_{Instr}		min E_{Data}		
	E_{Total} (uJ)	I/D-cache	E_{Instr} (uJ)	I-cache	E_{Data} (uJ)	D-cache	
P224-verify	95.0	4KB	50.2	8KB	28.7	2KB	78.9
P256-verify	152.9	8KB	77.2	16KB	39.4	2KB	116.6
P384-verify	209.0	4KB	114.4	8KB	75.2	2KB	189.6
P521-verify	736.3	4KB	391.0	8KB	250.2	2KB	641.2
P25519-verify	13.1	8KB	4.9	8KB	4.0	1KB	8.9
P224-sign	83.7	4KB	43.7	8KB	25.0	2KB	68.7
P256-sign	134.1	8KB	68.5	16KB	36.2	2KB	104.7
P384-sign	192.5	4KB	103.1	8KB	65.6	2KB	168.7
P521-sign	645.3	4KB	343.1	8KB	219.3	2KB	562.4
P25519-sign	6.7	2KB	3.4	16KB	1.9	2KB	5.3

VI. CONCLUSIONS AND FUTURE WORK

This paper showed that, in microarchitectures devoid of hardware accelerators, the starting point towards energy-efficient ECC solutions is the adoption of an instruction cache of around 8KB. The use of a data cache is beneficial only if its capacity is at least 4 times as small or if a non-associative memory of comparable size is used to accommodate the data working set, which is smaller than the code working set. For the same level of security, the simple adoption of the curve P25519 running on a 32-bit RISC processor leads to 10 times less instructions, 15 times less dynamic energy consumption, and half the required cache capacity as compared to P256. Thus, the use of more elaborated elliptical curves and the exploitation of instruction caches are likely to reduce the need for hardware accelerators in public key cryptography.

As future work, we plan to rely on a compiler optimization for further reducing energy consumption via instruction prefetching. Besides, we intend to evaluate its impact on the full security protocol (asymmetric, symmetric, hash and non-cryptographic tasks) for distinct transaction sizes.

REFERENCES

[1] N. R. Potlapally, S. Ravi, A. Raghunathan, and N. K. Jha, "A study of the energy consumption characteristics of cryptographic algorithms and security protocols," *IEEE Transactions on Mobile Computing*, vol. 5, no. 2, pp. 128–143, Feb 2006.

[2] N. Koblitz, "Elliptic curve cryptosystems," *Mathematics of computation*, vol. 48, no. 177, pp. 203–209, 1987.

[3] A. D. Targhetta, D. E. Owen, and P. V. Gratz, "The design space of ultra-low energy asymmetric cryptography," in *2014 IEEE International Symposium on Performance Analysis of Systems and Software (ISPASS)*, March 2014, pp. 55–65.

[4] A. D. Targhetta, D. E. Owen, F. L. Israel, and P. V. Gratz, "Energy-efficient implementations of gf (p) and gf(2m) elliptic curve cryptography," in *2015 33rd IEEE International Conference on Computer Design (ICCD)*, Oct 2015, pp. 704–711.

[5] IEEE, *IEEE Std 1363-200, Standard Specifications for Public-Key Cryptography*, Aug 2000.

[6] NIST, "Fips pub 186-4 digital signature standard (dss)," 2013.

[7] D. J. Bernstein, *Public Key Cryptography - PKC 2006: 9th International Conference on Theory and Practice in Public-Key Cryptography, New York, NY, USA, April 24-26, 2006. Proceedings.* Berlin, Heidelberg: Springer Berlin Heidelberg, 2006, ch. Curve25519: New Diffie-Hellman Speed Records, pp. 207–228.

[8] A. S. Wander, N. Gura, H. Eberle, V. Gupta, and S. C. Shantz, "Energy analysis of public-key cryptography for wireless sensor networks," in *Third IEEE International Conference on Pervasive Computing and Communications, 2005. PerCom 2005*, March 2005, pp. 324–328.

[9] E. Wenger, T. Unterluggauer, and M. Werner, *Progress in Cryptology – INDOCRYPT 2013: 14th International Conference on Cryptology in India, Mumbai, India, December 7-10, 2013. Proceedings.* Cham: Springer International Publishing, 2013, ch. 8/16/32 Shades of Elliptic Curve Cryptography on Embedded Processors, pp. 244–261.

[10] J. Fan, O. Reparaz, V. Rožić, and I. Verbauwhede, "Low-energy encryption for medical devices: Security adds an extra design dimension," in *2013 50th ACM/EDAC/IEEE Design Automation Conference (DAC)*, May 2013, pp. 1–6.

[11] T. O. Project. (2015) Openssl: The open source toolkit for SSL/TLS. [Online]. Available: https://www.openssl.org

[12] N. Binkert, B. Beckmann, G. Black, S. K. Reinhardt, A. Saidi, A. Basu, J. Hestness, D. R. Hower, T. Krishna, S. Sardashti, R. Sen, K. Sewell, M. Shoaib, N. Vaish, M. D. Hill, and D. A. Wood, "The gem5 simulator," *SIGARCH Comput. Archit. News*, vol. 39, no. 2, pp. 1–7, Aug. 2011. [Online]. Available: http://doi.acm.org/10.1145/2024716.2024718

[13] N. Muralimanohar, R. Balasubramonian, and N. P. Jouppi, "Cacti 6.0: A tool to model large caches," *HP Laboratories*, pp. 22–31, 2009.

Cluster-based Architecture Relying on Optical Integrated Networks with the Provision Of a Low-latency Arbiter

Felipe Göhring de Magalhães[†*], Fabiano Hessel[*],
Odile Liboiron-Ladouceur[‡] and Gabriela Nicolescu[†]

[†]Ecole Polytechnique de Montreal, Canada - [*]PPGCC/PUCRS, Porto Alegre, Brazil - [‡]McGill University, Canada

Contact: felipe.magalhaes@acad.pucrs.br

Abstract—State-of-art Multiprocessor Systems-on-Chip (MP-SoC) struggle to respect the increasing requirements of high performance interconnects for high throughput communications. Optical integrated Networks (OIN) represent currently one of the most promising paradigms for the design of such next generation MPSoC. They provide increased bandwidth and better reaction to electromagnetic noise while decreasing latency and power consumption. In this paper we propose a new cluster-based architecture, the Hybrid Torus MPSoC (HTM). We propose also a low-latency arbiter allowing to exploit the full potential of the HTM architecture. Our experiments, based on FPGA prototyping and simulation, show the efficiency of the proposed architecture.

Index Terms—Optical Integrated Networks; Low-Latency Arbiter; Cluster-based Architectures

I. INTRODUCTION

Modern systems have their implementation based on multiple integrated processing elements, running at a lower clock frequency, due to energy consumption constraints. Such integrated system is called Multiprocessor System-on-Chip (MPSoC). Since the introduction of MPSoCs, one of the design's main concerns lies in how the communication between internal components is performed. Electrical networks-on-chip (eNoCs) provide good communication performance [37] while maintaining an improved energy efficiency and a high re-usability level [6]. However, as the number of possible integrated cores on a single chip continues to increase, metallic interconnects in eNoCs will become a bottleneck leading the ITRS (International Technology Roadmap for Semiconductors) [1] to point out the need for a new technology to overcome such restrictions.

In this design context, on-chip optical interconnects and 3D die stacking are currently considered to be the two most promising new paradigms. Optical Integrated Networks (OINs) have already been proven to be feasible for inter-chip communications [4][7], and previous work presented photonic architectures with low power consumption, low insertion loss (7.9 dB for an 8×8 structure) and a power penalty of less than 1 dB [25]. These works bring forward OINs as attractive candidates for high demanding architectures.

Several works presented cluster-based architectures that make usage of different communication infra-structures in order to extract the best of each, such as [39][12]. This type of architectures alleviates the clock skew [27] as the components are grouped, reducing the complexity to design the clock-tree. For the same reason, the employment of Global

Asynchronous Local Synchronous (GALS) design technique [9] is facilitated when using cluster-based architectures, as each cluster might execute under its own clock domain. As cluster-based architectures rely on more than one communication infrastructure in the same system, fast interfaces and control techniques are needed. This challenge was previously addressed for the electrical communication structures, but OIN-based architecture still lack better solutions.

This work presents two main contributions:

1) A cluster-based architecture, the Hybrid Torus MPSoC (HTM). The HTM architecture uses electrical components to perform in-cluster communications and optical components for the intra-cluster communications.
2) A low-latency arbiter used to control efficiently the OIN. This controller allows the exploitation of the full potential of the HTM architecture.

Results show the efficiency of the introduced architecture as well as the low-latency impact of its arbiter. Different traffic patterns and traffic injection configurations are used to evaluate the performance.

The remaining of the paper is organized as it follows. The next section brings the state-of-art revision, positioning our approach with existing works. In Section III, the proposed cluster-based architecture is presented, followed by Section IV in which the arbiter is illustrated. In Section V, the obtained results are presented and finally, Section VI concludes this work.

II. RELATED WORK

A. Cluster-Based Architectures

Two main approaches are currently employed for the design of embedded cluster-based architectures: (1) defining architectures using different communication approaches, such as buses, shared memory and NoCs and (2) architectures based on hardware modifications, like embedded virtualization.

The most common approach is the usage of buses for inter-cluster communications and NoCs for intra-cluster communications. In [39] the authors propose a cluster-based system formed by ARM processors grouped in 'n' clusters of variable size which communicate through an AMBA-AHB bus [2]. The communication between clusters is performed by a NoC with a cluster on each router. An architecture using a NoC as the communication infrastructure between clusters and a simple bus to communicate internal elements on each cluster is presented in [10]. In this model, each node is composed by IPs that may be hardware or soft cores and two modules to communicate with the NoC router.

Another class of architectures are the one based on shared memories for local communications. The architecture presented in [15] is composed of 17 processors organized into four clusters of four processors each, plus a central processor that controls the whole system. In [38] the authors present a model of dynamic clusters where the message exchange is performed using local memories. For both works, the internal modules of each cluster use a memory and the clusters are connected through a point-to-point connection.

Some previous works relied on modifications in the hardware to keep only one communication infrastructure. The authors in [33] presented a cluster-based system using a single NoC with modified routers. Unlike all previous works, the inner IPs of each cluster are directly connected to the NoC router. To do so, the NoC routers were changed to include three more local ports.

The work presented in [3] introduces a different approach for embedded cluster-based system. Virtualization techniques are used to expand the MPSoC capability, running 'n' virtual processors on each physical unit. This solution uses MIPS processors [30] and a central NoC to perform the communication between physical cores.

The works presented in [20], [23] present cluster-based architectures relying on OINs. The used approach and obtained results are very similar for both cases, where cluster composed for four IPs are connected using NoC and the clusters are connect through a 10×10 optical switch fabric. The work [36] uses hybrid routers (optical-electrical) in the system, where the clusters are connected using an OIN organized in a butterfly fashion. This work also uses 10×10 optical switch fabrics.

Finally, the works introduced in [32], [29] use a ring-based optical network to connect clusters, where wavelength division is employed to have an non-blocking OIN. Their approach is based on two layers: the electrical layer has the clusters and the optical layer connects the clusters using MR-based optical switches.

As it is possible to verify, systems relying solely on electrical interconnects are majority. Also, most works have predefined in-cluster organizations, decreasing the flexibility to use them.

Comparing with previous works, the originality of the proposed HTM architecture lies in the fact that a high-bandwidth OIN is used to connect clusters. Also, the clusters' constitution is not hard-limited or defined, which expands the design space exploration. Still, besides the internal cluster organization, the difference for the architectures that also utilize OINs lies in the employed OIN. This work relies in well established torus architectures, while the previous works use in-house OINs. Moreover, the ring-based OIN system has a constraint in the number of possible wavelengths used, thus jeopardizing its usage.

B. OIN Controlling Solutions

Most of the state-of-art controllers are topology - or architecture specific, thus optimizing at most the performances of a specific network.

In [31], the control unit used is based on the circuit-switched algorithm. A dimension order algorithm is applied on an electrical-layer and closes the paths for the optical layer. A similar technique is used by [22], where the controlling scheme is also based on circuit-switching. The latency was calculated to be around 3.5ns on a 8×8 network, where resonators and peripherals run at 5 GHz. An electrical-optical mixed approach is presented by [21], where Optical switches are in charge of transmitting data on a circuit-switching fashioned way, while electrical switches are in charge of closing the path by using package-switching techniques.

The control unit in [16] is based on wavelength-division multiplexing (WDM). Each I/O is assigned with one specific wavelength and might communicate at any time, without arbitration. Authors of [8] introduce a routing technique based on wavelength selection integrated with spatial routing. Circuit-switching technique is used along with WDM and each router is composed by a junction of a receiver bank and a modulator bank.

An asynchronous and variable-length packet switching is presented in [17]. Every IP is attributed with one exclusive label, which corresponds with each output fiber. While the message travels over the optical path, when it comes across a new network node, the message gets delayed while its label is computed by the electrical node. Further, a multi-cast scheduling control solution is proposed in [34], focusing input-queued switches based on the Weight Based Arbiter (WBA) and Time-division Multiplexing (TDM). The technique used by the authors is based on time sharing and aged-based weight calculations.

Finally, authors in [40] presented a controlling solution for contention handling based on optical-buffering, by introducing a three stages buffering method. This method uses electronically controlled wavelength routing switches in combination with optical delay lines to temporarily store data [11], [24], [35].

A good number of works make use of circuit-switching techniques, thus adding a control-latency that could shatter completely the benefits of using an OIN-based architectures. Also, most solutions are deeply attached to the controlled topology, which leverages their employment to different scenarios.

III. THE HYBRID TORUS MPSoC

The Hybrid Torus MPSoC is designed in three different layers: the electrical communication layer, the interface layer and the optical layer. Its organization might be deployed in regular 2.5D fabrication technologies, but it is designed for the future 3D integration processes. Also, each cluster can operate in a different frequency, in a Global Asynchronous Local Synchronous (GALS) fashion.

978-1-5090-2737-8/16 $31.00 © 2016 IEEE

The *electrical layer* is composed by generic IPs and communication infrastructures (networks-on-chip). Even though the HTM is not restricted to one specif NoC router, in this work, we relied on the HERMES router [14] for implementation and validation purposes. The HERMES is deployed as a mesh and is composed by routers, buffers and controllers of routers information. The router overview is presented in Figure 1. Still, the internal queue scheduling uses a priority round robin algorithm. The packet routing algorithm is the XY [28] and the packet flow protocol is credit-based. Each cluster is composed by 'x' IPs interconnected by a $n \times m$ NoC.

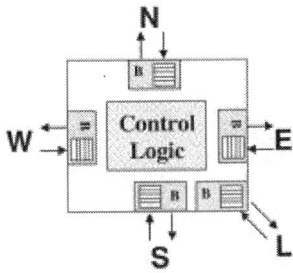

Fig. 1. HERMES router internal organization [14].

The *interface layer*, namely the *Cluster Interface (CI)*, performs the communication between each cluster and the optical layer. The CI consists of two circular queues that temporary store the data traveling between layers and a serializer/deserializer module. This module runs in parallel with the IPs executing in each cluster and in order to exchange a message between clusters, the message should pass through the CI first. On the receiver node, the CI module forwards the message to the destination IP unit.

In order to make all clusters execute independently from each other motivated the usage of buffers to temporarily store the messages. The main idea is to make the message exchange overhead smaller as possible in the application level, working in a pipeline fashioned way. Thus, when an IP is communicating with another IP that is not in the same cluster, it sends the message to the CI module and after continues its regular execution, while the CI module performs the rest of the message delivering. Figure 2 presents the fifo employed in the CI, where it is possible to see the circular infrastructure used.

Fig. 2. Cluster interface circular FIFO overview.

The optical layer is responsible for exchanging messages between each cluster and is organized in a torus topology. Its design is based on a 5×5, strictly non-blocking optical router [19]. Figure 3 illustrates the router internal organization where it is possible to see the 16 micro-ring resonators (MRs), six

waveguides, and two waveguide terminators. The MRs in the switching fabric are identical, and have the same on-state and off-state resonance wavelengths.

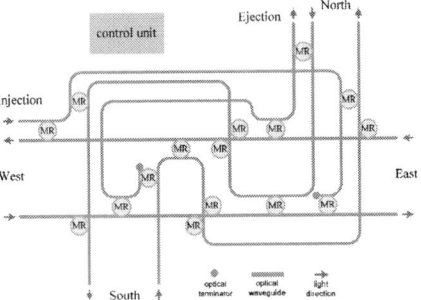

Fig. 3. Optical router organization [19]

Each cluster is composed by a variable number of IPs and each cluster can be configured with one frequency. Each cluster is connect to one optical router through the CI. Figure 4 presents the schematic overview of the HTM, where both optical and electrical layers are illustrated. In the Figure, each cluster is highlighted with a doted box and is composed by 25 routers, in a 5×5 NoC mesh-topology. Each electrical router in the electrical network is represented by a small circle. Still, the optical network is illustrated, where each optical router is represented as a big circle, with the CR inscriptions inside. For the sake of better viewing the CI was omitted in the Figure.

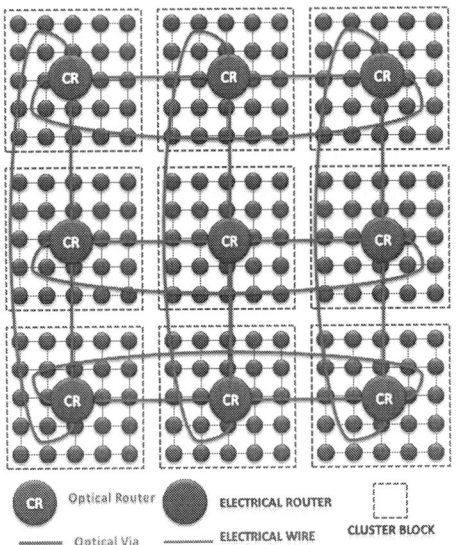

Fig. 4. HTM topology schematic overview.

IV. Low Latency Arbiter

The performance and efficiency of OIN-based architectures can be constrained by their controllers. Long setup time of circuit-switching techniques make them not practical and, at the same time, centralized controllers have been successfully demonstrated [13], thus this is the model we are going to use.

That being said, following the blocks that compose the arbiter are presented:

- **conflicts resolution block (CRB)**: this block is responsible for detecting destination conflicts and solving them by the usage of a given algorithm;
- **memory (LUT)**: is used to store static data accessed by the controller during run-time. This memory is used mainly to reduce computation time, thus reducing control latency, and;
- **dynamic setup (DSB)**: block responsible for on-line calculations, like path attribution and memory addresses reading, by the usage of a real-time calculation (RTC) unit.

A. Path Analyzer and LUT Creation (PALC)

The Path Analyzer and LUT Creation (PALC) block is responsible for evaluating the path diversity of network topologies by analytically analyzing them and generate the memory-arrays to be used as addressing lists. In order to create the table, the Dijkstra algorithm [26] is used to compute the shortest possible routes. Every single possible communication scenario is evaluated in this stage and all of them are stored as look-up tables.

B. Conflict Resolution Block (CRB)

The CRB is a hardware block responsible for detecting conflicts (or contention) in the targeted IPs. A conflict is defined as any situation in which two or more source IPs are targeting the same destination IP, simultaneously.

The CRB works in two steps: first, it analyzes all requests, looking for a conflict, and; second, if a conflict is found, a Round-Robin (RR) algorithm is applied to define which IP will have its accessed granted. The matrix method works by checking all matrix's columns and, for the cases where more than one output is marked as one, it sets a conflict. Following, two matrices are presented : in the matrix to the left **A** is targeting **C**, **B** targets **A**, **C** targets **D** and **D** targets **B**, hence no conflicts. in the matrix to the right, **B** targets **A** but **D** is also targeting **A**, so a conflicting situation is found.

$$
\begin{bmatrix} 0 & 0 & 1 & 0 \\ 0 & 0 & 1 & 0 \\ 0 & 0 & 1 & 0 \\ 0 & 0 & 1 & 0 \end{bmatrix}, \begin{bmatrix} 0 & 0 & 0 & 1 \\ 0 & 0 & 1 & 0 \\ 0 & 1 & 0 & 0 \\ 1 & 0 & 0 & 0 \end{bmatrix}
$$

After, all columns j of the matrix are checked to find conflicting points, such as:

$$
\forall j \in \mathcal{M}, \quad \neg XOR(j) \wedge OR(j) \implies conflict(j) = 1.
$$

Furthermore, the request computation runs parallelized, in which each possible request port is considered as one running process. It made possible for the arbiter to receive requests, solve conflicts and grant access to the network with a low latency.

C. Dynamic Setup Block (DSB)

The LUT is created based on the possible paths a message may take on the network. It stores the route each message should follow through the network in order to reach its destination. As it would be very costly to store every single combination of requests and targets, the Dynamic Setup Block (DSB) is used. The DSB realizes real-time calculation of paths configuration based on a minimized LUT version, the LITE-LUT. The LITE-LUT stores only portions of the network paths, like the configuration of one switch, so the LUT usage does not turn into an overhead.

V. RESULTS

This section presents the results obtained using the arbiter and its application in the HTM architecture. Firstly, only the arbiter was simulated and also FPGA prototyped. The FPGA board was integrated with fabricated switches in order for the arbiter to be validated in realistic scenarios. Later, its latency time was compared against state-of-art works in order to analyze their impacts on the system execution. Finally, it was integrated with the HTM architecture and the performance evaluated.

A. Arbiter Validation and Comparison

In order to validate and evaluate the arbiter, FPGA prototyping was performed where the FPGA board was integrated with fabricated MZI-switches[1] [25]. Also, it was validated through VHDL-based simulations, where we used values extracted from fabricated devices. The arbiter was tested under the injection of different traffic patterns, like all-to-all, all-to-one and compliment. The design was synthesized using the proposed flow for the STMicroelectronics 65nm technology process.

The synthesis process was used in order to verify the minimum time period possible. To do so, different network sizes' arbiters were configured: 8 inputs, 16 inputs, and 32 inputs. The average minimum delay obtained for this step was $\approx 1.4ns$.

Figure 5 presents the simulation waveform of a 64×64 topology. In the presented scenario, the input ports are configured to target output ports using a compliment pattern, except for the input port 1 that is targeting output port 1[2]. This configures a conflict, as two ports target the same output. In the Figure, it is possible to see that it takes one clock cycle for a request (rx) to be acknowledged (ack). Still, the conflicted port waits for the end of previous communication (tail), which leads to a no conflict situation, and then has its request granted as well.

By having the latency measured, the real impact of the controller latency on a system was verified and compared with state of art work. For a fair comparison, the well known 8×8 Beneš [5] topology was used. Also, taking as base a fabricated optical switch, the latency for each optical bit to pass through the network was rounded to 200 ps. The comparison was

---

[1]*Mach-Zehnder Interferometer* (MZI) is a device used to control the amplitude of an optical wave by diving it in two, applying a given delay and then merging the two beans of light into one [18].

[2]the following mapping is performed, illustrated in the Figure by the signal DEST: *1 → 1, 2 → 63, 3 → 62 ... 62 → 3, 63 → 2, 64 → 1*

978-1-5090-2737-8/16 $31.00 © 2016 IEEE

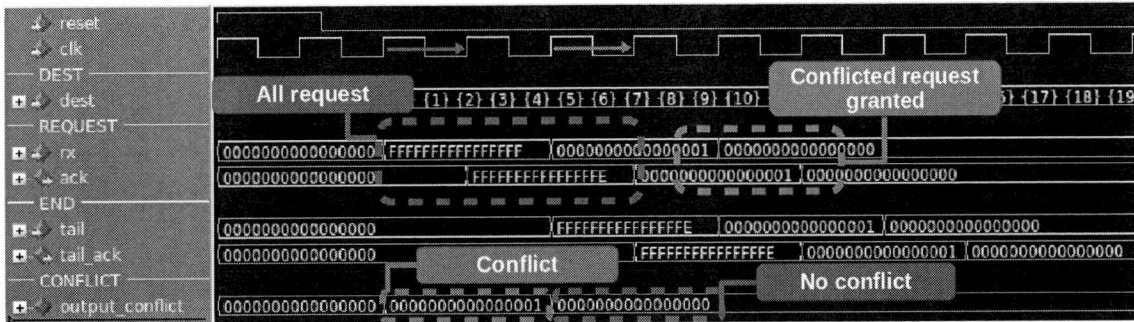

Fig. 5. Arbiter simulation waveform.

performed by analyzing the total time it takes for a message to be arbitrated and pass trough the network, such as:

$$TotalTime = CL + Nob * TD, \quad (1)$$

where CL stands for control latency, Nob stands for number of bits transmitted and TD stands for transmission delay. Still, four different message sizes (128 B, 256 B, 512 B, 1 Kb) were used.

Figure 6 presents the latency comparison with four state of the art solutions [22], [21], [17], [34]. The Figure shows that the arbiter latency is comparable to the fastest presented ones. However, it still contains differences that put them apart from each other. The solution presented in [22] claims to use an operation frequency of 5 GHz, which is not realistic, so much that all validations were under simulations only. Further, the provided solution in [17] was validated through FPGA prototyping, with similar latency to LUCC. Nevertheless, its usage imposes modifications on the application network layer, which is not always possible, thus reducing its applicability. Finally, the approach used in [34] uses the same time division technique as this work, and obtained fairly similar results. The solution is suited for a specific topology, jeopardizing its usage for other cases.

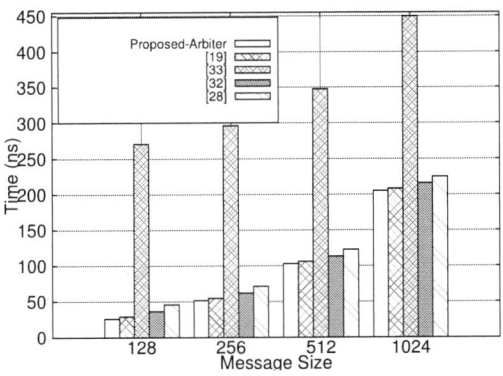

Fig. 6. Latency comparison.

B. HTM Validation and Comparison

The Hybrid TORUS MPSoC was validated through simulation and different traffic patterns were adopted. In order to evaluate the HTM performance, different cluster configurations were used, presented on Table I. In the Table, it is possible to see four columns: *Total I/Os* defines the total number of in/out ports in the network; *OIN Nodes* shows the number of optical routers, which is equal to the number of clusters in the system; *Cluster Nodes* holds the number of inner nodes for each cluster, and; *NoC Size* determines the intra-cluster NoC size for each cluster.

TABLE I
HTM SIZES

Total I/Os	OIN Nodes	Cluster Nodes	NoC Size
36	4	9	3x3
100	4	25	5x5
196	9	49	7x7
324	9	81	9x9
576	9	144	12x12
900	9	225	15x15

The traffic injection was configured to insert data using different traffic patterns, such as all-to-all and compliment. Also, the injection rate was configured to the maximum frequency allowed by the components, in order to obtain the maximum available throughput. Still, different messages sizes were adopted. By analyzing the traffic scenarios, we extracted the average latency for messages to be delivered, where the latency is measured as the time between the moment the first flit is inserted into the sender and the last flit leaves the receiver.

Results show that the communication capability of the HTM matches state of art solutions, where the average bit latency is found to be $\approx 0.91ns$ and the worst-case latency $\approx 1.09ns$. Based on the bits latency, the total throughput for each channel, and lastly for the entire network was measured.

The Figure 7 presents the obtained results for the HTM under different traffic patterns and messages sizes. The results are presented in a normalized form and are based on the measured latency and three different traffic patterns: compliment, local and non-uniform. It is possible to see that the HTM suits better for the compliment pattern, due the fact that the HTM is designed to improve the distant communications, leaving the local traffic for the eNoCs.

978-1-5090-2737-8/16 $31.00 © 2016 IEEE

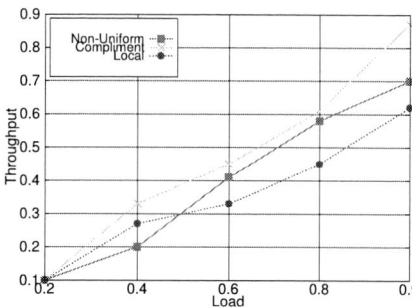

Fig. 7. Normalized throughput graph.

VI. CONCLUSION

This work presented the Hybrid TORUS MPSoC, a cluster-based architecture for future multiprocessor systems. It takes advantage of Optical Integrated Networks' high bandwidth in order to perform long communications and the already well deployed Networks-on-chip for shorter communications. Also, an arbiter to be used in the proposed architecture was introduced. Obtained results showed a fast response time when employing the arbiter in OINs. Also, the impact on latency was tested against state-of-art works and showed that the proposed solution proved to be more efficient in most cases. Finally, the HTM's performance was presented, showing the high bandwidth obtained.

REFERENCES

[1] International technology roadmap for semiconductors, http://www.itrs.net/ - last access on 12/2014.

[2] Amba ahb reference, url: http://alturl.com/88d98, last access April, 2014.

[3] A. Aguiar et al,. Embedded virtualization for the next generation of cluster-based mpsocs. In *Rapid System Prototyping (RSP)*, 2011.

[4] A.V. Rylyakov , et al. Silicon Photonic Switches Hybrid-Integrated With CMOS Drivers. *IEEE Journal of Solid-State Circuits*, 2012.

[5] V. Beneš. On rearrangeable threestage connecting networks. pages 1481–1492. Bell Syst. Tech. J., 1962.

[6] L. Benini and G. De Micheli. Powering networks on chips: energy-efficient and reliable interconnect design for socs. In *Proceedings of the 14th international symposium on Systems synthesis*, New York, NY, USA, 2001. ACM.

[7] B.G. Lee , et al. Monolithic Silicon Integration of Scaled Photonic Switch Fabrics, CMOS Logic, and Device Driver Circuits. *Journal of Lightwave Technology*, 2014.

[8] J. Chan and K. Bergman. Photonic interconnection network architectures using wavelength-selective spatial routing for chip-scale communications. *Optical Communications and Networking, IEEE/OSA Journal of*, 2012.

[9] D. M. Chapiro. *Globally-asynchronous Locally-synchronous Systems (Performance, Reliability, Digital)*. PhD thesis, Stanford, CA, USA, 1985. AAI8506166.

[10] D. Melpignano et al,. Platform 2012, a many-core computing accelerator for embedded socs: Performance evaluation of visual analytics applications. In *Design Automation Conference (DAC), 2012 49th ACM/EDAC/IEEE*, 2012.

[11] D.K. Hunter et al,. Buffering in optical packet switches. *Lightwave Technology, Journal of*, 1998.

[12] F. Magalhaes et al,. Embedded cluster-based architecture with high level support - presenting the hc-mpsoc. In *Rapid System Prototyping (RSP)*, 2014.

[13] Fei Lou , et al. Towards a centralized controller for silicon photonic MZI-based interconnects. In *Optical Interconnects Conference - paper WD4*, 2015.

[14] Fernando Moraes et al,. Hermes: an infrastructure for low area overhead packet-switching networks on chip. *Integr. VLSI J.*, 2004.

[15] Geng Luo-feng et al,. Performance evaluation of cluster-based homogeneous multiprocessor system-on-chip using fpga device. In *Computer Engineering and Technology (ICCET)*, 2010.

[16] H.A. Khouzani et al,. Fully contention-free optical NoC based on wavelenght routing. In *Computer Architecture and Digital Systems (CADS)*, 2012.

[17] Haijun Yang , et al. Design of Novel Optical Router Controller and Arbiter Capable of Asynchronous, Variable length Packet Switching. In *Photonics in Switching, PS*, 2006.

[18] P. Hariharan. *Basics of interferometry*. Elsevier Academic Press.

[19] Huaxi Gu et al,. A low-power low-cost optical router for optical networks-on-chip in multiprocessor systems-on-chip. In *ISVLSI'09 - 2009*.

[20] Hui Li et al,. A hierarchical cluster-based optical network-on-chip. In *Future Computer and Communication (ICFCC), 2010 2nd International Conference on*, May 2010.

[21] Junhui Wang , et al. A Highly Scalable Butterfly-Based Photonic Network-on-Chip. In *Computer and Information Technology (CIT)*, 2012.

[22] Z. Li and T. Li. ESPN: A case for energy-star photonic on-chip network. In *Low Power Electronics and Design (ISLPED), 2013 IEEE International Symposium on*, 2013.

[23] Luying Bai et al,. A cluster-based reconfigurable optical network on chip design. In *Photonics and Optoelectronics (SOPO), 2012 Symposium on*, May 2012.

[24] M. Renaud , et al. Transparent optical packet switching: The European ACTS KEOPS project approach. In *IEEE Lasers and Electro-Optics Society Annual Meeting*, 1999.

[25] M.S. Hai et al. MZI-based non-blocking soi switches. In *Asia Communications and Photonics Conference 2014 - paper ATh3A.147*.

[26] N. Jasika , et al. Dijkstra's shortest path algorithm serial and parallel execution performance analysis. In *Proceedings of the 35th International Convention MIPRO*, 2012.

[27] P. Ramanathan et al,. Clock distribution in general vlsi circuits. *Circuits and Systems I: Fundamental Theory and Applications, IEEE Transactions on*, May 1994.

[28] S. Pasricha and N. Dutt. *On-Chip Communication Architectures: System on Chip Interconnect*. Morgan Kaufmann Publishers Inc., San Francisco, CA, USA, 2008.

[29] S. Pasricha and N. Dutt. Orb: An on-chip optical ring bus communication architecture for multi-processor systems-on-chip. In *Design Automation Conference, 2008. ASPDAC 2008. Asia and South Pacific*, pages 789–794, March 2008.

[30] S. Rhoads. Mips plasma, url: http://opencores.org/project, last access July, 2014.

[31] Ruiqiang Ji et al,. Five-port optical router based on microring switches for photonic networks-on-chip. *Photonics Technology Letters, IEEE*, 2013.

[32] S. Le Beux et al,. Optical Ring Network-on-Chip (ORNoC): Architecture and design methodology. In *Design, Automation Test in Europe Conference Exhibition (DATE), 2011*, 2011.

[33] M. Seifi and M. Eshghi. A clustered noc in group communication. In *TENCON*, 2008.

[34] M. Shoaib. Selectively weighted multicast scheduling designs for input-queued switches. In *Signal Processing and Information Technology, 2007 IEEE International Symposium on*, 2007.

[35] T. Sakamoto et al. Variable optical delay circuit using wavelength converters. *Electronics Letters*, 2001.

[36] X. Tan, M. Yang, L. Zhang, X. Wang, and Y. Jiang. A hybrid optoelectronic networks-on-chip architecture. *Lightwave Technology, Journal of*, 32(5):991–998, March 2014.

[37] Tota, S. et al,. A multiprocessor based packet-switch: performance analysis of the communication infrastructure. In *Signal Processing Systems Design and Implementation, 2005. IEEE Workshop on*, 2005.

[38] M. Tudruj and L. Masko. Dynamic smp clusters with communication on the fly in soc technology applied for medium-grain parallel matrix multiplication. In *Parallel, Distributed and Network-Based Processing, 2007. PDP '07. 15th EUROMICRO International Conference on*, 2007.

[39] Xin Jin et al. Fpga prototype design of the computation nodes in a cluster based mpsoc. In *Anti-Counterfeiting Security and Identification in Communication (ASID), 2010 International Conference on*, 2010.

[40] Y. Liu , et al. All-optical buffering using laser neural networks. *Photonics Technology Letters, IEEE*, 2003.

A Security Aware Routing Approach
for NoC-based MPSoCs

Ramon Fernandes[1], César Marcon[2], Rodrigo Cataldo[3], Jarbas Silveira[4], Georg Sigl[5], Johanna Sepúlveda[6]

[1,2,3]Pontifícia Universidade Católica do Rio Grande do Sul (PUCRS)

[5,6]Institute for Security in Information Technology, Technical University of Munich (TUM)

[5]Fraunhofer Research Institution for Applied and Integrated Security (AISEC)

[4]LESC-DETI, Federal University of Ceará (UFC)

Email: {ramon.fernandes[1],rodrigo.cataldo[3]}@acad.pucrs.br, cesar.marcon@pucrs.br[2],
jarbas@lesc.ufc.br[4], sigl@tum.de[5], johanna.sepulveda@tum.de[6]

Abstract—Malicious applications target Multi-Processors System-on-Chip (MPSoCs) to capture sensitive information or disrupt normal operation; therefore, security is now a design requirement for MPSoC design. Network-on-Chip (NoC) is a key communication structure to aid in the overall MPSoC protection. Firewall-based NoC protection allows data exchange monitoring and controlling according to the MPSoC security policy. Secure NoCs enable to detect and prevent a broad range of software-based attacks. However, complex security policies may turn firewalls costly. This paper proposes a protection technique based on the NoC routing algorithm. By manipulating the routing of packets, security zones can be built. Our routing algorithm prioritizes communication among paths deemed secure while guaranteeing deadlock freedom. We evaluate the scalability of the proposed technique using synthetic and real application scenarios, as well as the security of the proposed technique.

Keywords—MPSoCs; NoCs; routing algorithm; security;

I. INTRODUCTION

The next generation of Multi-Processors System-on-Chip (MPSoCs) will integrate hundreds of Intellectual Property (IP) modules, such as processors, memories and other application specific components into a single chip. MPSoCs promise to achieve high performance and flexibility [1]. The high communication parallelism of several applications targeting MPSoC architectures has turned the Network-on-Chip (NoC) as the more suitable interconnection structure [2][3], providing a reliable and scalable communication infrastructure [4].

Data is exchanged in the NoC as packets, which are transmitted by a set of routers and links. The router is the underlying communication fabric of the system, switching packets from its inputs ports to the output ports. The links perform the interconnection between routers, and the routing algorithm defines the selection of the output port of the router. The final configuration of the NoC should satisfy the performance and cost requirements of the system while preventing undesirable behaviors, such as deadlocks. IPs are linked to the routers through the Network Interface (NI), responsible for packing and unpacking data sent or received through the NoC. Each IP module has a unique address in the system used by other IPs for message exchanging.

Since MPSoCs have become a trend in the industry as a major design solution for the increasing demands of mobile

platforms, such as the Internet of Things (IoT) [5], the concern for security has also gained increasing relevance in MPSoC design. Such devices may perform critical tasks, as well as store and process sensitive and private information [6].

Attacks at MPSoC aim to extract sensitive data, modify the system behavior or denial the system operation (*Denial-of-Service*, *DoS*) [7]. Software attacks account for 80% of security incidents in embedded systems, often preceded by abnormal communication events [8]. Software attacking techniques are getting more complex and effective. The resource sharing in MPSoC is widely exploited; thus, the computation and communication of sensitive data must be isolated [7]. However, to enhance the performance of the MPSoC, the designer usually spreads the applications on the computation resources, forcing the sensitive traffic through the NoC. Aiming to protect the MPSoC, security services can be implemented in the computation structure by using encryption techniques to avoid plain data transmission among IP modules [9]. Security may also be integrated into the communication structure, employing firewalls to monitor the traffic, detect abnormal communication behavior in the system and isolate sensitive traffic [10].

Security zones at NoCs are used to wrap IPs and protect the sensitive flows from attackers. Isolation of traffic avoids that attackers capture data-dependent traffic characteristics of critical tasks, thus revealing sensitive data [7]. Firewalls are used in [8] to create security zones. However, complex security policies may turn firewalls costly [11]. This work proposes for the first time the utilization of NoC routing algorithm for implementing security zones. Considering the security configuration of applications mapped in the MPSoC, the routing algorithm is able of establishing safe communication paths. We show that the routing algorithm can be an efficient and scalable alternative for promoting the data protection in the NoC. The novelties of this work are:

- Implementation of a routing algorithm technique, based on *Region-based Routing* algorithm (*RBR*) method [12], which is aware of the security requirements of the system, defined by security zones;
- The exploration of the proposed routing technique regarding scalability and secure communication paths, consid-

978-1-5090-2737-8/16 $31.00 © 2016 IEEE

ering the *Segment-based Routing* (*SBR*) deadlock prevention method [13].

This papers organization is as follows: Section II presents related work on NoC-based MPSoC security. Section III discusses the considered threat model. Section IV presents the proposed secure routing mechanisms and algorithms. Section V shows the evaluation criteria. Section VI analyzes the results. Finally, Section VII concludes our work.

II. RELATED WORK

Previous works have shown that the integration of security services at NoCs may aid in the overall MPSoC protection. Firewall-based protection is the most common strategy to implement NoC-based MPSoC security. Such hardware structures are integrated into the NI and NoC components to filter data according to the security policy of the system. According to the updating capabilities of the security tables that store the policy, firewalls can be classified as static or dynamic. Static firewalls are used in [6], while reconfigurable firewalls, able to update the security policy, are used in [8][14][15].

In [6][14][15][16], firewalls are integrated at the initiator and destination NIs. Those firewalls check different information embedded into the packet. Source and privileges are checked in [6][14][16]. Address ranges and read/write operation requested by IPs are checked in [15][16]. Firewalls can be integrated into the NoC, as for instance, the work of [8] that proposes 3D-ACeNoC, integrating firewalls at the NIs and between consecutive routers. By restricting the NoC traffic, the firewalls wrap IPs inside a security zone. Components inside the same security zone are trusted and considered secure among them. The concept of security zones is adopted in this paper.

Communication scheduling for security is explored in the three-dimensional NoC of [16]. The vertical interconnection based on *Through-Silicon-Vias* (*TSVs*) greatly suffer from coupling effects, turning this technology vulnerable to integrity and operational problems. Malicious traffic flows could promote electro-migration effects that may alter packets in adjacent TSVs. Thus, the packets are scheduled in the different TSVs to avoid sensitive data interference.

Despite the good results regarding protection of the sensitive content, none of the approaches mentioned above attempt to deal with system protection at NoC routing level, which is the approach of this work that applies routing strategies to implement the security zones in the MPSoC.

III. THREAT MODEL

The MPSoCs considered in this work are composed of a set of IPs interconnected by a shared NoC. Parallel applications are split into smaller pieces of code, called tasks, and mapped into different and several IPs to enhance system performance. For sensitive applications, such strategy forces the communication of sensitive data through the NoC. The sensitive communication between a pair of IPs, that execute the critical application, is called sensitive path. Fig. 1 shows an MPSoC

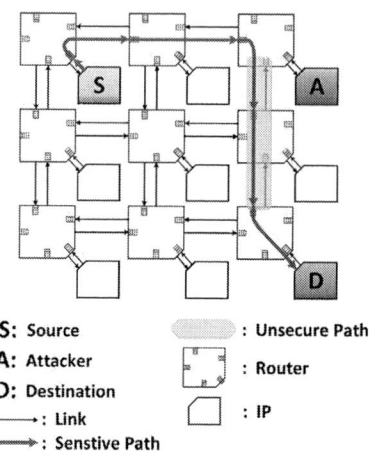

S: Source
A: Attacker
D: Destination
———▸ : Link
═══▸ : Sensitive Path
▭ : Unsecure Path
▢ : Router
▢ : IP

Fig. 1. A source process S communicates with a destination D over an insecure path due to a DoS or Timing Attack from an attacker A to D.

that includes 9 IPs, highlighting a sensitive communication between the IPs *S* (source) and *D* (destination).

The attacker performs a software attack infecting an IP through a malicious code. We assume that the NoC is secure. The malicious program is loaded into the MPSoC and executed by an IP. It can be done by downloading malwares directly from the Internet to the chip or by modifying the external memory used by the MPSoC to store applications. Depending on the attack type (extraction of sensitive data), it may be desirable to retrieve data from the MPSoC, which requires that the attacker infects an IP with the right to use the peripherals. An infected IP *A* (attacker) is shown in Fig. 1.

To increase the efficiency of the attack it is desirable that the infected IP be located inside the sensitive path. The attacker must know the MPSoC mapping strategy and the NoC routing algorithm. This latter requirement is not mandatory for some attacks of *Denial-of-Service*, whose goal is to disrupt the system operation by overloading the resources.

The attacks considered in this work are the timing and *DoS* attacks, previously described by [7][17][18][10]. Timing attacks use the communication collision between the sensitive traffic and the attacker A request in order to reveal a secret. Data dependences of critical applications are reflected in the traffic pattern [18]. DoS attacks are performed by flooding the NoC resources that are used by the sensitive path with useless communications.

Both attacks exploit the collisions and interferences of the sensitive traffic. Security zones arise as an alternative for isolating sensitive traffic inside protected environments. IPs that belong to the security zone are considered secure. Selecting a path that avoids the router linked to the malicious A core and the possible malicious paths aids to mitigate the attack. In this paper, we propose a deadlock-free routing algorithm that maximizes the encapsulation of the sensitive path inside a security zone. Such strategy aids to decrease the attacks.

978-1-5090-2737-8/16 $31.00 © 2016 IEEE

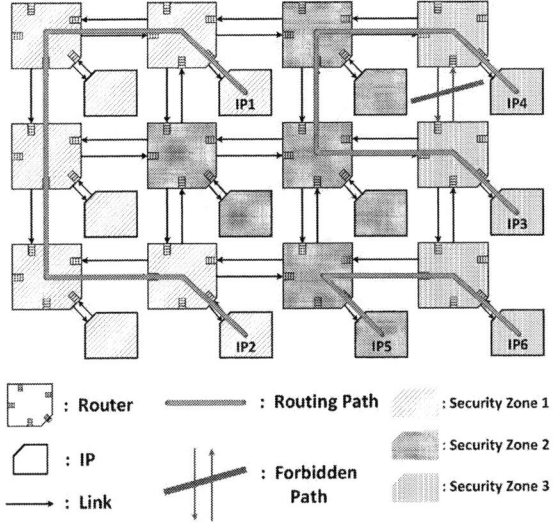

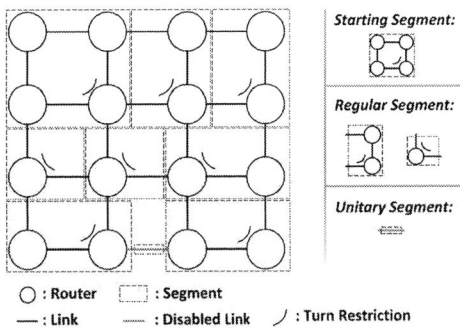

Fig. 3. Segments and turn restrictions computed by the *SBR* algorithm in a 4x4 2D Mesh NoC (based on [13]).

Fig. 2. The three communication scenarios with security zones: *FIZ* (IP1 to IP2); *PIZ* (IP3 to IP4); and *IZ* (IP5 to IP6).

IV. SECURITY AWARE ROUTING

This work proposes *SBR Security Zone Awareness (SBR-SZA)*, an alternative segment computation heuristic that is aware of the security characteristics of the system. The secure routing approach is based on two concepts: (i) *Security Zones*; and (ii) *Routing Algorithm*. These concepts and their utilization are described in the following two Subsections.

A. Security Zones

A security zone SZ is a physical space (continuous or disrupted) that wraps the IPs that execute critical applications. IPs that belong to the security zone are considered trusted among them [11]. The task mapping of critical applications inside the MPSoC defines the shape of the security zone.

A set of IP cores (IP) that executes a critical application defines a security zone SZ, such that the elements $p_i, p_j \in IP$ are considered secure and trusted. A transaction from p_i to p_j, where $i, j \in [0, N-1]$ with N representing the total amount of IP blocks in the system, is called sensitive and must be performed inside SZ. However, typical NoC routing algorithms may force the route of the sensitive path outside the SZ. Three communication scenarios are shown in Fig. 2:

- **Full intra-zone communication (FIZ)**: S and D are in the same SZ. The sensitive path is **completely** inside the SZ, e.g., the path from *IP1* to *IP2*;
- **Partial intra-zone communication (PIZ)**: S and D are in the same SZ. However, the sensitive path is **partially** inside the SZ. **PIZ** takes place in disrupted security zones and for irregular SZ shapes, when typical routing algorithms force out the sensitive traffic, e.g., the path from *IP3* to *IP4*;
- **Inter-zone communication (IZ)**: S and D are in different SZ, e.g., the path from *IP5* to *IP6*.

FIZ communication is the most secure situation, as sensitive communications are contained in secure elements of the

system. Whenever possible, traffic flows should adhere to this model. **PIZ** occurs when the security zone is fragmented, or the routing algorithm forbids a communication path, usually to avoid a route that would lead to a routing deadlock. Lastly, **IZ** should occur by communications among distinct applications in the system.

B. Routing Algorithm

Searching for secure paths on MPSoCs demands greater flexibility [11]. As shown in Fig. 2, establishing secure paths inside the SZ requires in some cases non-minimal paths, e.g., the path from *IP1* to *IP2*. It heavily depends on the shape of the security zone.

Several routing algorithms have been studied before in the areas of high performance and fault-tolerant MPSoCs. *Segment-based Routing (SBR)* [13] and *Region-based Routing (RBR)* [12] have been used in conjunction to efficiently find non-minimal paths. *SBR* is responsible for deadlock prevention while *RBR* computes the routing entries.

SBR is composed of two phases: (i) segment computation; and (ii) placement of routing restrictions. At segment computation, *SBR* partitions the NoC into segments comprising routers and links. As shown in Fig. 3, each segment is characterized by a turn restriction that avoids routing deadlocks. *SBR* classifies segments into three types: (i) *starting*, which starts and ends at the same router forming a loop; (ii) *regular*, which starts at a link, contains at least one router, and ends on another link; and (iii) *unitary*, which contains a single link that does not allow traffic. *SBR* aims to create segments that minimize the number of elements per segment in order to reduce the occurrence of *unitary* segments.

When placing routing restrictions, *SBR* defines that each segment can contain a localized turn restriction that best suits the routing of elements in that segment. Globally, *SBR* guarantees deadlock freedom and connectivity among all elements in the network; i.e., the turn restrictions still allow communication among all source-destination pairs.

RBR takes the turn restrictions, computed by *SBR*, to find paths between all origins and destinations in the NoC. As a result, the routing entries for each router are generated. The main advantage of *RBR* is that a single routing entry

978-1-5090-2737-8/16 $31.00 © 2016 IEEE

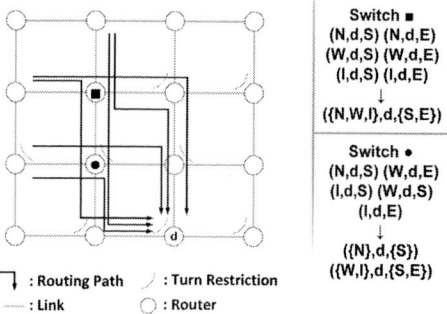

Switch ■
(N,d,S) (N,d,E)
(W,d,S) (W,d,E)
(I,d,S) (I,d,E)
↓
({N,W,I},d,{S,E})

Switch ●
(N,d,S) (W,d,E)
(I,d,S) (W,d,S)
(I,d,E)
↓
({N},d,{S})
({W,I},d,{S,E})

⌐ : Routing Path   / : Turn Restriction
---- : Link   ○ : Router

Fig. 4. Paths computed with *RBR* algorithm to a destination d from two different source switches (based on [12]).

can represent a path to more than one destination, which can reduce the size of routing tables significantly.

RBR computation occurs in three steps. The first step is the *routing computation* from each NoC router to every other router. The path-selection is performed according to the designer goals and system requirements. The found paths are stored in the routing table of the router. They are represented as a 3-tuple (N, d, S), where N is the packet input port, d the destination address, and S is the output port.

The second step is the *region computation*, where multiple entries are joined based on the input and output port values. Fig. 4 shows the paths computed by *RBR* algorithm to a destination **d** from two different source routers. The entries (N, d, S), (W, d, S) and (I, d, S) of router ■, that have the same set of output ports for the same destination, can be grouped. As a result, a single routing entry $(\{N, W, I\}, d, S)$ is stored. Analogously, the 3-tuples with the E output port can be packet into $(\{N, W, I\}, d, E)$. Further packing can be done within the grouping results, leading to a single entry $(\{N, W, I\}, d, \{S, E\})$. This result represents an adaptive routing, as more than one output port exists to reach the same destination.

The third step, *region merge*, merges overlapping regions in order to reduce the amount of routing entries.

RBR defines regions in the NoC by grouping entries based on their destination. A single entry can represent routing entries that have the same set of input and output ports to distinct destinations.

Regarding *the computational cost* of *SBR*, the segment computation has a cost of $O(m)$, while placing routing restrictions has a cost of $O(s)$, where m is the number of links in the NoC and s the number of segments. Moreover, for *RBR*, the cost of checking all sources to every possible destination is $O(n^2)$, where n represents the amount of routers in the NoC. It becomes important to use an efficient pathfinding algorithm to alleviate these costs, such as *Dijkstra*'s shortest path or *A**.

C. Security Zone Awareness (SBR-SZA)

SBR capabilities can be used to compute the segments and turn restrictions based on the security zones of the MPSoC. Fig. 5 shows two cases of segment computation for the same six routers of the MPSoC. Depending on the NoC segments

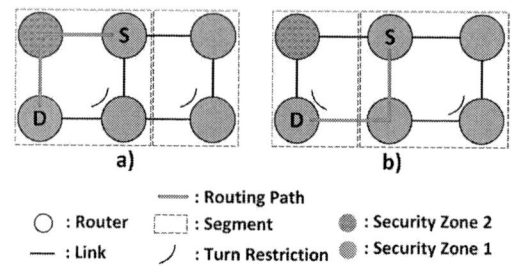

····· : Routing Path
○ : Router   [___] : Segment   ● : Security Zone 2
---- : Link   ⌐ : Turn Restriction   ◉ : Security Zone 1

Fig. 5. Two *SBR* segment computations: a) the path among S and D goes through an insecure element due to a routing restriction; b) a set of restrictions in the segments enables a secure path between S and D.

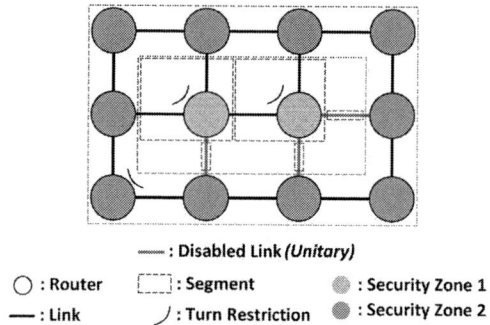

---- : Disabled Link *(Unitary)*
○ : Router   [___] : Segment   ◉ : Security Zone 1
---- : Link   ⌐ : Turn Restriction   ● : Security Zone 2

Fig. 6. Segmentation example with *SBR-SZA*. While *PIZ* scenarios do not occur, the segmentation results in *unitary* segments.

computed by *SBR*, the communication path between S and D can be either *PIZ*, as shown in Fig. 5a; or *FIZ*, as shown in Fig. 5b.

SBR-SZA uses the *SBR* algorithm in order to compute NoC routes that favor the occurrence of *FIZ* scenarios. The segments are tailored to a security zone so that *SBR-SZA* creates the smallest possible segments that contain elements from the same security zone.

While *SBR-SZA* should favor the occurrence of *FIZ* scenarios, a performance impact is a possibility since the segmentation can lead to a greater occurrence of *unitary* segments, as Fig. 6 illustrates.

D. Modeling of Security Zones

We model the MPSoC as a graph. Vertexes correspond to routers and their associated IPs, while edges to links between routers. Each vertex belongs to a security zone, and each edge has a positive weight that is relative to the pathfinding iteration.

When computing the path from a source router to a destination, the weight of edges adjusts to favor paths to other routers from the same security zone as the source. In Fig. 7, there are three paths from S to D. The topmost path traverses an insecure element in the context of S, so it has a higher edge cost than edges that lead to a router of the same security zone of S. The middle path traverses two secure routers, reaching D. Meanwhile the bottommost path has a routing restriction that forbids traffic from S to D.

Determining the edge cost is about the system requirements. It could represent the cost to protect a package while traversing

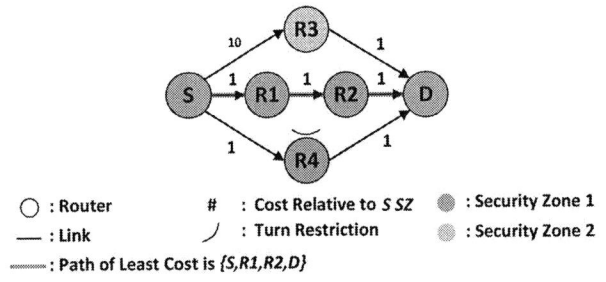

○ : Router   # : Cost Relative to *S SZ*   ● : Security Zone 1

── : Link   ⌡ : Turn Restriction   ● : Security Zone 2

┄┄ : Path of Least Cost is *{S,R1,R2,D}*

Fig. 7. Modeling of the paths from *S* to *D*. Traversing a different security zone leads to a higher path cost.

TABLE I
SUMMARY OF OBTAINED RESULTS

		Scenario		
	Algorithm	NAS	Synth1	Synth2
Increase in Routing Entries per Router	SBR	16.50%	9.42%	15.57%
	SBR-SZA	16.56%	4.52%	14.30%
Coefficient of Variation in Routing Entries per SBR Seed	SBR	1.01%	2.47%	2.42%
	SBR-SZA	1.02%	5.37%	2.83%
Minimum PIZ Scenarios	SBR	230	0	0
	SBR-SZA	198	0	0
Coefficient of Variation in PIZ Scenarios per SBR Seed	SBR	18.21%	117.95%	89.95%
	SBR-SZA	18.63%	115.14%	76.22%

to an insecure zone by a cryptographic module, or the cost of passing a firewall. The spectrum of possibilities to determine the edge cost is too wide and not a focus of study in this paper. Meanwhile, it suffices to say that our model defines the cost of traversing an insecure element in such a way that it is always preferable to route inside a secure zone.

V. EVALUATION

In this work, we are mainly concerned with two aspects of our proposed secure routing approach: (i) the scalability of the *RBR* routing tables; and (ii) to minimize *PIZ* routing scenarios. For these purposes, we compute some *SBR* segmentation configurations by varying the initial router (seed) used for segment computation. This approach results in different turn restriction placement (shown in Fig. 5) along the NoC and therefore different routing tables for each configuration and *FIZ/PIZ* distributions.

We consider two synthetic scenarios and another based on a real application benchmark mapped to an MPSoC. The first synthetic scenario (*Synth1*) consists of a 6x4 NoC with four security zones; the second synthetic scenario (*Synth2*) has a 10x6 NoC with ten security zones. Both scenarios only contain continuous security zones, and aim to reproduce the situations depicted in Fig. 2, Fig. 5 and Fig. 6.

The real application consists of the *NASA Numerical Aerodynamic Simulation* (*NAS*) Benchmark [19]; a 13x13 NoC contains the mapping of five applications used by the *NAS* benchmark using a *Simulated Annealing* mapping algorithm. Each application contains 32 tasks, and we consider that each task fully occupies an IP core in the system. Tasks from an application characterize the security zones, totaling five zones for the *NAS* scenario.

Additionally for the *NAS* scenario, we also estimate the latency impact of using security zones with *SBR* and *SBR-SZA*. Assuming that packets which are traversing an insecure zone require encryption, we evaluate for the six block ciphers techniques in [20] as the protection mechanism.

VI. RESULTS

Due to the modeling of security zones with non-minimal distance routing paths, it is expected that additional *RBR* routing entries are necessary to accommodate routes within security zones. Fig. 8 illustrates the average distribution of routing entries per router in each scenario. Considering a *Base-NoC* with no security zones, there is an average increase of 16.50% and 16.56% of routing entries per router for the *NAS* scenario with *SBR* and *SBR-SZA*, respectively. Scenarios *Synth1* and *Synth2* also incur in an overhead for the required routing entries to accommodate security zones, as Table I shows.

Regarding the *PIZ* and *FIZ* communication scenarios, Fig. 9 illustrates the distribution of *PIZ* communications for each scenario per *SBR* seed. *FIZ* and *PIZ* scenarios are inversely proportional. A configuration with zero *PIZ* scenarios means all communication occurs as *FIZ*.

As Fig. 9 and Table I show, there is no ideal secure configuration for the *NAS* scenario for any seed used in *SBR*; however, *SBR-SZA* yields more secure communication paths than *SBR*. Both *Synth1* and *Synth2* can be configured in a way that guarantees only *FIZ* routing with either *SBR* or *SBR-SZA*.

In both evaluations changing the seed used by *SBR* or *SBR-SZA* yields different results, as expected. While the routing tables present a small coefficient of variation when changing the seed, the occurrence of *PIZ* and *FIZ* scenarios can vary greatly due to the segmentation process. The proposed *SBR-SZA* model can also result in smaller routing tables and increase *FIZ* routing scenarios in some situations, although its benefits depend on the shape of the security zones.

Considering the *NAS* scenario, Fig. 10 illustrates the variation in latency when compared to a NoC without security zones. The choice of the encryption technique can have a significant impact on the communication latency, with **HIGHT** encryption having $< 1.0\%$ performance penalty, while **AES-128** can more than double the average communication latency. As shown in Fig. 10, *SBR-SZA* also incurs in greater communication latency than standard *SBR*, which is expected as *SBR-SZA* can generate more *Unitary* segments, decreasing link availability and creating longer communication paths.

VII. CONCLUSION

This work proposes a routing technique aware of security requirements in the system. As information security and availability become a new design requirement in MPSoCs, the exploration of different system elements for security purposes, such as the routing algorithm, presents alternative approaches to aid in the overall MPSoC security.

The proposed technique is sufficient to address the identified threat model with a small to moderate overhead in *RBR* routing

978-1-5090-2737-8/16 $31.00 © 2016 IEEE

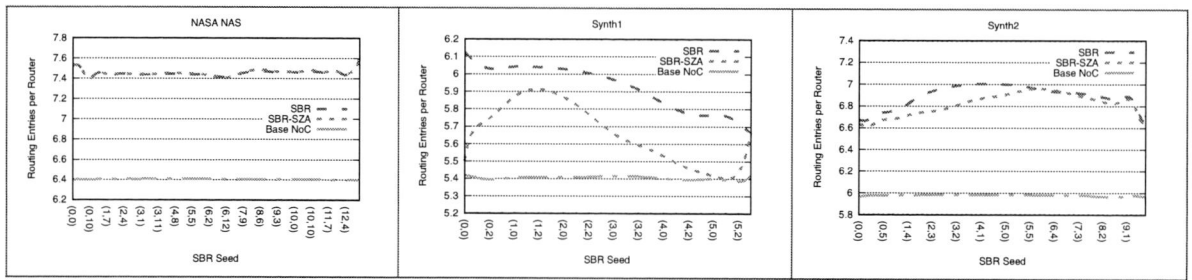

Fig. 8. Average distribution of routing entries per router per *SBR* Seed, using *SBR* and *SBR-SZA*, compared to a Base-NoC without security zones.

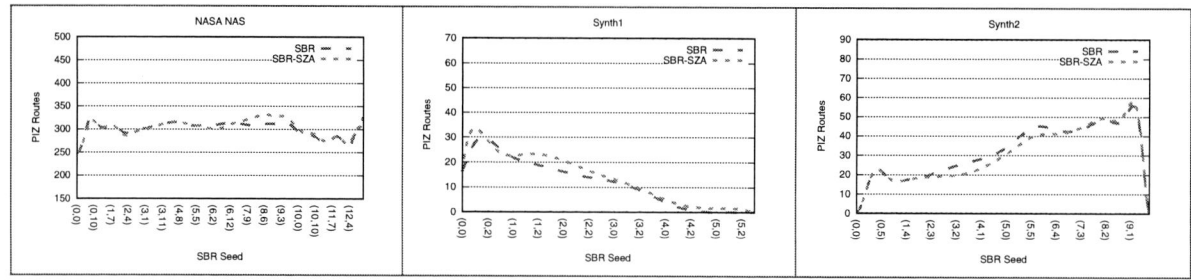

Fig. 9. Average distribution of **PIZ** routes per *SBR* Seed, using *SBR* and *SBR-SZA*. When zero **PIZ** routes occur, it characterizes a configuration with only **FIZ** communication.

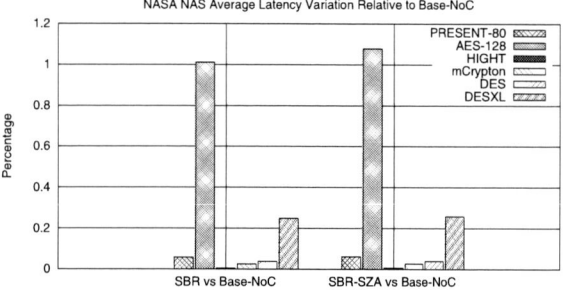

Fig. 10. Average latency variation for *NAS* scenario with different data protection mechanisms.

tables. *SBR-SZA* can also benefit certain security scenarios while providing configurations with routing contained to a security zone, albeit at a slightly larger application performance impact. Further studies regarding different applications can also yield a broader understanding of the proposed model.

REFERENCES

[1] ITRS. *International Technology Roadmap for Semiconductors*. 2015. URL: http://www.itrs.net/reports.html (visited on 11/29/2015).

[2] M. Sgroi et al. "Addressing the System-on-a-chip Interconnect Woes Through Communication-based Design". In: Design Automatic Conference (DAC). 2001, pp. 667–672.

[3] Terry Tao Ye et al. "Packetization and Routing Analysis of On-chip Multiprocessor Networks". In: *J. Syst. Archit.* 50.2-3 (Feb. 2004), pp. 81–104.

[4] L. Benini et al. "Networks on chips: a new SoC paradigm". In: *Computer* 35.1 (Jan. 2002), pp. 70–78.

[5] P. Greenhalgh. "big.LITTLE Processing with ARM Cortex-A15 and Cortex-A7". In: (2011).

[6] L. Fiorin et al. "Secure Memory Accesses on Networks-on-Chip". In: *IEEE Transactions on Computers* 57.9 (Sept. 2008), pp. 1216–1229.

[7] M. J. Sepúlveda et al. "NoC-Based Protection for SoC Time-Driven Attacks". In: *IEEE Embedded Systems Letters* 7.1 (Mar. 2015), pp. 7–10.

[8] J. Sepulveda et al. "Elastic security zones for NoC-based 3D-MPSoCs". In: *Electronics, Circuits and Systems (ICECS)*. Dec. 2014, pp. 506–509.

[9] D. M. Ancajas et al. "Fort-NoCs: Mitigating the threat of a compromised NoC". In: *Design Automation Conference (DAC)*. June 2014, pp. 1–6.

[10] MJ Sepúlveda et al. "An Hybrid Switching Approach for NoC-based Systems to avoid Denial-of-Service SoC Attacks". In: *16th Iberchip Wksp (IWS 2010)* (2010), pp. 23–25.

[11] J. Sepulveda et al. "Reconfigurable security architecture for disrupted protection zones in NoC-based MPSoCs". In: *Reconfigurable Communication-centric Systems-on-Chip (ReCoSoC), 2015 10th International Symposium on*. June 2015, pp. 1–8.

[12] J. Flich et al. "Region-Based Routing: An Efficient Routing Mechanism to Tackle Unreliable Hardware in Network on Chips". In: *Networks-on-Chip (NOCS)*. May 2007, pp. 183–194.

[13] A. Mejia et al. "Segment-based routing: an efficient fault-tolerant routing algorithm for meshes and tori". In: *Parallel and Distributed Processing Symposium (IPDPS)*. Apr. 2006, pp. 105–115.

[14] Ramon Fernandes et al. "A Non-intrusive and Reconfigurable Access Control to Secure NoCs". In: IEEE International Conference on Electronics, Circuits and Systems (ICECS). Nov. 2015, pp. 316–319.

[15] M. D. Grammatikakis et al. "Security Effectiveness and a Hardware Firewall for MPSoCs". In: *High Performance Computing and Communications (HPCC)*. Aug. 2014, pp. 1032–1039.

[16] J. Sepulveda et al. "3D-LeukoNoC: A dynamic NoC protection". In: *ReConFigurable Computing and FPGAs (ReConFig)*. Dec. 2014, pp. 1–6.

[17] Yao Wang et al. "Efficient Timing Channel Protection for On-Chip Networks". In: *IEEE/ACM Sixth International Symposium on Networks-on-Chip*. 2012, pp. 142–151.

[18] J. Sepulveda et al. "Dynamic NoC Buffer Allocation for MPSoCs Timing Side Channel Attack Protection". In: *Latin American Symposium on Circuits and Systems (LASCAS)* (2016), p. 4.

[19] NASA. *NAS Parallel Benchmarks*. 2016. URL: http://www.nas.nasa.gov/publications/npb.html (visited on 03/31/2016).

[20] A. Bogdanov et al. "PRESENT: An Ultra-Lightweight Block Cipher". In: *Cryptographic Hardware and Embedded Systems (CHES)* (2007), pp. 450–466.

MagPDK: an Open-Source Process Design Kit for Circuit Design with Magnetic Tunnel Junctions

Raphael M. Brum and Gilson I. Wirth
Programa de Pós-Graduação em Engenharia Elétrica (PPGEE)
Universidade Federal do Rio Grande do Sul (UFRGS)
Av. Osvaldo Aranha, 103 – 90035-190 Porto Alegre, RS, Brazil
brum@ufrgs.br, gilson.wirth@ufrgs.br

Abstract—In this paper, we introduce MagPDK, an open-source process design kit that represents a magnetic tunnel junction fabrication process. Based on the widely known open design kit FreePDK, MagPDK extends it by adding magnetic tunnel junction devices on top of the standard CMOS transistors. The core of this design kit includes physical design verification rules, connectivity and parasitic extraction decks and a set of parametric cells representing variations of the MTJ device. Electrical simulations can be performed by integrating MagPDK with external MTJ compact models. For this purpose, wrappers are provided for three open-source models available in the literature. A basic set of non-volatile standard-cells and memory bit-cells is supplied with the package to illustrate its capabilities. All of the elements in the PDK were developed using de-facto standard formats, such as OpenAccess, LEF and Spice, and can be seamlessly employed within industry-standard digital, analog or mixed-signal design flows. In addition, MagPDK is backward-compatible with any FreePDK-based layout. With MagPDK, which is the only open-source design kit for magnetic processes available to date, we provide a common environment for researchers, circuit designers and EDA developers, so that they can explore the MTJ-based circuit design space not only in the behavioral level, but in the physical implementation level as well.

I. Introduction

The tech industry has evolved in a non-stopping, exponential pace for the past decades, as predicted by Gordon Moore fifty years ago. Though the semiconductor industry is still able to keep this rate, the end of Moore's Law quickly approaches, as transistors dimensions are just a few generations away from atomic scale.

In parallel with the efforts to move the CMOS technology forward, industry and academia are also investing on alternative technologies, which promise to excel where CMOS now struggles to deliver the required performance-to-power ratio. Among numerous contenders, the magnetoresistive random-access memories (MRAM) appear as a suitable replacement for Static (SRAM) and Dynamic (DRAM) technologies, presenting fast read and write access times, near-zero leakage power and practical integration with CMOS processes. On top of that, MRAM is non-volatile, a characteristic not currently present in the first levels of memory hierarchy; exploiting this new feature is being subject of several publications [1].

The basic building block of MRAM is the Magnetic Tunnel Junction (MTJ), a current or magnetic-field driven

programmable resistor. Apart from memory arrays, MTJ applications span to magnetic sensors, hard-disk read heads and even logic circuits. Furthermore, research on magnetoresistive technology crosses several domains, such as (a) developing the manufacturing process on itself, (b) uncovering the physics behind the device, (c) developing innovative circuits that explore the features of MTJs and (d) developing new tools to automate the design of circuits containing thousands to millions of these devices. Bridging the gap between these fields is difficult, as the manufacturing processes are experimental and often undisclosed to the community in general.

In this work, we introduce the first open-source, freely distributed process design kit (PDK) dedicated to hybrid CMOS/Magnetic processes. We devised a magnetic process that mimics the features of state-of-the-art magnetic tunnel junctions published in the literature. With this kit, we expect to contribute to the efforts of designing such circuits by providing:

- a framework for physical design exploration of magnetic memory arrays and general-purpose, magnetoresistive-based logic circuits;
- a reference technology for comparing non-volatile circuit topologies;
- a reference design kit for testing new technology-specific design, simulation and verification methodologies;
- a step towards detecting shortcomings of current tools and standardizing the implementation of magnetic process design kits;
- compatibility with the CMOS FreePDK design kit, widely used by the community to validate circuits and tools targeted to standard CMOS technologies.

This paper is organized as follows: section II presents the MTJ device, along with the compact electrical models and design kits published so far in the literature; section III presents the proposed open-source design kit, detailing its current features; section IV provides an example of a read circuit designed and simulated using the reference flows provided with the design kit; finally, section V draws some conclusions and discusses a roadmap for future developments.

978-1-5090-2737-8/16 $31.00 © 2016 IEEE

II. BACKGROUND AND RELATED WORK

Typical electronic devices manipulate the charge of electrons to represent and process information. Spintronic devices, on the other hand, manipulate not only the electron charge, but also the spin associated with it. Magnetic Tunnel Junction (MTJ) are spintronic devices made of a stack of ferromagnetic layers separated by a thin insulator (Fig. 1a). They exploit a physical phenomenon called tunnel magnetoresistance (TMR), whereby electrons are able to tunnel through the insulator when the structure is subjected to a voltage.

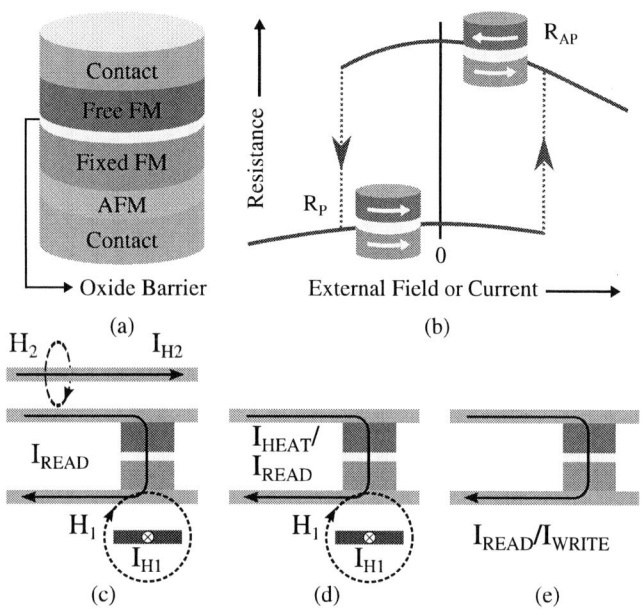

Fig. 1. Magnetic Tunnel Junctions: (a) A typical MTJ stack; (b) Characteristic resistance curve (the quantity on the x-axis depends on the MTJ variant); (c)FIMS-MTJ; (d) TAS-MTJ; (e) STT-MTJ.

Depending on the relative magnetization of these ferromagnets, MTJs may present more or less resistance to the current flowing through, as electrons flowing through the junction interact with the layers' magnetization. Practical implementations usually employ a reference ferromagnet (or fixed FM), whose magnetization cannot be easily reversed, and a free ferromagnet, which can be reoriented by an external input. This, in turn, results in two possible states (as in Fig. 1b): the anti-parallel state, which offers maximal resistance to the current flow (R_{AP}), and the parallel state, which offers minimal resistance (R_P).

From an electrical standpoint, MTJs are thus programmable resistors, which can assume either one of the two states mentioned above. Programming the MTJ can be done in a number of ways, which must be engineered in the device. The first generation (Fig. 1c) employed a field-induced magnetic switching (FIMS) approach, requiring two field-inducing currents (I_{H1}, I_{H2}) of about 15 mA. Thermally-assisted switching (TAS) appeared shortly thereafter; in this variant, the free layer is heated (Fig. 1d) by a heating current (I_{HEAT}), in order to become more sensitive to the external field induced by I_{H1}.

A third family, referred to as Spin-Transfer Torque (STT), exploits the magnetic momentum induced by the switching current (I_{WRITE}) passing through the MTJ (Fig. 1e) to reverse the free layer orientation.

Regardless of the variant, several works demonstrated that standard CMOS circuits can be fully integrated with MTJs [2], [3]. This is commonly done by depositing the MTJ stack over the CMOS metalization, in a magnetic post-process that takes over after the CMOS back end of line (BEOL) reaches the top metal. Integrating these processes is still a challenge, as the copper BEOL is performed under high temperatures, whereas MTJs may be degraded when subjected to temperatures above 400°C [4].

As the fabrication process evolves from experimentation, more and more magnetic devices can be integrated altogether. Tools used in the design flow must be capable of understanding the process limitations and abide to well-defined rules in order to produce a manufacturable circuit. In what follows, we review the process design kits that appeared so far in the literature.

A. Process Design Kit

A Process Design Kit (PDK) is a set of files that encode technology-specific information in a format that Electronic Design Automation (EDA) tools can understand and process. This information spans to the three design domains (physical, structural and behavioral), and may comprise, for instance, geometric design rules that must be obeyed to produce valid masks, layout-to-netlist recognition rules, electrical, thermal or electrical models for behavioral simulation and pre-designed and pre-characterized blocks.

In the past years, a number of complex MTJ-based circuits have consistently been reported as successfully manufactured. Whereas the layout of single devices can be handcrafted by designers, many-device integration requires automation and, thus, an established design flow, which surely employs hybrid CMOS/Magnetic PDKs. Some publications on practical prototypes or innovative circuits [5]–[7] briefly mentions the physical implementation flow and/or the electrical simulation tools and models used in the process. However, details on the implementation of these design kits, in general, remain undisclosed to the scientific community. Table I compares the features of the present work with the only three hybrid CMOS/MTJ design kits currently disclosed to the public. Each criterion will be explained in more detail in section III.

B. Electrical models

One of the most prominent features of a design kit is to provide an accurate model for electrical simulations. Virtually all CMOS PDKs include parameters for the BSIM4 model, currently the *de facto* standard for bulk CMOS technologies. No such standard exists for the MTJ devices, though. For this reason, compact electrical models with varying degrees of complexity and accuracy are being actively studied in the last years. Table IV (in the last page) presents some of these

978-1-5090-2737-8/16 $31.00 © 2016 IEEE

TABLE I
MAGNETIC PDKs PUBLISHED IN THE LITERATURE

Ref.	[8]	[9]	[10]	This work
CMOS node (nm)	130	130	28	45
MTJ node (nm)	90	120	200	50
MTJ family	STT	TAS	P-STT	STT
Design rules	✓	✓	✓	✓
MTJ parameter ext.	?			✓
Connectivity ext.	✓	✓	✓	✓
MTJ parasitics ext.	?			✓
Electrical model	[11]	[9]	[10]	*
Manufacturable	✓	✓	✓	
Publicly available				✓
Open-source				✓

* Extracted devices can be mapped to models [12]–[14].

works, along with their main features, which are explained in what follows.

Compact models can be created by one of the following *methods or languages*: (a) building an equivalent circuit, using known passive and active elements; (b) implementing the behavior in an analog description language, such as Verilog-A or VHDL-AMS, or any other customization interface provided by a given simulator; and by (c) coding the MTJ behavior directly into the simulator. This will impact on whether the model is *simulator-independent* or not.

The most accurate compact models rely on directly implementing the Landau-Lifshitz-Gilbert-Slonczewski (LLGS) differential equation, which is a model for the magnetization of the free layer under the influence of external magnetic fields and, in the case of STT, spin-polarized currents. Such an implementation captures the MTJ's *dynamic behavior*. When this is not the case, these models neglect the dynamic behavior and use an approximation for the transition time, in what could be called a semi-dynamic approach.

Some of these models account for the *thermal activation* effect, whereby the junction can randomly switch its magnetization depending on the temperature. The MTJ conductance is also influenced by the *temperature* and by the *voltage* applied to its terminals.

MagPDK was made to be compatible with three of the publicly available compact models already published in the literature. In the next section, we present the MagPDK 45/50 design kit and its main features.

III. DESIGN KIT FEATURES

MagPDK 45/50 is built on top of the widely-known NCSU FreePDK 45nm design kit [15]. All CMOS devices remain exactly the same, maintaining full backward-compatibility with FreePDK. The magnetic elements are built between the second and the third metal layers, and we assume that the further metalization is possible up to the tenth layer. The process cross-section is shown in Fig. 2a.

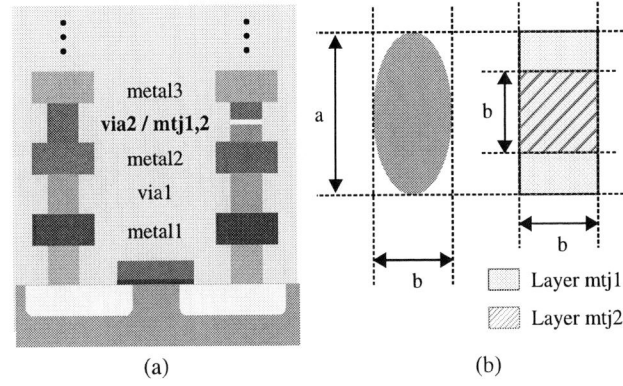

Fig. 2. (a) MagPDK 45/50 cross-section (up to metal layer 3); (b) Actual MTJ geometry and its practical two-layer layout representation.

TABLE II
LIST OF MAGPDK 45/50 MTJ-SPECIFIC DESIGN RULES

Rule	Description	Value (nm)
mtjn.1	Min. width mtjn	50
mtjn.2	Min. space mtjn	110
mtjn.3	Min. enclosure of metal2 around mtjn	10
mtjn.4	Min. enclosure of metal3 around mtjn	10
mtjn.5	Min. spacing between mtjn and via2	60
mtj1.6	mtj2 must be contained by mtj1	-

A. MTJ Parameter Extraction

Our target magnetic device is a STT-MTJ, being in line with most of the presented in table IV. We chose MTJ dimensions and characteristics similar to those published by Hitachi [16] in 2007. We assume an in-plane STT device, implying that the device's shape plays a critical role in its stability. An ellipsoidal shape (Fig. 2b) with major and minor axes a and b, respectively, is adopted. These dimensions may be exploited to provide further stability, at the expense of a larger switching current. When oversized MTJs are laid out using MagPDK, their dimensions can be captured by the netlist extraction tool, and later used in simulation analyses.

Due to limitations in the layout database formats, we chose to represent these shapes by a square of size b in layer mtj2, centered on a rectangle with sizes a and b drawn in layer mtj1. In an actual, manufacturable process, a mask post-processing procedure would then transform these shapes in the actual mask layouts used in the lithography process. For now, these procedures are not part of the design kit.

B. Design Rules

A basic set of MTJ-specific design rules was devised for this process. A graphical representation is provided by Fig. 3. The critical dimension of the device is enforced by the minimum width rules. Spacing rules were arbitrated according to the minimum metal-to-metal distances. The remaining values are presented in table II.

978-1-5090-2737-8/16 $31.00 © 2016 IEEE

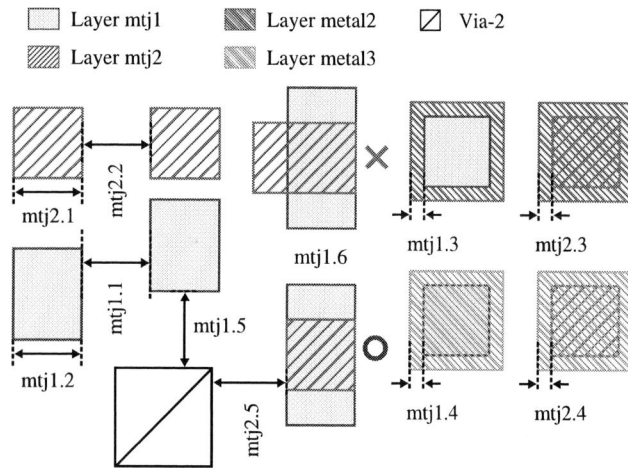

Fig. 3. Graphical representation of MagPDK 45/50 MTJ-specific design rules.

C. Connectivity extraction

Connectivity extraction is the procedure that recognizes the devices from the layout shapes. It is critical for the layout-vs-schematic (LVS) check, as well as for the parasitic extraction (PEX) flow. The LVS and PEX rule files include the rules for extracting the connectivity.

MTJs are seen as two-terminal devices: the topmost electrode corresponds to the positive terminal, and the bottommost electrode corresponds to the negative terminal. As switching the STT state depends on the current sense, these terminals cannot be interchanged. The device is formed by stacking shapes in layers metal2, mtj1, mtj2 and metal3. A DRC-correct MTJ layout will necessarily issue an well-formed device in the netlist.

D. MTJ Parasitic Extraction

The original FreePDK design kit provides parasitic extraction (PEX) rules for the basic CMOS device. Extraction rules are formulae to generate the parasitic RLC components from the layout masks of a particular design. The Calibre xRC deck supplied with the original FreePDK package has more than 50.000 lines. This file is not manually written, though: tools such as Cadence QRC Techgen and Mentor xCalibre generate these rules from a detailed technology description, known as stack-up file. This description is similar to the one shown in Fig. 2, but it provides the thickness, height and resistivity of conductive layers, as well as the thickness, dielectric constant and height of each insulating material in between.

Planar capacitors naturally rise from two metal shapes separated by a dielectric material, for instance. A complex layout may have billions of these interactions between its features. PEX-generation tools cross the stack-up information with predefined geometries and run them through a electrical field solver, which split the geometries into a mesh and estimates the electrical field on each point.

From this estimate, simpler rules, related to the layer length and width, absence of neighboring features in the layers above or below and others are derived. These are the rules found in the PEX deck usually supplied by the foundries.

MagPDK adds the MTJ layers to the standard FreePDK stack-up, introducing new parasitics that will relate to the intermediate steps introduced by the magnetic process. The current version of this PDK adds lateral dielectrics to the side of MTJs, but we intend to refine it further as more information regarding current manufacturing process becomes available. A single extraction corner (under typical conditions) is supplied with the design kit for now. With the stack-up descriptions, the user may generate new extraction corners as needed.

E. Electrical model and parameters

MagPDK is compatible with models Spinlib-STT [12], Harms et al. [13] and NVM-Spice [14], all of which are highlighted in table IV. This compatibility is ensured by a set of model-specific netlist extraction decks, which are able to convert the MTJ layout masks into the actual device, supplying the parameters needed by each model.

Modelcards supplied with MagPDK are based on the default parameters of Spinlib-STT model, which in turn were calibrated to match the Hitachi STT-MTJ process [17]. We then worked backwards to produce equivalent parameters for the other two models. For instance, Spinlib-STT calculates the parallel resistance (R_P) as a function of the resistance-area product (RA) of the material, the oxide thickness (tox), the energy barrier height (E_b) and the MTJ dimensions (a, b). The other two models simply require this quantity as a parameter. We then employ the formula used within the Spinlib-STT model to produce the equivalent parameter, such that

$$R_P = k_1 \frac{\text{tox} \cdot \text{RA}}{a \cdot b \cdot \sqrt{E_b}} e^{k_2 \text{tox} \sqrt{E_b}} \qquad (1)$$

where $k_1 = 3.009 \cdot 10^{-6}$ and $k_2 = 1.025 \cdot 10^{10}$ are fitting constants. A full list of parameters can be found in each of the respective model publications (see table IV). For the sake of simplicity, other model conversion rules are omitted from this publication. Further information regarding the conversion rules can be found in the PDK package.

While simulation plays an important role on the optimization of any circuit, attacking a sizing problem often starts by providing an educated guess, obtained through hand calculations over a simplified model. Similarly to what happens in the transistor realm, all of the compatible MTJ models present dozens of physics, process-related and geometric parameters. A more suitable set of parameters would relate directly to the effective resistance of the MTJ while being read and to the effective write current waveform required to successfully perform the operation. We refer to this set as *design parameters*. They can be derived as previously described in [7]. Table III presents the resulting design parameters for the minimum-sized MagPDK 45/50 device.

IV. DESIGNING WITH MAGPDK: BLACK & DAS CELL

To illustrate the capabilities of the MagPDK design kit, we detail here the implementation of the Black & Das circuit [18] under this technology. Its schematic is shown in Fig. 4.

TABLE III
RESULTING DESIGN PARAMETERS

Name	Description	Unit	Value
Vbd	Breakdown voltage	V	1.0
Vr	Read operation voltage	V	0.4
Rp(Vr)	Parallel resistance for V=Vr	Ω	2408.2
Rap(Vr)	Anti-parallel resistance for V=Vr	Ω	5881.5
t_{PW}	Minimum write pulse width	s	10n
$I_{W,P \to AP}$	Minimum write current $P \to AP$	A	353 μ
$I_{W,AP \to P}$	Minimum write current $AP \to P$	A	586 μ

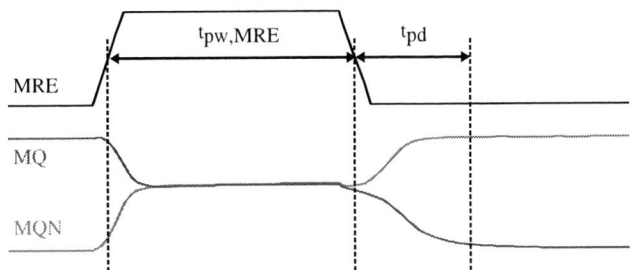

Fig. 5. Black & Das read pulse simulation, using Cadence Spectre and Spinlib-STT model.

It consists of a pair of cross-coupled inverters, whose outputs are unbalanced by a pair of MTJs holding complementary data. It is the self-referenced cell to be used as a mean to convert the magnetic information into a valid logic level. Activating MRE moves the circuit back to a metastable state, which brings the voltages of nodes MQ and MQN close to each other. Allowing the structure to fully reach the metastable state is essential to ensure proper operation. Once MRE is relaxed, MQ and MQN will converge to the stable state favored by the MTJ configuration.

A mask layout was drawn using Cadence Virtuoso and verified using Calibre DRC. A netlist was extracted using the deck appropriated for the Spinlib-STT model. A single read pulse simulation was performed over the extracted netlist, using Cadence Spectre. Graphical results are shown in Fig. 5. The input pulse width ($t_{PW,MRE}$) was set to 100 ps. Under typical conditions (Vdd=1.2, T=25°C), the response time (t_{pd}) after releasing the MRE pulse was 45.78 ps.

V. CONCLUSIONS AND PERSPECTIVES

This paper presented the first open-source, freely distributed hybrid CMOS/Magnetic process design kit. By using this tool, we expect to empower researchers in the field, by allowing them to (a) perform physical-level design exploration of spintronic circuits, (b) benchmark commercial and research-grade tools for this purpose and (c) to develop innovative

circuits even without having access to expensive experimental fabrication processes.

By sharing a common platform for the development of new EDA tools and new MTJ-based circuits, we expect to attract more interest to this area and to help enhance the design flows currently available. We believe that the design kit might have a role in the educational aspect as well, allowing students to get in touch with emerging technologies early on.

As soon as the MTJ manufacturing processes become more mature, designers will have a clearer picture of the new challenges impose by these devices. We expect to update MagPDK to better reflect these requirements, as well as to incorporate better electrical models and simulation tools in upcoming versions.

ACKNOWLEDGMENT

The authors would like to acknowledge the support of the Brazilian National Council for the Improvement of Higher Education (CAPES).

REFERENCES

[1] L. Torres, R. M. Brum, L. V. Cargnini, and G. Sassatelli, "Trends on the Application of Emerging Nonvolatile Memory to Processors and Programmable Devices," in *Circuits and Systems (ISCAS), 2013 IEEE International Symposium on.* IEEE Computer Society, May 2013, pp. 101–104.

[2] CMP, "CMOS / Magnetic Integration: Development and results at CMP," in *CMP annual users meeting*, January 2014. [Online]. Available: cmp.imag.fr/aboutus/slides/Slides2014/07_Magnetic_CMOS_2014.pdf

[3] K. Ikegami, H. Noguchi, C. Kamata, M. Amano, K. Abe, K. Kushida, E. Kitagawa, T. Ochiai, N. Shimomura, A. Kawasumi, H. Hara, J. Ito, and S. Fujita, "A 4ns, 0.9v write voltage embedded perpendicular stt-mram fabricated by mtj-last process," in *VLSI Technology, Systems and Application (VLSI-TSA), Proceedings of Technical Program - 2014 International Symposium on*, April 2014, pp. 1–2.

[4] K. Mizunuma, S. Ikeda, H. Sato, M. Yamanouchi, H. Gan, K. Miura, H. Yamamoto, J. Hayakawa, F. Matsukura, and H. Ohno, "Tunnel magnetoresistance properties and annealing stability in perpendicular anisotropy mgo-based magnetic tunnel junctions with different stack structures," *Journal of Applied Physics*, vol. 109, no. 7, pp. –, 2011. [Online]. Available: http://scitation.aip.org/content/aip/journal/jap/109/7/10.1063/1.3554092

[5] M. Natsui, D. Suzuki, N. Sakimura, R. Nebashi, Y. Tsuji, A. Morioka, T. Sugibayashi, S. Miura, H. Honjo, K. Kinoshita, S. Ikeda, T. Endoh, H. Ohno, and T. Hanyu, "Nonvolatile logic-in-memory array processor in 90nm MTJ/MOS achieving 75% leakage reduction using cycle-based power gating," in *Solid-State Circuits Conference Digest of Technical Papers (ISSCC), 2013 IEEE International*, Feb 2013, pp. 194–195.

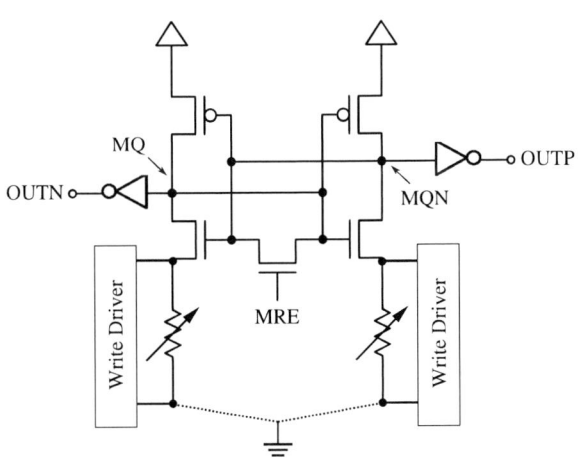

Fig. 4. Black & Das read circuit. [18]

TABLE IV
MTJ COMPACT ELECTRICAL MODELS PUBLISHED IN THE LITERATURE

Ref.	[12]	[13]	[19]	[14]	[20]	[21]	[22]	[9]	[10]
Identifier	Spinlib-STT	Harms	Mukherjee	NVM-Spice	NS-Spice	Madec	Vatankhah-ghadim	Spintec-TAS	Spinlib-PMA
MTJ family	STT	STT	STT	STT	STT	STT	STT	TAS	P-STT
Method or language	Verilog-A	Equivalent circuit	Equivalent circuit	Built-in LLGS	Built-in LLGS	VHDL-AMS	Verilog-A	C (CMI)	Verilog-A
Simulator-independent	✓	✓	✓			✓	✓		✓
Dynamic behavior				✓	✓	✓	✓	✓	
Thermal activation	✓	✓	✓	✓	✓		✓		✓
Voltage-dependent resistance	✓			✓	✓			✓	✓
Temperature-dependent resistance					✓			✓	
Publicly available	✓	✓		✓					✓
Open-source	✓	✓							✓

[6] H. Koike, T. Ohsawa, S. Ikeda, T. Hanyu, H. Ohno, T. Endoh, N. Sakimura, R. Nebashi, Y. Tsuji, A. Morioka, S. Miura, H. Honjo, and T. Sugibayashi, "A power-gated MPU with 3-microsecond entry/exit delay using MTJ-based nonvolatile flip-flop," in *Solid-State Circuits Conference (A-SSCC), 2013 IEEE Asian*, Nov 2013, pp. 317–320.

[7] B. Jovanovic, R. Brum, and L. Torres, "Comparative analysis of mtj/cmos hybrid cells based on tas and in-plane stt magnetic tunnel junctions," *Magnetics, IEEE Transactions on*, vol. 51, no. 2, pp. 1–11, Feb 2015.

[8] G. Di Pendina, G. Prenat, B. Dieny, and K. Torki, "A hybrid magnetic/complementary metal oxide semiconductor process design kit for the design of low-power non-volatile logic circuits," *Journal of Applied Physics*, vol. 111, no. 7, pp. –, 2012. [Online]. Available: http://scitation.aip.org/content/aip/journal/jap/111/7/10.1063/1.3680013

[9] M. El Baraji, V. Javerliac, W. Guo, G. Prenat, and B. Dieny, "Dynamic compact model of thermally assisted switching magnetic tunnel junctions," *Journal of Applied Physics*, vol. 106, no. 12, pp. –, 2009.

[10] Y. Zhang, W. Zhao, Y. Lakys, J. Klein, J.-V. Kim, D. Ravelosona, and C. Chappert, "Compact Modeling of Perpendicular-Anisotropy CoFeB/MgO Magnetic Tunnel Junctions," *Electron Devices, IEEE Tr. on*, vol. 59, no. 3, pp. 819–826, 2012.

[11] W. Guo, G. Prenat, and B. Dieny, "A novel architecture of non-volatile magnetic arithmetic logic unit using magnetic tunnel junctions," *Journal of Physics D: Applied Physics*, vol. 47, no. 16, p. 165001, 2014. [Online]. Available: http://stacks.iop.org/0022-3727/47/i=16/a=165001

[12] L.-B. Faber, W. Zhao, J. O. Klein, T. Devolder, and C. Chappert, "Dynamic compact model of Spin-Transfer Torque based Magnetic Tunnel Junction (MTJ)," in *Design Technology of Integrated Systems in Nanoscal Era, 2009. DTIS '09. 4th International Conference on*, April 2009, pp. 130–135.

[13] J. Harms, F. Ebrahimi, X. Yao, and J.-P. Wang, "SPICE Macromodel of Spin-Torque-Transfer-Operated Magnetic Tunnel Junctions," *Electron Devices, IEEE Transactions on*, vol. 57, no. 6, pp. 1425–1430, June 2010.

[14] Y. Shang, W. Fei, and H. Yu, "Fast simulation of hybrid CMOS and STT-MTJ circuits with identified internal state variables," in *Design Automation Conference (ASP-DAC), 2012 17th Asia and South Pacific*, Jan 2012, pp. 529–534.

[15] J. Stine, I. Castellanos, M. Wood, J. Henson, F. Love, W. Davis, P. Franzon, M. Bucher, S. Basavarajaiah, J. Oh, and R. Jenkal, "FreePDK: An Open-Source Variation-Aware Design Kit," in *Microelectronic Systems Education, 2007. MSE '07. IEEE International Conference on*, June 2007, pp. 173–174.

[16] T. Kawahara, R. Takemura, K. Miura, J. Hayakawa, S. Ikeda, Y. Lee, R. Sasaki, Y. Goto, K. Ito, T. Meguro, F. Matsukura, H. Takahashi, H. Matsuoka, and H. Ohno, "2Mb Spin-Transfer Torque RAM (SPRAM) with Bit-by-Bit Bidirectional Current Write and Parallelizing-Direction Current Read," in *Solid-State Circuits Conference, 2007. ISSCC 2007. Digest of Technical Papers. IEEE International*, Feb 2007, pp. 480–617.

[17] K. Tsuchida, T. Inaba, K. Fujita, Y. Ueda, T. Shimizu, Y. Asao, T. Kajiyama, M. Iwayama, K. Sugiura, S. Ikegawa, T. Kishi, T. Kai, M. Amano, N. Shimomura, H. Yoda, and Y. Watanabe, "A 64Mb MRAM with clamped-reference and adequate-reference schemes," in *Solid-State Circuits Conference Digest of Technical Papers (ISSCC), 2010 IEEE International*, Feb 2010, pp. 258–259.

[18] W. C. Black and B. Das, "Programmable logic using giant-magnetoresistance and spin-dependent tunneling devices," *J. Applied Physics*, vol. 87, no. 9, pp. 6674–6679, 2000.

[19] S. S. Mukherjee and S. Kurinec, "A Stable SPICE Macro-Model for Magnetic Tunnel Junctions for Applications in Memory and Logic Circuits," *Magnetics, IEEE Transactions on*, vol. 45, no. 9, pp. 3260–3268, Sept 2009.

[20] N. Sakimura, R. Nebashi, Y. Tsuji, H. Honjo, T. Sugibayashi, H. Koike, T. Ohsawa, S. Fukami, T. Hanyu, H. Ohno, and T. Endoh, "High-speed simulator including accurate MTJ models for spintronics integrated circuit design," in *Circuits and Systems (ISCAS), 2012 IEEE International Symposium on*, May 2012, pp. 1971–1974.

[21] M. Madec, J. Kammerer, F. Pregaldiny, L. Hebrard, and C. Lallement, "Compact modeling of magnetic tunnel junction," in *Circuits and Systems and TAISA Conference, 2008. NEWCAS-TAISA 2008. 2008 Joint 6th International IEEE Northeast Workshop on*, June 2008, pp. 229–232.

[22] A. Vatankhahghadim, S. Huda, and A. Sheikholeslami, "A Survey on Circuit Modeling of Spin-Transfer-Torque Magnetic Tunnel Junctions," *Circuits and Systems I: Regular Papers, IEEE Transactions on*, vol. 61, no. 9, pp. 2634–2643, Sept 2014.

Efficient Hardware Implementation of the Richardson-Lucy Algorithm for Restoring Motion-Blurred Image on Reconfigurable Digital System

Oscar Anacona-Mosquera, Janier Arias-García, Daniel M. Muñoz, and Carlos H. Llanos

Abstract—This work presents the hardware implementation of the RLA (Richardson-Lucy Algorithm) for image restoration task, in which the images are blurred by relative motion between camera and the scene. In this case the RLA was implemented in an FPGA-based platform using the hardware description language VHDL, and assuming the absence of additive noise in the capturing image system. The overall architecture is scalable from 3×3 to 9×9 mask sizes for the convolution steps of the RLA. The quality evaluation of the collected images was achieved using the SR-SIM (Spectral Residual Based Similarity) metric as well as by a visual verification of the images. The synthesis results and respective testing with real images are also presented in order to give support to video applications.

Keywords—*Richardson-Lucy Algorithm, Motion-Blurred Image Restoration, FPGA*

I. INTRODUCTION

Image degradations can occur due to mechanical movements of the camera during the acquisition process in a short exposure time of the scene and/or by the simultaneous movement of several objects in the scene (*motion blur*), in which each object can have different speeds; or due to the incidence of several effects involving lens characteristics, shutter mechanisms for controlling the exposure time, or even from effects of environment conditions [1].

From a theoretical perspective the blurring effects can be modeled by a mathematical operation called *convolution* in $2D$ [2]. In this kind of problems the convolution is done between a real image and a particular function called PSF (*Point Spread Function*), defining, in this case, a degradation [2]. The inverse process of convolution is applied to obtain the original image, which is known as *deconvolution*, being a mathematical process which reverts the effects of the PSF, given the estimation of a real image [1]. Generally, this process (the deconvolution) is part of restoration techniques.

Algorithms for image restoration are widely used in images that have been degraded during the capture process [2]. The

Oscar Anacona-Mosquera, Daniel M. Muñoz and Carlos H. Llanos are with the Department of Mechanical Engineering, University of Brasilia, Brasília, D.F., Brazil, 70910-900, Email:{oscar, damuz, llanos}@unb.br

Janier Arias-Garcia is with the Department of Electronic Engineering, Federal University of Minas Gerais, Belo Horizonte, MG, Brazil, 31270-901, Email: janier-arias@ufmg.br

Manuscript received Month xx, 2015; revised Month xx, 2015.

use of these algorithms has produced several research works in computer vision, above all due to the fact of the expensive computational cost for this kind of tasks [2]. The restoration process of images throughout the use of deconvolution operation presents a high computational cost, however the same allow the system to obtain images with fine structural details [1]. Otherwise, the same restoration process can be developed by other techniques such as: (a) previously knowing the PSF (*non-Blind deconvolution*), (b) by ignoring completely the PSF (*blind deconvolution*), and (c) knowing partially the PSF (*Myopic deconvolution*) [1].

The process of image restoration can be developed in an only stage (non-iterative methods) or in several stages (iterative methods). Among the possible approaches for solving the restoration problem iterative algorithms present a better estimation of the real image. Among the iterative algorithms there is one proposed by Richardson e Lucy in the 1970s, called RLA (*Richardson-Lucy Algorithm*), which is extensively used in astronomy and microscopy, distinguishing for being simple and easy to be implemented on software [1], [2]. The RLA is derived from the ML (*maximization likelihood principle*), assuming a Poisson model for noise distribution [2]. However, the RLA converges slowly and diverges after a certain number of iterations, in which the noise is increased during each new iteration, as a side effect of the restoration process [1], [2].

In this context, techniques to speed-up the RLA have been discussed in [1], [2], [3], which consist of reducing the number of iterations, and therefore increasing the convergence's speed of restoration task in each iteration. It is important remark that the RLA does not have a predefined criterion to restrict the number of iterations before the beginning of the restored image divergence [2]. Taking into account the performance aspect, the RLA uses two convolutions in each iteration. Therefore, RLA presents a high computational cost for embedded systems [1].

This paper presents an overall hardware architecture for the RLA implementation. The main contributions of this work are the following: (a) an scalable architecture for different size of masks for implementing the convolution operation; (b) the scalable convolution architectures allows for an efficient hardware implementation of the RLA for restoring images with different blurring level; (c) a quality evaluation of a restoration image example using SR-SIM metrics; (d) synthesis and performance results as well as comparisons with previous

978-1-5090-2737-8/16 $31.00 © 2016 IEEE

works implementing the RLA.

This work is organized as follows. Sec. II briefly presents the previous works of RLA hardware implementations. Sec. III presents theoretical aspects related to both image restoration and image quality metrics used in this paper. Sec. IV shows the proposed hardware architecture and before the conclusions Sec. V discusses the achieved results.

II. Related Works

There are some previous papers reporting implementations of $2D$ deconvolution, mainly for embedded system applications with real time requirements. Some of them have implemented the RLA to restore images addressing different computational platforms. In the case of mapping RLA on FPGAs (Field Programmable Gate Arrays) several approaches have been proposed based on the extraction of two separable filters in order to obtain a result corresponding to a non-separable filter response (see reference [2]). That method is based on the SVD (*Singular Value Decomposition*) and can be applied to an experimental PSF in order to calculate the coefficients of two FIR filters. Equation (1) shows the mathematical model used in [2] to implement the architecture.

$$I^{n+1} = I^n \cdot \left(PSF * \frac{B}{PSF * I^n} \right)^\beta, \qquad (1)$$

where PSF is the point spread function, B is the blurred image, I^n is the restored image in the n iteration, β is a constant for speeding up the convergence, and $*$ is the convolution operator. The exponent β is used with values among 1 to 3 (at the first iteration), yielding a decrement of the iteration number to be executed. That solution uses a 11×11 pixel mask and 640×480 pixels images at 30 fps achieving a throughput of 60 MP/s (millions of pixels per second) and a maximum frequency of 63 MHz in a Xilinx Virtex 2 XC2VP50.

In [1] the implementation of an image restoration algorithm, based on a sparse convolution matrix for time-variant PSFs [4], is proposed and mapped on a Stratrix V FPGA device. The deconvolution process uses two convolutions; therefore, it depends on the image sizes and on the PSF, having a high-level computational complexity. In [3] a DSP-based architecture using a Virtex-4 FPGA as coprocessor is proposed for image restoration. The algorithm is based on the equation (1) with $\beta = 1$. However, the hardware implementation was developed for frequency domain (using FFT-IFFT).

A method to improve the performance in software is proposed in [5], which estimates the movement vector using (1). This technique permits to calculate an iteration acceleration parameter (see β in (1)), increasing the convergence speed of restored image. The image processing is achieved by using a format given by the camera for an image resolution of 128×128 pixels. Nevertheless, an affirmation that the proposal does not present reliable results given a deviation of the trajectory estimation is presented in [6]. In that work, an implementation of the following algorithms is done in software: (a) RLA, (b) SGP (Scaled Gradient Projection), (c) OSEM (Ordered Subset Expectation Maximization). Apart from these developments, the results reported in [6] downscale the effects of edge problems in restored images. In this case, the implementations were developed using a GPU (CUDA) as well as a CPU.

Taking into account the related works about RLA, it can be observed that: (a) there is not reported in the literature a completed and scalable implementation of the RLA directly in hardware; (b) taking into account that the RLA is an iterative technique, there is not a proposal for an efficient architecture addressed to solve the RLA with its iterative characteristics and sensitivity to the motion-blur; (c) most of previous reports about restoration techniques have been addressed to GPUs and/or CPUs based platforms.

III. IMAGE RESTORATION

The image degradation model can be expressed as shown in (2).

$$g(x,y) = h(x,y) * f(x,y) + n(x,y), \qquad (2)$$

in which $f(x,y)$ is the original image, $g(x,y)$ is the noised and degraded image, $h(x,y)$ is the PSF, $n(x,y)$ is the additive noise, and $*$ is convolution operator [2]. Given that the noise always would be present, several simplifications are assumed over the basic restoration problem, in this case ignoring the additive noise (white and impulsive noises). In this context, image restoration uses an *a priori* information about degradations in such a way to make possible to apply the inverse process in order to remove the distortion.

A. Filtering by convolution

Convolution filtering is a basic step of the RLA, being used to modify the spatial frequency characteristics of an image. In this case, a matrix (called as *mask*) is applied to an image, and integer mathematical operations are achieved. The convolution operation works by determining the value of a central pixel by taking into account an image neighborhood and a mask. The output of the convolution is a new modified filtered image [7]. In order to apply a convolution, a neighborhood must be defined over the original image which depends on the sizes of the image and the mask. The mask is dislocated over the image area and when the end of an image row is reached the mask is dislocated a pixel down, reaching the beginning or the next row. The discrete image convolution is mathematically described as in (3)

$$g(x,y) = f(x,y) * h(x,y) = \sum_{s=\frac{-m}{2}}^{\frac{m}{2}} \sum_{t=\frac{-n}{2}}^{\frac{n}{2}} f(s,t) \cdot h(x-s,y-t),$$

$$(3)$$

where f is the input image, h is the mask, g is the filter output, and $*$ is the convolution operator. Typical mask sizes are 3×3 and 5×5, although higher mask sizes can be also used [2]. Almost any filtering function can be programmed in a mask, for instance smooth, correlation, border detection, correlation, etc, [7].

In this case, the RLA is a deconvolution method that consist of calculating an image close to the real one, starting from the observed image [2]. Notice that the image blurring produced by movement can be also modeled as a convolution between the image and a PSF [2].

In RLA is assumed a normalized PSF which gives an equation that can be integrative solved as presented by Wang [3] (see Equation 4).

$$O^{n+1}(x,y) = \left[\frac{I(x,y)}{(P * O^n(x,y))} * P^T(x,y) \cdot O(x,y) \right], \quad (4)$$

in which $P^*(x,y) = P(-x,-y)$, P is the PSF, P^T is the transpose of the PSF, I is the observed image, O^n is the estimate of the real image, and $*$ is the convolution operator [2].

It can be observed that the additive noise is amplified when each iteration is executed. A stop criteria can be established for the iteration number in order to decrease the quantity of amplified noise [3]. For instance, the quality of the restored image comparing with the original image could be a suitable criteria.

B. SSIM (Structural Similarity Index)

The SSIM metrics is defined as an objective metrics that estimates the image quality by using several aspects of the human perception. This metrics takes into account structural characteristics such as deviation, mean, and correlation [8]. Distortions can be originated from structural and nonstructural phenomena [9]. Among structural distortions it can be cited (a) blurring, (b) contamination by noise, (c) side effect from compression techniques such as JPEG, among others [8]. Nonstructural distortions are: (a) illumination changes, (b) contrast effects, (c) gamma distortion and (d) spatial changes [9]. A measurement of the structural change points out an approximation of the distortion degree of the acquired image [8]. SSIM comprises the use of three main components such as: (a) luminance, (b) contrast, and (c) structure, which are compared and combined to generate a similarity score between two images [8], [9].

The general expression of the SSIM index is given as shown in (5).

$$SSIM = \frac{(2\mu_x\mu_y + C_1)(2\sigma_{xy} + C_2)}{(\mu_x^2 + \mu_y^2 + C_1)(\sigma_x\sigma_y + C_2)}, \quad (5)$$

in which μ_x, μ_y, σ_x^2, σ_y^2 and σ_{xy} are the mean of x e y (the evaluating and reference images), the variance of x e y, and the cross-covariance of x e y respectively. The terms C_1 e C_2 are constants. The SSIM have different extensions, for instance it can be cited the MS-SSIM (*the multiscale SSIM index*), the IW-SSIM (*the information content weighted SSIM index*), the FSIM (*the feature similarity index*), and the SR-SIM (*Spectral Residual Based Similarity*) [10]. The SR-SIM has a low computational complexity, being based on the *Spectral Residual Visual Saliency*-SRVS [10]. The SR-SIM comprises two functions, namely: (a) to evaluate the

quality of the image regions, and (b) to create a ponderation to evaluate the prominence of an image area for the humans visual system. The SR-SIM permits object detections from the image background by using a specific visual saliency model map (spectral residual visual saliency). The expression for SR-SIM is defined as shown in equation 6.

$$SRSIM = \frac{\sum S_V \cdot [S_G]^\alpha \cdot max(R_1, R_2)}{\sum R_m}, \quad (6)$$

where R_1 e R_2 are the SRVS maps of the images, S_V and S_G are used for obtaining a similarity between two images. In [11] the SR-SIM metrics is considered a better evaluation, with low computational cost, being suitable to be implemented in an embedded system. Consequently, the SR-SIM can be used for real time applications such remote sensing, air recognition, computer vision in satellites, among others. Because this, the SR-SIM metrics will be used in our approach.

IV. HARDWARE ARCHITECTURE

The architecture is based on the RLA showed in 1, using masks and neighborhood circuits from 3×3 until 9×9. This algorithm starts with the information of the observed image (*B*), in a gray-level with a $m \times n$ size, and also with the (*PFS*), producing as output the restored image (I^{n+1}) from a given (I^n) image.

The masks used for the convolution with the image were calculated offline using the predefined *fspecial* function of the Matlab image processing toolbox, which produce bi-dimensional filters from the desired type of filter and the input parameters (for the case of blurring, x displacement and θ angle). Also, the additive noise (noise generated by the capture system) of the image model (see equation 2) was disregarding.

The RLA architecture is shown in Fig. 1, in which the input image is given such as a streaming in a gray scale, with a rate of one pixel for clock cycle. The input image is pre-multiplied by an α constant, providing the representation values to have the same integer format for the subsequent blocks. Then, the used mask values are stored in the internal memories of the FPGA, using also an integer representation. Finally, the result is divided by the same constant used in the pre-multiplication step, achieving the restoration image (see the input of the pos-division block in Fig. 1)

A. FPGA Implementation

Fig. 2 shows the approach used for loading the neighborhood to the convolution operation via multiplexing. The proposed system has to deal with the possibility of working with different mask sizes (3×3, 5×5, 7×7, and 9×9) due to the fact of the RLA needs to choose the mask size taking into account the motion-blur. For instance, if the blurring was produced by a motion-blur of 3 pixels the mask size must be of 3×3.

Being the RLA an iterative and a sequential algorithm, the proposed architecture was mapped into a pipeline of N stages in order to increase the parallelism level (see Fig. 3).

In this work, the output image size is equal to the input image size, which does not happen in other similar works [7],

978-1-5090-2737-8/16 $31.00 © 2016 IEEE

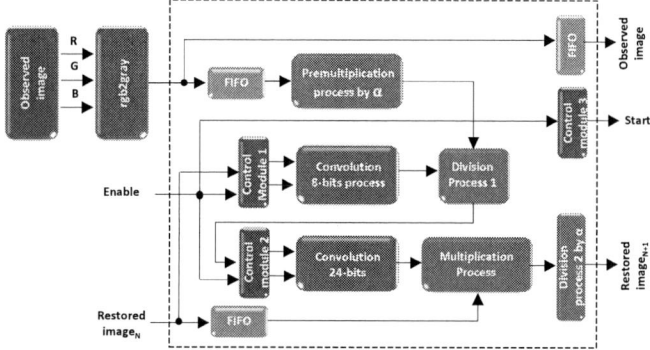

Figure 1: General Architecture of RLA

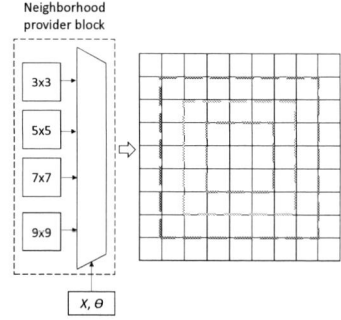

Figure 2: First approach for setting masks for PSF

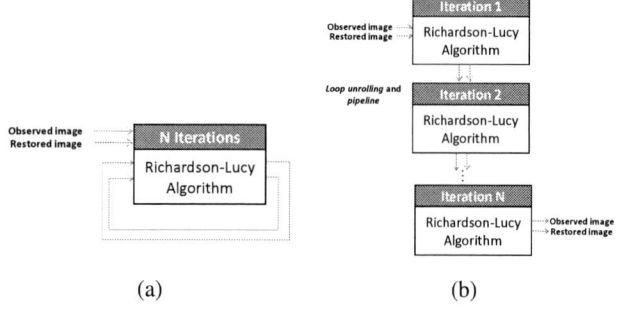

Figure 3: An unrolling loop of the RLA. (a) Software implementation and (b) hardware implementation

2) **The neighborhood-provider step:** it has to provide the values of the pixels of the image that will be processed, given a defined mask size.

3) **The convolution step:** it achieves the convolution operations, where the neighborhood values are multiplied by their respective mask weights, and added for obtaining an output pixel.

Fig. 4 shows the implemented convolution module for achieving the previously described steps, where each one of them were implemented with respective hardware blocks.

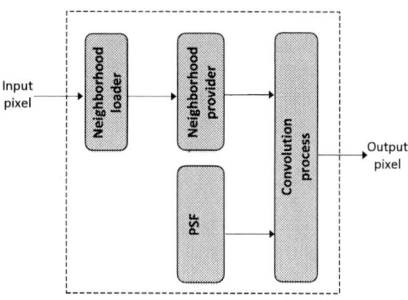

Figure 4: Convolution Filtering

The architecture used for the windowing operation is implemented by the *neighborhood-loader* block and shown in Fig. 5. The architecture is based on the mathematical definition of convolution filtering. However, instead of the mask displacement over the image the last is displaced through a pipeline. To perform this process in hardware it is necessary the use of structures for temporary storage of the image lines, which is accomplished by several shift-registers (see Fig 5).

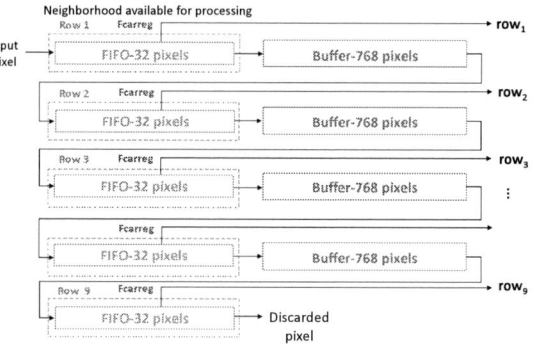

Figure 5: Architecture for windowing operation to an image of 800×525 pixels (the *neighborhood-loader* block)

The green blocks in the Fig. 5 correspond to the FIFOs, which work such as a shift register bank, allowing the access to the desired pixels of the neighborhood (in the Fig. 5 it is referred to as *Fcarreg*). After the pixel comes out of the *Fcarreg* register, the same is temporarily loaded in the line *buffer* and at every clock cycle the pixels are shifted to the same direction, being the oldest pixel discarded and the

where edges are added to the original image, depending on the mask size used in the convolution. As the RLA uses two convolution steps during processing, distortions by dimensionality created by the convolution of traditional architectures are interpreted as noise. In this case, three necessary steps of the convolution filter were developed (see Fig. 4), without dimensionality distortions which are presented below:

1) **The neighborhood-loader step:** it was designed for processing a neighborhood with a flexible handling, providing all of the possible surrounding data.

978-1-5090-2737-8/16 $31.00 © 2016 IEEE

youngest is loaded (blue block highlighted, see Fig. 5) [7]. The line buffer is implemented by using *block-RAMs* of the FPGA.

Fig. 6 depicts the *neighborhood-provider* block, shown in Fig. 4. Generally, the loading process provides the whole neighborhood (9×9, that is, 81 pixels). Instead of that, in this work, the provision was divided in a pixel per line, where pixels are sent simultaneously from the ninth position of the shift registers. Given that pipeline must be filled to develop the first convolution, this process has an initial latency for a image of $M \times N$ and a mask of $L \times P$ [7]. Also, in this work, images of 800×525 are being processed using a mask of 9×9. Therefore, the initial latency of the process of loading all the data for the first convolution is 3205 clock cycles.

To avoid distortions in the dimension of the image, after the first convolution, a special circuit was designed to separate the pixels of the image edges (see Fig. 6), providing a correct neighborhood in all situations during the convolution process. The problem occurs specifically when the mask is slipping over an edge of the image. The implementation of the architecture for providing the neighborhood is shown in Fig. 6, in which the MUX_1 and MUX_2 control the edge pixel processing. In the same module each pixel of the ninth position of *Fcarreg* is loaded into a FIFO of nine positions (*Fdisp*) and it is multiplexed to two sets of registers ($BReg_1$ e $BReg_2$) of equal size than *Fcarreg*. In this case, each register called as *neighborhood-provider row i* (Fig. 6) stores each row of *neighborhood-provider* block of Fig. 2. Additionally, the *frame-selection* block (Fig. 6) separates the frames from the pixel streaming coming from the camera.

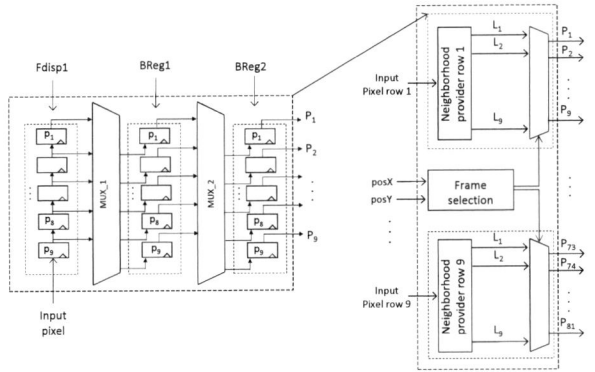

Figure 6: Architecture for the provision of pixels between image lines and video frames (the *neighborhood-provider* block)

In this context, the pixels form a multiple-stream where the pixels for each line-neighborhood are sent simultaneously for achieving nine simultaneous convolutions. The 2D convolution process, implemented in the overall architecture, is based on the definition of equation 3 and depicted in Fig. 7, which was described in Fig. 4 as the *convolution process* block. The processing occurs simultaneously sweeping across the loaded neighborhood, thus providing an output pixel at each

clock cycle (see Fig. 7). The release of the neighborhood is represented by the area (a) of Fig. 7, where implementations of multipliers and adders were omitted (see the areas (b) and (c) respectively). Note that the circuit provides the necessary hardware resources, namely 81 multipliers for multiplying each neighborhood pixel by each PSF value.

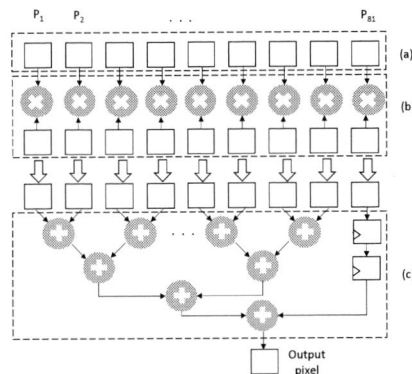

Figure 7: Architecture for convolution (the *convolution-process* block)

All multiplications are done in parallel between the neighborhood and the PSF (see Fig. 7b). After the parallel multiplication, results are added using a binary tree add (see Fig. 7c). The additions are performed in pairs and in a *pipeline* structure, and the result of these additions is the processed output-pixel. Finally the neighborhood circuit is scalable, providing also convolutions of 3×3, 5×5, and 7×7 (see Fig. 2) by using two control bits. In this case, the system automatically inserts zero-values in the external edges out from the chosen mask dimension, in both the neighborhood and in the PSF hardware structure.

V. RESULTS

Synthesis results are shown in Table I for 2 and 10 iterations for Cyclone IV (EPC4C115F29C7) and Stratix V (5SGXMB9R1H43C2) devices, respectively. It can be observed the consumption of LEs (*Logic Elements*) for Cyclone IV and ALMs (*Adaptive Logics Modules*) for Stratix V, the embedded DSPs and maximum frequency for the overall architecture.

TABLE I: Hardware resources consumption in FPGA

Board	LEs/ALMs	DSP	Fmax (MHz)	Iterations
Stratix	202.932 (64%)	352 (100%)	61,32	10
Cyclone IV	96.981 (65%)	532 (100%)	69,12	2

Given that the algorithm uses two convolutions it can be observed in Table I that the architecture demands a high resource consumption for each iteration. In this case the use of DSP reaches 100% as from the third iteration. Therefore, it is necessary the use of LEs for implementing the multiplication operations, leading to a loss of performance, due to the data

978-1-5090-2737-8/16 $31.00 © 2016 IEEE

representation (24 bits) which was chosen for achieving the precision of the restored image. Note that the camera frequency is achieved in all cases (50 MHz).

Tables II and III shown a comparison among several works discussed in the related work section. References [2] and [3] have synthesis results with high hardware consumption, even optimizing the convolution. Additionally, references [1] and [3] use PSNR (Peak Signal-to-Noise Ratio) and BSNR (Blurred Signal-to-Noise Ratio) metrics for quality image evaluation, respectively, whereas we have used the SR-SIM metrics which is more adapted to the human vision system, and therefore more suitable for verifying the restoration results. Finally, it can be observed that our implementation use images larger than other proposals.

TABLE II: Comparison between our proposed method and previous works (algorithmic aspects)

Ref	Iterations	Image size	PSF-dim	PSF-type	Device
[1]	15	640x480	< 10	SV	Stratix V
[2]	2	640x480	11x11	SI	Virtex 2
[3]	60	64x64	13x13	SI	Virtex 4
Our proposal	10	800x525	9x9	SI	Stratix V

SV: Space-variant. SI: Space-invariant.

TABLE III: Comparison between our proposed method and previous works (implementation aspects)

Ref	Proc. time (ms)	FPS	Domain	Freq. Max. (MHz)
[1]	40	25	Space	Unknown
[2]	Unknown	Unknown	Space	63
[3]	78	12	Frequency	100
Our proposal	80	12	Space	61

FPS: Frames per Second

Fig. 8 shows the restauration results of an image which was degraded by movement. In this case, the original image (8a) suffered motion-blur (8b, with $x = 9$ and $\theta=60°$) being restored by using RLA (8c). It can be observed that the movement of the image is not significant if compared with [1]. However, the similarity between the two images was high being a SR-SIM=0.9563. In our work the larger motion-blur implies in a lager PSF, spending larger hardware consumption.

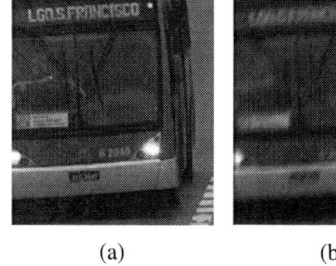

(a) (b) (c)

Figure 8: Image restoration using our architecture. (a) original image, (b) motion-blurred image, (c) restored image

VI. CONCLUSIONS

The hardware architecture developed for the RLA is scalable for mask sizes of 3×3, 5×5, 7×7, and 9×9, which is efficient for restoration image caused by motion-blur effect. In fact, the PSF may be defined for various types of degradations (optical degradations, air turbulence, etc.). The synthesis results were presented and discussed as well as image restoration effects with a real image. The restoration quality was evaluated using the SR-SIM metrics obtaining satisfactory results. The system was designed to be capable to restore images in real time (in VGA format), which can be complemented in an embedded system, for instance adding a motion measure system. A disadvantage of this architecture is the high DSP element usage, which depends on the number of iterations to be implemented. A future work is the implementation of equation 1 with β different of 1.

VII. ACKNOWLEDGEMENTS

The authors would like to the acknowledge the Brazilian institutions: FUNAPE and Petrobras for supporting the present study and PRH-PB 223. The supporting material were provided by Altera Corp. under the Altera University Program.

REFERENCES

[1] S. Carrato, G. Ramponi, S. Marsi, M. Jerian, and L. Tenze, "FPGA implementation of the lucy-richardson algorithm for fast space-variant image deconvolution," in *Image and Signal Processing and Analysis (ISPA), 2015 9th International Symposium on*, Sept 2015, pp. 137–142.

[2] O. Sims, "Efficient implementation of video processing algorithms on FPGA," EngD thesis, University of Glasgow, 2007.

[3] A. C. Atoche, O. P. Marrufo, and L. R. Castellanos, "Aggregation of parallel computing and hardware/software co-design techniques for high-performance remote sensing applications," in *Geoscience and Remote Sensing Symposium (IGARSS), 2011 IEEE International*, July 2011, pp. 217–220.

[4] S. H. Chan, "Constructing a sparse convolution matrix for shift varying image restoration problems," in *2010 IEEE International Conference on Image Processing*, Sept 2010, pp. 3601–3604.

[5] D. S. C. Biggs and M. Andrews, "Acceleration of iterative image restoration algorithms," *Appl. Opt.*, vol. 36, no. 8, pp. 1766–1775, Mar 1997.

[6] Prato, M., Cavicchioli, R., Zanni, L., Boccacci, P., and Bertero, M., "Efficient deconvolution methods for astronomical imaging: algorithms and IDL-GPU codes," *Astronomy & Astrophysics manuscript*, vol. 539, p. A133, October 2012.

[7] J. Y. Mori, C. Sánchez-Ferreira, D. M. Muoz, C. H. Llanos, and P. Berger, "An unified approach for convolution-based image filtering on reconfigurable systems," in *Programmable Logic (SPL), 2011 VII Southern Conference on*, April 2011, pp. 63–68.

[8] Z. Wang, A. Bovik, H. Sheikh, and E. Simoncelli, "Image quality assessment: from error visibility to structural similarity," *Image Processing, IEEE Transactions on*, vol. 13, no. 4, pp. 600–612, April 2004.

[9] A. C. Brooks, X. Zhao, and T. N. Pappas, "Structural similarity quality metrics in a coding context: Exploring the space of realistic distortions," *IEEE Transactions on Image Processing*, vol. 17, no. 8, pp. 1261–1273, Aug 2008.

[10] L. Zhang and H. Li, "SR-SIM: A fast and high performance IQA index based on spectral residual," in *Image Processing (ICIP), 2012 19th IEEE International Conference on*, Sept 2012, pp. 1473–1476.

[11] Z. Wang and A. Bovik, "Mean squared error: Love it or leave it? a new look at signal fidelity measures," *Signal Processing Magazine, IEEE*, vol. 26, no. 1, pp. 98–117, Jan 2009.

A systematic design approach for nanoscale inductor-less regulated cascode stages

C. Talarico
Department of Electrical and Computer Engineering
Gonzaga University
Spokane, WA 99258
talarico@gonzaga.edu

G. D'Amato, G. Avitabile, G. Piccinni, G. Coviello
Dipartimento di Ingegneria Elettrica e dell'Informazione
Politecnico di Bari
70125 Bari, Italy
{giulio.damato, gfa, giovanni.piccinni,
giuseppe.coviello}@poliba.it

Abstract—This paper presents a framework for the systematic design of inductor-less regulated cascode (RGC) stages. Targeting high-speed fiber optic data receiver front-ends, the technique reported combines the symbolic solution of the small-signal model of the RGC and the use of g_m/I_D based lookup tables to efficiently explore and optimize the resulting design space. A practical design is discussed and implemented in a 180 nm six-metal-layer CMOS process with 1.8V supply. The accuracy and viability of the proposed approach is validated through circuit simulation.

Keywords—regulated cascode (RGC); small-signal model; transfer function; root locus; non-dominant poles optimization.

I. INTRODUCTION

The increasing demand for low-cost, low power, high-speed and high-gain fiber optic data receivers has resulted in extensive research for broadband CMOS transimpedance amplifiers (TIAs). The main purpose of the TIA is to interface the large photodetectors present at the front-end of any fiber optic receiver. Among the many CMOS TIAs topologies present in literature [1]-[5], the regulated cascode (RGC) (Fig. 1) has the advantage to enhance the effective bandwidth of the TIA through a negative feedback induced input resistance reduction that desensitize the system from the large photodetector's capacitance present at its input.

The purpose of this paper is to introduce a systematic design approach for nanoscale inductor-less regulated cascode (RGC) stages, to be used in absence of closed-form MOSFET equations. The proposed technique is based on a g_m/I_D methodology as well as a computer-aided symbolic solution of the small-signal model of the stage. The role of the g_m/I_D methodology is to efficiently explore the large design space resulting from the symbolic solution of the RGC stage. The accuracy of the approach is validated through circuit simulation. The design of the RGC circuit is implemented in a 180nm six-metal-layer CMOS process with 1.8V supply.

II. GM/ID FRAMEWORK

When CMOS devices approach nanoscale size geometries, traditional "square-law" equations are too inaccurate to model transistors behavior. For deep submicron devices, second- and third-order effects (such as short-channel effect, channel length

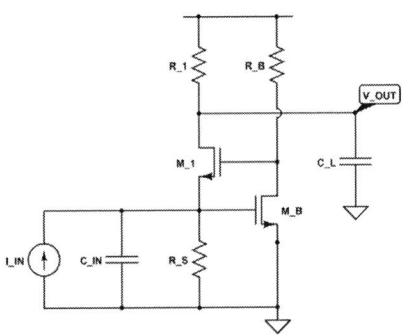

Fig. 1 - RGC input stage

modulation, drain induced barrier lowering, gate tunneling, etc.) become so significant that design methodologies based on closed-form expressions require too many iterations before a designer can converge toward a circuit that meet the desired specifications. On the contrary, g_m/I_D based methodologies provide a very effective and systematic framework to meet and optimize the desired design specifications [6][7]. In the g_m/I_D methodology the transistors are characterized and described through a set of look-up tables (or equivalently a set of charts) rather than closed form equations. As long as the second- and third-order effects afflicting the process are integrated in the SPICE models used to build the look up tables their effect is accurately captured and accounted for. In addition, the g_m/I_D framework is based on a set of design parameters (e.g. g_m, f_T and I_D) and figures of merit (transconductance efficiency g_m/I_D, transit frequency f_T, intrinsic gain $g_m r_o$ and current density I_D/W) that are more closely related to the design specifications, than the physical parameters (e.g. μC_{ox}, V_{th} and V_{dsat}) used in the closed form equations [8].

The charts illustrated in Fig. 2 have been calculated for the actual 180 nm six-metal-layer CMOS process used to design the RGC circuit. The plots provided in Fig. 2 are expressed with respect to the overdrive voltage $V_{OV} = V_{GS} - V_{th}$. A low g_m/I_D leads to devices biased in their strong inversion region, thus exhibiting a high transit frequency but poor power efficiency and a decreased output voltage swing. Instead, a high g_m/I_D leads to devices biased in their weak inversion region, thus exhibiting good power efficiency but an increased width size and thus lower speed. Assuming applications with

978-1-5090-2737-8/16 $31.00 © 2016 IEEE
121

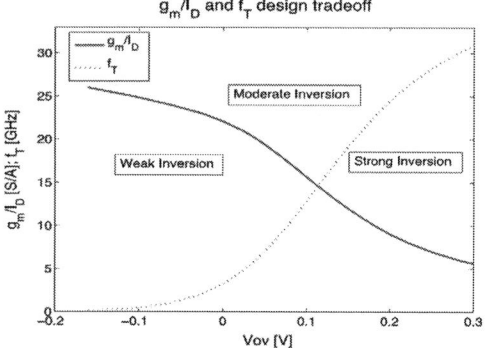

Fig. 2 - Plot of g_m/I_D and f_T at different levels of inversion V_{ov} for a 180nm nMOS transistor

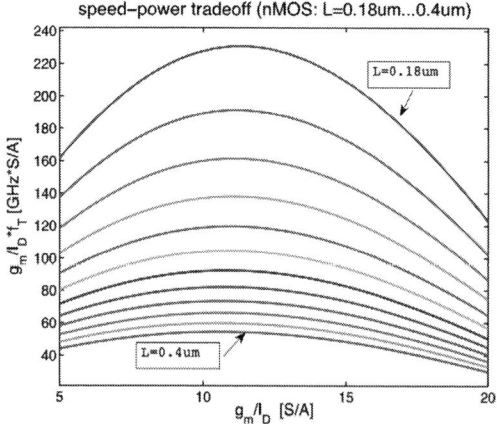

Fig. 3 – Relationship between the speed-power product and the level of inversion (g_m/I_D) for a 180 nm process as channel length varies

speed and power consumption having the same level of importance, biasing the devices in the moderate inversion region allows the best speed-power trade-off (see Fig. 3).

III. CIRCUIT ANALYSIS

The analytical closed-loop transfer function of the RGC is attained by symbolically solving the small signal model of the circuit through the use of the software package SapWin [9]. The symbolic solution is then imported in MATLAB to optimize the design with respect to bandwidth and power consumption.

The parasitic capacitances at the various nodes are respectively:

$$C_X = C_{in} + C_{gsB} + C_{sb1} \tag{1}$$

where C_{in} is the parasitic capacitance of the photodiode driving the circuit, C_{gsB} is the parasitic capacitance between gate and source of transistor M_B and C_{sb1} is the parasitic capacitance between source and bulk of transistor M_1.

$$C_Y = C_L + C_{db1} \tag{2}$$

where C_L is the capacitive load driven by the TIA and C_{db1} is the parasitic capacitance between drain and bulk of transistor M_1.

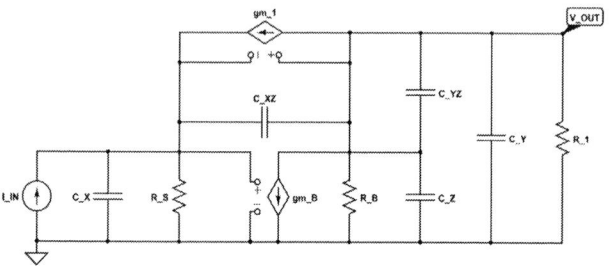

Fig. 4 - RGC small-signal model

$$C_Z = C_{dbB} \tag{3}$$

where C_{dbB} is the parasitic capacitance between drain and bulk of transistor M_B.

$$C_{ZY} = C_{gd1} \tag{4}$$

where C_{gd1} is the parasitic capacitance between drain and drain of transistor M_1.

$$C_{ZX} = C_{gs1} + C_{gdB} \tag{5}$$

where C_{gs1} is the parasitic capacitance between gate and source of transistor M_1 and C_{gdB} is the parasitic capacitance between gate and drain of the transistor M_B.

The transfer function has two zeros, one simple pole and one pair of complex conjugate poles. The zeros are at a significantly higher frequency than the poles so they are usually neglected [10]. Neglecting the zeros the transfer function of the RGC is of general the form:

$$A(s) \cong \frac{A_0}{\left(1 + \frac{s}{\omega_{p1}}\right)\left(1 + \frac{s^2}{\omega_0^2} + \frac{s}{\omega_0 Q}\right)} \tag{6}$$

and the poles can be written as:

$$p_1 = -\omega_{p1} \tag{7}$$

$$p_{2,3} = -\frac{\omega_0}{2Q}\left(1 \mp \sqrt{1 - 4Q^2}\right) \tag{8}$$

Typically RGC circuits are designed so that the bandwidth can be approximately determined by considering only the time constants at the nodes X and Y [10]. This corresponds to assume that the poles position is relatively close to the case $0.5\omega_0/Q \gg \omega_{p1}$ and the system tends to behave like a first order system with a dominant pole. The two time constants at the nodes X and Y are given by:

$$\tau_X = C_X R_X \quad and \quad \tau_Y = C_Y R_Y \tag{9}$$

Assuming the body effect of the transistor M_1 and the channel length modulation for both M_1 and M_B are negligible we can derive R_X and R_Y as follows:

$$R_X \cong R_S || \frac{1}{g_{m1}(1 + g_{mB}R_B)} \quad and \quad R_Y \cong R_1 \tag{10}$$

Given the assumptions just mentioned, as long as $g_{m1}R_S (1+g_{mB}R_B) \gg 1$ the DC gain of the TIA can be approximated as follows:

978-1-5090-2737-8/16 $31.00 © 2016 IEEE

$$A_0 \cong R_1 \frac{g_{m1}R_s(1 + g_{mB}R_B)}{g_{m1}R_s(1 + g_{mB}R_B) + 1} \cong R_1 \quad (11)$$

In practice the RGC circuit can be seen as a common gate (CG) input stage (composed by M_1, R_1 and R_S) to which a local feedback loop consisting of a common source (CS) amplifier (composed by R_B and M_B) is added with the goal of reducing the input impedance (or in other words reducing the otherwise large time constant τ_X associated to the input node X) by a "boosting" factor approximately equal to the gain of the common source stage ($A_{CS} \approx g_{mB}R_B$). Due to this property the RGC is one of stages most widely used for designing high-speed optical front ends [1][2][11], and as a result developing systematic optimization techniques for the design of RGC stages has become of significant interest. The main aim of this work is to illustrate and validate a design optimization technique to be used as an alternative to the common intuitive approach illustrated in [10]. Although the intuitive technique has the advantage of being extremely simple and it usually grants acceptable results, it is suboptimal. The main drawback in the intuitive approach is to assume that one of the time constants τ_X and τ_Y dominate and that the other time constants associated with the parasitic capacitances C_Z, C_{ZY} and C_{ZX} are all negligible. The consequence of these assumptions is that the approach is accurate only when the simple pole occurs at a much lower frequency than the complex conjugates poles. In reality, provided the designer takes care of "flattening" the potential peaking in the frequency response of the circuit, by positioning the complex conjugate poles at a lower frequency than the simple pole allows to further enhancing the bandwidth. Fig. 5 and Fig. 6 show the effect of poles location on bandwidth for two representative examples. The first example is achieved through the intuitive design strategy proposed in [10], while the second example is achieved using our optimization strategy. Table I and Table II summarize and compare the sizing of the devices required to implement the two designs.

Using the design and optimization strategy developed we are able to enhance the bandwidth of the RGC circuit of about 28.7% while simultaneously saving about 64.8% in power consumption. These improvements are at the expense of a minimal loss of 0.1 dBΩ in DC gain. Fig. 7 validates the viability and accuracy of the methodology by comparing the analytical frequency response obtained using our g_m/I_D based MATLAB framework and HSPICE simulation.

The percentage error in the analytical computation of the gain is about +0.6% while the percentage error in the analytical computation of the bandwidth is about −1.86%. The gain and bandwidth obtained through the analytical MATLAB framework and HSPICE simulation are summarized in Table 3.

IV. OPTIMUM DESIGN APPROACH

In order to design an optimized nanoscale inductor-less RGC stage, symbolic analysis is combined with a systematic synthesis procedure based on the g_m/I_D methodology. The approach is illustrated by means of an example targeting a gain of 61dBΩ, a bandwidth of at least 4GHz and power consumption less than 1.35mW, while considering a load C_L

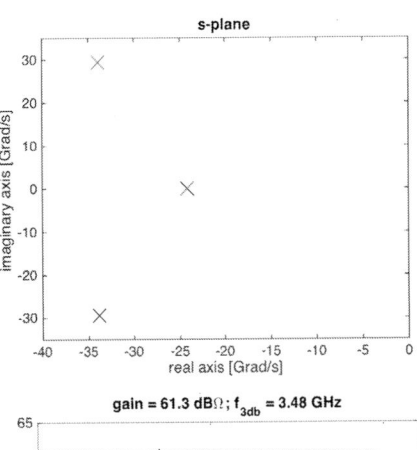

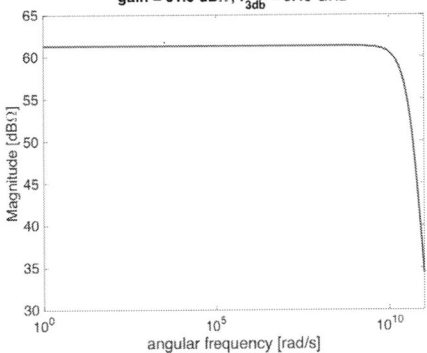

Fig. 5. – Poles location and frequency response achieved with the first design example (sub-optimal).

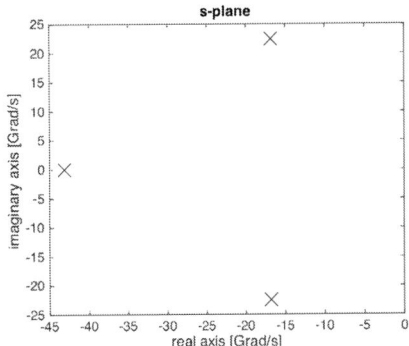

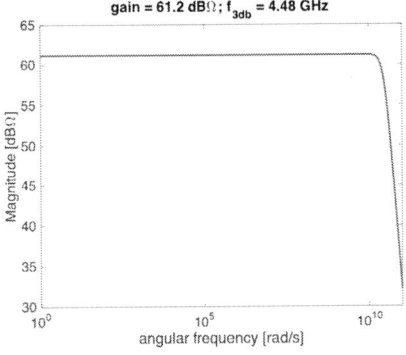

Fig. 6. – Poles location and frequency response achieved with our design strategy.

equal to 20fF and a photodiode's parasitic capacitance C_{in} equal to 200fF. Under the proposed framework, an initial solution is computed using the intuitive approach proposed in [10]. The initial solution is computed biasing both the transistors in the moderate inversion region ($g_m/I_D \approx 10$ V^{-1} to 15 V^{-1}) and selecting the shortest channel length (L=0.18μm) allowed by the technology. Although suboptimal, the above solution guarantees a good compromise between output signal swing, gain, bandwidth and power consumption. The design parameters for the transistors used in the initial solution are summarized in Table 4. The value of the resistors used are $R_1=R_S=1200\Omega$ and $R_B=800\Omega$. As noted in section III the gain is approximately set by R_1 (that is about 61.58 dBΩ). Although this solution is a convenient starting point, the bandwidth achieved is only 3.43GHz and power consumption is about 1.77mW, so the target performances are not met.

TABLE 1. DESIGN PARAMETERS FOR THE SUBOPTIMAL DESIGN

R_B [Ω]	R_1 [Ω]	R_S [Ω]	W_1/L_1 [$\mu m/\mu m$]	W_B/L_B [$\mu m/\mu m$]	I_{D1} [μA]	I_{DB} [μA]
800	1200	1200	18.2/0.18	26.18/0.18	511.02	472.94

TABLE 2. DESIGN PARAMETERS FOR THE OPTIMAL DESIGN

R_B [Ω]	R_1 [Ω]	R_S [Ω]	W_1/L_1 [$\mu m/\mu m$]	W_B/L_B [$\mu m/\mu m$]	I_{D1} [μA]	I_{DB} [μA]
2435	1200	1250	7.7/0.18	13.56/0.18	457.84	138.97

TABLE 3. COMPARISON BETWEEN ANALYTICAL AND HSPICE SIMULATION

Design Parameter	MATLAB Framework	HSPICE Simulation	% Error
A_0 [kΩ]	1.1529	1.1460	+0.6
BW [GHz]	4.3935	4.4768	−1.86

TABLE 4. DESIGN PARAMETERS FOR THE INITIAL GUESS

Parameter	M_1	M_B
g_m [mS]	5.17	6.07
g_{DS} [μS}	166.51	144.61
I_D [mA]	511.02	472.94
g_m/I_D [V^{-1}]	10.12	12.83
W [μm]	18.2	26.18
L [μm]	0.18	0.18
V_{GS} [mV]	616.23	808.42
C_{gtot} [fF]	53.25	36.90
C_{dtot} [fF]	20.91	29.82
f_T [GHz]	97.09	164.50

As discussed in section III, to reach an optimal solution that meets target specification we need to sweep the design parameters affecting the position of the poles. In order to make the optimization process fast, it is important to limit the design space exploration as much as possible. In order to reduce the number of solutions that the synthesis framework considers, it is crucial to constrain the optimization process using the physical insight we have about the behavior of the stage. For example, using transistors with longer channel lengths would lead to improved output resistances but would penalize gate and drain capacitance. Reducing the input resistance of the stage would require increasing the amount of feedback provided. This would require increasing g_{mB}, R_B, or both. Increasing g_{mB} would require increasing the current used

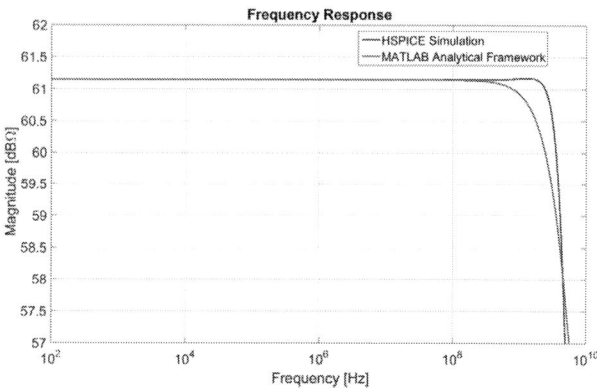

Fig. 7 – RGC Frequency Response: MATLAB vs. HSPICE comparison

to bias M_B, which unless we reduce R_B makes more difficult to maintain M_B in saturation for the desired signal swing. The best trade off in terms of speed and power consumption is to keep M_B in the moderate inversion region (10V^{-1} ÷ 15V^{-1}) and use an amount of feedback $g_{mB}R_B$ in the range 5 ÷ 8.

The automated design procedure developed has been fully implemented in MATLAB. It takes as inputs a model of the actual technology (in the form of look-up tables) and the desired circuit specifications (gain, bandwidth, load capacitance, photodetector's parasitic capacitance and maximum power consumption), and provides as output an optimal solution vector consisting of R_{S_opt}, R_{B_opt}, W_{1_opt}, W_{B_opt}, gain$_{\_opt}$, BW$_{\_opt}$ and Pdiss$_{\_opt}$. An iterative procedure is used to explore the design space. The procedure sweep a small number of primary design variables (R_S, $g_m/I_D(M_1)$, $I_D(M_1)$ and $I_D(M_B)$) and based on the symbolic solution of the small-signal model of the RGC stage compute gain, bandwidth, and power consumption. The design space exploration engine starts from an initial guess computed using the approach presented in [10] and then moves on by exploiting both design constrains and physical insight to reduce the solution space explored. The key is to avoid evaluating solutions that are known in advance to violate some design rules (e.g. transistors must be in saturation, the g_m/I_D ratio for M_B must be in the range 10÷15 V^{-1}, the product $g_{mB}R_B$ cannot be outside the range 5÷7). Fig. 8 summarizes visually the design exploration process. It is worthwhile to point out that all the MATLAB calculations related to the synthesis procedure are achieved in the order of minutes. The following steps describe the optimization algorithm implementing the design of the RGC:

1. Set L=0.18μm for all transistors, set R_1 to the target gain, assume $R_S=R_1$, set $I_{D1} = 500\mu A$, $(g_m/I_D)_1 \approx 10V^{-1}$, and $(g_m/I_D)_B \approx 13V^{-1}$. Given the DC analysis equations and the constraint that $I_{DB} \leq I_{D1}$ compute an initial feasible solution (i.e., find the values of W_1, W_B, and R_B required).

2. Find a viable interval of DC operating points (e.g. based on maximum power consumption specification, and desired output signal swing) and a number of design rules based on physical insight (e.g. both transistors must operate in saturation with sufficient margin, the current consumed by M_B should be at least

50% smaller than the current consumed by M_1) and set them as constraints.

3. Explore the design space by sweeping the primary design parameters R_S, I_{D1}, and $(g_m/I_D)_1$ around the initial guess. Discard "a priory" any solution that violates the constraints. In this way we separate the design space in two regions (acceptable and not), and avoid computing any not acceptable solution.

4. For each acceptable solution, compute performances (e.g. gain, bandwidth, power consumption) and record both performances and associated parameters (e.g. R_S, W_1, R_B, W_B)

5. Identify the design vector that optimizes performances

6. Extract from the solution subspace the optimum transistor sizes W_1 and W_B and the optimum values of R_S and R_B.

The results obtained through HSPICE simulation are in close agreement with the analytical results obtained through the use of the proposed framework. The results are reported in Table 5 and Table 6.

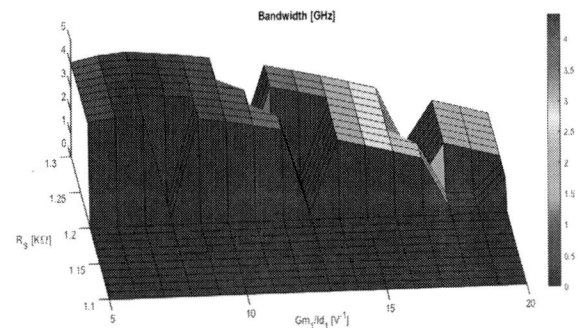

Fig. 8 – RGC design optimization: bandwidth vs. g_m/I_D of M_1 and source resistance R_S.

V. CONCLUSIONS

This paper presents a systematic design framework for the design of nanoscale inductor-less regulated cascode (RGC) stages that can be used in absence of closed-form MOSFET equations. The proposed technique illustrates the benefits of combining symbolic analysis and g_m/I_D methodology. The proposed approach has been validated through HSPICE post-layout simulation of a practical RGC stage implemented in a 180nm six-metal-layer CMOS process with 1.8V supply. Our design approach not only showed the limitations of the intuitive analysis presented in [10], but provided further insights and procedures missing to the previous approach. The close agreement between simulation results and the analytical results underlines the effectiveness of combining symbolic analysis and g_m/I_D methodology. In addition, the time efficiency of the MATLAB-based synthesis algorithm developed suggests that the framework can be easily augmented to enable further quick technology comparisons and selection or parameter sensitivity analysis. Finally, if further bandwidth extension is required the RGC synthesized

TABLE 5. ANALYTICAL RESULTS (OPTIMIZATION FRAMEWORK)

Parameter	M_1	M_B
g_m [mS]	2.8	2.9
I_D [mA]	400	214
g_m/I_D [V^{-1}]	7	13.55
W/L [μm/μm]	7.69/0.18	13.43/0.18
Gain [dBΩ]	61.24	
BW [GHz]	4.40	
R_S [Ω]	1250	
R_B [Ω]	2434.9	

TABLE 6. HSPICE SIMULATION RESULTS

Parameter	M_1	M_B
g_m [mS]	2.88	2.20
I_D [mA]	457.84	138.97
g_m/I_D [V^{-1}]	6.30	15.83
W [μm/μm]	7.70/0.18	13.44/0.18
Gain [dBΩ]	61.18	
BW [GHz]	4.48	
R_S [Ω]	1250	
R_B [Ω]	2435	

with our tool, can be used as starting point for more advanced topologies that relies on techniques like series inductive peaking and transformer based negative feedback [11][12].

REFERENCES

[1] S.M. Park, H.-J. Yoo, 1.25-Gb/s Regulated Cascode CMOS Transimpedance Amplifier for Gigabit Ethernet Applications. IEEE J. Solid State Circuits 39(1), 112–121 (2004)

[2] Y.-H. Kim, S.-S. Lee, A 72 dBΩ 11.43 mA novel CMOS regulated cascode TIA for 3.125 Gb/s optical communications, in IEEE International SOC Conference, pp. 68–72 (2013)

[3] H.M. Lavasani et al., A 76 dBΩ 1.7 GHz 0.18μm CMOS Tunable TIA using broadband current preamplifier for high frequency lateral MEMS oscillators. IEEE J. Solid State Circuits 44(1), 224–235 (2011)

[4] C.-F. Liao, S.-I. Liu, 40 Gb/s Transimpedance-AGC Amplifier and CDR Circuit for Broadband Data Receivers in 90 nm CMOS. IEEE J. Solid State Circuits 43(3), 642–655 (2008)

[5] B. Razavi, Design of Integrated Circuits for Optical Communications, 2nd edn. (Wiley, Hoboken, New Jersey, 2012)

[6] B. Murmann, in Analysis and Design of Elementary MOS Amplifier Stages (NTS Press, 2013)

[7] F. Silveira, D. Flandre, P.G.A. Jesper, A gm/ID based methodology for the design of CMOS analog circuits and its application to the synthesis of a silicon-on-insulator micropower OTA. IEEE J. Solid State Circuits 31(9), 1314–1319 (1996)

[8] C. Talarico, G. Agrawal, J. Roveda, and Hani Lashgari, Design Optimization of a Transimpedance Amplifier for a Fiber Optic Receiver, Springer J. Circuits, Systems, and Signal Processing (CSSP), 34(9), 2785-2800 (2015).

[9] SapWin, site: "http://cirlab.det.unifi.it/SapWin/"

[10] L. B. Oliveira, C.M. Leitao, M. Medeiros Silva, Noise Performance of a Regulated Cascode Transimpedance Amplifier for Radiation Detectors. IEEE J. Trans. Circuits and Systems 59(9), 1841-1848 (2011)

[11] C. Li, S. Palermo, A Low-Power 26 GHz Transformer-Based Regulated Cascode SiGe BiCMOS Transimpedance Amplifier. IEEE J. Solid State Circuits 48(5), 1264 – 1275, (2013)

[12] M.-J. Wu, Y.-H. Lee, Y.-Y. Huang, Y.-M. Mu, J.-T. Yang, A CMOS multi-band low noise amplifier using high-Q active inductors. Int. J. Circuits Syst. Signal Process. 2(2), 199–202 (2007)

Characterization and Nonlinear Modeling of MASMOS® Transistor in Order to Design Power Amplifiers for LTE applications

Frédérique Simbélie, Sylvain Laurent, Pierre Medrel,
Michel Prigent, Raymond Quere
XLIM C2S2
7 rue Jules Vallès
19100 Brive-la-Gaillarde, France
frederique.simbelie@xlim.fr

Myrianne Regis, Yann Creveuil
ACCO semiconducteur
36-38 rue de la princesse
78430 Louveciennes
Myrianne.Regis@acco-semi.com

Abstract — **This paper reports on the first experimental characterizations and modeling process of a MASMOS® transistor with a classical model, largely used for the modeling of other transistors. From DC IV and S-parameters measurements a large signal model (LSM) has been carried out. The great interest of this model is to allow a simulation time reduced by a factor of 100 compared to foundry model, like BSIM3 model, in classical one-tone HB simulation and to perform multi-tones simulation which is not possible with the foundry model due to prohibitive simulation times. The LSM has been validated through extensive multi-tones and load pull measurements. A good agreement between LSM simulation results and measurements fully validates the proposed modeling methodology. This LSM will serve to design a power amplifier (PA) for LTE applications.**

Keywords— MASMOS, transistors, Modeling, linearity, HF characterization, multitones signal, IM3, linearity

I. INTRODUCTION

The Radio Frequency Power Amplifier (RFPA) is a key component in modern communication system, because it impacts significantly the amount of consumed DC power and the overall linearity of RF transmitters. All the steps toward the final implementation of the RFPA are largely conditioned by the accuracy of the developed nonlinear transistor model. Moreover, depending on the RF application, the design of RFPA is always subjected to a trade-off between efficiency, linearity and gain overall performances.

III-V technologies (e.g. GaAs, GaN) are the preferred technologies for high power and/ or high frequency solid-state applications, because of the inherent high breakdown voltage and high saturation velocity of charge carriers. However most of these devices are still expensive and incompatible with the full integration in silicon platform, and thus suffer from a prohibitive cost for mass market. In order to use CMOS as a technology for implementing power amplifiers blocks, most attempts have focused on complex structures (e.g. the use of multiple stage drivers, each involving the stacking of multiple devices) to overcome the inherent reduced voltage swing across the individual device that ultimately degrades power density and

VSWR tolerance in RF-oriented applications. Most of the reported approaches require complex design procedures that ultimately degrade the time and cost to market. As a promising alternative, ACCO has patented a new device that solves the inherent CMOS breakdown voltage limitation RF application in which is referred to as MASMOS®.

This paper presents, for the first time, the characterization and large signal modeling (LSM) of a MASMOS® transistor which can be used for PA design for LTE applications. There are several methods reported in the literature for the transistor modeling: physical (good accuracy but prolonged simulation time), empirical and behavioral (demands long time measurements to elaborate the model). The proposed modeling process is based on an empirical method, allowing the best compromise between development time and accuracy. The nonlinear electrical model contains both linear and non-linear elements. Non-linear elements are modeled by mathematical expressions depending on transistor internal voltages, which ensures reduced simulation time and rapid convergence of simulators. The model parameters are extracted using I-V and S-parameters measurements. In order to validate the model, many load pull and multi-tones measurements are performed. Note that in our model, the breakdown effect will not be taken into account but some equations exist and it can be simply added if necessary.

The paper is organized as follows. In Section II, we describe the various measurements performed to extract the model parameters. I-V measurements dedicated to extract non-linear current source parameter GAMM [1] are discussed. Low frequency S-parameters measurements are dedicated to extract non-linear capacitance as a function of internal voltage. In Section III, we present the test bench used to make load pull measurements over different load impedance conditions. Also, a comparison between LSM simulations and measurements is presented. In Section IV, we discuss about an innovative multi-tones characterization methodology proposed to highlight the non-linearity of the transistors or amplifiers. We compare results between two-tone and eight-tone measurements and the

MASMOS® model envelope simulations. Finally, Section V concludes the paper.

II. MASMOS® CHARACTERIZATION AND ELECTRICAL MODELING

This section explains the DC and low frequency (LF) characterizations used for developing the MASMOS® electrical model.

A. I-V measurements

The MASMOS® device used has a total periphery of 0.55 mm x 0.27 mm. DC measurements were performed using the semiconductor device parameter analyzer (B1500A) equipped with two SMUs, one for high power and the other for medium power generation.

Although the MASMOS® component is built with a cascode combination of 2 different transistors, we model it using only one electrical circuit as illustrated in Figure 1, always in order to limit simulation time. This large signal equivalent model includes extrinsic and intrinsic components. Extrinsic components C_{PG}, C_{PD}, R_g, R_d, R_s, L_g, L_d, L_s are independent of applied bias voltages and the intrinsic part of this circuit includes nonlinear elements (current source, nonlinear capacitances, R_i and R_{gd}). All modeling steps are described here after.

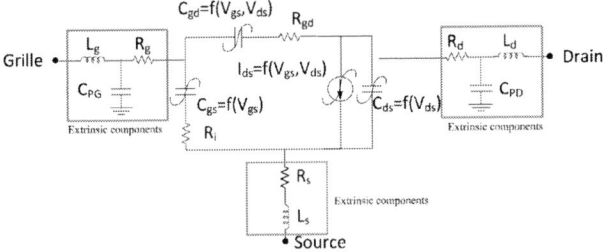

Figure 1: Large signal equivalent model of MASMOS®

B. The current source GAMM

The nonlinear drain current source of MASMOS® model is based on modified Tajima's equations [1] and named as GAMM model.

In order to optimize the 18 parameters of the GAMM model, we have to take into account the parasitic access resistances, R_s and R_d. In this work, we determine the access resistances using the linear region of the measured DC-IV characteristics. R_{on} resistance can be expressed as the sum of channel (R_c), parasitic source (R_s) and drain (R_d) resistances.

Figure 2 shows the comparison between measurement data and simulated results of GAMM current source model. The characteristic $I_{ds}=f(V_{ds})$ was measured for V_{ds} ranging from 0 to 6V in steps of 0.1 V and the gate-source voltage V_{gs} varying from 0 to 1V in steps of 0.1V. A good agreement between the measurement and the model is observed.

C. S-parameters measurements and model extraction

Multibias S-parameter measurements have been performed using Vector Network Analyzer (E5061B VNA). This instrument has an internal DC source which provides a maximum supply voltage of ± 40 V and have a current compliance of 100 mA. The measurement frequency ranges from 5 kHz to 3 GHz.

Depending on the application – in our case for power amplifier design – a fictive load-line cycle was chosen on I(V) characteristics (Figure 2), that corresponds to 12 pairs of values (V_{gs}, V_{ds}). The nonlinear capacitances are thus extracted and optimized along this I-V trajectory in order to fit the model to measurements.

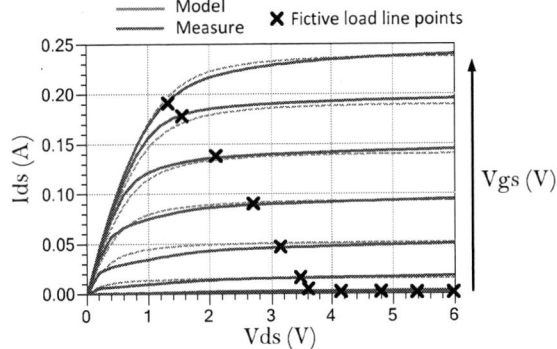

Figure 2: Comparison Ids-Vds for V_{ds}=0 to 6V, step: 0.1 V and V_{gs}=0 to 1 V, step: 0.1 V

Low frequency (LF) S-parameters measurements allow a straightforward extraction of the intrinsic nonlinear elements such as transconductance (G_m), conductance (G_d), gate-source capacitance (C_{gs}), gate-drain capacitance (C_{gd}) and source-drain capacitance (C_{ds}) as they remove the influence of the extrinsic reactive effects of the device.

The measured LF S-parameters (Figure 4 and Figure 5) are converted into equivalent Y-parameters and the capacitances of MASMOS® model are extracted using the following expressions:

$$C_{gs} = \frac{\text{Im}(Y_{11})+\text{Im}(Y_{12})}{2\pi f} \quad (1)$$

$$C_{gd} = \frac{-\text{Im}(Y_{12})}{2\pi f} \quad (2)$$

$$C_{ds} = \frac{\text{Im}(Y_{22})+\text{Im}(Y_{12})}{2\pi f} \quad (3)$$

Thus, C_{gs}, C_{gd} and C_{ds} are extracted as a function of frequency and for each applied bias voltages (V_{gs}, V_{ds}). The extracted capacitances values remain constant over the 5 kHz to 3 GHz frequency range. Therefore, these capacitances are described with respect to the voltages applied at the device terminals.

In order to simplify the modelling process, we use the same mathematical expression for all capacitances. In Figure 3, are plotted the C_{gs} and C_{gd} capacitances as a function of applied voltages and moreover, the capacitances can be divided into three regions. In the first region capacitances demonstrate

978-1-5090-2737-8/16 $31.00 © 2016 IEEE

quasi-constant values for low V_{gs} and V_{gd}. In the second region, C_{gs} and C_{gd} increase and finally, at high voltages the capacitances decrease significantly. Consequently, we use a suitable mathematical function which can capture the increase and the decrease of the values.

The corresponding mathematical function that can effectively fits the measurement results can be expressed as [2]:

$$C = C_0 + \frac{C_1 - C_0}{2}\left[1 + tanh\big(A * (V + V_m)\big)\right]$$

$$-\frac{C_2}{2}\left[1 + tanh\big(B * (V + V_p)\big)\right]$$

$$+\frac{C_3}{2}\left[1 + tanh\big(C * (V - V_n)\big)\right] \quad (4)$$

This above expression has 10 fitting parameters. The parameters, C_0, C_1, C_2, C_3, A, B, C and V_m, V_n, V_p which can be optimized for each capacitance. According to the capacitance being modelled, the variable V stands for the command voltage of C_{gs}, C_{gd} and C_{ds}, respectively V_{gs}, V_{gd} and V_{ds}. Figure 3 demonstrates good agreement between model capacitances and measurements.

Once a satisfactory MASMOS® current source and nonlinear capacitances have been achieved, we attempt to choose the set of extrinsic parameters C_{PG}, C_{PD}, R_g, L_g, L_d, Ls that ensures the matching between S-parameters simulations and measurements. In this step, the parasitic resistances value of R_s and R_d remains unaltered. Since these resistance values have been already optimized during the current source modeling.

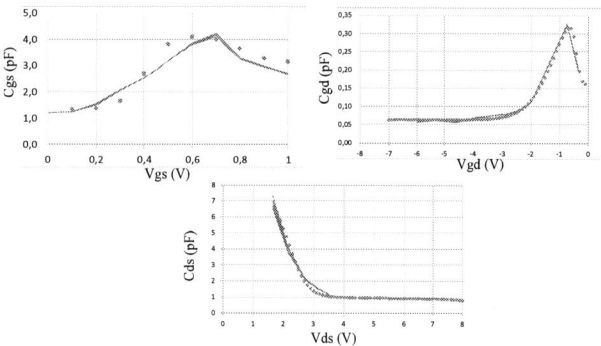

Figure 3: Comparison of capacitance values obtained using LF S-parameters measurements (red lines) and model extraction along the dynamic load line (blue points)

Comparison examples between S-parameters measurements and modeling for two different bias points selected from the fictive load line are shown in Figure 4, Figure 5 and Figure 6.

Comparison results show a good agreement between measurements and model simulations for all bias points along the chosen fictive load line (Figure 2). This load line points are the most critical since they are closely related to the final device operating conditions. Moreover for MASMOS® transistor we performed S-parameter measurements beyond 3 GHz to

validate transistor modeling. Figure 6 shows the S-parameters for a frequency range of 100 MHz to 10 GHz and for the same bias point as shown in Figure 5.

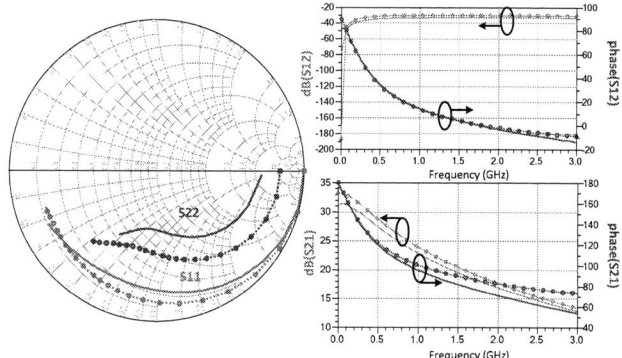

Figure 4: Comparison between measured (lines) and modeled (symbols) S-parameters at V_{gs}=0.7 V and V_{ds}=2.7 V

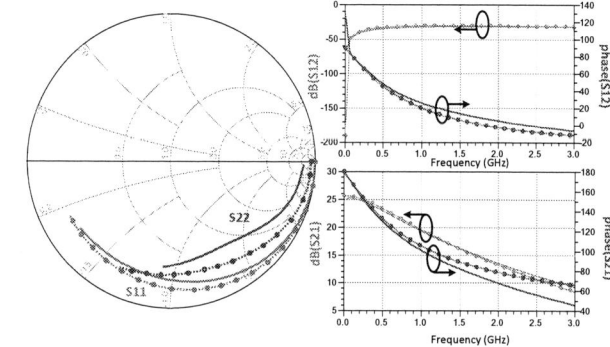

Figure 5: Comparison between measured (lines) and modeled (symbols) S-parameters at V_{gs}=0.5 V and V_{ds}=3.45 V

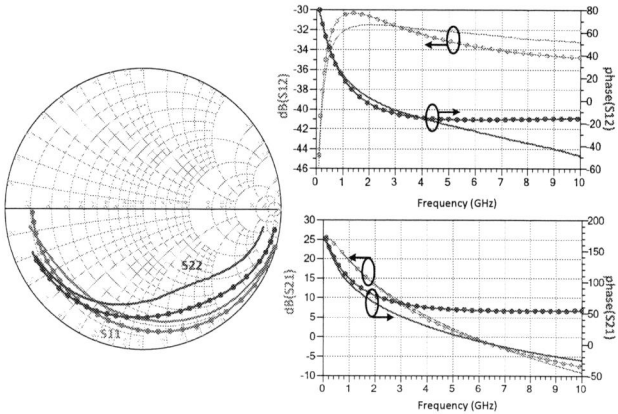

Figure 6: Comparison between measured (lines) and modeled (symbols) S-parameters at V_{gs}=0.5 V and V_{ds}=3.45 V

Thus, Section III is dedicated to load pull measurements and Harmonic Balance simulation results comparison in order to check the large signal model validity.

III. LOAD PULL MEASUREMENTS AND MODEL VALIDATION

In this section, load pull experiments are explained and the model is tested using similar conditions in Advanced Design System Keysight EEsof.

A. Load pull measurement bench

The description of the experimental load-pull test-bench is given in Figure 7. Large signal characterizations have been carried out at a fundamental frequency (f_0) of 2 GHz, and for different load impedances at this frequency and its corresponding harmonics. Each harmonic frequency load has been also measured to take it into account in the model simulations. To perform this kind of measurements, a Synthetized Signal Generator (68367C Anritsu 10MHz-40GHz) has been used. In order to control source impedance an ICCMT 1816-2C tuner is used while to control load impedance a MPT 1808 tuner is used. During the measurements, the source impedance was tuned to 50Ω. The load tuner enables the control of the f_0 (fundamental), $2f_0$ and $3f_0$ impedances. To perform the incident and reflected power waves collection and analysis, a Large Signal Network Analyzer (SWAP X402) is used, combined with two couplers, and labelled "wave probes" [3].

All the measurements have been conducted at a single bias point (V_{gs}=0,5 V , V_{ds}=3,5 V) at the center of the fictive load line. The wave probes couplers are limited to 18 GHz, therefore the measurements include data at f_0 and its eight first harmonics.

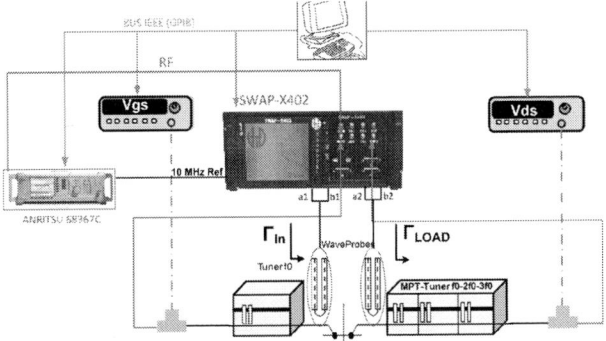

Figure 7. Load pull measurement bench

B. Measurements and modeling results

The fundamental and harmonic load impedances for optimum Power Added Efficiency (PAE) performances have been obtained with the following procedure. First, a measurement at a 50Ω f_0 impedance condition is performed (Figure 8). In a second step, the optimal PAE f_0 load impedance is found by scanning numerous impedances in the Smith chart (Figure 9). Then the f_0 load impedance is fixed at the previous value determined in step 2 and the impedance at $2f_0$ is chosen to maximize PAE.

Once the fundamental and harmonic load impedances are determined, we can plot power transfer characteristics. The complete large signal MASMOS® model has been simulated with the harmonic balance simulator. Comparisons have been made for three different load impedance configurations in Figure 8, 9 and 10, such as:

- A: Z_{LOAD} = 50 Ω @ f0 and harmonics frequencies

- B: Z_{LOAD} = (28+j21) @ f0, 50 Ω to harmonics frequencies

- C: Z_{LOAD} = (28+ j21) @ f0, Z_{LOAD}= (4+j55) @ 2f0, 50Ω everywhere else

The plotted characteristics are the PAE, the output power and the gain in function of input power for the three load impedance configurations. In addition, we plotted on smith chart load and input reflection coefficient defined in Figure 7. These figures demonstrate that the MASMOS® model chosen can effectively predict results in large signal simulations for different load impedances. This MASMOS® model takes only 0.85 second in Keysight EEsof ADS software to perform load pull simulation for 24 points of power level. It should be noted that output power characteristics have a good agreement, also for the input reflection coefficient.

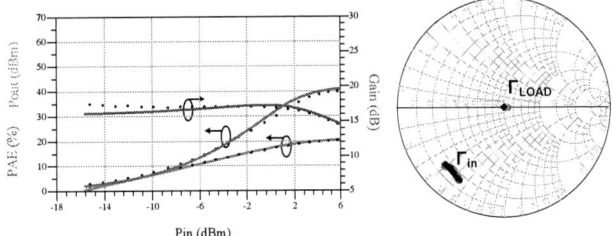

Figure 8. Output power, gain and PAE comparison between measurements (black dots) and model simulation (lines) with V_{gs}=0.5 V and V_{ds}=3.5 V; Configuration A

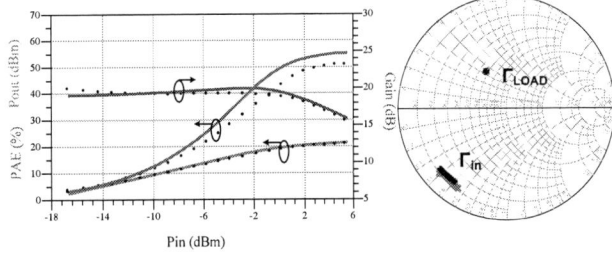

Figure 9. Output power, gain and PAE comparison between measurements (black dots) and model simulation (lines) with V_{gs}=0.5 V and V_{ds}=3.5 V; Configuration B

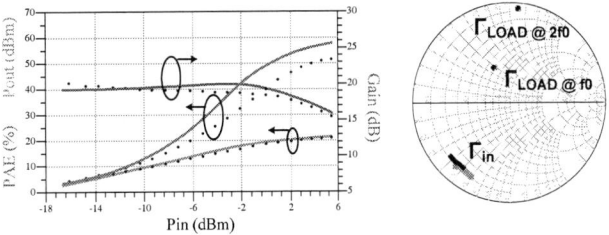

Figure 10: Output power, gain and PAE comparison between measurements (black dots) and model simulation (lines) with V_{gs}=0.5 V and V_{ds}=3.5 V; Configurations C

In order to realize more comparisons and to confirm the model accuracy, we propose new large signal measurements which are detailed in the Section IV.

IV. LINEARITY MEASUREMENTS

To characterize transistors/PA linearity performances under modulated signals conditions, numerous measurements and related test benches have been proposed. Multi-tones signals have been used to perform the same kind of measurements [4] [5] [6] [7]. In these proposed methods, tones are equally spaced. In this section, a new multi-tones load pull (MLTP) set up is detailed using non equally spaced tones [8].

A. Multitones measurement bench

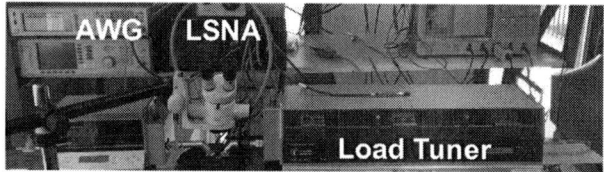

Figure 11: Multitones measurement bench

To perform this new measurement a multi-tones signal and a tailored algorithm in LSNA are used. It can be used to perform classical linearity measurements on MASMOS® transistor close to real modulated signal conditions, without a need of expensive dedicated instruments. The multi-tones signal is describe as follow:

$$f_k = f_1 + (k-1)*\Delta_f + \mathcal{E}_k \quad 1 \le k \le n \quad (5)$$

The parameter f_k referred to the different frequencies. In equation (5) k is the rank frequency, Δ_f is the frequency spacing between the tones, $\mathcal{E}_k$ represents the frequency shift of the kth frequency. In order to accurately measure the response of the device at the input frequencies and at IM frequencies, all the input frequencies are chosen on a frequency grid, which resolution is f_ε. Thus equation 5 can be written as:

$$f_k = [l + (k-1)*m + p_k]*f_\varepsilon \quad 1 \le k \le n \quad (6)$$

Where l, k, m, p_k are integers. The shifting vector p_k ensures that all IM3 products are distinct from the input frequencies and from each other. Moreover, it has been shown [9], [10] that by injecting 8 tones with correctly selected amplitudes and phases, the signal statistics in term of spectral density, CCDF (Complementary Conservative Density Function) and PAPR (Peak to Average Power Ratio) are almost perfect (like a continuous spectrum signal). The setup is tuned for a generation of a maximum of 8 independent tones as input signal and 224 resulting IM3 tones. The modulated signal is generated by an Arbitrary Waveform Generator (MXG N5162B 9 KHz - 6 GHz). In the first case, we measure 2 third order intermodulation (IM3) and the second case we measure 224 IM3. The main advantage of this bench is the instantaneous acquisition of incident and reflected waves at all frequencies for each incident power.

The MTLP set-up (Figure 11) is similar to the load pull bench. An oscilloscope Tektronix DPO7054 (500 MHz bandwidth, 20 GS/s, 8bit) was added to measure the IQ waves and the LF bias drain current.

B. Experimental results for 2-tones signal

From MTLP measurements, numerous information can be derived. The MTLP measurements are done at a bias point (V_{ds}, Ids) = (3.5V, 14mA) and a load impedance of 50 Ω. In Figure 12 a comparison between output power spectrums for two different input power levels is shown. It represents the measured output power spectrum. The two dominant lines correspond to the 2 injected frequencies tones, the close 2 lines are the IM3 frequencies. All the lines at a level less than -60 dBm correspond to the chosen frequency grid (224 tones) required for measuring IM3 power when an 8-tones signal is injected but in this case there is no power at these frequencies. Figure 12B represents the plot of output power versus input power at fundamental frequency (blue line) and at IM3 (red line). The crossing between the two trend lines gives the output IP3 power: OIP3=34 dBm.

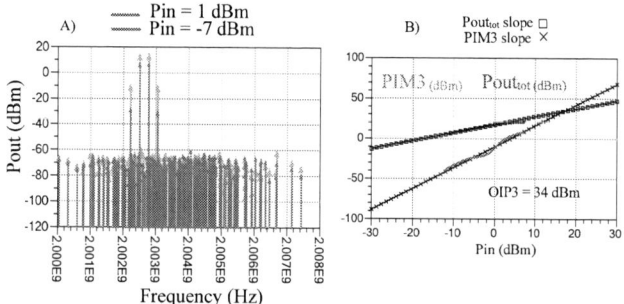

Figure 12: (A) Output power spectrum and (B) PIM3 and output power characteristics for two-tone signal

C. Experimental results for 8-tones signal

The spectrum in Figure 13 shows the experimental results for an 8-tones input signal (two frequencies are close to each other).

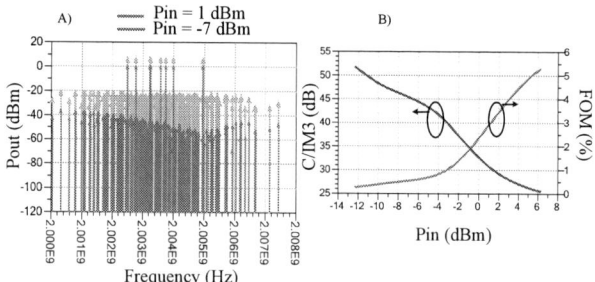

Figure 13: (A) Output power spectrum and (B) C/IM3 and Figure Of Merit (FOM) for eight-tones signal

In this case, we can compute C/IM3 (Carrier to 3rd Intermodulation Ratio) by

$$\frac{C}{IM3}(dB) = 10.\log(\frac{\frac{Pu}{8}}{\frac{P_{IM3}}{224}}) \quad (7)$$

Where P_u is the sum of the 8-tones signal power and P_{IM3} that of IM3 products. In addition, a Figure Of Merit (FOM) which, in certain conditions, can be identified as the Error Vector Magnitude (EVM), is defined as [9].

$$20.\log(\text{FOM}) \, (dB) = 40 - C/IM3 \qquad (8)$$

These two criteria are plotted in Figure 13B.

This MTLP bench allows to record the time-domain envelop signals and to compute the corresponding CCDF which allows to characterize the probability that the signal power is higher than a given envelop average power level. These CCDF curves allow to characterize the power statistics of different modulation schemes. The computed CCDF of an eight-tone signal is shown in Figure 14.

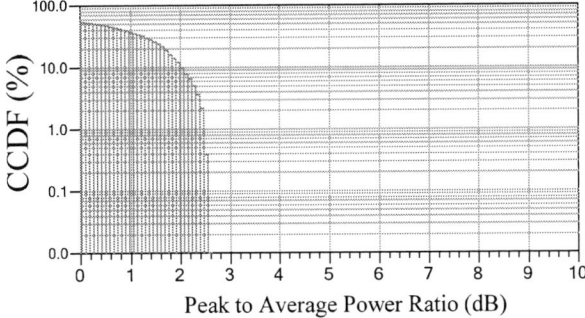

Figure 14: CCDF of an eight-tone signal

In order to make comparison between the developed LSM and measurements we performed envelope simulation on ADS software. Figure 15 illustrates the comparison between the LF drain current waveforms and time-domain envelope power measured and those obtained with the envelope simulation with the proposed nonlinear model, for an eight-tones signal at a load impedance equal to 50 Ω. The simulation time in a PC with a medium computation power is less than thirty seconds. With a typical foundry model like as BSIM3 or equivalent, this simulation exceed 1 day: a gain of factor of 3000. At the top of Figure 15 are plotted drain current waveforms for two average power levels. The simulated results are drawn in red line, and the measured ones in blue. At the bottom the time-domain output power envelopes are compared to LSM simulations. The MASMOS® model presents a good prediction. The simulated output power envelope evolves in the same way that the measurements.

V. CONCLUSION

A new model for the MASMOS® has been described in this paper. The model proposed reproduces I-V characteristics for one MASMOS® transistor. Also, it is demonstrated that LF S-parameters measurements are needed to determine nonlinear intrinsic capacitance. A one dimensional nonlinear model capacitance has been used in order to reduce simulation time. The developed model demonstrates an excellent agreement in both small-signal and large-signal operations.

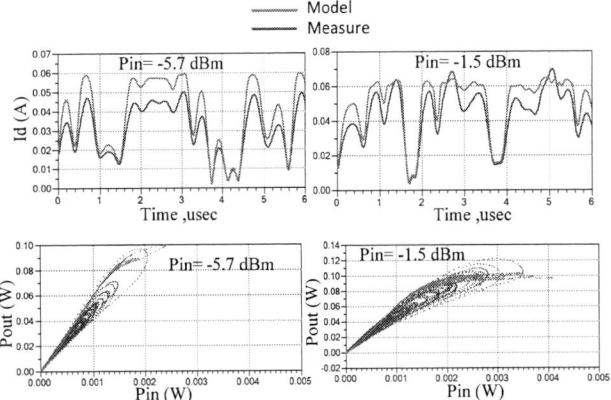

Figure 15: Drain current waveforms and time-domain output power envelope at two average power levels for eight-tone signals

VI. REFERENCES

[1] O. Jardel, G. Callet, C. Charbonniaud, J. Jacquet, N. Srazin, A. Morvan, R. Aubry, M.-A. Di Forte Poisson, J.-P. Teyssier, S. Piotrowicz and R. Quéré, "A new nonlinear HEMT model for AlGaN/GaN switch applications," *Proceedings of the 4th European Microwave Integrated Circuits Conference,* 28-29 September 2009.

[2] S. Forestier, T. Gasseling, P. Bouysse, R. Quere et J. and Nebus, «A new Nonlinear Capacitance Model of Millimeter Wave power PHEMT for Accurate AM/AM-AM/PM Simulations,» *IEEE microwave and wireless components letters,* vol. 14, January 2004.

[3] F. Ogboi et &. Al, «A LSNA configured to perform baseband engineering for device linearity investigations under modulated excitations,» *Microwave Conference (EuMC), 2013 European,* pp. 684-687, 6-10 Oct. 2013.

[4] S. Farsi et &. Al., «Characterization of intermodulation and Memory Effects Using Offset Multisine Excitation,» *Microwave theory and Techniques, IEEE Trans,* vol. 62, n° 3, pp. 645-657, March 2014.

[5] N. De Carvalho, «On the use of multitone techniques for assessing RF components intermodulation distorsion,» *IEEE trans.,* vol. 47, n° 12, 1999.

[6] M. A. Chaudhary, J. Lees, J. Benedickt et P. Tasker, «A sampling Oscilloscope based System with Active RF/IF Load-pull for Multi-tone Non-linear Device Characterization,» *International Journal of Microwave and Optical Technology,* vol. 8, n° 3, May 2013.

[7] R. Hajji, F. Beauregard et F. Ghannouchi, «Multitone Power and Intermodulation Load-Pull Characterization of Microwave Transistors Suitable for Linear SSPA's Design,» *IEEE Transactions on Microwave theory and techniques,* vol. 45, n° 7, July 1997.

[8] J. Teyssier, J. Sombrin, R. Quere, S. Laurent et F. et Gizard, «A test set-up for the anlysis of multi-tone intermodulation in microwave devices,» *Microwave Measurement Conference (ARFTG),* pp. 1-4, 2014.

[9] J. Sombrin, «On the Formal Identity of EVM and NPR Measurement Methods: Conditions for Identity of Error Vector Magnitude and Noise Power Ratio,» chez *Proceedings of the 41st European Microwave Conference,* Manchester, 2011.

[10] J. Sombrin, «Future test benches for the optimization of spectrum and energy efficiency in telecom nonlinear RF components and zmplifiers,» *ARFTG,* Montréal, 2012.

[11] Agilent Technologies, *Characterizing Digitally Modulated Signals with CCDF Curves, Application Notes,* 2000.

Focal-Plane Image Encoder with Cascode Current Mirrors and Increased Vector Quantization Bit Rate

Fernanda D. V. R. Oliveira, Tiago M. de F. Lopes, José Gabriel R. C. Gomes,
Fernando A. P. Barúqui and Antonio Petraglia
Universidade Federal do Rio de Janeiro – COPPE – PEE – Rio de Janeiro, RJ 21941-972
E-mails: fernanda.dvro@poli.ufrj.br, tiagomflopes@poli.ufrj.br, gabriel@pads.ufrj.br,
fbaruqui@pads.ufrj.br, petra@pads.ufrj.br

Abstract—**Focal-plane processing is the target of many studies due to its potential for enhancing the speed of the vision system flow. With focal-plane processing it is possible to perform parallel processing throughout the entire matrix. Usually, in vision systems, analog-to-digital conversion (ADC), transmission and storage represent a bottleneck. In order to alleviate these constraints, analog image compression is implemented at the focal-plane, thereby reducing the amount of data to be transmitted and the bandwidth requirements. The ADC is performed at the focal-plane as well, after the compression operation whose realization is based on differential pulse-code modulation (DPCM), linear transform and vector quantization (VQ) applied on every 4 × 4 pixel block using current-mode circuits. This paper presents experimental results obtained from the second-generation of an image sensor. The main contributions in comparison to the previous realization are: increase of the vector quantizer complexity, number of bits per pixel, pixel matrix size, and the use of cascode current mirrors in the linear transform matrix. The image sensor advanced in this paper was fabricated in a standard 180 nm CMOS process.**

I. INTRODUCTION

The CMOS image sensors framework applies to countless number of applications. Nowadays it is possible to find these sensors from simple cameras, such as the ones used for surveillance, to high quality cameras used by professional photographers. Industrial and academic fields invest resources on studying these sensors because of the image quality that is possible to achieve using the CMOS image sensors and due to their flexibility. An interesting feature of the CMOS image sensors is the possibility of introducing processing hardware in the same chip of the pixel matrix [1]. This allows for the design of an entire system on chip and has the potential for enhancing speed and reducing power of vision systems, very useful characteristics for embedded circuit applications. Furthermore, if we consider adding processing hardware inside the pixel matrix, every pixel would be part of a processing unit, and we can perform parallel processing throughout the entire matrix. This technique is called focal-plane processing and has lately been the topic of many publications [2]- [5].

Usually, in a vision system chain, all the pixels values from the pixel matrix must be converted to digital, sent out of the capture chip, stored in a intermediate memory and sent to a digital processor that will perform a desired task such as image compression, object recognition, face detection, among others. The bottleneck of this chain is the data transmission

from the pixel matrix to the intermediate memory. The analog to digital conversion, the transmission out of the chip and the storage in the memory are all steps that require significant amount of time in the vision system flow. In order to reduce time it is interesting to perform compression inside the pixel matrix chip, thereby also reducing the amount of data that need to be transmitted. Another advantage in doing compression inside the chip is to alleviate the bandwidth requirements of the system.

Our goal is to perform data compression at the focal plane. Current-mode analog circuits are used to implement differential pulse-code modulation, linear transform and vector quantization in every 4 × 4 pixel block [6]. Two generations of a compression imager have been designed, fabricated and tested. This paper presents experimental results for the second generation chip and compares them with results from the first generation ones [6]. Based on the experimental results from the first chip, improvements were included in the design of the second prototype. Among the modifications in the new design we highlight five main changes:

- VQ complexity: a new input component was added with the goal of being able to capture more details and improving the modulation transfer function. Five input dimensions are used now, instead of four, thus increasing the VQ complexity.
- Number of bits: due to the changes in the VQ, one additional bit is necessary to represent the component sign, and it was decided to add two bits for the representation of the new component absolute value.
- Cascode current mirrors: the linear transform matrix was implemented using cascode current mirrors, instead of simple current mirrors, as used in the first generation. A study was performed in order to verify that the cascode brings advantages to the implementation [7].
- Technology: the new chip was designed for fabrication in a 0.18 μm technology, while the previous one was designed for a 0.35 μm technology.
- Pixel matrix size: a 64 × 64 pixel matrix was designed for the second generation chip, in comparison with 32 × 32 for the first one.

Section II gives a brief explanation of the algorithm, highlighting the differences between the designs. The circuits that

978-1-5090-2737-8/16 $31.00 © 2016 IEEE

are required to implement the proposed data compression are shown in Section III. Section IV presents a qualitative comparison between the results obtained from both chips using similar targets. Section V closes the paper with a final discussion and ideas for future work.

II. BLOCK-BASED IMAGE COMPRESSION

The implemented image compression technique is block-based, performed in every 4×4 pixel block. This technique is explained in detail in [6]. The compression is divided into two parts: compression of the mean value of the block, also called DC component, and compression of the details of the block, also called AC component.

For the DC component representation, we use a DPCM encoder. The idea is to transmit, instead of the input signal itself, the difference between the input signal and a prediction of this signal [8]. With a good prediction, most of the values transmitted will be close to zero, which permits prioritizing the most likely values to be transmitted and even use less bits, if we can afford to lose the most unlikely values. For a natural image, there is a high probability of two neighbor pixels having close values, so a good prediction for a pixel is to use the value of a neighbor that was previously sent. This idea can be extended to the mean value of pixel blocks. That is, the mean value of a neighbor block is a good prediction for the mean value of the block we want to transmit.

In the following paragraphs, all algebric symbols refer to Fig. 1, which summarizes the operation of DPCM, linear transform and VQ performed inside a single 4×4 pixel block. For convenience, due to circuit simplifications, in our DPCM, instead of the mean value of a block, we use the sum of the pixels inside the block, s(n). Four bits are used for transmission, in which one, D_1, is for the sign and three, D_2, D_3 and D_4, are for the absolute value of the difference between the sum of a block and the sum of the previous block at the same row (e(n) = s(n) - ŝ(n)). The prediction value for the next block is given by the equation ŝ(n+1) = ê(n) + ŝ(n), where ê(n) is the decoded value that represents e(n) and ŝ(n) is the prediction for the current block. The maximum difference that can be encoded in our system is approximately equal to the pixel output signal range center value. This means that whenever there is a large transition from one block to the next, such as transitions from a white to a black block, the DPCM is not able to represent it properly. The choice of not representing steps higher than the middle of the signal range comes from the DPCM training, that considers the PSNR (peak signal to noise ratio) of the reconstructed images and the maximum number of desired bits. For focal-plane image compression, we use a low bit rate, which is enough for keeping the image quality at an acceptable minimum.

DPCM is performed for every row of blocks. A reference value is considered for the first block of every line. The main difference between the first-generation design DPCM and the second-generation DPCM is this reference value. On the first generation chip, this reference was equivalent to the lowest possible value from the signal range, while on the

new generation this reference was designed to be equivalent to the middle value of the signal range. This change in DPCM initialization was based on system-level simulations that indicated a PSNR increase associated with the new DPCM initialization. The consequence of considering the lowest value of the signal range is that the blocks at the beginning of every row became dark. This is not a problem when the mean of the blocks are closer to black, but if the first block is brighter, then we are not able to reach the correct average luminance value. On the other hand, when the reference is closer to the middle of the range it is easy to achieve darker or brighter values.

In the case of the AC component, we use linear transform and vector quantization for its compression and representation. As mentioned before, neighboring pixels usually have very close values, which means that inside a block the values are correlated. In other words, there is a significant amount of redundant data that does not need to be transmitted. The linear transform is responsible for changing the signal domain into a more efficient one, thus reducing redundancy and concentrating signal energy. The computation consists in multiplying the block of pixel values by a carefully chosen matrix that aims at maximum decorrelation. The output of this operation are transform coefficients. If these coefficients are multiplied by the 4×4 pixel-blocks that compose the transform basis, and the resulting pixel-blocks are added up, then the pixel-block texture details are reconstructed. The basis is composed by 1 DC component and 15 AC components of increasing frequency. The DC component was already previously encoded by the DPCM. Because of silicon area constraints, we encode a smaller set of AC components containing the highest-energy ones.

For the previous design, only four linear transform components were used. For the second design, we decided to use five linear transform components aiming at increasing the image quality and the ability to reconstruct finer image details. System-level simulations showed a significant PSNR increase with the inclusion of the new component [9]. After the linear transform, we compute the absolute value of the components, encode the component signs using one bit per sign (from S_1 to S_4 in the previous design and from S_1 to S_5 in the new design) and send the absolute values to be encoded by a vector quantizer.

The vector quantizer is a generalization of the scalar quantizer. It performs analog-to-digital transformations, but instead of considering each signal separately, it encodes the entire vector of components aiming at decreasing entropy and distortion [10]. Adding a new linear transform component has as direct consequence the bit rate increase by 1 bit, which is required to represent the sign of this component, and the VQ complexity increase. The VQ bit rate was also increased. Because of the inclusion of the new dimension, we have obtained better results when using 9 bits for the VQ (from B_1 to B_9), instead of 7 (from B_1 to B_7), as in the previous design. Reference [9] presents a theoretical comparison of both VQs and justifies the addition of a linear transform component at

978-1-5090-2737-8/16 $31.00 © 2016 IEEE

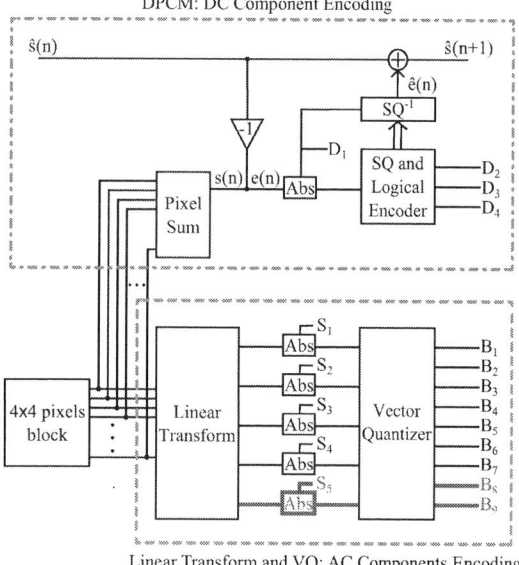

DPCM: DC Component Encoding

Linear Transform and VQ: AC Components Encoding

Fig. 1. Image compression algorithm block diagram. Inside the dash-dot green box: DC component encoding using the DPCM. Dashed blue box: AC components encoding using linear transform and VQ. The thicker lines indicate the second-generation chip changes with respect to the first.

the VQ input.

The DC and AC components data encoding flow can be seen separately in Fig. 1 highlighted by the boxes in dash-dot and dashed lines, respectively. The differences between the chip generations are indicated with the thicker lines. As it can be seen in the figure, in the previous design 15 bits were transmitted for each 4×4 pixel block. In the new design we have 18 bits per block. The bit rate increase is justified by the image quality improvement [9].

III. SCHEMATIC DIAGRAMS

Current-mode circuits were used to implement the compression algorithm. One of the main advantages of using the current-mode is that it allows for maintaining the signal range when it is necessary to change the power supply when scaling technologies. In our case, the current mode simplicity yields an important advantage over voltage-mode implementations with respect to signal summation. The sum of two currents can be performed simply by connecting both currents sources to the same node. To multiply a current by a fixed value, current mirrors with properly adjusted width and length are used. Those basic operations are necessary for the image compression algorithm implementation.

Fig. 2 (a) shows the photodiode readout circuit, which maps the photocurrent into a current that will be processed by the circuits that implement data compression. The readout circuit has three control signals, $Reset$, P_1 and P_2. When the reset is activated, the M_1 switch closes and the photodiode node, V_{ph}, is set to a initial value. As soon as the reset signal goes down and M_1 opens, we sample this value by opening the switch controlled by P_1. At the same time, the photodiode starts discharging the node V_{ph}, proportionally to the incident

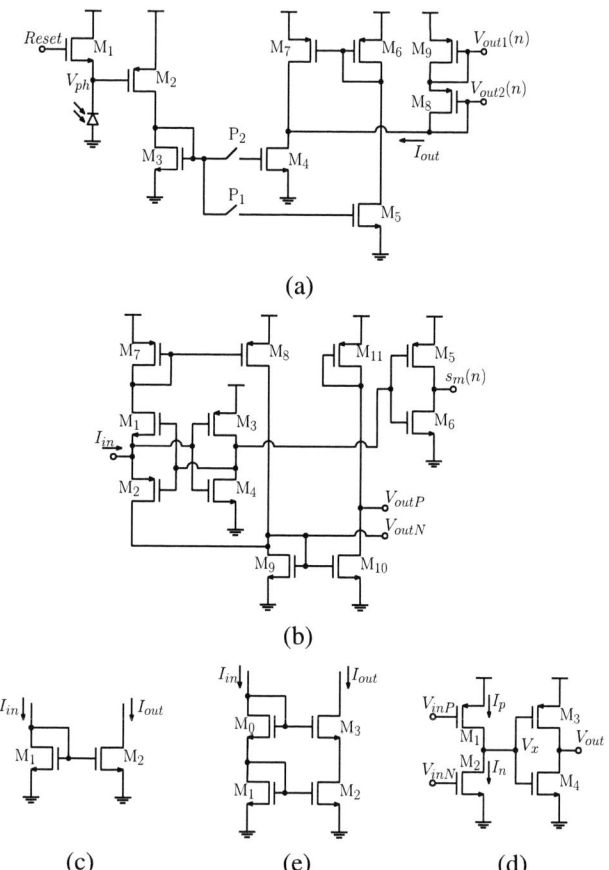

Fig. 2. Circuits for focal-plane image compression algorithm implementation in current mode: (a) pixel readout circuit; (b) absolute value circuit; (c) a simple current mirror; (d) a cascode current mirror; and (e) a current comparator.

light. After the integration period, which is the time interval during which the photodiode is kept working according to the incident light, we open the switch controlled by P_2 to obtain a second sample. The computation of the difference between the first and the second sample, which is a technique called CDS (correlated double sampling), is performed at the output node and is important to reduce fixed pattern noise. Transistors M_8 and M_9 transform the output current I_{out} into two output voltages, V_{out1} and V_{out2}, that are used by the processing circuitry to copy I_{out}.

The absolute value circuit is shown at Fig. 2 (b). As can be seen in Fig. 1, this circuit is required for DPCM and generation of the VQ input. Its input signal, I_{in}, is a current of either direction and its outputs are a bit that represents the direction of the input current, s_m, and two voltage references, V_{outP} and V_{outN}, from where the absolute-valued current can be copied in the desired direction.

Simple and cascode current mirrors are used in the new design. These circuits are presented in Fig. 2 (c) an (d), respectively. In both cases, the output current I_{out} will be equal to $\mathbf{K} \cdot I_{in}$, where $\mathbf{K}$ is the ratio between the dimensions of the output transistor (M_2) and the input transistor (M_1). The linear transform operation can be written as a sum of weighted pixel values. The current mirror is used to perform the multiplication of the current proportional to the pixel value. The first design used only simple current mirrors, but these circuits are not as accurate as the cascode current mirrors. The study presented in [9] showed that the use of cascode current mirrors for the linear transform operations is very beneficial. Current mirrors are also used for VQ implementation. To allow for the implementation of the VQ using simple analog circuits, we use a sub-optimal approximation consisting of another linear transform followed by a scalar quantizer. For this linear transform, simple current mirrors are used in both designs, since the advantages of using cascode at this part of the circuit were not significant [9].

The last circuit presented is a current comparator, Fig. 2 (e). The output of this circuit is a bit that indicates which current is higher. If I_p, which is given by the input voltage V_{inP}, is larger than I_n, given by V_{inN}, the node V_x will be charged and the output voltage V_{out} will be as low as to represent a logic zero bit. On the other hand, if I_n is larger than I_p, the node V_x will be discharged and the output bit will be one. This circuit is used for the analog-to-digital conversion inside the pixel block.

The new circuit was designed using a 180 nm CMOS technology. The layout of the 4×4 pixel block that is used throughout the pixel matrix is presented in Fig. 3. For data compression to be held inside the block, 833 transistors are necessary. On the previous design, that used only simple current mirrors and one less linear transform component, there were 607 transistors per block. Even with the increase in the number of transistors, using a 180 nm technology instead of the 350 nm technology from the first design, allowed for an increase in the fill factor, from 7.1 % to 13.5 %, and a reduction of the pixel pitch, from 37.5 μm to 27.2 μm.

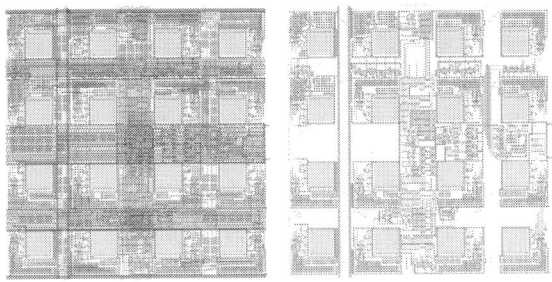

Fig. 3. Pixel block layout. On the left, complete layout, featuring all metal layers, and on the right, for better visualization, layout showing only the first metal layer, transistors and photodiodes.

IV. Experimental Results

Experimental tests are being performed with the purpose of characterizing the new chip and identifying whether the modifications were advantageous. In order to do that, an experimental setup has been designed so that a micro-controller could be used as the interface between a computer and the chip. The chips from both generations use the same lenses and structure shown in [6], but a new circuit board was designed to adapt for the requirements of the new chip. For example, the transistors from the first generation operate with a power

supply of 3.3 V, while the second generation ones work with 1.8 V.

Aiming at comparing the results from both chips, the same targets used for the first generation [11] were employed to test the new chip. Those are black and white images with geometrical shapes that allow for the visualization and first evaluation of the DPCM and VQ stages, as well as the final decoded image.

Figures 4 and 5 show the results for both generations when a white circle and a striped pattern are used as targets, respectively. It is important to note that although the images are presented with the same size, the resolutions are different. The images generated by the first chip have 32×32 pixels while the images from the second chip have 64×64 pixels. In both figures, the reconstructed partial image using only the DPCM bits can be observed at the first row, the reconstructed partial image employing only the VQ bits is at the second row, the final image is presented at the third row, and the mean image after 100 captures is shown at the fourth row. The reconstructed final images have being filtered to remove noise and enhanced for a better visualization. For all the presented figures, the DPCM row starts on the left side of the image.

If we compare the borders of the objects presented in the images of both chips, we can observe the improvement from the changes performed in the VQ. The transition between black and white is now more precise and better defined. On the images that show the VQ partial results the objects can be clearly seen. It is also interesting to note that on the darker regions there is much less noise in the results from the new design than from the former one.

On the other hand, observing the final reconstructed images, we can see that the beginning of the image is brighter than what should have been. This effect is produced by the DPCM. It seems that the reference value for the first block, that was supposed to be in the middle of the dynamic range, is higher than expected. As a consequence, it takes more steps to reach darker values than what was expected. In order to alleviate this effect, the DPCM is being carefully studied. Aside from this DPCM unexpected result that degrades the image quality, the overall result shows that the image quality has improved.

Figure 5 is interesting because it shows an image where the transition between white and black stripes are narrower than a DPCM block, which means that the VQ is the predominant signal in the image. In this figure, the new generation image has twice the number of black stripes compared to the first generation image. Since the resolution changes from the first design to the second, we depict these images to guarantee that a fair comparison is presented. That is, in a 32×32 section of the new generation image there is approximately the same number of stripes as that of the first generation chip. Comparing the images in the second row of Fig. 5, we see that the VQ is more efficient for the new design, since the stripes have a better definition.

More results can be seen in Fig. 6, where the VQ partial results and a snapshot image for various targets are displayed. In the snapshot images from the second row of Fig. 6, we

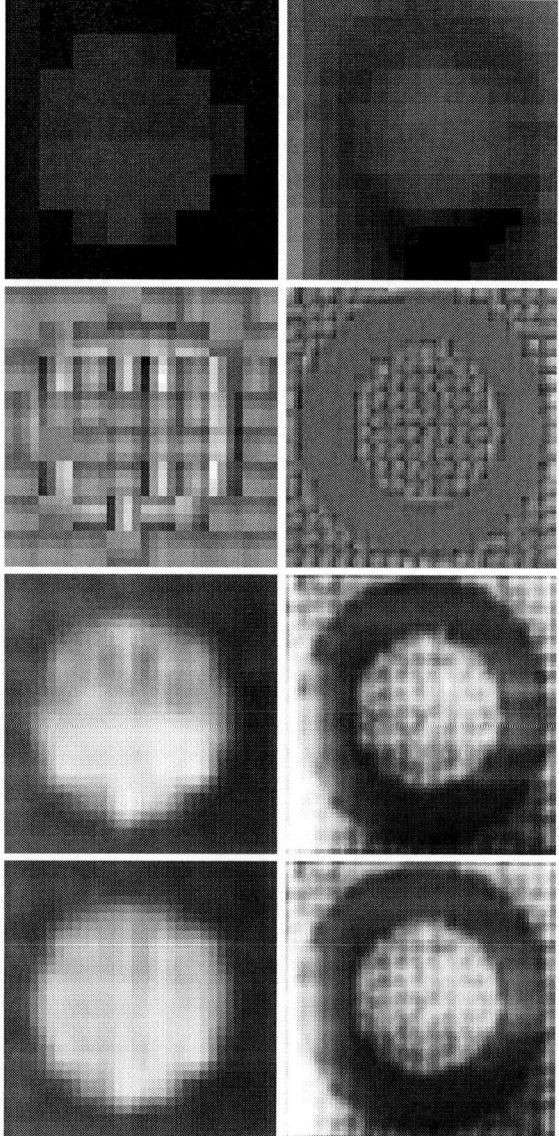

Fig. 4. Results when the target image is a circle. On the left column, results from the first generation chip, on the right column, second generation chip. From top to bottom: DPCM partial result, VQ partial result, decoded image after filtering and enhancement, and the result after computing the mean of 100 decoded images.

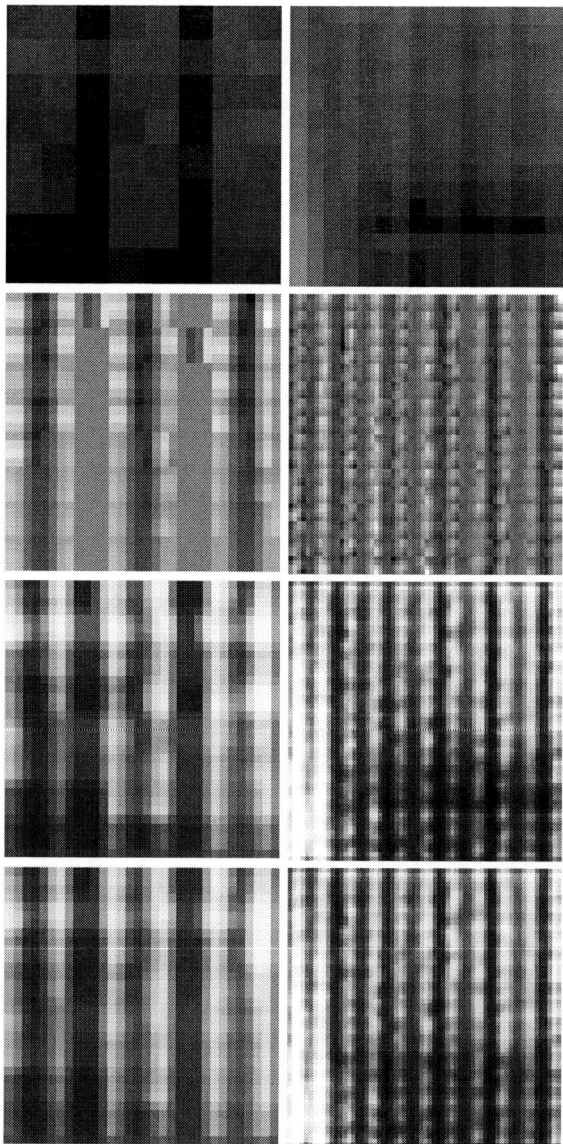

Fig. 5. Results when the target is a striped black and white pattern. On the left column, results from the first generation chip, with a target of 1.67 cycle/cm spatial resolution; on the right column, results from the second generation chip, with target of 3.33 cycle/cm spatial resolution. From top to bottom: DPCM partial result, VQ partial result, decoded image after filtering and enhancement, and the result after computing the mean of 100 decoded images.

can see two errors generated from the DPCM: the first one is the white blur that appears in the image from the second column, and the second one is the black blur that appears in the image from the fourth column. Both errors seem to have the same nature, since the DPCM noise assumes values that are higher or lower than what was supposed to be and the total noise causes the blur. The VQ results confirm the conclusion drawn from Figs. 4 and 5 that there was a significant quality improvement in this section of the compression algorithm, as in the arrow image, for example, shown in the fourth row and third column of Fig. 6. In the first generation VQ, the arrow can be hardly seen if we look just at the VQ result, while the VQ from the new design clearly represents the arrow.

V. CONCLUSION

This paper compared the performances of two chips, both capable of capturing and compressing at the focal plane, in terms of algorithm and experimental results. Table I summarizes the main differences between the chips. The results showed that there was a significant improvement from the first chip to the second one in the linear transform and vector quantization steps, since the borders of the images are better defined. On the other hand, the DPCM stage can still be improved. We can observe two main behaviors that reduce the image quality: some blocks at the beginning of the rows are brighter than they were supposed to be, and there is a visible gradient between successive DPCM rows. The first problem

TABLE I
Comparison between first and second generation chips.

	1st generation	2nd generation
Bit rate	0.94 bpp	1.13 bpp
Transform coeffs.	4	5
Sign bits	4	5
VQ bits	7	9
Fab. process	AMS 0.35 μm Opto	IBM 0.18 μm
Transistor count	607 per block	833 per block
Pixel area	37.5 μm $\times$ 37.5 μm	27.2 μm $\times$ 27.2 μm
Photodiode area	10 μm $\times$ 10 μm	10 μm $\times$ 10 μm
Fill factor	7.1 %	13.5 %
Chip area	2.4 mm $\times$ 2.1 mm	2.8 mm $\times$ 2.8 mm
Resolution	32 $\times$ 32	64 $\times$ 64
DPCM $\hat{s}(1)$	0.0	7.5
Power supply	3.3 V	1.8 V

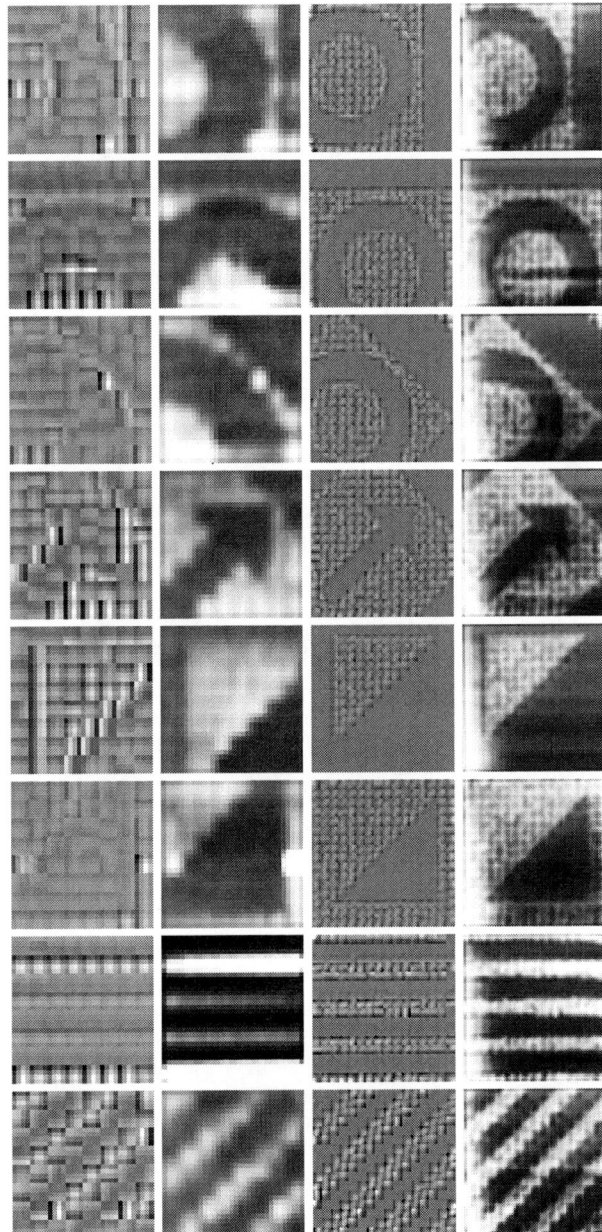

Fig. 6. From left to right columns: (1) VQ partial results from the first generation chip and (2) corresponding snapshot, (3) VQ partial results from the second generation chip and (4) corresponding snapshot.

is capable of perceiving. In the case of the FPN, it is important to identify the amount of noise in the images, compare it with the previous design and identify its sources.

ACKNOWLEDGMENTS

This work was supported by Brazilian higher education and research funding agencies: CAPES, CNPq, and FAPERJ. Special thanks to CMsatisloh for providing us with the optical setup necessary for the experimental tests.

REFERENCES

[1] J. Nakamura, *Image Sensors and Signal Processing for Digital Still Cameras*, 1st ed. EUA: CRC Press, Talyor & Francis Group, 2006.
[2] S. Vargas-Sierra, G. Linan-Cembrano, and A. Rodríguez-Vázquez, "A 151 dB high dynamic range CMOS image sensor chip architecture with tone mapping compression embedded in-pixel," *IEEE Sensors Journal*, vol. 15, no. 1, pp. 180–195, 2015.
[3] J. Fernández-Berni, R. Carmon-Galán, and A. Rodríguez-Vázquez, "Single-exposure HDR technique based on tunable balance between local and global adaptation," *Circuits and Systems II, IEEE Transactions on*, 2015.
[4] S. Chen, A. Bermak, and Y. Wang, "A CMOS image sensor with onchip image compression based on predictive boundary adaptation and memoryless QTD algorithm," *Very Large Scale Integration Systems (VLSI), IEEE Transactions on*, vol. 19, no. 4, pp. 538–547, 2011.
[5] W. D. Leon-Salas, S. Balkir, K. Sayood, N. Schemm, and M. W. Hoffman, "A CMOS imager with focal plane compression using predictive coding," *IEEE Journal of Solid-State Circuits*, vol. 42, no. 11, pp. 2555–2572, 2007.
[6] F. D. V. R. Oliveira, H. Haas, J. G. R. C. Gomes, and A. Petraglia, "CMOS imager with focal-plane analog image compression combining DPCM and VQ," *Circuits and Systems I: Regular Papers, IEEE Transactions on*, vol. 60, no. 5, pp. 1331–1344, 2013.
[7] F. D. V. R. Oliveira, J. G. R. C. Gomes, and A. Petraglia, "Influence of cascode and simple current mirrors in inner product implementations for CMOS imagers," in *Circuits Systems (LASCAS), 2015 IEEE 6th Latin American Symposium on*, Feb 2015, pp. 1–4.
[8] S. Haykin, *Communication Systems*. Quarta Edição. EUA: John Wiley & Sons, Inc., 2001.
[9] F. D. V. R. Oliveira, J. G. R. C. Gomes, and A. Petraglia, "Comparison of low-complexity image compression algorithms for analog circuit implementation," in *2014 14th International Workshop on Cellular Nanoscale Networks and their Applications (CNNA)*, July 2014, pp. 1–2.
[10] A. Gersho and R. M. Gray, *Vector Quantization and Signal Compression*. Massachusetts, EUA: Kluwer Academic Publishers, 1992.
[11] F. D. V. R. Oliveira, H. L. Haas, J. G. R. C. Gomes, and A. Petraglia, "Current-mode analog integrated circuit for focal-plane image compression," in *Integrated Circuits and Systems Design (SBCCI), 2012 25th Symposium on*, Aug 2012, pp. 1–6.
[12] G. D. Boreman, *Modulation Transfer Function in Optical and Eletro-Optical Systems*. Primeira Edição. Bellingham, Washington: SPIE, 2001.

seems to be due to the imprecision of the reference value for the first block in a row. The second problem is probably due to fabrication imperfections. Both are being studied using theoretical and practical approaches, with simulations and circuit analysis and with additional experimental tests.

Furthermore, it is important to perform experimental tests that allow us to characterize the chip in terms of figures of merit, such as modulation transfer function (MTF) and fixed pattern noise (FPN). The MTF defines the spatial frequency response [12] of the chip by measuring the contrast of images with white and black stripes, such as the ones presented in Fig. 5. Targets with increasing spatial frequency are used with the goal of determining the highest frequency that the sensor

978-1-5090-2737-8/16 $31.00 © 2016 IEEE

An Ultra Wide Band Analog-to-Digital Converter based on a Delta-Riemann architecture

Francois Rivet*, Elina Fiawoo*, Richard Montigny[†], Patrick Garrec[†], Yann Deval*
*University of Bordeaux, Bordeaux, France,[†]Thales Airborne Systems, Pessac, France
Email: francois.rivet@ims-bordeaux.fr

Abstract—**Wireless telecommunication network requires the use of new frequency bands to increase the data rate. The upcoming 5G standard will address in its lower band (sub 6GHz) at least 10 aggregated LTE carriers with 64 to 256QAM modulation scheme. Whether those bands are high in frequency, 20MHz wide and spread over several GHz, the analog-to- digital conversion becomes a critical technical bottleneck for any ultra low power receivers. Challenges are to convert simultaneously with a complete synchronization all the various bands while keeping in mind constraints of RFIC dedicated for mobile terminals. This paper presents a disruptive architecture of an analog-to- digital converter based on a delta quantization with a closed-loop including a Riemann Pump. A system overview is proposed with behavioural simulations results in MatLab and VHDL-AMS. A prospective of integration in 28nm CMOS technology will target a direct application for sub-6GHz 5G standard with a 4-bit resolution for a frequency range from 1.8 to 3.6 GHz.**

Keywords: analog-to-digital, RFIC, Riemann Pump, CMOS

I. INTRODUCTION

Digital technology takes a major place in electronics so much it seduces by its simplicity, its flexible programming flexibility, especially regarding the telecommunication standards, its reproducibility, less sensitive to the environment (EMC) and its quality. Although present in numerous domains, the domain of telecommunications was particularly affected by the digital era. Whether its influence is directly felt by the population or hidden in new features, the digital technology will be pushed in its last cuttings off to satisfy numerous domains such as the upcoming 5G standard.

However, in numerous applications, digital technology cannot completely be a substitute for analog [1]. Indeed for instance, in telecommunication RF signals require analog circuitry to be sent. As a consequence, Analog-to-Digital (AD) interface and vice versa are required. More commonly referred to as Digital-to-Analog Converter (DAC) or Analog-to-Digital Converter (ADC), these interfaces constitute a critical bottleneck. Technological breakthroughs of these last years have already helped to improve their speed, precision and integration. In parallel, their power and energy consumptions were decreased. However, their architectures confine them very often to a certain type of physical signal specific to the characteristics of standards (resolution, input frequency range, sampling frequency) [2] [3] [4].

In a perfect world such as presented in Fig. 1, ADC and DAC would be infinitely fast, would not change from original values or add distortion or noise, would have an infinite input frequency range and would have a negligeable power consumption. No such converters had been realized yet but many different converters exist, each of them have their application domains.

This paper introduces the realization of an ultra wide band ADC. It aims at designing a new conversion architecture which enables to realize more efficient ADC.

- Ultra wide band, especially within the frequency range $1.8GHz$ to $3.6GHz$ to address the sub-6GHz 5G standard with several carriers aggregated,
- Low-power consumption,
- Reduced number of bits,
- High SNR (more than 60dB).

The paper presents a model and its simulation using MATLAB to exhibit system performances. Then, an analysis of the ADC performances is done using a VHDL-AMS behavioral model. It verifies the validity of MATLAB simulation results at a level of analysis slightly closer to the realization of circuit thanks to the use of behaviourial models.

II. DELTA-RIEMANN ADC ARCHITECTURE

This concept of this ADC is based on a signal derivative conversion, using a limited panel of predetermined slopes. In other words, we propose to quantify the first continuous derivative of the input signal by means of a finite number of slopes. At the origin, the idea was to get back only the useful information of the signal by identifying its extrema, i.e. the points when its first derivative is null. But this principle showed itself little relevant as it was not sufficient

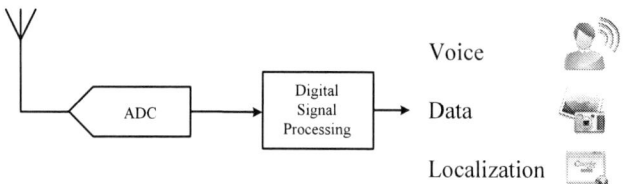

Fig. 1. A direct conversion of RF signals

978-1-5090-2737-8/16 $31.00 © 2016 IEEE

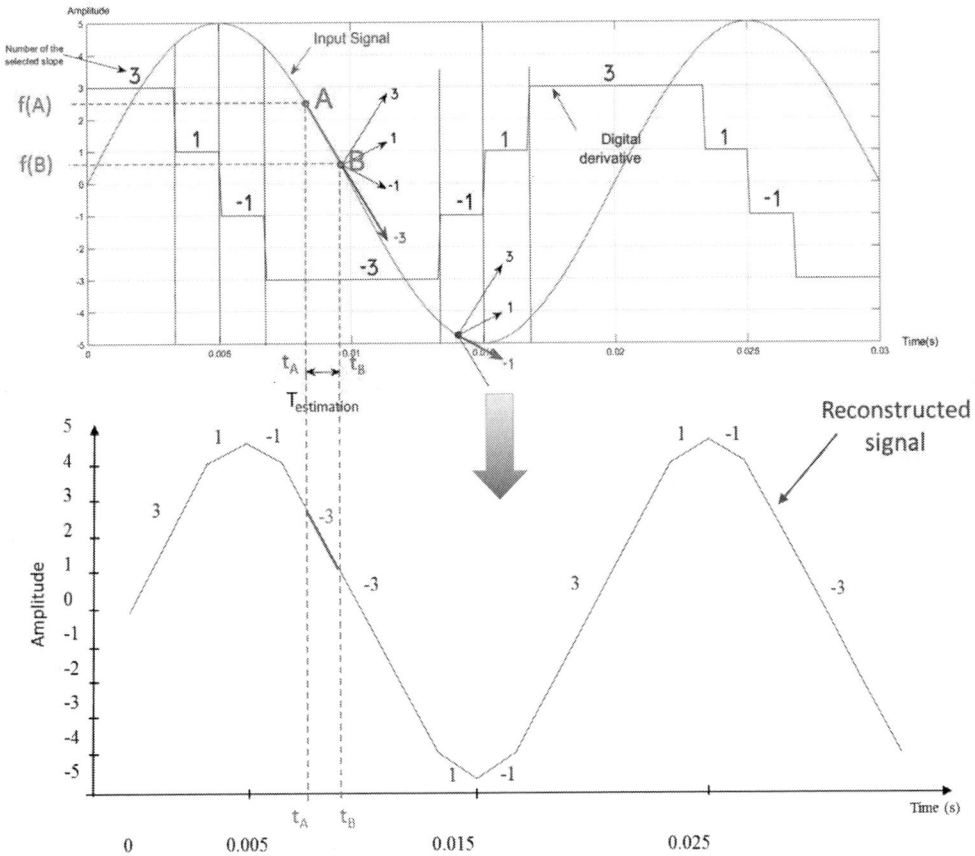

Fig. 2. Conversion principle and the corresponding reconstructed signal from the digital derivative

to reconstruct the signal and, in a way, evolved towards a quantification of the first derivative of the signal by returning to its definition.

Fig.2 depicts the principle of the conversion. Let us consider an analog input signal $f(t)$. By definition, its first derivative at any point A has for expression $f'(A) = \lim_{B \to A} \frac{f(B)-f(A)}{A-B}$. Thereby, for A and B close enough $\frac{f(B)-f(A)}{A-B}$ gives a certain estimation of the derivative. From another point of view, it gives an estimation of the slope of the signal between the points A and B. Taking this reasoning one step further, it turns out that when the evaluation of the slope is made in a given and known period $T_{estimation}$ the difference $f(B) - f(A)$ is sufficient to estimate the slope. A simple subtractor enables to retrieve this information.

The concept consists in comparing the difference supplied by the subtractor with a finite number of reference slopes or rather with the corresponding differences $f(B) - f(A)$ for the given estimation period, $T_{estimation}$. For each estimation, the closest reference slope was selected, as shown in Fig.2. Only the number of the selected slope is converted to digital. As a consequence, there is the possibility to realize an ADC

requiring few bits, which leads to a low-power consumption and to push signal processing in digital.

[5] presents the format of the slopes panel, the scaling of the slopes and the theory on the SNR. One can see in Eq. 1 that SNR depends on the number of bits N and the overspampling ratio (OSR) introduced by $f_s = 2^r.f_{max}$.

$$SNR_{dB} = 6,02N+9,03r-7,78+10\log_{10}(1 - \frac{1}{2^{N-1}} - \frac{1}{2^{2N}})$$
(1)

Nevertheless, a direct reconstruction is not as easy as it appears. The limited number of slopes, and thus conversion of the signal, introduces an error. This error on the derivative conversion could be compared to an offset on the reconstructed signal. We propose to add an integrator in a loop to compensate errors in real-time with mixed signals. The integrator added is a Riemann Pump presented in [5].

As shown in Fig. 3, the system is composed by 3 blocks: a subtractor, comparators and the Riemann Pump. The advantage of this architecture is that the ouput of the comparators is locked on the clock of the converter (f_s).

978-1-5090-2737-8/16 $31.00 © 2016 IEEE   139

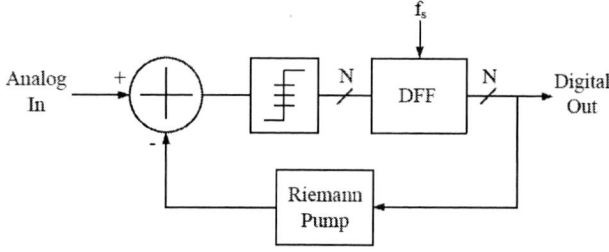

Fig. 3. Differentiating ADC based on the Riemann Pump. It is composed of a subtractor, comparators with N bits, DFF clocked at f_s and Riemann Pump

As a consequence, the delta-modulator gave the difference between the input signal and the point just reconstructed by the Riemann Pump, so the point that would be digitally reconstructed by the converter. The point of the signal just reconstructed by the Riemann Pump, thus at the instant t, is to match the point of the signal at the instant $t - T_{sampling}$. The system could take into account the errors of approximation which were introduced by both the comparators block, due to the fact that the reference slopes selected may not be equal to the slope of the output of the delta modulator, and the error of the sampling used to synchronize the DAC. We propose to name this conversion a Delta-Riemann ADC.

III. SIMULATION RESULTS

Simulations were performed to analyze the reconstructed signal (thus the output of the Riemann Pump) compared to the input signal.

A. Delta-Riemann model using MatLab

The model developped in Matlab do not take into account technlogical parameters or integrated circuits discrepancies. It is composed of a 4-bit Riemann Pump working at $f_s = 25GHz$. Simulation results are displayed in Fig. 5 exhibiting a proper behavior of Delta-Riemann ADC. Moreover, the spectrum of the reconstructed signal highlighted some parasitic harmonics which corresponded to the odd harmonic of the input signal. This latter could be explained by the fact that the main errors on the reconstructed

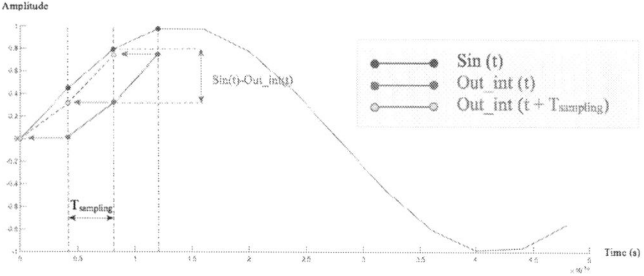

Fig. 4. Illustration of the principle of a Delta-Riemann ADC

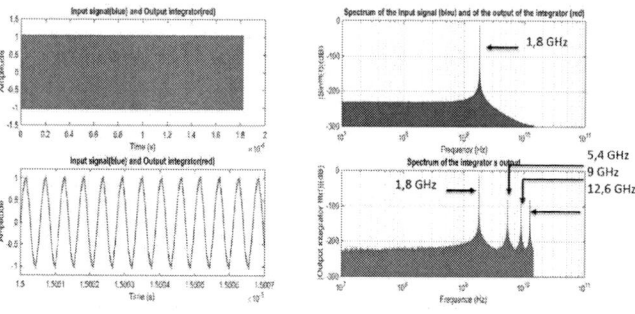

Fig. 5. Simulation results of Delta-Riemann ADC with 4 slopes, an input sinus at $1.8GHz$ and a sampling frequency f_s of $25GHz$. The integrator correspond to the Riemann Pump.

signal were done in wide input signal variation areas thus rounding the maximum and minimum values for a sinus. In these areas the slope selected were often once the highest then once the lowest. As a consequence, the reconstructed signal was close to the time signal of a sum of a sinus and its 3_{rd} and 5_{th} harmonics, as shown in Fig. 5.

B. Delta-Riemann model using VHDL-AMS

Delta-Riemann ADC is modelled in VHDL-AMS to include some electrical proprieties. The ADC is then modelled with:

- a digital generator ordering the switchs of the integrator (Riemann Pump. The pure ideal behavioral model being sufficient to validate the system this sub-system was described in pure VHDL language.
- a differentiator making the difference of the input signal and the output of the integrator
- an integrator based on [5]. It consists of:
  - a load capacity (10pF) corresponding to the output of the integrator. As the behaviour model of a capacity is relatively simple, it was then used in a netlist.
  - a circuit made of switches and current-generators. The switchs enabled to monitor current-generators in order to inject more or less current into the capacitor. By the charge or discharge of this capacitor at a certain speed, the slope of the output signal is generated. Switchs were controlled by digital signals but were also described through their current and voltage at their terminals according to electronics laws. Therefore, a mixed-model is given.

Fig. 6 displays simulation results. For reasons of homogeneity, the signals obtained with a transient simulation were extracted from VHDL-AMS and analysed with Matlab, following a strict protocol. Thus, the reconstructed signal extracted was the output of the integrator and in order to take into account the propable errors introduced by VHDL-AMS the input signal was extracted during the same simulation, both with a sample at the sampling frequency of the digital

978-1-5090-2737-8/16 $31.00 © 2016 IEEE

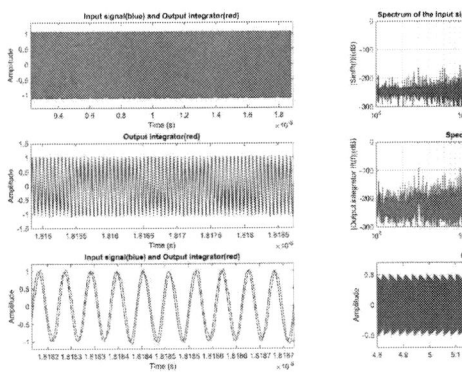

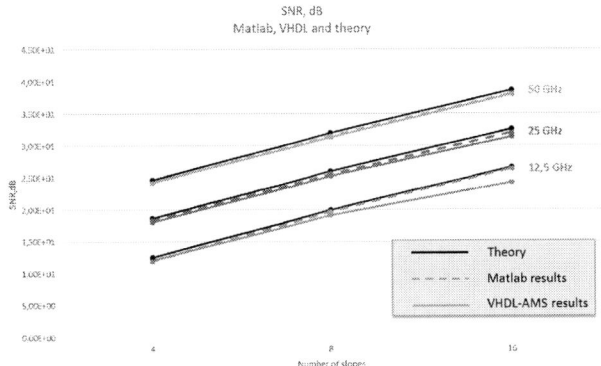

Fig. 6. Simulation results of Delta-Riemann ADC in VHDL-AMS with 4 slopes, an input sinus at $1.8GHz$ and a sampling frequency f_s of $25GHz$. The integrator correspond to the Riemann Pump.

Fig. 7. Comparaison between theory, MatLab and VHDL-AMS behavioral model

generator. Concerning the spectrum, VHDL-AMS clearly added some noise floor. This noise floor was more important for the reconstructed signal. Furthermore, the odd harmonics of the fundamental could slightly be glimpsed amongst the noise floor. However both the noise floor and the harmonics presented no potential perturbations for the functioning of the analog to digital converter as they barely overpassed $-100dB$ while the fondamentale were close to $-15dB$.

IV. MatLab and VHDL-AMS behavioral model results comparaison to theory

Fig. 7 presents SNR_{dB} extracted from MatLab and VHDL-AMS behavioral model results and theory in function of the number of bits N. As the SNR_{dB} estimation is concerned, the same number of point were taken in order to compare as fair as possible Matlab and VHDL-AMS results. The VHDL-AMS results globally confirmed the theoritical SNR tendance. Indeed, it seemed that increasing the number of reference slopes and/or the sampling frequency linearly increased the SNR_{dB} although a non negligeable difference for 16 reference slopes was highlighted. At 16 reference slopes, it seemed that the sampling frequency was critical. Indeed, at $12.5GHz$ the result was $2.5dB$ lower than the theory whereas it was $0.5dB$ lower at $50GHz$ thus almost the same range of error as for other numbers of slopes. This could be due to electronics delay phenomenons which could have shifted the different signal, thus introduce a mistake in the coorrespondance between each point of the input and of the reconstructed signal.

In the light of these final results, it seemed that the architecture could really be used as ADC. As the power consumption may increase with the numbers of reference slopes (one comparator for one reference slope), and that only 4 slopes might be too limiting for more complex inputs, the next step of this architecture will be to study higher number of reference slopes (such as 8) and at least a $25GHz$ sampling frequency.

V. Conclusion

Considering the strong requirements in wireless communications and the constraints on some converters, a Delta-Riemann is a potential candidate to overcome technological bottlenecks. Indeed, the principle idea could lead to the realization of low power ultra-wide band converters requiring fewer bits. The ideal behavioral models developed on MATLAB and VHDL-AMS demonstrated the feasibility and the relevance of the architecture. The Delta-Riemann ADC which strength lies in the fact that the loop takes into account the error made on the estimation of the previous sample but also the error made by the sampling stage of the converter.

However, this architecture needs a high sampling frequency to be relevant which is not so easy to fulfil at a transistor level especially considering all the different blocks needed. At present, the transistor level study is about to be completed in 28nm FDSOI CMOS technology which is a pre-requisite to high frequency system.

References

[1] F. Rivet, Contribution ltude et la ralisation dun frontal radiofrquence analogique en temps discrets pour la radio-logicielle intgrale, , PhD dissertation, Universit de Bordeaux, p.27, 2009.

[2] Jiangfeng Wu, A 5.4GS/s 12b 500mW Pipeline ADC in 28nm CMOS, Symposium on VLSI Circuits Digest of Technical Papers, 2013.

[3] Frank van der Goes, Chris Ward, Santosh Astgimath, Han Yan, Jeff Riley, Jan Mulder, Sijia Wang, Klaas Bult, A 1.5mW 68dB SNDR 80MS/s 2 Interleaved SAR Assisted Pipelined ADC in 28nm CMOS, ISSCC, 2014.

[4] David Bellasi, Luca Bettini, Thomas Burger, Qiuting Huang, A 1.9GS/s 4-bit Sub-Nyquist Flash ADC for 3.8GHz Compressive Spectrum Sensing in 28nm CMO, IEEE, 2014.

[5] Y. Veyrac, F. Rivet, Y. Deval, D. Dallet, P. Garrec and R. Montigny, "A 65-nm CMOS DAC Based on a Differentiating Arbitrary Waveform Generator Architecture for 5G Handset Transmitter," in IEEE Transactions on Circuits and Systems II: Express Briefs, vol. 63, no. 1, pp. 104-108, Jan. 2016.

A balanced logic routing block for Factorial-DLL based frequency generation

Yann Deval and François Rivet

IMS Lab - UMR5218
University of Bordeaux, Bordeaux Institute of Technology, CNRS
France
Email: yann.deval@ieee.org

Abstract— This paper presents the design of a balanced logic block that can act as either a delay line or a ring oscillator, with the addition of a voltage controlled element to pave the way for flexibility and tunability. This block is the key element of a Factorial Delay Locked Loop (F-DLL), and its balanced nature allows to significantly reduce the reference frequency-related spurious that are generated within the spectrum of a frequency generation unit when the latter is based on DLL. Once compared to the classical implementations of F-DLL logic block a Boolean routing is substituted to the switches. This allows a constant loading to the delay elements regardless the selected functionality of the logic block, namely delay line or ring oscillator. Doing so, the classical discrepancy that occurs in propagation delays in the two different modes is cancelled or at least minimized. The reference frequency-related error is therefore repealed.

Keywords—Delay Locked loop, voltage controlled delay line, frequency generation, RFIC, ultra-low power lectronics.

I. INTRODUCTION

As part of the highly expected Internet of Things (IoT) revolution, a tremendous number of object is likely to wirelessly connect themselves to the internet. Apart from the Internet Protocol (IP) software issue that will occur with the necessity of distinctive addresses for trillions of unique objects, from the hardware prospective the issue is different and determined by either the cost of a given object or its power consumption. Indeed, since tons of objects are anticipated it is mandatory that the cost of a single one is minimized to ensure wide diffusion. Also object power consumption has to be minimized as well, especially when wireless capability is of concern. Indeed it will reduce power supply maintenance, paving the way for an optimal object availability.

In this paper we focus on the frequency generation of an object, i.e. an IoT node. This subsystem is used to create the local oscillator of the node wireless interface. It has to be flexible and programmable, but among all as discussed it has to be low cost and ultra-low power.

Low cost behavior is achievable thanks to the reduction of the silicon footprint of Frequency Generation Unit (FGU). Actually most of the silicon area in classical FGUs is used by the passive part of the sinewave oscillator, the varactor and the inductor, with an emphases on the latter [1]. It yields to the adoption of an inductorless oscillator for our FGU. There are different possible options for inductorless oscillator implementations, such as relaxation [2] or phase-shifted

topologies [3], however the most efficient one in terms of silicon footprint is the ring oscillator since the latter does not rely on neither an inductor nor a capacitor, further reducing the circuit area and thus the cost required for its integration [4, 5]. Our FGU has thus been built around a ring oscillator, i.e. a loop of logic gates including an odd number of inverters. It leads to an astable configuration, unstable in either state so it continually switches from one state to the other, providing oscillations at the end of the day.

Low power behavior is another story. Once again a possible approach is to drive the circuit toward digital as much as possible. Indeed, as soon as a signal is analog it requires both a voltage and a current while a digital signal, due to its binary nature, can rely on solely a voltage or a current, independently. Consequently the overall power consumption of the considered circuit can be nulled assuming zero current when a voltage is used or zero voltage when a current is used instead. This paradigm is perfectly suited to the above-discussed ring oscillator as the latter is made of logic gates, and thus fully digital. Together with the addition of a highly advanced silicon technology such as the 28nm fully-depleted silicon on insulator (FD-SOI) ultra-thin body and buried oxide (UTBB) CMOS technology from STMicroelectronics [6], ultra-low power operation is then feasible.

This paper is based on the above mentioned principle to design an ultra-low power frequency generation unit based on a factorial DLL. The paper is organized as follow: in the first section the delay locked loop synthesizer mode of operation is detailed, and the factorial version of the DLL is discussed to demonstrate its key advantage in terms of flexibility and tunability. In the second section the issue of Duty Cycle Distortion (DCD) is presented and this periodic jitter that raises the harmonic spurs at frequency synthesizer output spectrum, corrupting the wanted signal, is reviewed. The third and last section presents a balanced logic routing block which, when implemented within the ring oscillator, dramatically reduces the unwanted DCD-related spurious.

II. DLL-BASED FREQUENCY GENERATION

A. Delay Locked Loop principle

When it was invented back in the 60's [7] the primary purpose of a Delay Locked Loop was to ensure synchronization of both clock and data within large digital integrated circuits. Indeed, when a circuit is large, significant delays can occur on

978-1-5090-2737-8/16 $31.00 © 2016 IEEE

either the clock or the data, mostly due to parasitic capacitors connected to the long wires upon which the signals are distributed within a given silicon die. Since these delays on clock and data are unlikely to match, it can generate asynchronous phenomenon, and thus functional failures. Synchronizing delays solves the issue, and this is what a DLL is supposed to do.

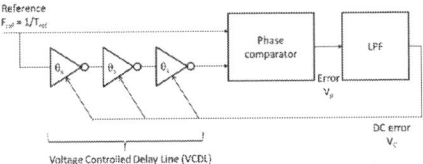

Fig. 1. DLL architecture.

The classical DLL topology is depicted in Fig. 1. A reference frequency Fref is injected together on a phase detector and a series of tunable inverters. The latter is the Voltage Controlled Delay Line (VCDL). Both the direct signal and its delayed version are compared at the inputs of the detector, and the error signal at its output is low pass filtered to extract its DC value. The loop is then closed as this DC error is used to tune back the VCDL. If the loop is a feedback and can lock, a steady state is obtained in which the DC error is a constant. It means the delay between the direct signal and its delayed version is a constant thanks to the loop operation.

To highlight the advantage of the DLL, we have to consider its linear model of Fig. 2.

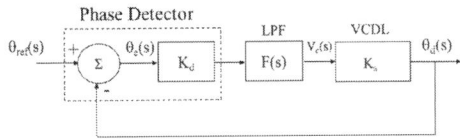

Fig. 2. DLL linear model.

When solved from the direct signal θref(s) to the delayed version θd(s) it comes the transfer function H(s) such as:

$$H(s) = \frac{\theta_{ref}(s)}{\theta_d(s)} = K_d K_s \cdot \frac{F(s)}{1 + K_d K_s F(s)} \qquad (1)$$

When the loop filter F(s) is made a ma mere capacitor loading a charge pump (1) becomes simply

$$H(s) = 1 / \left[1 + s \left(\frac{C}{K_d K_s} \right) \right]$$

yielding to ω_0, the pole of this first order system which is defined by the phase detector gain, the VCDL coefficient and the capacitor of the loop filter:

$$\omega_0 = K_d K_s / C \qquad (2)$$

Consequently a DLL with a loop filter reduced to a mere capacitor is a first order system, and thus unconditionally stable. Using a tiny capacitor not only the solid-state integration of the DLL is feasible but its bandwidth can be made very large at the same time, without the risk of instability one faces with a Phase Locked Loop in the same circumstances [8, 9]. Therefore, assuming the reference signal is robust an accurate, its delayed version is made accurate as well on a wideband basis since (1) has a unity DC gain.

The delay of the inverters with which the VCDL is made of is thus almost perfectly controlled, and directly linked to the

period of the reference signal and the number of inverters in the VCDL.

B. The digital paradigm : fatorizing the VCDL of the DLL.

The major issue of the DLL in Fig. 1 is its lack of flexibility. Indeed, since the delay of an inverter is linked to the number of element within the VCDL, the range in which we can control this delay is quite limited. This is dramatically emphasized if the reference frequency is made fix, in order to pave the way for highly accurate albeit low cost crystal oscillator reference.

The factorial DLL introduced in [10] offers an alternative for flexibility. The F-DLL architecture is depicted in Fig. 3, where the VCDL is decomposed of a delay element, a switch and a programmable counter.

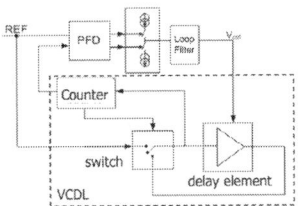

Fig. 3. Factorial DLL architecture.

When it starts a cycle, the switch connects the delay element to the reference. The rising edge of the reference goes through the switch and clocks the counter which, in return, modify the switch position for the latter to loop the delay element on itself, forming a ring oscillator of its own. The signal then circles in this ring as long as needed for the counter to reach its programmed value, being incremented at each and every turn. When the counter overflows, it generates an output rising edge for the phase and frequency detector (PFD) and changes the switch position one more time to allow the next rising edge of the reference to reinitiate the overall cycle. Consequently, the VCDL is equivalent to a series of identical delay elements with a length determined by the programmed value entered in the counter.

When the F-DLL is locked, the delay of the VCDL is now controlled by that programmed value and the reference signal period. Therefore, it is highly flexible and based on a digital paradigm as stated in the introduction for targeting both low cost and low power behaviors.

C. Application to frequency generation

The F-DLL in Fig. 3 can easily perform frequency synthesize. Indeed, the output of the delay element provide a frequency Fout which can be expressed as:

$$F_{out} = N \cdot F_{ref} \qquad (3)$$

with Fref the reference frequency and N the programmed value entered in the counter. Consequently, as soon as the programmable counter is an integer one, the frequency step of this synthesizer is fixed by Fref.

In addition, as this system is made of a DLL, the advantage of jitter cancellation which is a known specificity of DLL [11] is observed. It further reduces the low frequency noise of the synthesizer. Indeed at the beginning of every new cycle of the system, the rising edge of the reference is substituted to the rising edge of the ring oscillator, and the system is resynchronized. It induces the cancelling of jitter accumulation. This is a key advantage when compared to the usual PLL-based synthesizer.

978-1-5090-2737-8/16 $31.00 © 2016 IEEE

III. Limitations – Duty Cycle Distorsion

When dealing with frequency generation, DLL suffers from a major issue. As discussed, the rising edge of the reference is actually intended to generate the first rising edge of the series of pulses generated within the VCDL when the latter is looped on itself as a ring oscillator thanks to the switch. The issue is then to synchronize all the series of rising edges, one after the other, and of course this is a very challenge for the DLL.

Figure 4 is a simplified illustration of this limitation, which generates a duty cycle distortion (DCD) on the very last period of the series of rising edges.

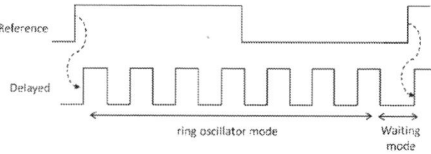

Fig. 4. Duty cycle distortion due to unperfect synchronization of the series of rising edges..

Based on the architecture of Fig. 3 the reference frequency initiate the first rising edge of the delayed signal within the VCDL. The VCDL is then turned into a ring oscillator and generates as much rising edges as necessary for the counter to reach overflow. As a ring oscillator, the duty cycle of the series of generated rising edges is 50%. At the next step, the system has to wait for the next rising edge of the reference to start a new cycle. If this reference rising edge is not perfectly synchronized with the last falling edge of the series, the duty cycle of the last period of the delayed signal is no longer 50%, as depicted in Fig. 5 where the waiting mode is a bit too long when compared to previously generated low level at the output of the VCDL.

That DCD occurs periodically at the frequency of the reference, and thus it generates spurs within the spectrum of the synthesized signal with a frequency offset fixed by multiples of the reference frequency. Consequently, the larger the frequency multiplication realized by the F-DLL the closer the spurious are from the carrier. Then, it is impossible to clean the spectrum by the mere addition of filters, and another option is to be proposed.

Actually DCD is not the only phenomenon that generates spurs when a DLL is used to synthesize a continuous wave carrier [12]. Delay mismatch of different elements – including the two paths of the phase detector – are also of concern. However, while these mismatched delays can be counteracted thanks to a calibration loop the DCD cannot. Indeed DCD is mostly due to an unbalanced time delay of the delay element in Fig. 3 when acting as a delay block or a ring oscillator, and thus this limitation is intrinsic to the topology of Fig. 3. The unbalanced time delay is a consequence of the switch, as the latter position modifies the parasitic capacitors at both the input and the output of the delay element, and propagation delays at first order are linked to these capacitors and the ability of the delay element to charge and discharge them [13].

Without taking care of that critical issue in DLL-based synthesizer, highly polluted spectrum can be obtained for a carrier instead of the expected pure single tone which is required for wireless communications. As an illustration, in Fig. 5 is depicted the first version of the circuit in [14], designed in a 130nm Partially Depleted Silicon on Insulator (PD-SOI) CMOS technology.

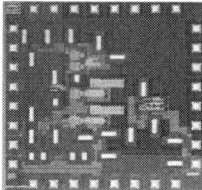

Fig. 5. F-DLL for a 1-6 GHz frequency synthesizer.

This prototype was fully functional, demonstrating the feasibility of a F-DLL based frequency synthesizer with a tremendous frequency range, but since at this time no real attention was taken in balancing the delay element capacitors a lot of spurs where observed within the spectrum of the synthesized signal. A wideband measurement of this continuous wave carrier is depicted in Fig. 6 where a Spur-to-Carrier Ratio (SCR) as high as -6 dBc is observed, which is clearly unacceptable. In addition spurious are observed even at very large offset form the carrier, which is unacceptable as well especially to provide a local oscillator for a wireless data link.

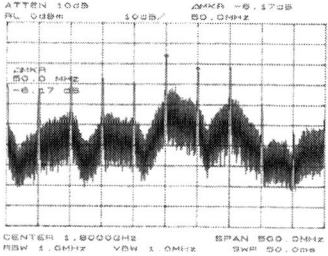

Fig. 6. F-DLL-based frequency generation and reference-related spurious.

While the circuit was redesigned with a careful balanced layout and the addition of buffers at critical nodes in order to reduce spurious at an acceptable level, yielding to the results published in [14], an alternate to the topology in Fig. 3 was yet to be proposed to offer a robust and technology-independent solution to the issue of unbalanced propagation delay.

IV. A Balanced Logic Routing Block

A. Boolean gates instead of switches

While the issue of Fig. 3 topology is due to the switch, the solution relies in suppressing that device. A possible replacement is a multiple-input latch as in the original architecture of the F-DLL for digital design purposes [10], but the latter is also unbalanced due to asymmetrical paths.

The here proposed option is to take advantage of simple Boolean logic gate, and to make the combination as symmetric as possible. For that purpose one has to consider the two possible modes of the delay element on which the F-DLL rely to create the VCDL: the element can act as (a) a delay line or (b) a ring oscillator. Since the counter in Fig. 3 is in charge of selecting either the delay mode or the ring mode, it has to internally create a 'ring' logic signal to allow the oscillating mode and then the delay element has to be in its delay mode when 'ring' is low AND the rising edge of the reference arrives, OR in its ring oscillator mode when 'ring' is high AND the rising edge of the element itself arrives. It yields to the following Boolean equation:

$$A = \overline{ring} \cdot Reference + ring \cdot Delayed \quad (4)$$

where A is the Boolean output of the logic block intended to replace the switch. Equation (4) can be rewritten based on NAND gate as the latter is easier to implement in CMOS. It gives:

$$A = \overline{\overline{\overline{ring}.Reference}.\overline{ring.Delayed}} \qquad (5)$$

B. Balancing delays

Based on (5) it brings the overall VCDL in Fig. 7 with the balanced logic routing block highlighted. This VCDL is to be substituted to the one in Fig. 3, solving the issue of DCD since the propagation delay for the reference rising edge to reach the input of the counter is the same as the propagation delay of the delayed signal rising edge –after the three controlled inverters - due to the symmetrical nature of the logic routing.

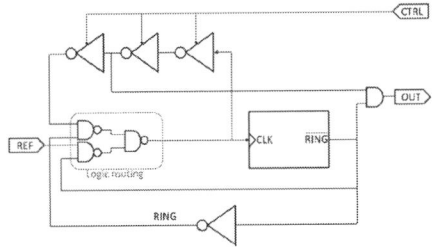

Fig. 7. The balanced logic routing block within the new VCDL

C. Simulation results

The circuit in Fig. 7 was implemented in the 28nm Fully-Depleted Silicon on Insulator (FD-SOI) CMOS technology from STMicroelectronics, and propagation delay discrepancies of barely 1% was observed, validating the balanced nature of the logic routing block when compared to the switch.

Powered by a 1 V power supply with a 2.5 MHz reference frequency and a 10-bit counter programmed at a value in the thousand range the here proposed F-DLL can synthesize an almost spurs-free 2.5 GHz continuous wave carrier to be used in IoT at ultra-low power levels. Since the circuit in Fig. 7 is the major contributor its power consumption was observed and reported in Table I.

TABLE I. FREQUENCY AND POWER CONSUMPTION AS A FUNCTION OF CONTROL VOLTAGE

Control Voltage	Carrier Frequency	VCDL Power Consumption*
700 mV	1.4 GHz	2 µW
800 mV	2.4 GHz	4.6 µW
900 mV	3.26 GHz	5 µW

Excluding the counter

Consequently, an overall power consumption in the vicinity of 100 µW for the complete synthesizer can be targeted, paving the way for a microwatt radio implementation within a wireless node of the IoT.

V. CONCLUSION

A new VCDL allowing a significant reduction of spurious generation in Factorial DLL based frequency generation is presented. A balanced logic routing block has been substituted to the classical switch in order to ensure a constant parasitic loading of the controlled delay element, cancelling the major contributor to duty cycle distortion. Unbalance in the range of one percent has been observed on propagation delay, validating the approach. When inserted within a F-DLL this new block allows synthesizing a 2.5 GHz continuous wave carrier for just a couple of microwatt, excluding the counter, allowing the power budget devoted to the frequency generation unit of a IoT node to be as low as 100µW.

REFERENCES

[1] N. M. Nguyen and R. G. Meyer, "Si IC-compatible inductors and LC passive fillers", IEEE Journal of Solid-State Circuits vol. 25, no 8, August 1990, pp. 1028-1031

[2] Y. Deval, J. Tomas, J-B. Begueret, H. Lapuyade and J-P. Dom, "1-V low-noise 200 MHz relaxation oscillator," Solid-State Circuits Conference, 1997. ESSCIRC '97. Proceedings of the 23rd European, Southampton, UK, 1997, pp. 220-223.

[3] G. Hoffmann de Visme, "High-frequency phase shifter and phase-shift oscillator," in Electrical Engineers, Proceedings of the Institution of, vol. 111, no. 11, pp. 1831-1832, November 1964.

[4] Ü Güler and G. Dündar, "Modeling CMOS Ring Oscillator Performance as a Randomness Source," in IEEE Transactions on Circuits and Systems I: Regular Papers, vol. 61, no. 3, pp. 712-724, March 2014.

[5] T. Kwasniewski, M. Abou-Seido, A. Bouchet, F. Gaussorgues and J. Zimmerman, "Inductorless oscillator design for personal communications devices-a 1.2 µm CMOS process case study," Custom Integrated Circuits Conference, 1995., Proceedings of the IEEE 1995, Santa Clara, CA, 1995, pp. 327-330

[6] P. Flatresse, "Process and design solutions for exploiting FD-SOI technology towards energy efficient SOCs," Low Power Electronics and Design (ISLPED), 2014 IEEE/ACM International Symposium on, La Jolla, CA, 2014, pp. 127-130.

[7] J. J. Spilker and D. T. Magill, "The Delay-Lock Discriminator-An Optimum Tracking Device," in Proceedings of the IRE, vol. 49, no. 9, pp. 1403-1416, Sept. 1961.

[8] H. de Bellescize, "La réception synchrone", L'Onde Electrique, vol. 11, June 1932, pp. 230-240.

[9] Best, E. B., Phase locked loops, McGraw-Hill, New York, NY, USA, 1984.

[10] M. Combes, K. Dioury and A. Greiner, "A portable clock multiplier generator using digital CMOS standard cells," in IEEE Journal of Solid-State Circuits, vol. 31, no. 7, pp. 958-965, Jul 1996.

[11] G. Chien and P. R. Gray, "A 900-MHz local oscillator using a DLL-based frequency multiplier technique for PCS applications", IEEE J. Solid State Circuits, vol. 35, no. 12, pp. 1996-1999, 2000

[12] A. Ojani, B. Mesgarzadeh and A. Alvandpour, "Modeling and Analysis of Harmonic Spurs in DLL-Based Frequency Synthesizers," in IEEE Transactions on Circuits and Systems I: Regular Papers, vol. 61, no. 11, pp. 3075-3084, Nov. 2014.

[13] Hirata, H. Onodera and K. Tamura, "Estimation of propagation delay considering short-circuit current for static CMOS gates," in IEEE Transactions on Circuits and Systems I: Fundamental Theory and Applications, vol. 45, no. 11, pp. 1194-1198, Nov 1998.

[14] C. Majek, Y. Deval, H. Lapuyade and J. B. Begueret, "The factorial Delay Locked Loop: a solution to fulfill multistandard RF synthesizer requirements," Research in Microelectronics and Electronics Conference, 2007. PRIME 2007. Ph.D., Bordeaux, 2007, pp. 185-188

Energy-aware Scheduling in Transactional Memory Systems

Ademir Marques Junior and Alexandro Baldassin
UNESP – Univ Estadual Paulista, Rio Claro, Brazil
ademirmj@fc.unesp.br, alex@rc.unesp.br

Abstract—Transaction scheduling is a relatively new technique for transactional memory systems responsible for deciding which transactions to run in a given moment. Current transactional schedulers are designed with performance in mind, leaving unattended other important metrics such as energy consumption. In order to address this important concern, this paper presents a novel heuristic, called *Dynamic Serializer* (DS), for scheduling transactions with the aim of reducing energy consumption and improving the energy-delay product (EDP) of transactional applications. DS serializes the execution by choosing either a spinlock or mutex according to a dynamic profile based on abort rates. The idea is to use a spinlock when a transaction is likely to succeed in the near future, since spinning while waiting for the lock is faster than blocking as in the case of a mutex. On the other hand, a mutex is used in case transactions need to wait a long time before resuming because in this case energy can be saved by blocking a thread and enabling the processor Dynamic Voltage and Frequency Scaling (DVFS) to act. DS exploits this tradeoff to improve the EDP of transactional programs. Experimental results with an Intel Core i7 with 8 logical threads and the STAMP benchmark show an EDP improvement of up to 17% and an average of 4.9% compared to LUTS-Dynamic, a state-of-the-art scheduling heuristic for transactional systems.

I. Introduction

Energy consumption is a growing concern in all areas of computing systems, ranging from small embedded systems to large data centers [1]. At the same time, the popularization of multicore processors requires programmers to write parallel code in order to fully exploit the available performance. Parallel software is usually developed with performance in mind and typically does not take into consideration energy concerns in its design. Such is the case of most current implementations of Transactional Memory (TM) systems [2], [3]. TM is a programming model that adopts the concept of transactions. When transactions conflict with each other a contention manager is required to decide what to do. Although most of the first contention policies aimed at increasing performance [4], [5], [6], [7], [8], a number of later works have addressed energy consumption by creating energy-aware policies [9], [10], [11], [12], [13], [14]. These policies tend to use dynamic voltage and frequency scaling (DVFS) techniques in order to provide a good energy/performance tradeoff. Devising energy-aware policies is particularly interesting as transactional memory is also viable for embedded applications. Indeed, the first policies in this category were designed to be used in embedded systems [9], [10], [12].

There are, however, a number of important drawbacks with the use of contention managers. Most importantly, they take

action only *after* a conflict has happened. In order to avoid that, researchers have proposed scheduling-based approaches, wherein the decision of executing or not a transaction is taken *before* the appearance of the conflict. Two of these transaction schedulers are the Adaptive Transaction Scheduling (ATS) of Yoo and Lee [15] and Lightweight User-Level Transaction Scheduling (LUTS) of Nicacio et al. [16]. While ATS serializes the execution in high contention scenarios, LUTS provides an API that can be used to develop different scheduling heuristics. The most recent heuristic developed by Nicacio et al. is LUTS-Dynamic [17], which provides different execution behaviors according to a transaction length: for short transactions a fast heuristic based on contention level is employed, whereas a past conflict history is used to decide whether a long transaction should be scheduled.

Both ATS and LUTS-Dynamic have been designed to improve performance. Indeed, to the best of our knowledge, there is no work on transaction scheduling targeting energy consumption reduction. In this work we present *Dynamic Serializer* (DS), a novel heuristic aimed at providing a good performance/energy tradeoff for TM systems. DS is built over the LUTS infrastructure and is the first scheduling-based approach to be energy-aware. The main idea behind DS is to serialize the execution by dynamically choosing between a spinlock and a mutex. A spinlock is used for transactions that are likely to succeed in the near future whereas a mutex is used for transactions that require a longer waiting time (because the contention level is too high). Since a mutex is usually blocking, at most one thread will be executing at one time, creating an opportunity for the system to implicitly use DVFS techniques to reduce the energy consumption [18].

This paper makes the following contributions:

- It proposes *Dynamic Serializer* (DS), a new energy-aware scheduling heuristic for transactional memory systems. The key idea of DS is to dynamically decide between using a spinlock or mutex in order to provide energy savings (see Section III);
- It shows the efficiency of the DS heuristic by performing an experimental evaluation using the STAMP benchmark [19]. Results show an improvement in the energy-delay product (EDP) of up to 17% and 61%, with an average of 4.9% an 15.8%, when compared to LUTS-Dynamic and ATS, respectively (see Section IV).

This paper starts by discussing the main concepts and

978-1-5090-2737-8/16 $31.00 © 2016 IEEE

related works in Section II. After that, Section III presents the *Dynamic Serializer* algorithm and the choices we have made, while Section IV discusses the results. Finally, the conclusions are drawn in Section V.

II. BACKGROUND AND RELATED WORK

Transactional Memory (TM) [3] provides a high level programming model wherein transactions are the main units of concurrency. A transaction is guaranteed to execute atomically, in isolation, and preserve consistency. A conflict happens when multiple transactions access the same piece of data and at least one of the accesses is a write operation. In such a case, transactional systems resort to contention managers in order to resolve the conflicts.

The first contention policies have been designed in order to improve performance [4], [5], [6], [7], [8]. Later on, policies were also created to tackle energy consumption [9], [10], [11], [12], [13], [14]. While Ferri et al. [9], [12] investigated methods to improve energy savings in hardware implementations of TM (HTM), Klein et al. [10] and Baldassin et al. [11], [13] focused on software implementations (STM). All of these works have used a simulated embedded environment for their analysis and relied on the DVFS mechanism provided by the platform to devise heuristics for reducing energy consumption. More recently, Issa et al. [14] have investigated the design of energy-aware policies on contemporary processors by mixing spin-based and sleep-based implementations of the backoff mechanism. Their approach is similar to the one present here, but was applied in the context of contention management instead of transaction scheduling.

The main issue with contention managers is that they take action only *after* the appearance of a conflict. Therefore, modern systems employ scheduling-based approaches wherein a decision about allowing a transaction to execute is taken *before* the conflict shows. Yoo and Lee pioneered the design of scheduling techniques by presenting a heuristic called Adaptive Transaction Scheduling (ATS) [15]. ATS keeps track of the system's contention level and, when a given threshold is reached, ATS serializes the execution of transactions. Other scheduling techniques followed, such as the works of Dolev et al. [20], Dragojevic et al. [21], Maldonado et al. [22], and Nicacio et al. [16].

The design of the Lightweight User-Level Transaction Scheduler (LUTS) framework by Nicacio et al. [16] differs from the previous works in the sense that LUTS provides a scheduling API that can be used to design different scheduling strategies. The dynamic heuristic designed by the authors, called LUTS-Dynamic [17], is able to detect short and long transactions and applies different scheduling strategies: for short transactions a contention intensity threshold is used and the system simply reschedule transactions if above the threshold; for long transactions a history is kept of past conflicts and the system uses this information to decide which transaction to schedule.

Common to all scheduling-based approaches is the fact that they were developed to improve performance. To the best of

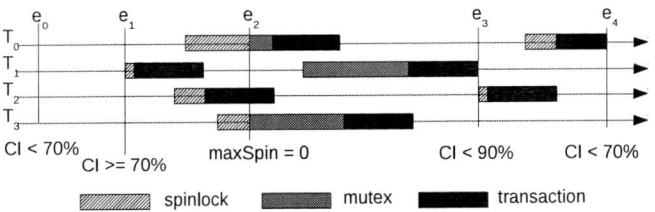

Fig. 1. Example of *Dynamic Serializer* in action with 4 threads.

our knowledge, there is no work addressing the energy issue in transaction scheduling. Therefore, in this paper we propose *Dynamic Serializer* to remedy the situation, as presented next.

III. THE DYNAMIC SERIALIZER APPROACH

In this section we present our energy-aware heuristic, *Dynamic Serializer* (DS). The core idea of DS is to implicitly use DVFS as described by Wamhoff et al. [18]. Basically, the levels of performance and power consumption of current processors are controlled by two operational states: *P-states* implement DVFS, whereas *C-states* are used to save energy when cores are idle. Using the terminology presented by Wamhoff et al. [18], the nominal P-state is referred as P_{base}. P-states above P_{base} are known as turbo modes and denoted by P_{turbo}. Turbo modes are usually entered automatically if certain conditions hold, for instance if not all cores are in C0 (which is the normal operational state). P_{slow} refers to the P-state with the slowest speed configuration.

It is possible to manage the P and C states manually to a certain degree, although some processors (such as Intel ones) are more hardware-centric. Since we do not want to tie our technique to a specific manufacturer, the approach employed by DS is to enable different DVFS values by forcing transactions to block through the use of mutexes. This, in turn, will allow the processor to enter boosted P-states. Thus, we strive to let the mutex owner to run at P_{turbo}, other threads to wait at P_{slow}, and otherwise execute transactions in parallel at P_{base}. Entering a turbo mode is usually costly, so DS firstly attempts to serialize the execution through a spinlock, which spins until the lock is free. DS automatically replace the spinlock by a mutex in case the system is under high contention. The idea of using implicit DVFS management is not new itself [18], [14], [23], but our heuristic is the first one to apply it in a scheduling-based setting.

A. Running Example

Figure 1 shows an example of the DS heuristic in a system with four threads, T_0, T_1, T_2, and T_3 (y axis). Time runs from left to right (x axis). The events e_0, e_1, e_2, e_3, and e_4 describe time intervals where the system behavior changes, as described next. CI denotes the Contention Intensity, first introduced by ATS [15] to decide when to start serializing the code. In the example a value of 70% is used, meaning that transactions are serialized when at least 70% of them are aborting.

As shown in Figure 1, during the time interval $e_0 \rightarrow e_1$ CI is lower than 70% and therefore no serialization is applied.

978-1-5090-2737-8/16 $31.00 © 2016 IEEE

This changes when e_1 is reached and CI is now greater or equal than 70%. At this point the transaction at T_1 needs to grab a spinlock before start executing. Transactions at T_0, T_2, and T_3 spin waiting for T_1 to relinquish the lock. Up until e_2 the system is running at the nominal P-state (P_{base}). Internally, DS stores the maximum number of retries for acquiring the spinlock in a $maxSpin$ variable. When this variable reaches 0 at e_2, DS starts using a mutex instead. Transactions in the interval $e_2 \rightarrow e_3$ are also serialized, but now they contend for the ownership of the mutex.

Since acquiring a mutex usually requires a system call and may result in blocking threads, mutex operations are much more costly than spinlocks. However, since at most one transaction will be executing during $e_2 \rightarrow e_3$, it can run at full speed (P_{turbo}). Meanwhile, the remaining threads will use a low power state (P_{slow}) because they are blocked waiting for the mutex to be released. When using a mutex, DS uses a threshold to determine when to switch back to a spinlock. In this example the value 90% is used and a lower value is reached at e_3. During the interval $e_3 \rightarrow e_4$ the system is using a spinlock again and the operation mode is back to P_{base}. After e_4, the system is under low contention and no transactions are serialized.

B. Algorithm

The main parts of the *Dynamic Serializer* heuristic is presented in Algorithm 1. The algorithm is made up of extra actions that need to be performed when a transaction starts (lines 1–21), commits (lines 22–36) or aborts (lines 37–38). Typical software implementations provide hooks to plug these actions into the main transactional algorithm.

The `globalMode` variable holds the lock mode being used by DS (either spinlock or mutex). In order to reduce contention on this global variable, a transaction-local version of the mode is also held in `lockType`. The locks `spinLock` and `mutexLock` are the synchronization variables employed by DS. The threshold used by DS to decide when to switch from spinlock to mutex is given by `maxSpin`. Each transaction also stores locally whether a lock was acquired (`lockAcquired`) and the contention intensity (`CI`). The decision of making `CI` a local variable is to avoid contention on the variable. This is also how the idea of contention intensity is implemented in ATS [15].

When a transaction starts (or restarts after aborting) it checks whether `CI` is greater or equal to 70% and no lock has been acquired (line 2). We opted for using this threshold (70%) since it provided the best results. If the lock type is spinlock (line 3), DS stays in a while loop until one of the following three scenarios occur (lines 5–9): (i) the spinlock is acquired; (ii) the number of retries exceeds the threshold; (iii) `globalMode` is changed to mutex by another transaction. If after leaving the loop the spinlock was not acquired, it changes the mode to mutex (lines 10–12), or just make `lockAcquired` true otherwise (lines 13–15). Before finishing, it still needs to check if the lock type is now mutex in line 17 (this could be caused because the spin threshold

---

Algorithm 1: The *Dynamic Serializer* algorithm

System-wide variables:
`globalMode:` /* `is_spinLock or is_mutexLock` */
`maxSpin:` /* `Spin threshold` */
`spinLock:` /* `The global spinlock` */
`mutexLock:` /* `The global mutex` */

Transaction-local variables:
`CI:` /* `Contention Intensity` */
`lockAcquired:` /* `True if a lock was acquired` */
`lockType:` /* `is_spinLock or is_mutexLock` */

```
 1  upon start:
 2  if CI >= 0.7 and lockAcquired == false then
 3  │   if lockType == is_spinLock then
 4  │   │   retries ← maxSpin;
 5  │   │   while (not spinLock.Lock() and
 6  │   │       retries != 0 and
 7  │   │       globalMode == is_spinlock) do
 8  │   │   │   retries ← retries −1;
 9  │   │   end
10  │   │   if spinLock not acquired then
11  │   │   │   lockType ← is_mutexLock;
12  │   │   │   globalMode ← is_mutexLock;
13  │   │   else
14  │   │   │   lockAcquired ← true;
15  │   │   end
16  │   end
17  │   if lockType == is_mutexLock then
18  │   │   mutexLock.Lock(lock);
19  │   │   lockAcquired ← true;
20  │   end
21  end

22  upon commit:
23  Calculate new maxSpin based on past number of aborts per
    second (only done by the main thread);
24  Update CI;
25  if lockAcquired == true then
26  │   if lockType == is_spinLock then
27  │   │   spinLock.Unlock();
28  │   else
29  │   │   mutexLock.Unlock();
30  │   end
31  │   lockAcquired ← false;
32  end
33  if CI < 0.9 and lockType == is_mutexLock then
34  │   lockType ← is_spinLock;
35  │   globalMode ← is_spinLock;
36  end

37  upon abort:
38  Update CI;
```

---

has been reached or another transaction changed the mode to mutex). Either way, the transaction acquires the mutex and sets `lockAcquired` to true (lines 18 and 19).

After a transaction is committed, `maxSpin` is recalculated (line 23). The way it is done is similar to how the Dynamic heuristic in LUTS-Dynamic works: a special thread (the main thread, for instance) keeps track of how many times its transactions have aborted in the past 100 executions and the time taken. It then divides the measured time by the number of aborts. This value is used to derive `maxSpin` by applying it to a logarithmic function obtained empirically after profiling the applications in the STAMP benchmark. The contention intensity is then updated (line 24) similarly to how it is done by ATS. If the committed transaction was holding a lock, it relinquishes the ownership of the respective lock and sets `lockAcquired` to false (lines 25–32). If the contention intensity is below a given threshold (90% in this case) and the lock type is mutex, it changes the mode to spinlock (lines 33–36). Finally, transactions that abort update their contention intensity value (line 38).

C. Discussion

Dynamic Serializer borrows some ideas from both ATS and LUTS-Dynamic. The contention intensity (`CI`) threshold is used to decide whether to apply the scheduling heuristic or not. As originally introduced by ATS [15], `CI` is given by the following equation:

$$CI_n = \alpha \times CI_{n-1} + (1 - \alpha) \times CC$$

`CI` is maintained in a decentralized manner (each thread has its own contention information). The value of `CI` is initially zero, and the presented equation is evaluated every time a transaction commits (line 24) or aborts (line 38). This evaluation is performed based on the previous value of `CI` and the current contention (CC), which is set to 0 if the transaction committed, or 1 if the transaction aborted. A weight variable (α) is used to give more priority either to the past history or the current contention. Note that for large values of α the equation biases toward past history, whereas small values favor the current contention.

From LUTS-Dynamic, *Dynamic Serializer* borrows the concept of using a thread to periodically summarize runtime statistics. More precisely, DS uses the number of aborts per time to infer a new value for `maxSpin`. The way it is done currently is based on a profiling stage where we exploit the correlation between abort rate and `maxSpin`. This correlation is transformed into a function that is used at runtime to determine the new value of `maxSpin`. Ideally, we would like to use a machine learning technique to avoid the profiling stage, as done by other works [14], [24], but we leave this feature for future work.

Dynamic Serializer also uses two thresholds: one to decide when to enter the serialization mode (70% as in line 2) and another to decide when to switch back to spinlock mode from a mutex mode (90% as in line 33). These values were obtained empirically through a sensitivity analysis and provided the best results for the transactional applications from the STAMP benchmark. Notice that this tuning methodology was also used by ATS and LUTS-Dynamic. Again, automatic tuning is left for future work.

One last observation about Algorithm 1 is that it allows different transactions to run under different lock modes simultaneously, albeit for a short time. For instance, assume two transactions t_1 and t_2 are running the spinlock loop (lines 5–8) and t_1 is able to acquire the spinlock. Meanwhile, t_2 may leave the loop because it exhausted the number of retries, thus acquiring the mutex (line 18). Notice, however, that all transactions from this point on will correctly use the mutex mode, since `globalMode` was changed by t_2 (line 12). Notice, also, that having two locking modes active for a short time cannot cause incorrect execution. Therefore, the changes to the algorithm in order to guarantee "pure" exclusive modes will only add overhead and are not worth the effort.

IV. EXPERIMENTAL RESULTS

In this section we seek to evaluate the proposed heuristic, *Dynamic Serializer*, by measuring the execution time and energy consumption of transactional programs. For that purpose, the Stanford Transactional Applications for Multi-Processing (STAMP) [19] benchmark suite was chosen given its wide use in transactional systems. STAMP is comprised of 8 applications covering a wide variety of domains such as security and machine learning. We make use of 7 of the 8 available applications: `Genome`, `Intruder`, `Kmeans`, `Labyrinth`, `SSCA2`, `Vacation`, and `Yada`. The remaining one, `Bayes`, was not used due to nondeterministic behavior as pointed out by others [25].

We compare DS against other two well-known scheduling techniques: ATS [15] and LUTS-Dynamic [17], as discussed earlier. We implemented the DS algorithm using the C language and adapted it as part of the TinySTM library [26]. The same procedure was used for ATS and LUTS-Dynamic, providing a common code base for fair comparisons. DS also makes use of the scheduling infrastructure of LUTS [16]. For each application, we show the average of ten executions when using 8 threads, which is the maximum number of logical cores in the system. The processor used is a quadcore Intel Core i7 2600K with 8GB of RAM running Linux with kernel 2.6.32. All applications were compiled using the GCC compiler version 4.47. Intel's RAPL (Running Average Power Limit) [27] interface was used to collect energy measurements.

A. Evaluation

Figure 2 summarizes our results by presenting execution time (a), energy consumption (b), and EDP (c) for the 7 studied applications, with ATS and LUTS-Dynamic normalized with regard to DS (values greater than 1 indicate that DS presented better results). The last cluster of bars in each subfigure shows the geometric mean.

Looking at execution time (Figure 2a), it is noticeable that ATS presents higher values. The reason for this is mostly due to the fact that only mutexes are used for serialization and, as a consequence, applications with high abort rates (e.g., `Intruder`, `Kmeans`, and `Yada`) tend to have worse execution times with ATS. Also notice that, with 8 threads, the system is using all available logical cores (through simultaneous multithreading), creating a larger overhead when using mutexes. Since DS first attempts to use spinlocks instead of mutexes (resorting to mutexes only in case of extreme contention), it does not suffer from these effects and performs slightly better than LUTS-Dynamic. ATS does perform better than both LUTS-Dynamic and DS for `Genome` and `SSCA2`.

978-1-5090-2737-8/16 $31.00 © 2016 IEEE

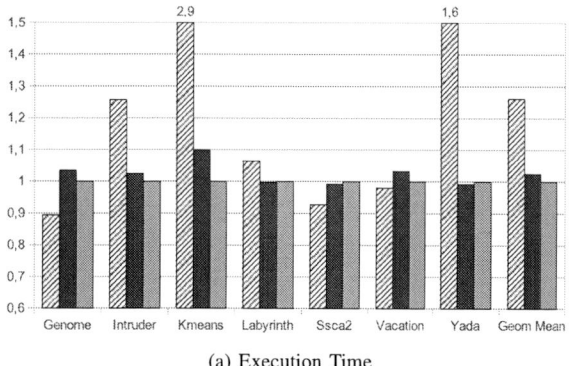

(a) Execution Time

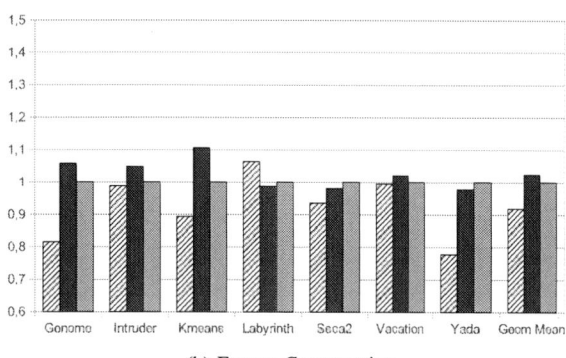

(b) Energy Consumption

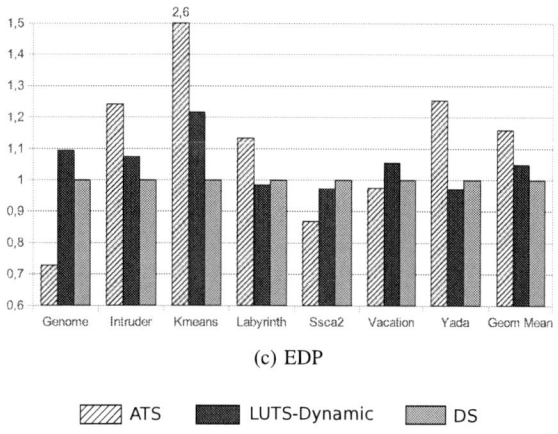

(c) EDP

ATS   LUTS-Dynamic   DS

Fig. 2. Execution time (a), energy consumption (b), and EDP (c) numbers for ATS, LUTS-Dynamic, and DS with 8 threads. Results are normalized with regard to DS.

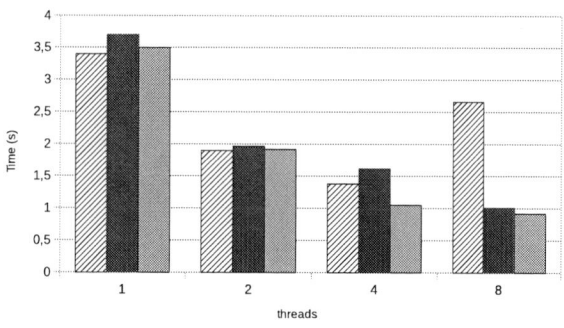

(a) Execution Time

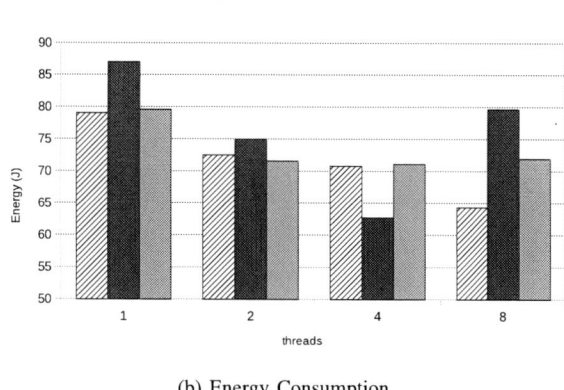

(b) Energy Consumption

ATS   LUTS-Dynamic   DS

Fig. 3. Execution time (a) and energy consumption (b) for the `Kmeans` application.

For these applications, the abort rate is very low and therefore serialization is rarely applied. Thus, LUTS-Dynamic and DS are affected by the overhead of the scheduling infrastructure of LUTS, without any benefits.

On the other hand, the energy results (Figure 2b) show that ATS provided an overall better energy consumption. It is particularly interesting to notice that for `Kmeans` and `Yada`, where ATS performed poorly, it provided the best energy savings. This is because it serialized the execution using mutexes which, despite increasing execution time, allowed the system to put 3 of the 4 physical cores in a low energy state. DS' approach of using spinlocks and mutexes, in general, provided

better energy savings than Dynamic-LUTS, particularly for `Genome`, `Intruder`, and `Kmeans`.

When the combined effect is considered using the energy-delay product (Figure 2c), it can be noticed that DS provided good results compared to both ATS and LUTS-Dynamic, providing an average improvement of 15.8% and 4.9% and a maximum of 61% and 17% (`Kmeans`), respectively. The mixed approach of using spinlocks and mutexes allowed DS to obtain both good execution times and energy savings in the majority of the applications.

In order to better understand the benefits of DS, Figure 3 shows the execution time (a) and energy consumption (b) for `Kmeans`. Results are shown for 1, 2, 4, and 8 threads and the 3 considered scheduling techniques. It is worth noting that ATS' execution time increased from 4 to 8 threads because the high contention level forced this technique to serialize almost the entire execution. Since transactions in `Kmeans` are very short, the time spent during the abort operation is very fast and the use of spinlocks makes more sense. However, if the contention level reaches critical levels, using mutexes is a good idea since they will quickly force the abort rate to decrease. Thus, the mixed approach of using both spinlocks and mutexes as done by DS pays off.

B. Discussion

In general, the results indicate that DS is more useful for applications displaying average to high contention levels and short transactions. For applications with longer transactions, such as Yada and Labyrinth, the proposed heuristic did not provide any improvement over LUTS-Dynamic, although it was able to achieve better results than ATS. Lastly, for applications with low contention levels, such as Genome and SSCA2, ATS was able to get better overall EDP numbers due to its low implementation overhead.

Although our evaluation was not conducted in an embedded setting, we believe that the proposed scheduling technique could be applied in such environment with similar results. For instance, the work on an embedded TM system conducted by Ferri et al. [9] also considers the tradeoffs of using spinlocks, mutexes and TM. DS could potentially be adapted to use the transactional hardware support presented in that work.

V. CONCLUSION

In this paper we proposed *Dynamic Serializer* (DS), a novel scheduling technique for transactional systems. DS employs both spinlocks and mutexes and chooses between the two types dynamically according to the system's contention level. Experimental results performed with the STAMP benchmark suite shows that DS is particularly useful for applications with moderate to high abort rates and with relatively short transactions. As future work we intend to adapt an online algorithm to dynamically tune some of the DS parameters and expand our evaluation to other systems.

ACKNOWLEDGMENT

This work is supported by FAPESP (2011/19373-6) and CNPq (446160/2014-8).

REFERENCES

[1] L. A. Barroso and U. Holzle, "The case for energy-proportional computing," *IEEE Computer*, vol. 40, no. 12, pp. 33–37, Dec. 2007.

[2] M. Herlihy and J. E. B. Moss, "Transactional memory: Architectural support for lock-free data structures," in *Proceedings of the 20th Annual International Symposium on Computer Architecture*, Jun. 1993, pp. 289–300.

[3] T. Harris, J. Larus, and R. Rajwar, *Transactional Memory*, 2nd ed. Morgan & Claypool Publishers, Jun. 2010.

[4] W. N. Scherer and M. L. Scott, "Advanced contention management for dynamic software transactional memory," in *Proceedings of the 24th Annual Symposium on Principles of Distributed Computing*, 2005, pp. 240–248.

[5] R. Guerraoui, M. Herlihy, and B. Pochon, "Toward a theory of transactional contention managers," in *Proceedings of the 24th Annual Symposium on Principles of Distributed Computing*. ACM Press, Jul. 2005, pp. 258–264.

[6] ——, "Polymorphic contention management," in *19th International Symposium on Distributed Computing*, Sep. 2005, pp. 303–323.

[7] M. F. Spear, L. Dalessandro, V. J. Marathe, and M. L. Scott, "A comprehensive strategy for contention management in software transactional memory," in *Proceedings of the 14th Symposium on Principles and Practice of Parallel Programming*, Feb. 2009, pp. 141–150.

[8] G. Sharma, B. Estrade, and C. Busch, "Window-based greedy contention management for transactional memory," in *24th International Symposium on Distributed Computing*, Sep. 2010, pp. 64–78.

[9] C. Ferri, A. Viescas, T. Moreshet, R. I. Bahar, and M. Herlihy, "Energy efficient synchronization techniques for embedded architectures," in *Proceedings of the 18th ACM Great Lakes symposium on VLSI*, May 2008, pp. 435–440.

[10] F. Klein, A. Baldassin, G. Araujo, P. Centoducatte, and R. Azevedo, "On the energy-efficiency of software transactional memory," in *Proceedings of the 22nd Annual Symposium on Integrated Circuits and System Design*, Sep. 2009, pp. 1–6.

[11] A. Baldassin, F. Klein, G. Araujo, R. Azevedo, and P. Centoducatte, "Characterizing the energy consumption of software transactional memory," *IEEE Computer Architecture Letters*, vol. 8, no. 2, pp. 56–59, 2009.

[12] C. Ferri, S. Wood, T. Moreshet, R. I. Bahar, and M. Herlihy, "Embedded-TM: Energy and complexity-effective hardware transactional memory for embedded multicore systems," *Journal of Parallel and Distributed Computing*, vol. 70, no. 10, pp. 1042–1052, Oct. 2010.

[13] A. Baldassin, J. P. L. de Carvalho, L. A. G. Garcia, and R. Azevedo, "Energy-performance tradeoffs in software transactional memory," in *Proceedings of the 24th International Symposium on Computer Architecture and High Performance Computing*, Oct. 2012, pp. 147–154.

[14] S. Issa, P. Romano, and M. Brorsson, "Green-CM: Energy efficient contention management for transactional memory," in *Proceedings of the 44th International Conference on Parallel Processing*, Sep. 2014, pp. 550–559.

[15] R. M. Yoo and H.-H. S. Lee, "Adaptive transaction scheduling for transactional memory systems," in *Proceedings of the 20th Annual ACM Symposium on Parallel Algorithms and Architectures*, Jun. 2008, pp. 169–178.

[16] D. Nicacio, A. Baldassin, and G. Arajo, "LUTS: A lightweight user-level transaction scheduler," in *Proceedings of the 11th International Conference on Algorithms And Architectures For Parallel Processing*, Oct. 2011, pp. 144–157.

[17] D. Nicacio, A. Baldassin, and G. Araujo, "Transaction scheduling using dynamic conflict avoidance," *International Journal of Parallel Programming*, vol. 41, no. 1, pp. 89–110, 2012.

[18] J.-T. Wamhoff, S. Diestelhorst, C. Fetzer, P. Marlier, P. Felber, and D. Dice, "The TURBO diaries: Application-controlled frequency scaling explained," in *Proceedings of the 2014 USENIX Annual Technical Conference*, Jun. 2014, pp. 193–204.

[19] C. C. Minh, J. Chung, C. Kozyrakis, and K. Olukotun, "STAMP: Stanford Transactional Applications for Multi-Processing," in *Proceedings of the IEEE International Symposium on Workload Characterization*, Sep. 2008, pp. 35–46.

[20] S. Dolev, D. Hendler, and A. Suissa, "CAR-STM: Scheduling-based collision avoidance and resolution for software transactional memory," in *Proceedings of the 27th Annual Symposium on Principles of Distributed Computing*, Aug. 2008, pp. 125–134.

[21] A. Dragojevic, R. Guerraoui, A. V. Singh, and V. Singh, "Preventing versus curing: Avoiding conflicts in transactional memories," in *Proceedings of the 28th Annual Symposium on Principles of Distributed Computing*, Aug. 2009, pp. 7–16.

[22] W. Maldonado, P. Marlier, P. Felber, A. Suissa, D. Hendler, A. Fedorova, J. L. Lawall, and G. Muller, "Scheduling support for transactional memory contention management," in *Proceedings of the 15th Symposium on Principles and Practice of Parallel Programming*, Jan. 2010, pp. 79–90.

[23] D. Lo and C. Kozyrakis, "Dynamic management of TurboMode in modern multi-core chips," in *Proceedings of the 20th International Symposium on High-Performance Computer Architecture*, Feb. 2014, pp. 603–613.

[24] N. Diegues and P. Romano, "Self-tuning Intel transactional synchronization extensions," in *Proceedings of the 11th International Conference on Autonomic Computing*, Jun. 2014, pp. 209–219.

[25] D. Christie, J.-W. Chung, S. Diestelhorst, M. Hohmuth, M. Pohlack, C. Fetzer, M. Nowack, T. Riegel, P. Felber, P. Marlier, and E. Riviere, "Evaluation of AMD's advanced synchronization facility within a complete transactional memory stack," in *Proceedings of the 5th European Conference on Computer Systems*, Apr. 2010, pp. 27–40.

[26] P. Felber, C. Fetzer, and T. Riegel, "Dynamic performance tuning of word-based software transactional memory," in *Proceedings of the 13th Symposium on Principles and Practice of Parallel Programming*, Feb. 2008, pp. 237–246.

[27] H. David, E. Gorbatov, U. R. Hanebutte, R. Khanna, and C. Le, "RAPL: Memory power estimation and capping," in *Proceedings of the 2010 ACM/IEEE International Symposium on Low Power Electronics and Design*, Oct. 2010, pp. 189–194.

978-1-5090-2737-8/16 $31.00 © 2016 IEEE

A Parallel Motion Estimation Solution for Heterogeneous System on Chip

Mateus Melo, Gustavo Smaniotto, Henrique Maich, Luciano Agostini, Bruno Zatt, Leomar Rosa Jr, Marcelo Porto

Group of Architectures and Integrated Circuits (GACI)

Federal University of Pelotas – UFPel

Pelotas, Brazil

{msdmelo, ghsmaniotto, hdamaich, agostini, zatt, leomarjr, porto}@inf.ufpel.edu.br

Abstract—This paper presents a parallel Motion Estimation (ME) solution for video coding on heterogeneous System-On-Chip (SoC), with two Implementation Versions: an OpenCL-based version targeting embedded GPGPUs and a hardware design targeting an embedded FPGA device. The current work considers a heterogeneous SoC composed of a variety distinct processing units such as CPU, DSP, Memory, GPGPU, and FPGA, where the FPGA component has support for dynamic reconfiguration. These two versions implement a parallelism-oriented algorithm and provide two performance/energy operation points allowing flexibility for dynamic power management according to runtime scenarios. The solution presented in this paper uses a scheme to reduce the number of operations required to perform the Sum of Absolute Differences (SAD) for the evaluated candidate blocks. This scheme is based on the accumulation of previously calculated SADs, considering the 8x8 Prediction Unities (PU) as base blocks, to generate the SAD for larger PUs. The proposed solution was evaluated in two platforms, (1) an Odroid XU-3, with a Samsung Exynos 5422 SoC, featuring a 64-core Mali-T628 MP6 GPGPU, and (2) an FPGA device. The performance and energy consumption results shows the FPGA implementation are able to process 49 HD 1080p fps with 1000x increased in energy efficiency when compared to the GPGPU implementation.

Keywords—Heterogeneous System; SoC; GPGPU; FPGA; Video Coding; Motion Estimation

I. INTRODUCTION

State-of-the-art mobile systems are becoming more heterogeneous in order to meet severe performance and energy constraints imposed by current applications. The high level of integration enabled by submicron technologies made possible the realization of SoCs [1] featuring multiple CPU cores (typically ARM cores), DSP processors, GPUs (Graphic Processing Unit) capable of general purpose processing (GPGPUs) and dedicated hardware accelerators for highly demanding applications, such as 4G communication and video coding and decoding. The availability of heterogeneous processing units guarantees better fitting between application requirements and hardware resources allocation, leading to improved energy efficiency through dynamic power management (DPM).

However, the heterogeneity brings multiple challenges related to both, system design and runtime management. On the one hand, it is necessary a proper design-time system specification (number of cores, characteristics of each core, communication, memory hierarchy, etc.) to ensure processing power to attend applications demanded by the consumer along the upcoming years. On the other hand, there is a need for efficient runtime management algorithms able to deal with distinct usage scenarios through implementing dynamic task binding, DPM and dynamic thermal management whereas considering dark silicon-related challenges.

FPGA-based hardware acceleration has been considered for embedded applications in multiple research works [6] due to its higher flexibility in relation to ASIC and to its superior performance/energy-efficiency in relation to software solutions. Still, FPGAs have never become an industrial reality for high-end mobile devices. However, in the current scenario where the consumer uses an endless variety of applications, including some with tight performance/energy constraints, FPGAs are a promising option.

In the current market, the most part of heterogeneous systems have only eGPU. However, the FGPA + GPGPU solution can be seen in few devices, such as, Zynq Ultrascle+ device [4]. However, the acquisition of Altera by Intel [3] indicates that FPGAs may appear in SoCs released by Intel, similarly to the Atom E600c [5] processor.

The popularization of teleconferencing and video sharing platforms (YouTube, Netflix, Twitch, etc.) made video (de)coding a mandatory capability for mobile devices. Multimedia applications, especially video (de)coding, are processing/memory intensive tasks and represent huge challenge for system designers. For this reason, dedicated hardware accelerators [6] or complete hardware codecs [7] are implemented within the main devices available in the market. Such implementations, however, are designed to support specific coding standards being unable to process in accordance to other ones. This limitation is undesirable in an ecosystem under constant change where multiple standards – MPEG-2, H.264, HEVC, VP9, AVS, AVS2 – coexist simultaneously.

In the heterogeneous system field, there are many works using CUDA, OpenCL, and HDL languages to deal with problems as reconfigurable acceleration and dynamic self-reconfigurable in general proposed system [13], digital signal processing [14] and video coding [2]. Therefore, there is a need to provide standard-independent video (de)coding support whereas efficiently exploiting the available hardware resources.

Among the video coding standards in the current market share, High Efficiency Video Coding (HEVC) standard [8] represents the state of the art and brings the largest challenges. It promotes 50% of video compression increase maintaining the same subjective video quality when compared to previous standard, the H.264/AVC (Advanced Video Coding) [9]. In fact, the most efficient technique of HEVC and its predecessors, in

978-1-5090-2737-8/16 $31.00 © 2016 IEEE

terms of compression gains, is the Motion Estimation (ME) [10], the main step of the inter-prediction process. The ME uses previously processed frames as references to remove temporal redundancies in the current encoding frame. However, ME computational effort corresponds to 62%-94 % of total encoding time in the HEVC standard [12].

In this context, the current paper presents a ME solution targeting HEVC (but not limited to) video coding with two Implementation Versions featuring a parallelism-oriented ME algorithm. The first version is implemented using OpenCL targeting embedded GPGPUs. The second version is a hardware design described using VHDL targeting eFPGA fabrics. Although the task allocation/binding is not part of this work, the two versions allow the system to adapt according to the runtime scenario and provide two performance/energy operation points for DPM algorithms.

This paper is organized as follows: Section II presents the ME in the HEVC standard. Section III describes heterogeneous systems and related works solutions. Section IV.A describes the ME solution proposed in this work, and the two Implementation Versions. Section IV.B presents the GPGPU implementation and Section IV.C presents the FPGA implementation. Finally, Section V presents results and Section VI presents conclusions of this work.

II. MOTION ESTIMATION IN THE HEVC STANDARD

As mentioned before, the HEVC standard promoted 50% of video compression gain maintaining the same subjective video quality when compared to previous standard, H.264/AVC. However, the tools used in the HEVC standard increase the complexity and hardware resources when compared to previous standards. The HEVC reached important coding efficiency improvements at the cost of a high increase in the computational effort for the coding process, between 9.1% and 502.2% higher than H.264/AVC [8]. In fact, the HEVC standard has complex stages, and it demands a huge computational effort, especially to process high resolution videos. Thus, a parallel approach is desirable for some HEVC coding steps.

The HEVC standard uses several coding tools to improve the data compression, such as intra and inter prediction, followed by transform, quantization and entropy coding. The intra prediction is used to identify the redundancies present within a frame. The inter prediction is responsible to identify temporal redundancies, i.e., redundancies between different frames. The transform is responsible to translate the prediction errors signal to the frequency domain, after that, the quantization step reduces the amplitude of the signal considering human visual system characteristics and facilitating the entropy encoding process.

However, ME is the most important technique of HEVC in terms of compression gains. It divides the current encoding frame into smaller blocks as applies a block matching algorithm to find the most similar block, between the candidate blocks, within the reference frames. Thus, it is not necessary to represent again the current block, only the Motion Vector (MV), reference frame index and prediction errors are transmitted.

In the HEVC, the largest block is 64x64, called Coding Tree Unit (CTU). The CTU is recursively partitioned according to a quadtree structure [8] with up to four levels, where the CTU is

the highest level. The sub-partitions of the CTU are called Coding Units (CU), ranging from 64x64 (level 1) down to 8x8 block size (level 4). The CUs can be further divided into smaller blocks for prediction and transform steps. The blocks used in the prediction step are named Prediction Units (PU) and can be divided in 7 blocks sizes – for tree levels 1 to 3 (Fig. 1) – or 3 block sizes – for tree level 4 (Fig. 1.a) – adding up 24 PU sizes.

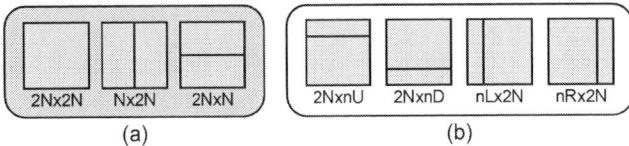

Fig. 1. PU partitioning: Symmetric (a) and (b) Asymmetric

The ME typically limits the search to a Search Area (SA), smaller than frame dimensions, around the co-located block position on the reference frame. This strategy is applied because normally the best matching block is localized nearby co-located position of the current block.

The FS (Full Search) algorithm finds the best matching within the SA, since it compares all candidate blocks. On the other hand, fast algorithms, such as Diamond Search (DS), Hexagon Search (HS), and Test Zone Search (TZS) [15][16], do not compare all candidate blocks present in the SA. These fast algorithms use heuristic to speedup the search process, achieving a suboptimal result with a considerable reduction in the number of candidate blocks evaluated.

Similarity criteria are used to identify the best matching, such as Sum of Absolute Differences (SAD) and Sum of Squared Error (SSE), where the differences are calculated between samples of the original block and the candidate block. The SAD [22] is a lower-complexity distortion metric and represents the most commonly used option in the literature. Note, the SAD presents no data dependencies between distinct candidate blocks, as can be seen in Equation (1), where **w** and **h** represent width and height of the block, respectively, **RB** represents the reference block, and **CB** represents the current block.

$$SAD = \sum_{j=0}^{w-1} \sum_{i=0}^{h-1} |RB_{i,j} - CB_{i,j}| \qquad (1)$$

Thus, distortion metric due to the high amount of data to be processed and the data independence between the candidate blocks, ME represents a key module to be optimized using an efficient parallel implementation targeting energy consumption reduction and performance increase for heterogeneous SoCs.

III. HETEROGENEOUS SYSTEMS RELATED WORKS

As presented in Section I, employing heterogeneous systems is an efficient way to deal with performance and energy constrains of current applications. Aware of that, many works propose systems using the hardware dynamic reconfiguration and/or dynamic task migration between heterogeneous processing units such as CPUs, GPGPUs, and FPGAs.

Designing the project using the heterogeneous and reconfigurable concepts, [17] presents a new approach for general-purpose systems with reconfigurable hardware acceleration and self-reconfiguration. A compiler C-to-VHDL generates the hardware configuration (bitstream) and the host processor sends these bitstreams to reconfigurable device to perform the reconfiguration at runtime. This work demonstrates that it is possible to project a heterogeneous system with CPU and FPGA components, where the host processor controls the reconfigurable device, applying dynamically reconfiguration to set the hardware with specific function when it is necessary, for energy optimization and to make an efficiently general-purpose system.

When high performance parallel execution is demanded, GPGPUs have been adopted in multiple solutions. Many works [18]-[21] use GPGPUs to increase the system processing performance. The referred works employ CUDA language to implement the proposed solutions. In [18] and [19] the GPGPU and CPU are in the same system and the processing is divided between the CPU and GPU, balancing the workload between heterogeneous components, in order to achieve maximum performance. Authors of [20] and [21] propose optimizations for HEVC steps using GPU and CPU. In the first paper has been developed a four-layer parallelized ME in these platforms. The second focuses on making the HEVC encoding speed acceptable for a cloud video service using a cooperatively CPU and GPU multi-core system.

All previous works presented in this section only have been developed with heterogeneous system composed by CPU plus FPGA or CPU plus GPGPU, i.e., none considers accelerating the application in systems with both FPGA and GPGPU. It means the related solutions consider software-only or hardware-only solutions. In addition, these works did not develop for embedded or mobile heterogeneous system.

Therefore, our parallel ME solution focuses heterogeneous systems where it is possible to perform both, runtime task migration and hardware reconfiguration. It is worth to mention that the dynamic task migration itself is out of the scope of this work. Here we present two Implementation Versions in order to allow the system to adapt according to the runtime scenario and provide two performance/energy operation points. Apart from that, we demonstrate that ME can be implemented in heterogeneous systems and benefit from this heterogeneity. This paper proposes FPGA and GPGPU solution for ME step targeting HEVC standard – but not limited to – running on heterogeneous SoCs.

IV. PARALLEL ME SOLUTION FOR HETEROGENEOUS SoCs

This section presents the proposed parallel ME solution for video encoding on heterogeneous SoC. Two implementations were developed, an OpenCL version targeting embedded GPGPUs and a hardware design targeting eFPGA devices.

A. Parallelism-Oriented Motion Estimation Algorithm

In order to better exploit the parallelism allowed by GPGPUs and FPGAs, a parallelism-oriented ME solution is presented. It features a FS algorithm with a scheme to reuse SAD values from 8x8 blocks to group the SAD values for larger block sizes, considering a bottom-up approach. Similar approach was adopted by [22] and [23] for the H.264/AVC standard which uses lower block sizes in the ME step. Additionally, the asymmetric PUs (Fig. 1.b) and 4x8 and 8x4 PUs (symmetrical)

are not supported in our solution, since these PU sizes are rarely used and their removal represents low losses of efficiency encoding [11].

The Search Area (SA) was defined considering a Search Range (SR) of [-15, 15], around the co-localized block, in the Reference Frame, resulting in a SA of 38x38 pixels, for an 8x8 PU size. This SR results in 961 candidate blocks (offset of -15 to 15 vertically and horizontally of 8x8 candidate blocks), for all PUs to be processed, independent of the block size. FS algorithm is adopted in this work, because as cited before, it is the algorithm which leads to the best matching in a given search area. Additionally, the FS regularity allows the SAD grouping scheme.

In brief, the 8x8 PU size was adopted as base block, because this block size is present in all current video coders. Moreover, the use of 8x8 PUs allow the development of a multi-size ME, by the accumulation of the 8x8 SAD values to group the SAD values of the larger PU sizes used in the HEVC standard, i.e., all calculated SAD values of candidate blocks of 8x8 PUs will be used to group candidate blocks for all larger PUs.

Fig. 2 shows the grouping scheme for the SAD calculation of bigger PU sizes from the 8x8 PUs. The 8x8 SAD Calculation (grey box in the Fig. 2) is the first and most important/complex step of this solution as it demands SAD trees [12] to calculate the 8x8 SAD values. For the grouping flow (orange, blue, and green boxes in Fig. 2) it is only necessary to add two SAD values to compose a new SAD value for a larger PU size.

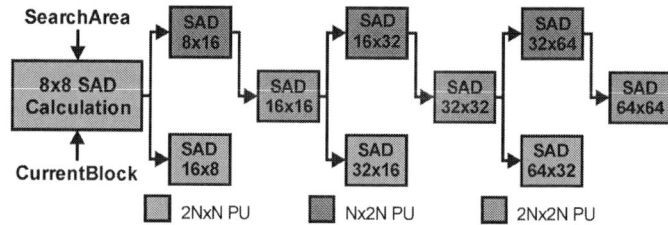

Fig. 2. Grouping using a bottom-up approach from 8x8 block size

After the SAD calculation for 8x8 PUs (NxN; where N represents the dimension of the base block), the candidate blocks of 16x8 ($2NxN$) and 8x16 ($Nx2N$) PUs will be grouped by adding two neighbor 8x8 PUs, as depicted in Fig. 3. Vertically neighboring 8x8 PUs are used for calculate the candidate blocks of $2NxN$ PUs, the blue box in the Fig. 2 (e.g., $16x8_{(0)} = 8x8_{(0)} + 8x8_{(1)}$ and $16x8_{(4)} = 8x8_{(8)} + 8x8_{(9)}$ in the Fig. 3); and horizontally neighboring 8x8 PUs are used for calculate the candidate blocks of $Nx2N$ PUs, the green box in the Fig. 2 (e.g., $8x16_{(0)} = 8x8_{(0)} + 8x8_{(8)}$ and $8x16_{(1)} = 8x8_{(1)} + 8x8_{(9)}$ in the Fig. 3).

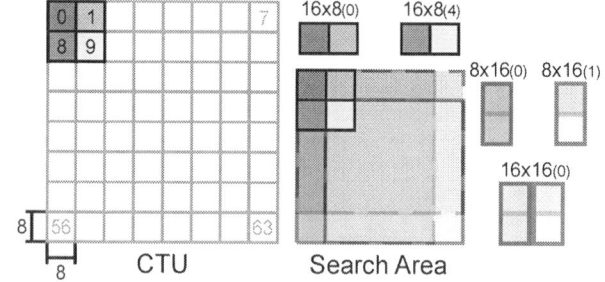

Fig. 3. SAD grouping scheme overview

Then, the candidates block of 16x16 PU ($2N$x$2N$) are grouped by two horizontally neighboring 8x16 PUs, the orange box in the Fig. 2 (e.g., $16x16_{(0)} = 8x16_{(0)} + 8x16_{(1)}$ in the Fig. 3). Finally, the first step of SAD grouping is concluded. After that, the candidates of $2N$x$2N$ PUs are considered as NxN PU size, to calculate the following PUs, as shown in the Fig. 2. This process is repeated until the candidate blocks for a 64x64 PU.

B. The GPGPU Implementation Version

The GPGPUs can be used as hardware accelerators with high memory bandwidth and a large number of programmable cores, with potentially thousands of threads executing a kernel according to single instruction multiple data (SIMD) fashion. Thus, due to no data dependence between the candidate blocks to calculate the SAD, this component can be used for a highly-parallel implementation for the proposed solution.

The first step, for this implementation, is to divide the current frame in CTUs and divide each CTU in 8x8 PUs. This division is done by a kernel responsible to calculate the SAD values for the candidate blocks of each 8x8 block present in the CTU (*CalcSAD8x8*, the grey box in the Fig. 4).

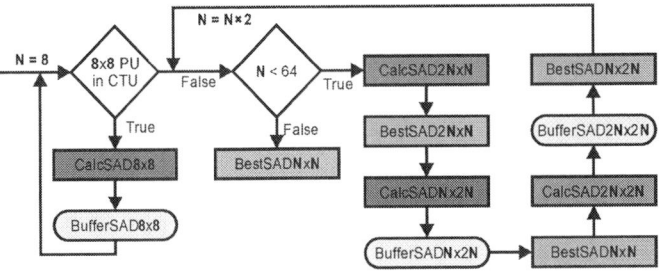

Fig. 4. GPGPU 8x8 SAD calculation and grouping flow

The GPGPU architecture has hundreds of computing units available in this architecture. Thus, it is possible to calculate SAD values in parallel, for all 961 candidate blocks present in the proposed SA. Each computing unit is responsible to calculate the SAD value (sequential) of an 8x8 candidate block. In this Implementation Version, the kernel used by a computing unit requires 64 absolutes, 64 subtractions, and 63 addition operations to calculate an 8x8 SAD value. After calculating SAD values of an 8x8 block size, it is necessary to store these values in a buffer (*BufferSAD8x8*) to use these values in the SAD grouping step.

The grouping step for the GPGPU Implementation Version presents the same idea presented before, i.e., it uses high-parallel approach, where the sums of base blocks to group candidate block for all PUs of a given size are done in parallel (orange, blue, and green box in the Fig. 2 and Fig. 4). For example, 961×32 adders to generate SAD values of candidates blocks for all 32 8x16 PUs from candidate blocks of 8x8 PUs.

The kernel to find the best candidate block is interleaved with the grouping kernels, as shown in the Fig. 4. The kernel to find the best candidate block of a PU is composed of comparator trees, where all SAD values of candidate blocks for all PUs of a given are processed in parallel. This kernel returns the best candidate block of each PU size present in the CTU.

C. The FPGA Implementation Version

The FPGA based hardware architecture Implementation Version is present in Fig. 5. As can be seen, the architecture has as input two sample lines of the search area (2×38) and two sample lines of the block to be processed (2×8). The output of this step is the Motion Vector (MV) of the best-matching candidate block. These architecture is mainly composed by six modules: *SADTree*, *Comp.TreeSAD8x8*, *BufferSAD8x8*, *BufferBest8x8*, *mux* and *decoder*.

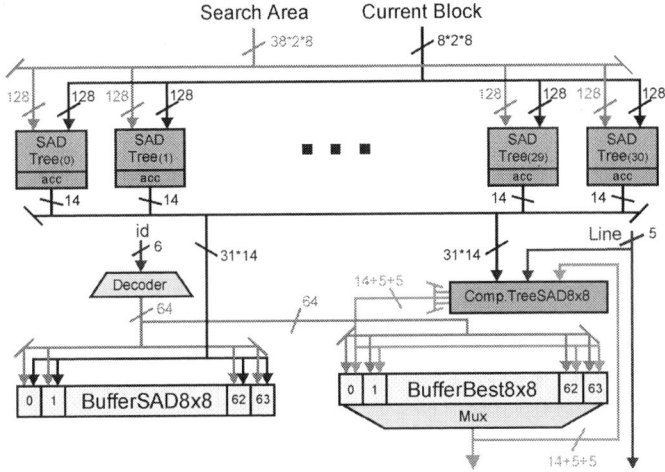

Fig. 5. SAD hardware architecture for 8x8 block size

The *SAD Tree*, the most important module in this architecture, is responsible for calculating the SAD value between two lines of the candidate block and two lines of the current block. The *SAD Tree* uses 16 absolutes, 16 subtractors and 15 adders. In addition, it is necessary to use accumulators, which are composed of one adder and one register, to obtain the SAD value of 8x8 PUs, during four cycles. In this architecture, there are 31 *SAD trees* to process the SAD values of a line of candidate blocks present in SA. The SA position of the candidate blocks is generated according to *SAD tree* id and the *line*.

The *Comp.TreeSAD8x8* has 32 candidate blocks as input, composed of the SAD value and its position in the SA, and the best SAD and its respective position as output. The 64:1 *mux* is responsible to select the best matching already calculated, to be processed by *Comp.TreeSAD8x8*. The *decoder* 6:64 generates the *load* signal to control the buffers (*BufferBest8x8* and *BufferSAD8x8*) to be used according to the current block (*id*).

Finally, the B*ufferBest8x8* stores the best matching of each 8x8 PU present in the CTU. The *BufferSAD8x8* stores the SAD lines to be processed by the grouping step. This buffer is responsible to store the last line of SAD values, represented by 31 SAD values of 14 bits, for each 64 8x8 PUs present in the CTU. These 31 SAD values are called SAD lines in this work.

Once the SAD line is complete, this line is used as input to the *BufferSAD8x8* and to the *Comp.TreeSAD8x8*. In the *Comp.TreeSAD8x8,* this SAD line and its position in the SA are processed with the best matching already processed (one SAD value and its position in the SA), which are stored in the B*ufferBest8x8*, selected using the *mux* according to the block being processed (*id* signal control).

978-1-5090-2737-8/16 $31.00 © 2016 IEEE   155

The *Comp.TreeSAD8x8* defines if the current best match is present in the current SAD line or if it was processed before. After that, it stores them in the *BufferBest8x8*.

The data flow used by the architecture is: divide the video into CTU (64x64) and divide the CTU into 8x8 PUs. After, the architecture will calculate the SAD line using first line of candidate block in the SA of the first 8x8 PU present in the CTU, after, will be processed first line of candidate block in the SA of the second 8x8 PU present in the CTU, this processes will be carried until the last 8x8 PU present in the CTU. After that, the same idea is used, but the processing is done using the second line of candidate blocks present in the search area of all PUs. This process will be carried for all (31) lines of candidate blocks present in the search area.

The grouping flow has data dependences according to the PU being processed. Fig. 6 presents the time dependencies in the SAD grouping process. The SAD values for candidate blocks of 8x8 block size are calculated in every four cycles (for all *id8x8*, that present the 8x8 block position in the CTU). However, to calculate the SAD value of 16x8 PU candidates, eight cycles are spent (every second 8x8 SAD processing, according to half of *id8x8*). On the other hand, to calculate the SAD value of 8x16 PU candidates it is necessary to wait for the processing of SAD values from eight 8x8 PUs, as shown in the Fig. 6.

The generic architecture for the grouping step is shown in Fig. 7. It uses the same architectural template from Fig. 5 but replaces the *SAD Trees* by *adders* of *n*-bits, according to the SAD line used as input. It employs nine of such architectures, where each one is used for one block size used in the grouping flow. Other difference is the elimination of *bufferSAD* for the 2*NxN* (blue boxes in the Fig. 2), because these SAD values are not used as base blocks for the following block size. Finally, the

number of *bufferSAD* and *bufferBest* are defined according to the number of PU sizes present in the CTU. For instance, there are 16 16x16 PUs and four 32x32 PUs in one CTU.

The whole architecture processes each SAD line with 31 SAD values using four clock cycles, for 8x8 PUs. In fact, it is necessary to spend 64 clock cycles to process the first SAD line of candidate blocks for each 8x8 PU of an CTU; also, it is necessary to process 31 SAD lines in the SA. Thus, 7684 cycles are necessary define the best candidate block of the first 8x8 PU of an CTU. After that, more four cycles are necessary to define the best candidate block of the second 8x8 PU, this processes will be carried until the last 8x8 PU present in the CTU. In fact, 7936 cycles are necessary to complete the whole CTU processing and define the best candidate block for all PUs. After that, the architecture is available to process a new CTU.

V. IMLEMENTATION RESULTS

The GPGPU Implementation Version was described in OpenCL 1.1 and the performance/energy results were obtained by real execution on the Odroid XU3 platform [24]. The Odroid XU3 is based on the Samsung Exynos 5422 (28-nm) SoC featuring the ARM® big.LITTLE™ architecture with Heterogeneous Multi-Processing (HMP). It is composed of a quad-core Cortex™-A15 @ 2.0 GHz, a quad-core Cortex™-A7 @1.4 Ghz, 2GB LPDDR3 RAM, and a GPGPU Mali-T628 MP6@600 MHz supporting OpenCL 1.1 Full Profile.

The proposed FPGA Implementation Version was described in VHDL and validated using the Mentor Graphics ModelSim tool. The power analysis and synthesis results were obtained targeting a Stratix V 5SGXMABN3F45I4 FPGA device (28-nm) using the Quartus Prime 15.1.0 tool. The FPGA core is powered by 0.85V and disposes of 359.2K ALMs, 1064 I/O pins, and 352 DSP Blocks.

The FPGA synthesis results are presented in Table I, which also presets the performance results for both implementations. A total of about 69K ALMs were used (19% of the total). As shown in Table I, the maximum frequency obtained by the FPGA Implementation Version is about 200 MHz. It is necessary 222,208 cycles to process a QWVGA frame (28 CTUs), and 4,047,360 cycles to process a HD 1080 (510 CTUs) using this architecture. Thus, the performance allows to process 898 QWVGA and 49 HD 1080 frames per second (fps). The GPGPU Implementation Version reaches only 0.7 QWVGA fps when processing at 600 MHz. These results demonstrate that this GPGPU implementation is not able to handle with real time applications, which requires a more powerful device of a simplified version of the ME algorithm, with reduced support for the different block sizes and/or providing some level of subsample to speed up the process.

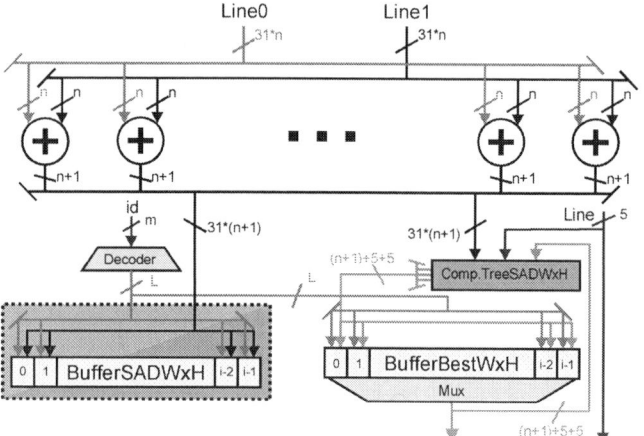

Fig. 7. Generic hardware architecture for the grouping flow

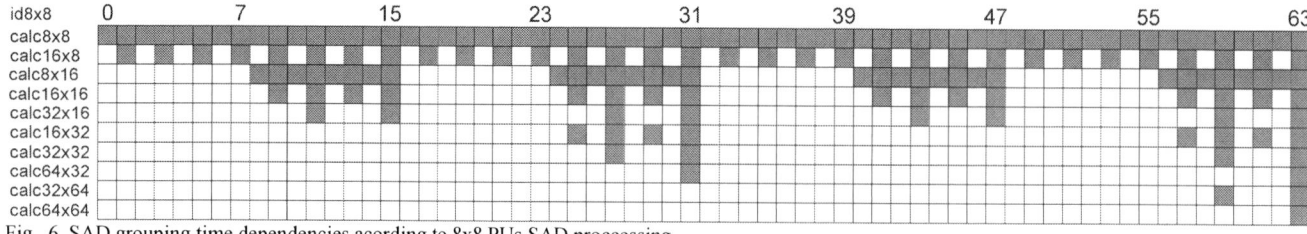

Fig. 6. SAD grouping time dependencies acording to 8x8 PUs SAD proccessing

TABLE I. SYNTHESIS AND PERFORMANCE RESULTS

	FPGA	Mali-T628
ALMs	69,481 (19%)	-
Register	63132	-
Pins	838 (79 %)	-
DSP Blocks	0 (0%)	-
Freq.	199.68 MHz	600MHz
FPS (QWVGA – 416x240)	898	0.7
FPS HD 1080p	49	-

The energy results for both Implementation Versions were obtained using real data from a QWVGA video (*RaceHorses*) and a SA of 38x38. FPGA power analysis was performed using the Altera PowerPlay tool whereas for GPGPU real measurements were performed using Exynos built-in sensors. Observe from Table II that in addition to the meaningful difference in term of performance, the FPGA Implementation Version increases about 1000x the energy efficiency.

TABLE II. ENERGY CONSUMPTION TO PROCESS ONE QWVGA FRAME

FPGA Energy (mJ)		Exynos 5422 Energy (mJ)	
-	-	GPU	623
Dynamic	0.054	Memory	101
Static	1.332	Cortex A15	170
I/O	0.036	Cortex A7	417
Total	**1.422**	**Total**	**1,311**

Firstly, it is possible to see that reconfigurable FPGA fabrics are highly desirable in order to improve energy efficiency in heterogeneous SoCs. Secondly, ME is a high-demanding block in the HEVC and its allocation in the FPGA should be prioritized. Thirdly, although providing much inferior energy efficiency and performance, the Implementation Version focusing for embedded GPGPUs provide a second energy/performance operation point that must be considered if the SoC does not feature an FPGA fabric or the FPGA is allocated to a higher priority task.

VI. CONCLUSIONS

This paper presented a parallel ME solution for video coding on heterogeneous SoC, with two implementation versions: a GPGPU Implementation Version, described in OpenCL, and a FPGA based hardware design, described in VHDL.

The performance and energy consumption results were presented for both implementations considering real data from an QWVGA video. The energy consumption results for the GPGPU implementation were obtained by Exynos built-in sensors, and the FPGA based energy consumption results were estimated with the Altera PowerPlay tool. The synthesis results for the FPGA implementation show an energy consumption of 1.422 mJ per frame on a QWVGA resolution, and a maximum performance of 49 HD 1080p per second.

ACKNOWLEDGMENT

We have a special acknowledgement to CNPq, CAPES and FAPERGS to support this work.

REFERENCES

[1] S. B. Furber, *ARM System-on-Chip Architecture*, Addison-Wesley Longman Publishing Co., 2000.

[2] F. Luo et al, "Multiple layer parallel motion estimation on GPU for High Efficiency Video Coding (HEVC)," *2015 IEEE International Symposium on Circuits and Systems (ISCAS)*, Lisbon, 2015, pp. 1122-1125.

[3] Intel. Intel Completes Acquisition of Altera [Online]. Available: http://intelacquiresaltera.transactionannouncement.com/.

[4] S. Ahmad et al, "A 16-nm Multiprocessing System-on-Chip Field-Programmable Gate Array Platform," in IEEE Micro, vol. 36, no. 2, pp. 48-62, 2016.

[5] Intel. Intel Introduces Configurabel Intel Atom – Based Processor [Online]. Available: https://newsroom.intel.com/press-kits/intel-introduces-configurable-intel-atom-based-processor/

[6] K. F. K. Wong et al, "Hardware accelerator implementation on FPGA for video processing," *Open Systems (ICOS)*, 2013 IEEE Conference on, Kuching, 2013, pp. 47-51.

[7] Dan Miao et al, "Layered screen video coding leveraging hardware video codec," *2013 IEEE Int. Conf. on Multimedia and Expo (ICME)*, San Jose, CA, 2013, pp. 1-6.

[8] G. J. Sullivan et al, "Overview of the High Efficiency Video Coding (HEVC) Standard," in *IEEE Trans. on Circuits and Systems for Video Technology*, vol. 22, no. 12, pp. 1649-1668, Dec. 2012.

[9] International Telecommunication Union. 2007. ITU-T Recommendation H.264: Advanced video coding for generic audiovisual services, 2007.

[10] A. Puri et al, "Video Coding Using the H.264/MPEG-4 AVC Compression Standard," *Signal Process.: Image Commun.*, Vol. 19, no. 9, pp. 793-849, Jul. 2004.

[11] H. Maich et al, "HEVC Fractional Motion Estimation complexity reduction for real-time applications,", In IEEE 5th Latin American Symp. on Circuits and Systems (LASCAS), Santiago, 2014, pp. 1-4.

[12] Afonso, Vladimir, et al. "Memory-Aware and High-Throughput Hardware Design for the HEVC Fractional Motion Estimation." *Proc. of the 28th Symp. on Integrated Circuits and Systems Design*. ACM, 2015.

[13] I. Sourdis et al, "Reconfigurable acceleration and dynamic partial self-reconfiguration in general purpose computing," *Field-Programmable Technology (FPT), 2011 Int. Conference on*, New Delhi, 2011, pp. 1-8.

[14] D. Rossi et al, "A Heterogeneous Digital Signal Processor for Dynamically Reconfigurable Computing," in *IEEE Journal of Solid-State Circuits*, vol. 45, no. 8, pp. 1615-1626, Aug. 2010.

[15] M. S. Porto et al, "An efficient ME architecture for high definition videos using the new MPDS algorithm," In *Symposium on Integrated Circuits and Systems Design*, João Pessoa, 2011, pp. 119-124.

[16] L. Xufeng et al, "Context-adaptive fast motion estimation of HEVC," *IEEE Int. Symp. on Circuits and Systems (ISCAS)*, Lisbon, 2015, pp. 2784-2787.

[17] I. Sourdis et al, "Reconfigurable acceleration and dynamic partial self-reconfiguration in general purpose computing," *Field-Programmable Technology (FPT), 2011 Int.Conference on*, New Delhi, 2011, pp. 1-8.

[18] R. Shi et al, "A scalable and portable approach to accelerate hybrid HPL on heterogeneous CPU-GPU clusters," *Cluster Computing (CLUSTER), 2013 IEEE International Conference on*, Indianapolis, IN, 2013, pp. 1-8.

[19] J. Zhao and H. Zhou, "Design and optimization of remote sensing image fusion parallel algorithms based on CPU-GPU heterogeneous platforms," *Image and Signal Processing (CISP), 2011 4th International Congress on*, Shanghai, 2011, pp. 1623-1627.

[20] W. Xiao et al, "HEVC Encoding Optimization Using Multicore CPUs and GPUs," in *IEEE Transactions on Circuits and Systems for Video Technology*, vol. 25, no. 11, pp. 1830-1843, Nov. 2015.

[21] F. Luo et al, "Multiple layer parallel motion estimation on GPU for High Efficiency Video Coding (HEVC)," *2015 IEEE International Symposium on Circuits and Systems (ISCAS)*, Lisbon, 2015, pp. 1122-1125.

[22] D. Cláudio, et al. "Comparative analysis of parallel SAD calculation hardware architectures for H. 264/AVC video coding." In *IEEE Latin American symp. on circuits and systems* (LASCAS). Foz do Iguacu, 2010.

[23] Wei and M. Gang, "A novel SAD computing hardware architecture for variable-size block motion estimation and its implementation with FPGA," ASIC, 2003. In *5th Int.Conference on*, 2003, pp. 950-953 Vol.2.

[24] Hardkernel. ODROID-XU3 platform description [Online]. Available: http://www.hardkernel.com/main/products/prdt_info.php.

A Digitally Tunable 4th-order Gm-C Low-Pass Filter for Multi-Standards Receivers

Mateus S. Oliveira*, Paulo C. de Aguirre*[†], Lucas C. Severo*, Alessandro G. Girardi* and Altamiro A. Susin[†]

*Federal University of Pampa (UNIPAMPA) - GAMA

Alegrete, Brazil

Email: mateus.oliveira@alunos.unipampa.edu.br and paulo.aguirre@unipampa.edu.br

[†]Federal University of Rio Grande do Sul (UFRGS) - PPGEE

Porto Alegre, Brazil

Abstract—This paper presents a digitally tunable 4th-order Gm-C low-pass filter (LPF) for multi-standards radio receivers. The cutoff frequency tuning is provided by changing the transconductance of a reconfigurable operational transconductance amplifier (OTA). Two control bits are employed to digitally control the OTA transconductance and also the power consumption. This LPF is designed in a 180 nm CMOS process and powered by a 1.8 V power supply. Post-layout simulation analysis indicate that this flexibility provides a cutoff frequency of 2.54/5.11/7.68/10.29 MHz with a power consumption ranging from 10.27 to 12.69 mW. The designed filter achieves an IIP3 of 4.14 dbm for a signal bandwidth of 10.29 MHz and electrical characteristics comparable to recent published works in the literature.

I. Introduction

Several communication standards are supported and employed in today mobile wireless communication systems. The use of flexible analog blocks is required to support these multiple standards in a single device [1]. Furthermore, the power consumption of flexible analog circuits should be proportional to the required performance level. It is necessary in order to minimize the power consumption in each operation mode, such as in software-defined radio (SDR) devices [2]. The specification of analog blocks can, for example, be reduced when the system is switched from a hard standard to a softer one. This enables the front-end circuitry to fulfill the specifications of each standard with a lower power consumption, similar to a single-mode device designed only for that application.

The direct-conversion receiver architecture has being employed in multi-standard wireless transceivers since it achieves a highest level of integration [3]. These receivers convert the signal directly to baseband. Thus, the continuous time active low-pass filter (LPF) is a key building block in direct-conversion integrated circuit (IC) front-ends [4]. In order to avoid an array of channel-select filters, it is more energy efficient to design a single tunable LPF to support different communication standards. So, the reconfiguration of these filters is performed in architectural and circuit levels [1].

Based in these considerations, this paper presents a digitally tunable Gm-C low-pass filter for multi-standard radio receivers. The tuning is performed by changing the filter transconductance by means of a digitally programmable transconductance operational amplifier (OTA). Besides of a digitally controlled transconductance, this technique also provides a scalable power consumption proportional to the transconductance.

This paper is organized as follows: Section II presents the tunable gm-C filter architecture. Circuit implementation is detailed in Section III. Post-layout simulation results of the designed filter and comparison with related works are presented in Section IV. Finally, conclusions are given in Section V.

II. Reconfigurable Gm-C Filter

When dealing with integrated active low-pass filter design, two common implementation approaches are employed: Gm-C [5] or active-RC [1]. To tune the cutoff frequency of an active-RC filter it is necessary to change the value of passive components (resistors and capacitors) in order to change the filter transfer function. It can be implemented by using digitally configurable banks of capacitors and resistors. However, integrated resistors are very sensitive over process variations - the absolute value of an integrated resistor varies up to 20 % -, impacting in a deviation in the cutoff frequency. In addition, the DC voltage gain and gain-bandwidth product (GBW) of the operational amplifiers in active-RC filters should satisfy the specification for the higher cutoff frequency. On the other hand, the cutoff frequency tunning of a Gm-C filter could be performed by changing the capacitors or the transconductance. It is less sensitive to component mismatches and the scaling of power consumption can be achieved by the transconductance reconfigurability. In addition, Gm-C filters have great potential of working in higher frequencies since the GBW of the transconductance amplifiers only needs to be slightly higher than the cutoff frequency of the filter [4]. Thus, due to the characteristics of Gm-C filters, they present higher power efficiency when compared to active-RC ones. So, they are more suitable to attend low-power receivers requirements.

In this work, a 4th-order Gm-C filter topology is employed to design a tunable LPF whereas the cutoff frequency is controlled by changing the OTA transconductance. The filter order was set in agreement with related multi-mode LPF for radio receivers presented in the literature [6]. The filter

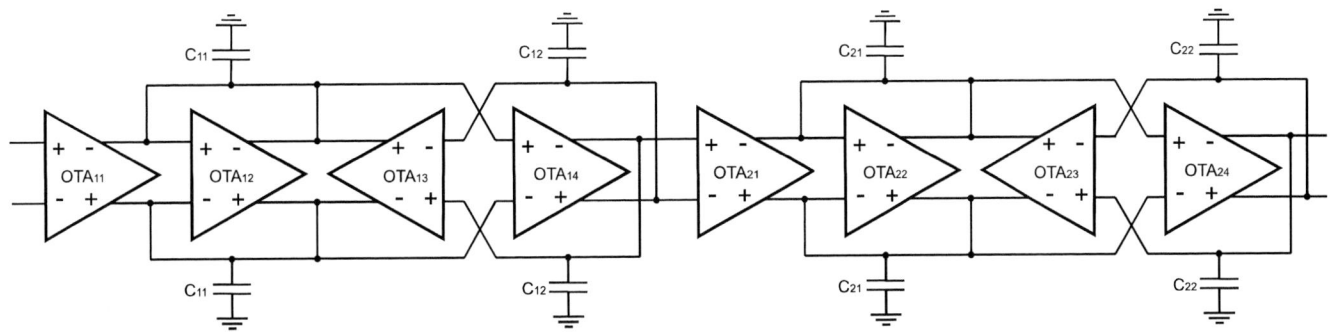

Fig. 1. Simplified schematic of a 4th order reconfigurable Gm-C filter.

schematic is shown in Fig. 1 [7]. The transfer function of the filter is split in two terms related to two biquad filters, as presented in (1).

$$F(s) = \frac{gm_{11}.gm_{14}/(C_{11}.C_{12})}{s^2 + \dfrac{gm_{12}}{C_{11}}s + \dfrac{gm_{13}.gm_{14}}{C_{11}.C_{12}}} \times$$
$$\frac{gm_{21}.gm_{24}/(C_{21}.C_{22})}{s^2 + \dfrac{gm_{22}}{C_{21}}s + \dfrac{gm_{23}.gm_{24}}{C_{21}.C_{22}}} \quad (1)$$

The cutoff frequency (fc) is controlled by the transconductances gm_{i3} and gm_{i4} of OTA$_{i3}$ and OTA$_{i4}$, respectively, as presented in (2), in which i indicates the biquad index. In addition, the quality factor (Q) of each biquad is defined according to (3).

$$fc_i = \frac{1}{2\pi}\sqrt{\frac{gm_{i3}.gm_{i4}}{C_{i1}.C_{i2}}} \quad (2)$$

$$Q_i = \sqrt{\frac{C_{i1}.gm_{i3}.gm_{i4}}{C_{i2}.gm_{i2}^2}} \quad (3)$$

The filter capacitors have a capacitance of 12.58 pF. In addition, to reduce the design complexity, only one reconfigurable OTA can be designed and replicated in the filter. Thus, each biquad in the designed filter has a quality factor equal to one, providing a frequency response similar to the Butterworth transfer function.

III. CIRCUIT IMPLEMENTATION

This 4th-order reconfigurable Gm-C filter was designed in a TSMC 180 nm CMOS process with a 1.8 V power supply employing only standard transistors and metal-insulator-metal (MiM) capacitors. The reconfiguration approach is based on a reconfigurable transconductance amplifier (OTA) while the filter capacitor values are kept constant.

A. Reconfigurable OTA

Reconfigurable operational transconductance amplifiers are fundamental blocks for multi-mode analog baseband circuits of radio receivers, such as integrated analog filters and analog-to-digital converters (ADCs). In addition, these reconfigurable

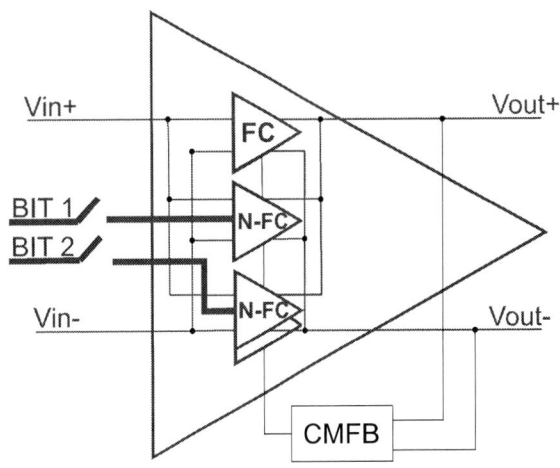

Fig. 2. Reconfigurable OTA architecture.

OTAs should fulfill the demand of flexibility and energy efficiency [8].

The reconfigurable OTA architecture used in this work is shown in Fig. 2 [1]. It is composed by an array of multiple switchable parallel amplifiers controlled by digital inputs. The main amplifier remains always active, whereas the switchable units are controlled by digital inputs that can be activated in order to modify the reconfigurable OTA transconductance. All switchable amplifier units are equal. Thus, by adding N amplifiers in parallel, the transconductance g_{m0} is multiplied N times while the reconfigurable amplifier DC gain is held constant. It provides a change in the filter cutoff frequency, as intended in software-defined radio applications. In addition, this approach provides a change in the OTA gain-bandwidth product and in the power consumption.

Four unit amplifiers are employed in the designed reconfigurable OTA. The first one is the main amplifier. It is a conventional fully-differential folded cascode (FC) amplifier due to the characteristics such as high DC voltage gain capability, wide input common-mode range (ICMR) and self-

978-1-5090-2737-8/16 $31.00 © 2016 IEEE

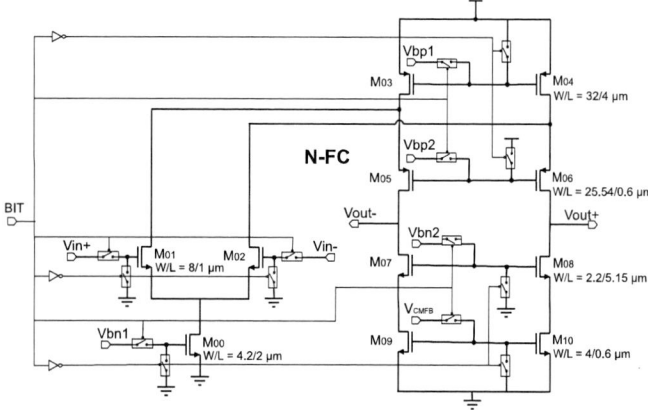

Fig. 3. Switchable OTA schematic (Bias and CMFB circuits not shown).

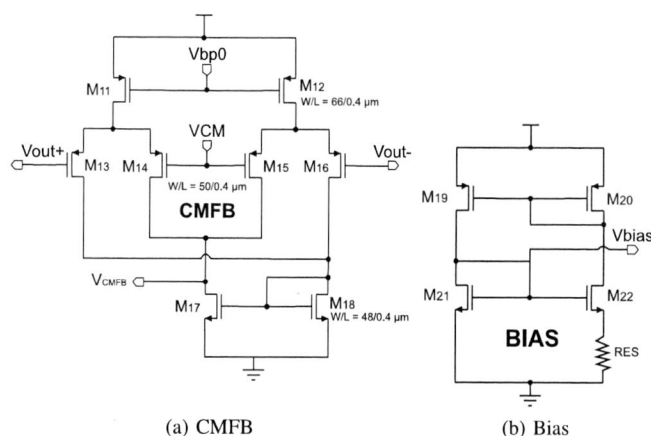

(a) CMFB

(b) Bias

Fig. 4. Common-mode feedback and bias circuitry employed in the reconfigurable OTA.

compensation [9]. This amplifier is always on. The another ones are switchable unit amplifiers, called in this paper as N-FC, whose schematic is shown in Fig. 3. These switchable units are also fully-differential folded-cascode amplifiers with switches that control the gate voltage of the transistors in order to switch them on or off. These switches are built with minimum-sized transistors in transmission gate configuration. In the "off" state, all switches are open and the supply sources Vbp1, Vbp2, Vbn1, Vbn2 and V_{CMFB} are electrically disconnected from the circuit. NMOS gates are grounded and PMOS gates are connected to VDD [1]. In this state, the static power consumption is minimized since the unit amplifier is not operating. In the "on" state all supply sources are connected and the inputs are connected to the $Vin+$ and $Vin-$ nodes, respectively.

Two control bits are employed in order to provide four operation modes associated with the following digital words: "00", "01", "10", "11". The first bit controls the second unit amplifier. The second bit controls both the third and fourth amplifiers. The "00" configuration represents the minimum operation mode, where just the non-switchable amplifier is active. This configuration provides a lower transconductance and also the lowest GBW and power consumption. On the other hand, the configuration "11" provides the maximum transconductance and also the highest GBW and power consumption, since all switchable unit amplifiers are on.

1) Common-mode feedback (CMFB) circuitry: The common-mode feedback (CMFB) circuit employed in the reconfigurable OTA is a PMOS differential-difference CMFB [10] and its schematic is depicted in Fig. 4a. This topology employs only CMOS transistors, eliminating the use of passive components in order to reduce area. The V_{CM} is 0.9 V and only one CMFB is employed for all unit amplifiers, in order to reduce area and power consumption since they are in parallel. Also, the diode connected M_{18} provides a higher output common-mode gain, improving the circuit stability.

2) Bias circuitry: Fig. 4b shows the beta-multiplier reference circuit [11] used to generate all bias voltages - Vbn1, Vbn2, Vbp1 and Vbp2 - in N-FCs, and Vbp0 in CMFB.

For each bias voltage a different cell was resized, changing transistors aspect ratios (W/L) and the resistor value in order to generate the desired bias value.

In this circuit, the resistor drains a current through M_{20}, which is diode-connected transistor, forcing the same current in M_{19}. As result, we have the same current in the NMOS current mirror composed by M_{21} and M_{22}. Since the current value is known, we can easily find the gate voltages by β relation obtained isolating it in the MOS current quadratic model.

B. Reconfigurable OTA post-layout simulation results

Since the reconfigurable OTA is the main analog building block of the tunable LPF, this subsection presents its complete post-layout analysis. In order to evaluate its behavior over process variation, a Monte Carlo (MC) simulation with 1,000 runs was performed. Fig. 5 presents the histograms for GBW considering the four different input configurations.

The Monte Carlo results obtained for GBW, DC voltage gain (Av) and phase margin (PM) are summarized in Table I. It can be observed that the mean values (μ) of the three specifications are very similar to the nominal results for all operation modes. Besides, through the achieved standard deviation (σ) values, it is possible to agree that the circuit is stable over process variations, since the parameters spread in a tight range.

The programmable OTA provides a GBW of 22.07/44.04/65.85/87.46 MHz with a power consumption of 1.28/1.39/1.51/1.62 mW for a 1 pF load connected at each output node, while the transconductance for each operation mode is 192.77/385.54/578.31/771.08 μS. Individually, each unit folded cascode cell consumes 109.15 μW, while 1.017 mW and 149.19 μW are the estimated power consumption for the CMFB block and the bias circuitry, respectively. The DC voltage gain is around 69 dB and changes less than 2 dB for all operation modes.

It can be noticed that the phase margin decreases when adding more basic OTA cells in parallel to increase the GBW. To keep a stable PM it is necessary to increase the PM of

978-1-5090-2737-8/16 $31.00 © 2016 IEEE

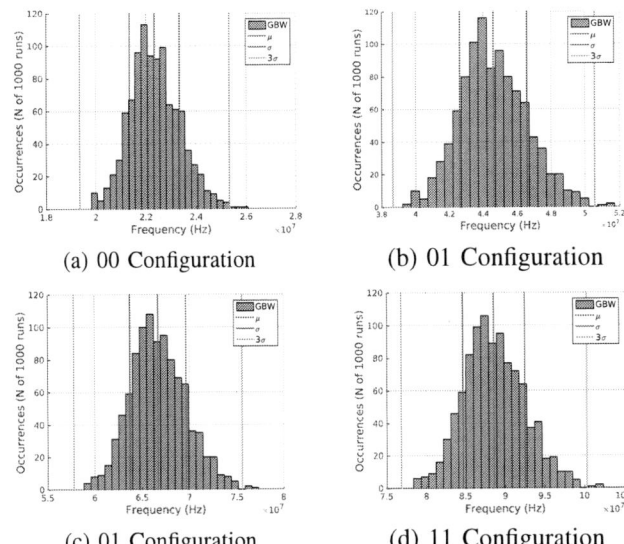

| (a) 00 Configuration | (b) 01 Configuration |
| (c) 01 Configuration | (d) 11 Configuration |

Fig. 5. Monte Carlo simulation for GBW specification under four operation modes.

TABLE I
MONTE CARLO SIMULATION RESULTS

Control bits	Av (dB)		PM (°)		GBW (MHz)	
	μ	σ	μ	σ	μ	σ
00	68.69	3.52	85.3	0.13	22.34	0.99
01	68.44	3.37	79.55	0.33	44.59	1.98
10	68.29	3.28	74.19	0.48	66.69	2.95
11	68.2	3.23	69.29	0.65	88.58	3.9

TABLE II
SUMMARIZED POST-LAYOUT SIMULATION RESULTS

Control bits	00	01	10	11
OTAs	1	2	3	4
Av0 (dB)	69.18	69.96	68.85	68.75
PM (°)	85.28	79.51	74.14	69.28
GBW (MHz)	22.07	44.04	65.85	87.46
gm (μS)	192.77	385.54	578.31	771.08
PWR (mW)	1.28	1.39	1.51	1.62
OS (V)	1.50	1.53	1.54	1.54
SR (V/μs)	15.78	26.16	32.79	37.22

the basic OTA cell. The drawback is that it also increases the power consumption of the unit cell, so increasing the total power of the complete programmable OTA. Moreover, the "N-FC" cells increase the amplifier output capacitive load, reducing the GBW.

The simulated output swing (OS) is near ± 0.75 V for all operation modes which is desirable for the design of analog continuous-time filters. However, the value of the slew-rate (SR) is affected for each operation mode, since it is related to GBW. The lower SR value is 15.78 V/μs for the "00" configuration and the higher SR value is 37.22 V/μs for the "11" configuration. The complete programmable OTA performance estimated by post-layout simulation is summarized in Table II.

IV. 4TH ORDER FILTER POST-LAYOUT SIMULATION RESULTS

In this section post-layout simulation results of the designed tunable Gm-C LPF are presented. The circuit layout is shown in Fig. 6, occupying a silicon area of 0.23 mm². During filter evaluation it was employed a load composed by a 1 pF capacitor connected at each output node (inverting and non-inverting). The frequency response of the filter is shown in Fig. 7. The filter DC gain is 0 dB at the pass-band with a small peak near the cutoff frequency due the chosen quality factor. The achieved cutoff frequencies, which are controlled

by changing the OTAs transconductance, are: 2.55, 5.11, 7.68 and 10.29 MHz. Thus, this filter could be employed in radio-receivers for multi carrier WCDMA signals [12]. The total power consumption of the filter for each operation mode is: 10.27, 11.18, 12.06 and 12.96 mW. Due to the power hungry CMFB employed on the reconfigurable OTAs in order do reduce de common-mode voltage error, the filter power consumption flexibility is not linear in relation to the cutoff frequency.

The in-band linearity of the designed filter was also evaluated by a two tone test. Fig. 8 presents the power spectrum density of the output filter for two input signals with differential peak amplitude of 40 mV and frequencies of 1.0 and 1.1 MHz. The third-order intermodulation distortion (IM3) amplitude is around -81 dB.

The input referred third order intercept points (IIP3s) of the filter, considering the extracted layout view, was also evaluated for each operation mode according to [3]. Since the reconfigurable OTA employed in the filter presents high output impedance, the IIP3 analysis was performed by applying a power signal at the filter input and measuring a voltage signal at the output instead of a power signal. However, for comparison purposes, the input and IIP3 voltage signal values are expressed as power signals referred to a 50 Ω load.

Fig. 9 presents the input power versus the filter output IM3 power for each operation mode. Two input tones with frequencies of 0.5 and 0.6 MHz and initial power of -40 dbm were applied at filter input. The values of IIP3 were estimated by the extrapolation of this points. The higher IIP3 is 4.14 dbm and it is achieved for the cutoff frequency mode of 10.29 MHz.

In order to compare this work with previous reported works in the literature, a figure of merit (FoM), which is independent to the tuning ratio, is evaluated by (4) [13].

$$\text{FoM} = \frac{P_{tot}}{N.f_c.\text{SFDR}.N^{\frac{4}{3}}} \qquad (4)$$

Here, P_{tot} is the LPF total power consumption, N is the number of poles and zeros, f_c is the cutoff frequency and SFDR is the signal-free dynamic range, which is expressed by (5).

$$\text{SFDR} = \left(\frac{\text{IIP3}}{P_N}\right)^{4/3} \qquad (5)$$

978-1-5090-2737-8/16 $31.00 © 2016 IEEE

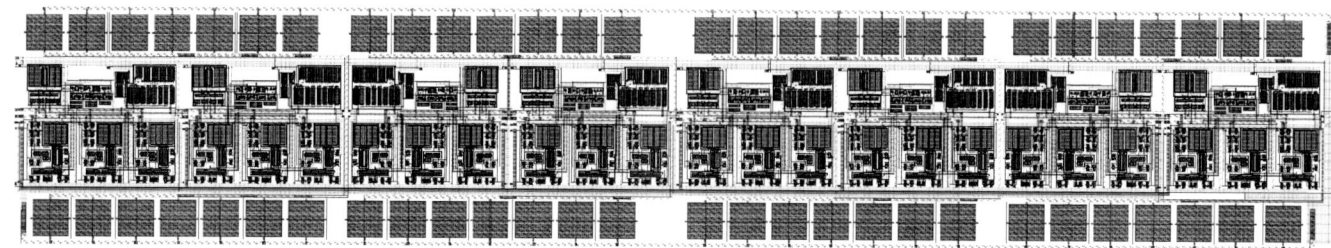

Fig. 6. Layout of the 4th order reconfigurable Gm-C filter (1137 x 199 μm).

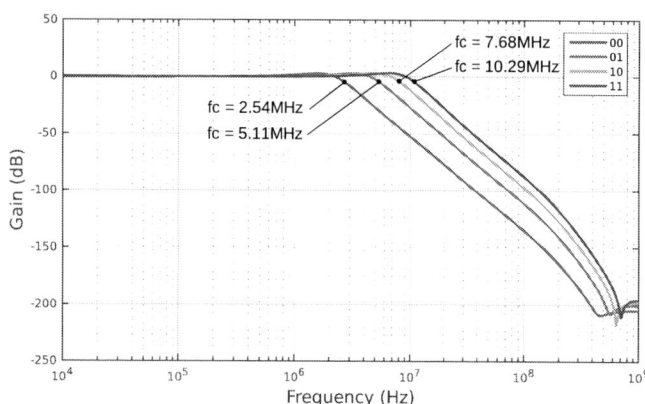

Fig. 7. Post-layout frequency response of the 4th-order reconfigurable filter.

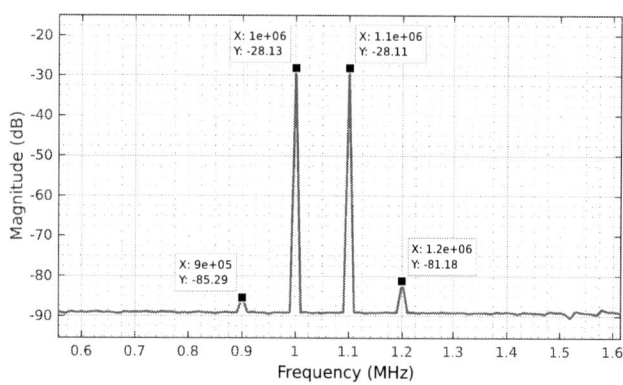

Fig. 8. PSD of the filter output for a two-tone test (10.29 MHz operation mode)

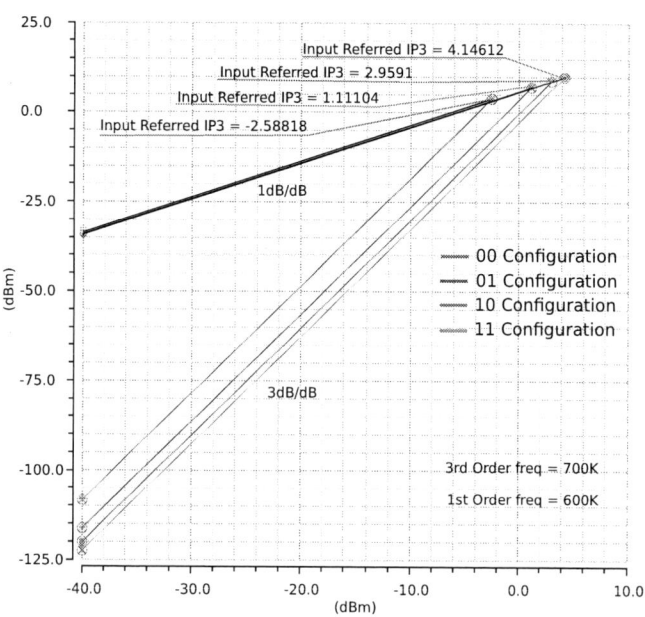

Fig. 9. In-band IIP3 for each cutoff frequency.

TABLE III
PERFORMANCE SUMMARY

Technology	180 nm CMOS			
Supply Voltage	1.8 V			
Filter Type	Biquad (Q=1)			
Order	4			
Silicon area	0.23 mm^2			
Cutoff Freq. (MHz)	2.55	5.11	7.68	10.29
IIP3 (dBm)	-2.59	1.11	2.96	4.14
Noise (nV/$\sqrt{Hz}$)	58.1	42.66	34.88	27.88
PWR (mW)	10.27	11.18	12.06	12.96
FOM (fJ)	7.30	2.371	1.2829	0.774

In this case, P_N is the noise power that was evaluated at the filter input by simulation. Table III summarizes the filter performance results including the FoM (expressed in fJ) and Table IV presents a performance comparison with related works reported in recent years. It is possible to verify that our tunable filter achieves a comparable FoM for the cutoff frequency of 10.29 MHz but it suffers due to the power consumption. However, we should remark that our results are based on post-layout simulations (extracted view with parasitics).

V. CONCLUSION

This paper presented the complete design and post-layout simulation results for a tunable 4th-order Gm-C filter in a 180 nm CMOS process. Four operation modes are achieved providing cutoff frequencies to cover radio receivers for multi-carrier WCDMA signals. The tunable capability is achieved by digitally reconfigurable OTAs that also provides a power consumption scaling. The filter provides a comparable FoM,

978-1-5090-2737-8/16 $31.00 © 2016 IEEE

TABLE IV
COMPARISON WITH PREVIOUSLY REPORTED WORKS

Parameters	This Work	[1] JSSC'07	[4] ESSCIRC'15	[14] CICC'07	[15] JSSC'11	[7] TCAS-I'14	[16] JSCC'09	[3] JSCC'09
Technology (nm)	180	130	180	180	90	180	130	190
Supply Voltage (V)	1.8	1.2	1.8	1.0	1.0	1.8	0.55	1.2
Filter Type	2xBiquad (Q=1)	Butterworth	Elliptic	Butterworth	Butterworth	Butterworth	Butterworth	Butterworth
Order	4	2-4-6	4	3	6	4	4	3
Cutoff Freq. (MHz)	2.5-10.3	0.35-23.5	7.4-27.4	0.135-2.2	8.1-13.5	0.3-12	11.3	0.5 - 20
Tuning ratio	4	67	3.7	16	1.66	40	No	40
IIP3 (dBm)	-2.59-4.14	22.97	20.75	16.3-20.1	21.7-22.1	8.7-18	10	19-22.3
Noise ($nV/\sqrt{Hz}$)	58.1-27.88	83.35-163	-	65	75	112-4780	33	425-12
PWR (mW)	10.27-12.96	0.72-21.6	3.8-13.6	1.57-1.92	4.35	1.08-4.68	3.5	4.1-11.1
Area (mm^2)	0.23	0.52	0.23	0.5	0.24	0.125	0.43	0.23
FOM (fJ)	7.30-0.774	0.013	-	0.207-0.179	0.0246	0.843	0.103	0.0345 - 3.055

for the higher cutoff frequency, and silicon area with recent related works reported in the literature.

As future work we intend to propose a new strategy to the CMFB circuitry implementation in order to reduce the total power dissipation.

ACKNOWLEDGMENT

The authors would like to thank FAPERGS process number 0460-2551/15-6 for the financial support for this work and IMEC Brasil for providing the process data and manufacturing through Mini@asic Program from Europractice.

REFERENCES

[1] V. Giannini, J. Craninckx, S. D'Amico, and A. Baschirotto, "Flexible baseband analog circuits for software-defined radio front-ends," *Solid-State Circuits, IEEE Journal of*, vol. 42, no. 7, pp. 1501–1512, July 2007.

[2] X. Zhang, Y. Xu, B. Liu, Q. Yu, S. Han, Q. Liu, Z. Zhang, Y. Gao, Z. Wang, and B. Chi, "A 0.1-5ghz flexible sdr receiver in 65nm cmos," in *Solid-State Circuits Conference (A-SSCC), 2014 IEEE Asian*, Nov 2014, pp. 249–252.

[3] T. Y. Lo, C. C. Hung, and M. Ismail, "A wide tuning range gm-c filter for multi-mode cmos direct-conversion wireless receivers," *IEEE Journal of Solid-State Circuits*, vol. 44, no. 9, pp. 2515–2524, Sept 2009.

[4] S. Ghamari, G. Tasselli, C. Botteron, and P. A. Farine, "A wide tuning range 4 th-order gm-c elliptic filter for wideband multi-standards gnss receivers," in *European Solid-State Circuits Conference (ESSCIRC), ESSCIRC 2015 - 41st*, Sept 2015, pp. 40–43.

[5] D. Chamla, A. Kaiser, A. Cathelin, and D. Belot, "A gm-c low-pass filter for zero-if mobile applications with a very wide tuning range," *IEEE Journal of Solid-State Circuits*, vol. 40, no. 7, pp. 1443–1450, July 2005.

[6] J. R. Ville Saari and S. Lindfors, *Continuous-Time Low-Pass Filters for Integrated Wideband Radio Receivers*. London: Springer-Verlag New York, 2012.

[7] T. M. S. Hori, N. Matsuno and H. Hida, "Low-power widely tunable gm-c filter employing an adaptive dc-blocking, triode-biased mosfet transconductor," *IEEE Trans. Circuits Syst. I Regul. Pap.*, vol. 61, no. 1, pp. 37–47, Jan 2014.

[8] A. Atac, C. Harder, R. Wunderlich, and S. Heinen, "A low power variable gbw opamp from 60mhz to 2ghz for multi-standard receivers," in *Electronics, Circuits and Systems (ICECS), 2012 19th IEEE International Conference on*, Dec 2012, pp. 1–4.

[9] B. Razavi, *Design of Analog CMOS Integrated Circuits*, 1st ed. New York, NY, USA: McGraw-Hill, Inc., 2001.

[10] R. Dehghani, *Design of CMOS operational amplifiers*. London: Artech House, 2013.

[11] R. J. Baker, *CMOS: Circuit Design, Layout, and Simulation*, 3rd ed. New York, NY, USA: John Wiley and Sons, Inc., 2010.

[12] J. Ryynanen, M. Hotti, V. Saari, J. Jussila, A. Malinen, L. Sumanen, T. Tikka, and K. A. I. Halonen, "Wcdma multicarrier receiver for base-station applications," *IEEE Journal of Solid-State Circuits*, vol. 41, no. 7, pp. 1542–1550, July 2006.

[13] S. Hori, T. Maeda, hitoshi Yano, N. Matsuno, K. Numata, N. Yoshida, Y. Takahashi, T. Yamase, R. Walkington, and H. Hikaru, "A widely tunable cmos gm-c filter with a negative source degeneration resistor transconductor," in *Solid-State Circuits Conference, 2003. ESSCIRC '03. Proceedings of the 29th European*, Sept 2003, pp. 449–452.

[14] T.-Y. Lo and C.-C. Hung, "Low-voltage multi-mode gm-c channel selection filter for mobile applications," in *IEEE Custom Integrated Circuits Conf. (CICC)*, Sep 2007, pp. 635 – 638.

[15] M. K. M. S. Oskooei, N. Masoumi and H. Sjoland, "A 4.35-mw +22-dbm iip3 continuously tunable channel select filter for wlan/wimax receivers in 90-nm cmos," in *IEEE J. Solid-State Circuits, vol.46, no. 6*, Jun 2011, pp. 1382–1391.

[16] M. D. Matteis, S. D'Amico, and A. Baschirotto, "A 0.55 v 60 db-dr fourth-order analog baseband filter," *IEEE Journal of Solid-State Circuits*, vol. 44, no. 9, pp. 2525–2534, Sept 2009.

978-1-5090-2737-8/16 $31.00 © 2016 IEEE

A new two-step $\Sigma\Delta$ architecture column-parallel ADC for CMOS image sensor

Pierre Bisiaux, Caroline Lelandais-Perrault,
Anthony Kolar and Philippe Benabes

GeePs | Group of electrical engineering - Paris, UMR CNRS 8507,
CentraleSupelec, Univ Paris-Sud, Universite Paris-Saclay
Sorbonne Universites, UPMC Univ Paris 06
3 &11 rue Joliot Curie, Plateau de Moulon
F91192 Gif sur Yvette CEDEX France

Email: Pierre.Bisiaux, Caroline.Lelandais-Perrault,
Anthony.Kolar, Philippe.Benabes@centralesupelec.fr

Filipe Vinci Dos Santos

Advanced Analog Design Group
CentraleSupelec
3, rue Joliot CuriePlateau de Moulon
91192 Gif-Sur-Yvette Cedex

Email : Filipe.Vinci@centralesupelec.fr

Abstract—The demand for high resolution CMOS image sensors (CIS) is rising. Analog-to-digital converters (ADC) represent one of the major bottleneck of CIS. One of the candidates to overcome the existing limits is the column-parallel ADC. Column-parallel extended counting ADCs (EC-ADC) are able to reach high resolution thanks to their two-step conversion. However the EC-ADC area increases due to the two-step design. A solution is to use the same hardware twice to perform both steps. This paper proposes a 14-b, 100 kHz Nyquist frequency, two-step incremental $\Sigma\Delta$ (I$\Sigma\Delta$) analog-to-digital converter suitable for column-parallel CIS. Several architectures with different modulator order are compared to determine the most promising one. The proposed architecture, compared to a one-step second order modulator, reduces the total oversampling ratio (OSR) from 150 to 60 to reach a resolution of 14-b. The operational transconductance amplifiers (OTA) is the most critical part in our ADCs. Its required DC-gain is around 80 dB for a 120 MHz gain-bandwidth product (GBW). The ideal DNL and INL of our two-step I$\Sigma\Delta$ ADC are respectively +0.55/-0.6 LSB and +0.5/-0.5 LSB. This work achieves a SNDR of 89 dB when a full scale sinusoid of 100 kHz is applied.

Index Terms—ADC, incremental, sigma-delta ($\Sigma\Delta$), two-step, CMOS Image Sensor, column-parallel ADC, second-order $\Sigma\Delta$

I. INTRODUCTION

High quality CMOS image sensors (CIS) have become indispensable in the aerospace field. This evolution is possible thanks to the progress in the integration of sensor, readout electronics and A/D conversion in a single chip. The increasing resolution of the image sensor restricts the pitch of the pixel. This also increases the rate of pixel data conversion for a given frame rate. Moreover, greater pixel sensitivity and dynamic range demand 12-bits or higher resolution conversion. The column-parallel conversion is seen as the best answer to these increasingly stringent requirements.

Among column-parallel ADCs, single-slope (SS-ADC) and successive approximation register (SAR) ADCs are usually chosen for low power consumption application [1] [2] [3] [4]. However, the SS-ADC requires a large amount

of clock cycles for high resolution conversion. The SAR ADC needs a capacitor DAC consuming large area due to the weight of the capacitor and is very sensitive to the capacitor mismatch. Cyclic ADCs are also widely used in the CIS field [5] [6] [7]. This ADC only requires n clock cycles for a n-bit resolution. However, cyclic ADCs require high DC gain OTA to avoid non-linearity. Matching problems limit the resolution achievable by cyclic ADCs. Sigma-Delta ($\Sigma\Delta$) ADCs are well known for their ability to achieve high resolution. However they are slower than Nyquist converters due to oversampling. $\Sigma\Delta$ ADCs have recently appeared in CIS field thanks to the scaling down technology [8] [9]. $\Sigma\Delta$ ADCs require smaller DC gain and sampling capacitance than cyclic ADCs because of the oversampling operation.

In order to achieve a better tradeoff between speed and resolution, extended-counting ADCs (EC-ADC) have been reported [10] [11]. The EC-ADC is an hybrid architecture that performs the conversion in two steps. First a coarse conversion is done using an incremental $\Sigma\Delta$ (I$\Sigma\Delta$) ADC. The first step generates an analog residue. Then the fine conversion of the residue is realized with a Nyquist converter. Thus the overall conversion time is reduced compared to a single step I$\Sigma\Delta$. However EC-ADCs have quite complex circuits due to the use of two different conversion principles. The two-step conversion can be pushed further and both conversions can be realized by the same I$\Sigma\Delta$ ADC. Such architectures are known as two-step I$\Sigma\Delta$ ADCs [12] [13]. The two-step I$\Sigma\Delta$ ADCs advantage is to use twice the same hardware for both conversions, while simultaneously decreasing the overall OSR. In order to obtain a higher resolution (compared to a single loop I$\Sigma\Delta$) and limit the complexity of the circuit, a two-step I$\Sigma\Delta$ ADCs can be preferable over EC-ADCs. In this paper, a two-step second order I$\Sigma\Delta$ ADC is proposed. The remainder of this paper is organized as follows : section II describes the I$\Sigma\Delta$ modulator. In section III a two-step I$\Sigma\Delta$ architecture and its characteristics are described. In section IV the high level modulator design is discussed. Then the section V presents

978-1-5090-2737-8/16 $31.00 © 2016 IEEE

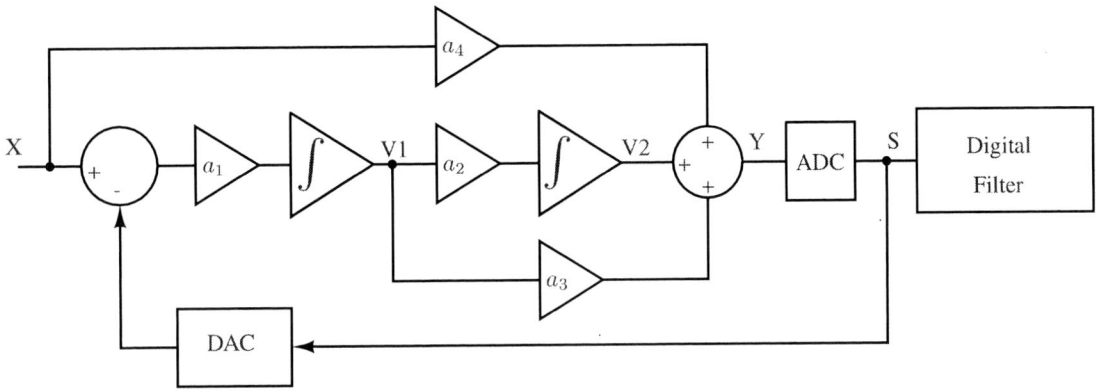

Figure 1. 2nd order modulator with feed-forward path

the simulated ADC characteristics.

II. INCREMENTAL ΣΔ ADC

The first IΣΔ ADC was published by Van De Plassche in 1978 [14]. It is composed of an integrator, an ADC (a comparator with 2 levels in most cases) and a digital-to-analog converter (DAC) in the feedback path. This architecture has the drawback to be very slow. Indeed it requires 2^n clock cycles to reach n-bits. To improve the conversion speed, one can increase the modulator order or the number of cascaded modulators. However, higher order modulators can lead to stability problems which are hard to solve. An example of a IΣΔ with a second order modulator is shown in fig.1. Here, the IΣΔ is studied in the time domain. In this way the residue value can be written depending on ΣΔ modulator parameters and the result is reused later for the two-step architecture. In this architecture the output of the integrators are given by

$$V_1[M] = a_1 \sum_{i=1}^{M-1} (X[i] - S[i]) \qquad (1)$$

$$V_2[M] = a_1 a_2 \sum_{K=1}^{M-1} \sum_{i=1}^{K-1} (X[i] - S[i]) \qquad (2)$$

where M is the OSR. After M clock cycles the error of conversion is given by

$$E = X - \hat{X} = \frac{2}{a_1 a_2 M(M-1)} . V_2[M] \qquad (3)$$

where $\hat{X}$ is the estimated reconstructed signal from the bitstream, defined as,

$$\hat{X} = \frac{2}{M(M-1)} \sum_{K=1}^{M-1} \sum_{i=1}^{K-1} S[i] \qquad (4)$$

One can observe that the residue value $V_2(M)$ is the image of the conversion error. With this modulator, 150 cycles are needed to reach 14 bits resolution [15].

III. TWO-STEPS IΣΔ ADC

A. Two-steps architecture

For several years, architectures with reused hardware have been investigated to decrease the size of the ADC [16] [17] . In a two-step IΣΔ ADC [12] [18], the conversion is divided in 2 phases and the hardware is used twice to perform the full conversion. The chosen architecture and its diagram timing are shown in fig.2.

The architecture is composed of a sample-and-hold (S/H), an IΣΔ modulator and a digital filter. At the beginning of the conversion, a reset is done on the integrators and the capacitors. During the first step ($\Phi1=1$) the pixel value is fed to the modulator. At the end of the coarse conversion ($\Phi_{EOC} = 1$ and $\Phi1=0$), the residue V_R (or $V_2[M]$) is loaded into the S/H. Then another reset is done and the fine conversion is performed. The timing diagram for a conversion is shown in fig. 2(b). To reach 100 frames/s (FPS) with a 1920x1080 array, the available row time is 9.2 μs. Two conversions are realized within the row time to do a correlated double sampling in digital domain. Since a few extra clock cycle are necessary to reorganize the switch and load the capacitance, a single step can last up to 2.1 μs.

B. Modulator scaling

As already mentioned the residue of the coarse conversion is used for the fine conversion. The goal is to fit the output of the first step to the input of the second step. Moreover the input range of a 2nd order modulator must be limited to avoid integrator overflow. So it is very important to limit the residue value. The chosen modulator is shown in fig.1. As one can see in eq.(3), a solution to modify V_R is to modify the parameter $a_{1,2}$ of the modulator. In fig.1, parameters a_1 and a_2 are arbitrary chosen at 0.5 and the maximum input at 75% of the reference. An optimization is performed by sweeping parameters a_3 and a_4 to minimize the swing of the signals into the integrators giving $a_3 = 2$ and $a_4 = 1$. Then a global scaling coefficient is applied to fit the output swing

978-1-5090-2737-8/16 $31.00 © 2016 IEEE   165

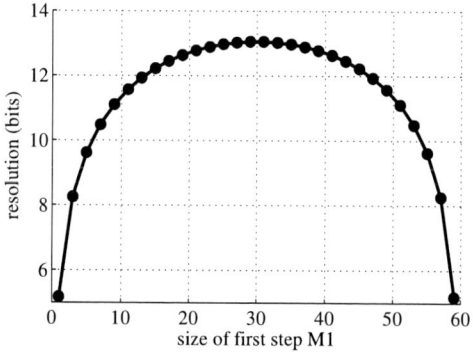

Figure 2. High-level view of (a) two-steps architecture with 2nd order modulator and (b) the timing diagram of the conversion

of the integrator with the input range of the modulator. The optimum coefficients are $a_1 = 0.75$, $a_2 = 0.5$, $a_4 = 1.5$ and $a_3 = 2$. However these optimal values are not very convenient to implement in switched capacitor circuit. Moreover a margin has to be taken from optimal coefficients to be insensitive to analog mismatches. At the end, the final values are $a_1 = a_2 = 0.5$, $b_1 = 1$ and $a_3 = 2$.

From eq.(3) a total of 27 cycles per step is needed to reach 7-b of resolution. In order to get some margin to deal with analog mismatches, the OSR is increased to 30 leading to a total OSR of 60 for a conversion giving 70 ns per cycle. Once again, a small margin is taken giving a clock frequency of 20 MHz.

C. OSR split of two-steps conversion

We previously supposed that the length of both steps were equal. This supposition is verified in the following part as the optimal choice. The parameters $k1 = a_1.a_2.M_1(M_1 - 1)/2$ and $k2 = a_1.a_2.M_2(M_2 - 1)/2$ are introduced with M_1 and M_2 respectively the OSR of the first and the second step. From eq.(3), one can derive the error of the conversion,

$$E_{TOT} = \frac{V_{FS}}{k1.k2} = \frac{4}{a_1^2.a_2^2.M_1(M_1 - 1).M_2(M_2 - 1)}.V_{FS}$$

$$E_{TOT} \approx \frac{4}{a_1^2.a_2^2.M_1^2.M_2^2}.V_{FS} \qquad (5)$$

Setting $M = M_1 + M_2$ and $\alpha = M_1/M_2$, eq.(5) can be written,

$$E \approx \frac{4.V_{FS}}{\frac{a_1^2 a_2^2 \alpha^2 . M^4}{(\alpha+1)^4}} \qquad (6)$$

From eq.(6), one can plot the curve shown in fig. 3. As one can see the resolution is maximum when the OSR is equal in both steps.

The choice of the modulator order for the two steps is now discussed. In recent two-step $I\Sigma\Delta$ ADC work, the modulator order can be different for the two-step [18]. For second and higher order modulator, the last integrator can be switched to a sample-and-hold for the second step. The modulator order of the second step is then decreased of one order. The analysis of the modulator order within the conversion is now studied. Four different architectures are compared : a third order modulator single loop ($I\Sigma\Delta3$), a

Figure 3. resolution of the conversion for different size of the first conversion

two-step with a second and first order modulator ($I\Sigma\Delta2$-1), a two-step with two second order modulator ($I\Sigma\Delta2$-2) and a two-step with a third order and second order modulator ($I\Sigma\Delta3$-2). The result of our simulations is shown in fig.4. The resolution of the different ADC is calculated from the conversion error, computed using eq.(5). In each case, integrators modulators coefficients and α in eq.(6) are optimized to minimize the conversion error, thus maximize the resolution. In the different architectures, input range is reduced to avoid integrator overflow. Considering the scaling of the different architectures and the aimed resolution, the proposed $I\Sigma\Delta2$-2 architecture requires the smallest OSR.

IV. MODULATOR SPECIFICATIONS

In this section, the switched capacitor circuit and the modulator characteristics are presented. The effect of non-idealities in the OTA (finite DC gain, gain bandwidth-product (GBW) and slew rate (SR)) are introduced and their effects simulated. In this part and for the rest of the paper, the resolution is calculated from the SNDR.

A. Switched-capacitor circuit

The switched-capacitor circuit implementation is shown in fig. 5. A single-ended version of the architecture is represented to simplify the circuit. The sampling capacitance

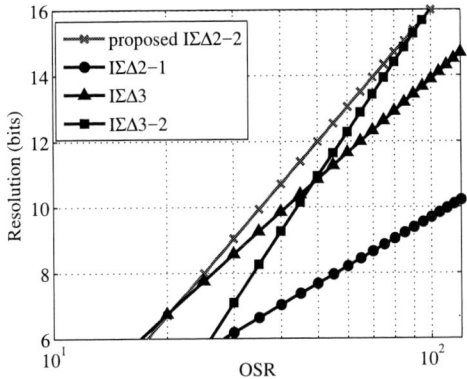

Figure 4. comparison of different architectures and modulator order

C_s is chosen to be 100 fF. The integrators realized with the OTA are the most critical part in a modulator. To minimize injection charge effect, two non overlapped clock p1 and p2, and their delayed signal p1d and p2d, are used

The characteristics of the OTA are now discussed. The DC gain OTA, the GBW and the slew-rate (SR) are important parameters in a switched capacitor circuit. These parameters are degrading the charge transfer and introduced errors at both input and output of the integrator.

B. Finite OTA DC gain

The finite DC gain OTA introduces an offset at the input of the integrators. The transfer function of the integrator with an finite OTA DC gain is expressed as follow

$$Vo(n) = \frac{(1+\mu)Vo(n-1) + \frac{C_s}{C_i}Vin(n-1)}{1+\mu(1+\frac{C_s}{C_i})} \quad (7)$$

with Vo the output of the integrator, Vin the input, $\mu = 1/Ao$, with Ao the finite DC gain, C_s and C_i respectively the sampling and the integrator capacitance. The simulation is shown in fig.6. One can see that the resolution drops from a DC gain of 5000, or 74 dB. Thus a minimum DC gain of 80 dB is chosen.

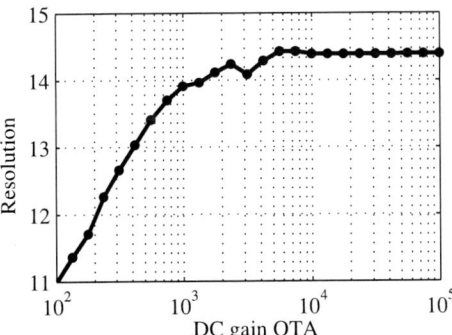

Figure 6. DC gain OTA versus ENOB

C. Finite GBW and SR

The charge transferring phase of a switched-capacitor circuit can be cut into a slewing and a settling part. Finite SR and GBW respectively influence the slewing and the settling time. The influence of the GBW on the system is first analysed. The transfer function of an integrator with a finite GBW is

$$Vo(n) = Vo(n-1) + \frac{C_s}{C_i}\left(1 - \exp^{-k_1}\left(\frac{C_s}{C_s + C_i}\right)\right) \quad (8)$$

with

$$k1 = \left(\frac{GBW.C_i}{C_i + C_s}\right)\left(\frac{T}{2}\right) \quad (9)$$

with Vo and the Vin respectively the output and the input of the integrator and T the available settling time. The result of the simulation is shown in fig 7. One can see that the

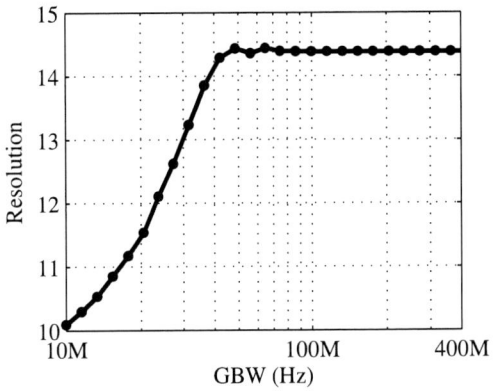

Figure 7. DC gain OTA versus ENOB

resolution quickly dropped for a GBW below 50 MHz. However this value is quite small considering the settling time. Thus a GBW of 120 MHz is chosen to be at least five times higher than the clock frequency. The slewing time is then determined to calculate the SR. A good trade-off is to set the slewing time at 25% of the available time. The remaining time is used for the settling time. The SR is expressed by

$$SR = \frac{\Delta V}{T_{slewing}} \quad (10)$$

with ΔV the maximum output swing of the integrator and $T_{slewing}$ the chosen slewing time. The minimum SR calculated is 220 V/μs. The values found previously are put together in table I.

V. RESULTS

An ideal model of the proposed ADC is used to compute the DNL and the INL of the ADC. The ideal DNL and the INL of the I$\Sigma\Delta$ are shown in fig.8 and fig.9. The DNL and INL of the ADC are respectively +0.55/-0.6 LSB and +0.5/-0.5 LSB. The ADC output spectrum is shown in fig.10. The SNDR is calculated with a full scale input signal of 100 kHz

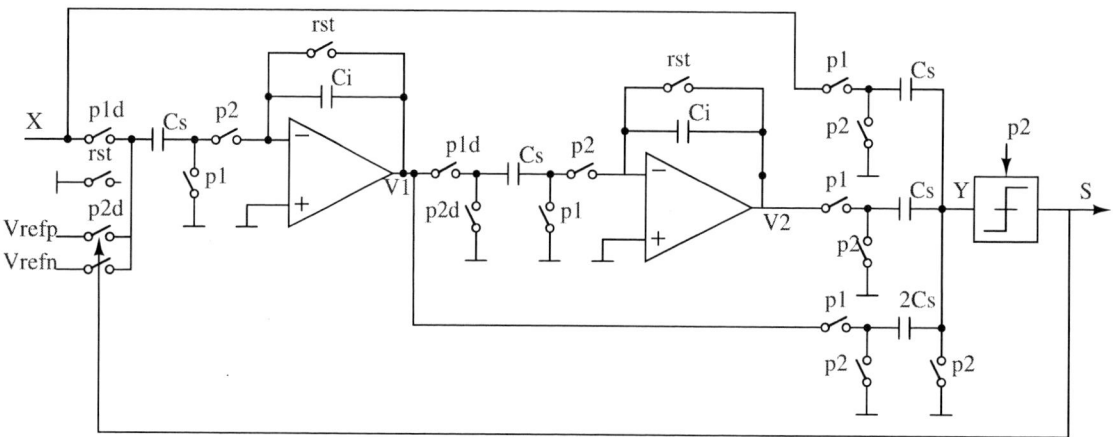

Figure 5. 2nd order modulator with feed-forward path

Table I
OTA CHARACTERISTICS

Slew rate	220 V/μs
GBW	120 MHz
Load capacitance	300 fF
DC gain	80 dB

and a sampling frequency of 20 MHz. This ADC achieves a SNDR of 89 dB. The power consumption is estimated from schematic simulation and is 450μW. The performance comparison with others EC-ADC is shown in table II.

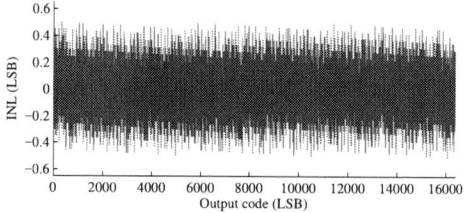

Figure 9. INL of proposed ADC

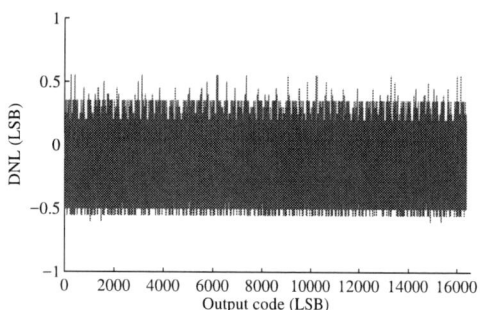

Figure 8. DNL of proposed ADC

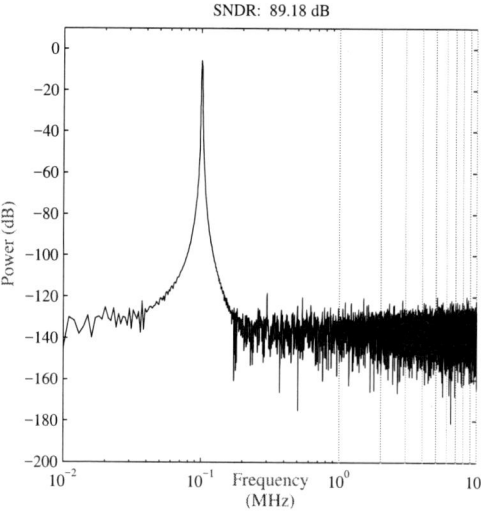

Figure 10. ADC output spectrum

VI. CONCLUSION

This paper presents a two-step I$\Sigma\Delta$ for column-parallel conversion of pixel arrays, allowing a monolithic integration of the whole CIS. First, several architectures are compared to analyze the one reaching the highest resolution with the lowest OSR. Compared to a single loop second order modulator, the proposed two-step ADC with second order modulator (I$\Sigma\Delta$2-2) can reduce the OSR from 150 to 60, to reach 14-b of resolution. Simulations are then realized to get OTA specifications that avoid lowering the resolution of the ADC. To finish simulations on ideal (I$\Sigma\Delta$2-2) ADC

Table II
PERFORMANCE COMPARISION

	[11]	[19]	[10]	[12]	This work
Technology (μm)	0.18	0.35	0.18	0.15	0.18
ADC architecture	$\Sigma\Delta$+Cyclic	$\Sigma\Delta$+SAR	$\Sigma\Delta$+Cyclic	2step $\Sigma\Delta$	2step $\Sigma\Delta$
Sampling rate (kS/s)	30	150	50	-	100
Resolution	17	14.3	10.2	12	14
Power (μW)	345	300	13	363	410
DNL (LSB)	-0.88/1.38	-0.79/0.97	-	-0.7/1.8	-0.55/0.6
INL (LSB)	-26.3/35	-1.7/2.79	-	-22/20	-0.5/0.5
SNDR (dB)	85	-	63	-	89

are performed giving DNL/INL characteristics and the ADC output spectrum.

REFERENCES

[1] MF Snoeij, AJP Theuwissen, and JH Huijsing. A low-power column-parallel 12-bit adc for cmos imagers. In *Proc. IEEE Workshop Charge Coupled Devices (CCDs) and Advanced Image Sensors (AIS)*, pages 169–172, 2005.

[2] Min-Seok Shin, Jong-Boo Kim, Min-Kyu Kim, Yun-Rae Jo, and Oh-Kyong Kwon. A 1.92-megapixel cmos image sensor with column-parallel low-power and area-efficient sa-adcs. *Transactions on Electron Devices*, 59(6):1693–1700, June 2012.

[3] Li Quanliang. A 12-bit compact column-parallel sar adc with dynamic power control technique for high-speed cmos image sensors. *Journal of Semiconductors*, 35(10):1–8, october 2014.

[4] D.G. Chen, Fang Tang, Man-Kay Law, Xiaopeng Zhong, and A. Bermak. A 64 fj/step 9-bit sar adc array with forward error correction and mixed-signal cds for cmos image sensors. *Circuits and Systems I: Regular Papers, IEEE Transactions on*, 61(11):3085–3093, Nov 2014.

[5] Mitsuhito Mase. A wide dynamic range cmos image sensor with multiple exposure-time signal outputs and 12-bit column-parallel cyclic a/d converters. *JSSC*, 40(12):2787–2795, December 2005.

[6] Jong-Ho Park, S. Aoyama, T. Watanabe, K. Isobe, and S. Kawahito. A high-speed low-noise cmos image sensor with 13-b column-parallel single-ended cyclic adcs. *Electron Devices, IEEE Transactions on*, 56(11):2414–2422, Nov 2009.

[7] Seunghyun Lim, Jimin Cheon, Youngcheol Chae, Wunki Jung, Dong-Hun Lee, Minho Kwon, Kwisung Yoo, Seogheon Ham, and Gunhee Han. A 240-frames/s 2.1-mpixel cmos image sensor with column-shared cyclic adcs. *Solid-State Circuits, IEEE Journal of*, 46(9):2073–2083, Sept 2011.

[8] A. Mahmoodi and D. Joseph. Optimization of delta-sigma adc for column-level data conversion in cmos image sensors. In *Instrumentation and Measurement Technology Conference Proceedings, 2007. IMTC 2007. IEEE*, pages 1–6, May 2007.

[9] Youngcheol Chae, Jimin Cheon, Seunghyun Lim, Minho Kwon, Kwisung Yoo, Wunki Jung, Dong-Hun Lee, Seogheon Ham, and Gunhee Han. A 2.1 m pixels, 120 frame/s cmos image sensor with column-parallel $\delta\sigma$ adc architecture. *Solid-State Circuits, IEEE Journal of*, 46(1):236–247, Jan 2011.

[10] Cencen Gao, Dong Wu, Hui Liu, Nan Xie, and Liyang Pan. An ultra-low-power extended counting adc for large scale sensor arrays. In *Circuits and Systems (ISCAS), 2014 IEEE International Symposium on*, pages 81–84, June 2014.

[11] Min-Woong Seo. A low-noise high-dynamic-range 17b 1.3mega-pixel 30-fps cmos image sensor with column-parallel two-stage folding-integration/cyclic adc. *Transactions on Electron Devices*, 59(12):3396–3400, november 2012.

[12] Y. Oike and A. El Gamal. Cmos image sensor with per-column $\sigma\delta$ adc and programmable compressed. *Solid-State Circuits, IEEE Journal of*, 48(1):318–328, Jan 2013.

[13] Mengyun Yue, Dong Wu, and Zheyao Wang. A 15-bit two-step sigma-delta adc with embedded compression for image sensor array. In *Circuits and Systems (ISCAS), 2013 IEEE International Symposium on*, pages 2038–2041, May 2013.

[14] R.J. van de Plassche. A sigma-delta modulator as an a/d converter. *Circuits and Systems, IEEE Transactions on*, 25(7):510–514, Jul 1978.

[15] Biao Wang. A 1.8-v 14-bit inverter-based incremental sd adc for cmos image sensor. In ASICON, editor, *ASICON*. ASICON, IEEE, 2013.

[16] Min-Woong Seo, Sung-Ho Suh, T. Iida, T. Takasawa, K. Isobe, T. Watanabe, S. Itoh, K. Yasutomi, and Shoji Kawahito. A low-noise high intrascene dynamic range cmos image sensor with a 13 to 19b variable-resolution column-parallel folding-integration/cyclic adc. *Solid-State Circuits, IEEE Journal of*, 47(1):272–283, Jan 2012.

[17] Jae hong Kim, Wun ki Jung, Seung hyun Lim, Yu jin Park, Won ho Choi, Yun jung Kim, Chang eun Kang, Ji hun Shin, Kyo jin Choo, Won baek Lee, Jin kyeong Heo, Byung jo Kim, Se jun Kim, Min ho Kwon, Kwi sung Yoo, Jin ho Seo, Seog heon Ham, Chi young Choi, and Gab soo Han. A 14b extended counting adc implemented in a 24mpixel aps-c cmos image sensor. In *Solid-State Circuits Conference Digest of Technical Papers (ISSCC), 2012 IEEE International*, pages 390–392, Feb 2012.

[18] C.-H. Chen, Y. Zhang, T. He, P.Y. Chiang, and G.C. Temes. A micro-power two-step incremental analog-to-digital converter. *Solid-State Circuits, IEEE Journal of*, PP(99):1–13, 2015.

[19] M.-S. Shin, J.-B. Kim, and O.-K. Kwon. 14.3-bit extended counting adc with built-in binning function for medical x-ray cmos imagers. *Electronics Letters*, 48(7):361–363, March 2012.

New Asynchronous Protocols for Enhancing Area and Throughput in Bundled-Data Pipelines

Jean Simatic, Abdelkarim Cherkaoui, Rodrigo Possamai Bastos, and Laurent Fesquet

Univ. Grenoble Alpes, TIMA Laboratory, F-38031 Grenoble, France

CNRS, TIMA Laboratory, F-38031 Grenoble, France

Email: {*First name*}.{*Last name*}@imag.fr

Abstract—This paper presents two new area-reduced controllers for bundled-data asynchronous pipelines in which the stages have long critical paths. The proposed protocols allow to reduce the number of required delay elements by using the falling edge of the asynchronous request to indicate data validity. For critical path lengths of 25 gates, the first presented scheme decreases the controller area by 48% and slightly increases the maximum throughput (2%) in comparison to a standard micropipeline implementation. The other more-concurrent scheme proposition leads to a 25% area reduction and a 40% improvement of the maximum pipeline throughput.

I. INTRODUCTION

Increasing the clock frequency to enhance the performance of digital circuits reached physical limits in the 20th century because of the high heat dissipation rates. Two complementary techniques have been exploited to continue the performance evolution of integrated systems: parallelization, which spatially separates a task and concurrently executes sub-tasks on different hardware; and pipelining, which decomposes a task into a succession of sub-tasks performed in sequential hardware stages.

Synchronous pipelines (Figure 1a) use a clock to notify all stages that it is time to forward their respective results. All stages must have the time to finish their computation within one clock cycle, i.e. the clock period has to be larger than the longest critical path of all stages. This timing constraint is global and determines the throughput of the pipeline.

Quasi-delay insensitive (QDI) pipelines (Figure 1b) require fewer timing assumptions [1]. QDI circuits use *1-of-N* encoding to indicate data validity: If all N wires are low, there is no data; if one of the wires is high, there is data. The value depends on which wire is high. This gives to QDI circuits a strong robustness against delay variations but requires more wires and more complex gates. The memory blocks use the encoding to detect the arrival of valid data. The acknowledgment indicates to the previous stage that data have been captured and that new data can be sent.

Bundled-data (BD) pipelines (Figure 1c), also known as micropipelines [2], are similar to synchronous pipelines but replace the global clock by local handshake signals. Each register (group of flip-flops) is controlled by an asynchronous controller. The controller handles the handshake signals: requests, which notify valid data; and acknowledgments, which indicate that new data can be captured. The notification of the data validity must occurs after the end of the data propagation

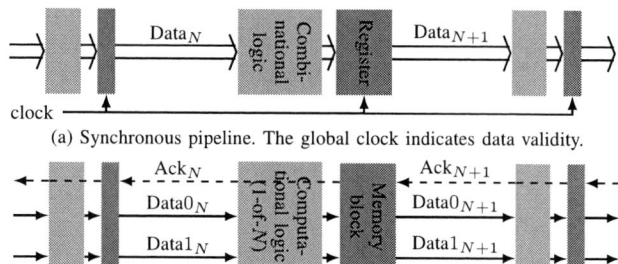

(a) Synchronous pipeline. The global clock indicates data validity.

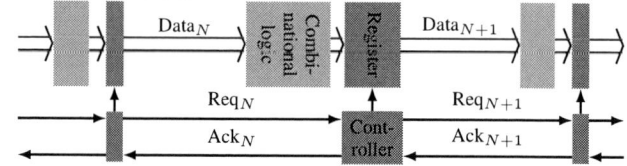

(b) Asynchronous QDI pipeline. The data encoding (dual-rail here) allows indicating data validity.

(c) Asynchronous Bundled-data pipeline. A local request signal indicates data validity. The notification must not arrive before the actual arrival of the data.

Figure 1. Synchronous and asynchronous pipelines. Three stages are represented. Stage N is in the center.

in the preceding combinational block. This timing constraint is the *bundled-data convention*.

In this paper, two new asynchronous pipeline controllers are proposed: the late capture and the maximus[1] controllers. Those controllers have been designed to reduce the cycle time and the circuit area of pipeline stages with long critical paths. Section II presents related works on protocols, especially the burst-mode controller [3] that introduces a strategy to halve the number of delay elements to be inserted in order to meet the bundled-data constraint. Presented in Section III, the late capture controller follows the same strategy but is slower, and thus, allows to gain area while meeting the same constraint. The maximus controller adds decoupling in the protocol to increase the pipeline throughput at the cost of additional logic. Section IV and V compare the proposed protocols and present a performance metric to identify the best operating conditions for the controllers.

---

[1]Maximus means "the largest" in latin. This attribute refers to the very large amount of decoupling in the protocol

978-1-5090-2737-8/16 $31.00 © 2016 IEEE

II. RELATED WORKS

All bundled-data controllers share a similar structure shown in Figure 2a. The controller is connected to two channels: the input channel L from which the controller receives data and the output channel R to which the controller sends data. A bundled-data channel is made of a data-path and two handshake wires: a request and an acknowledgment. The requests (L_i and R_i) indicate that new data arrive from the previous stage. On an abstract level, the requests are said to propagate *tokens* and to *fill* the channels. Conversely, the acknowledgments (L_i^a and R_i^a) indicate that the next stage is ready to receive new data. The acknowledgments are said to propagate *bubbles* and to *empty* the channels. When a controller has a token on its input channel and a bubble on its output channel, the controller must propagate forward the data and token from the input channel to the output one and propagate backward the bubble.

There are several possible ways for sequencing the request and acknowledgment events, and also for giving meaning to those events in term of tokens and bubbles. Several works in the literature specify constraints to ensure functional protocols and propose classifications of protocols. Furber and Day [4] impose a set of signal sequences and propose several protocols. Among them, the semi-decoupled protocol offers a good trade-off between decoupling and complexity. Lines [5] uses CHP reshuffling to list all 9 protocols complying with a set of rules. Focusing on QDI protocols, his work ignores data events and adds a constraint to simplify implementation. McGee and Nowick [6] added data events to also model dynamic protocols. They classify the protocols by their concurrency in a partial-order based framework. Birtwistle and Stevens [7] also proposed a framework based on a protocol offering a maximal concurrency. All protocols are then expressed as a subset of the possible states of the maximal protocol.

A. Micropipeline (WCHB)

In the 2-phase micropipeline protocol [2], each edge on the request indicates the availability of a new data and each edge on the acknowledgment indicates the possibility of moving a data forward. Thus, a channel is full if the values of the handshake signals are different and empty if they are the same. There is only one possible protocol. The controller is simple to implement but the data-path registers require flip-flops capturing new data for every edge (denoted + and −).

In order to use latches or D-flip-flops in the pipeline, the protocols must be level sensitive: only one every other edge indicates the arrival of a token or a bubble. Consequently, there are several possible protocols. They are called 4-phase protocols because for a channel X, four events are repeated in a loop: $X+$, X^a+, $X−$, and $X^a−$.

The weak condition half buffer (WCHB) protocol uses the same controller as Sutherland's micropipeline [2]. The protocol is also identical except that only rising edges are meaningful: the rise of the request indicates a new token and the rise of the acknowledgment indicates a new bubble. The falling edges are the Return-to-Zero (RZ) phase. Figure 2b

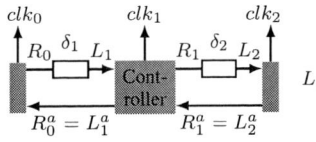
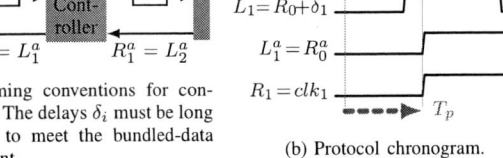

(a) Naming conventions for controllers. The delays δ_i must be long enough to meet the bundled-data constraint

(b) Protocol chronogram.

Figure 2. WCHB controller interface and protocol specification.

illustrates the WCHB chronogram. There are three important durations:

T_{BD} The *critical path delay* is length of the critical path in the combinational logic associated with the stage. As in synchronous circuits, the estimation of T_{BD} should take into account proccess varations.

T_p The *minimum propagation time* is the minimum time between the capture of a new data item by the previous stage (clk_0+) and the capture of this data item by the current stage (clk_1+). The bundled-data constraint imposes $T_p \geq T_{\mathrm{BD}}$.

T_c The *minimum cycle time* is the minimum time between two successive captures of new data items by a stage. The cycle time determines the throughput of the stage.

T_p and T_c are functions of the stage delay δ. Those function depend on the protocol. Chronograms alone do not allow a reliable analysis of the protocols because they do not represent the different ordering of the events in the controller. For example, Figure 2b shows simultaneous transitions on L_1^a and R_1. The chronogram does not specify what happens if they are not simultaneous.

Signal Transition Graphs (STGs) were especially introduced to model asynchronous circuits [8]. STGs are a subset of Petri-Nets (PNs) in which the transition marked $X+$ (resp. $X−$) represents a rising (resp. falling) edge on the signal X. To ensure determinism, the places have only one source transition and one destination transition. Consequently, the places are not represented between the transitions and only an arrow links two transitions. A dot is added on the arrow if the place is full. A signal transition only happens if all the source places (arrows pointed toward the transition) are full and all the destination places are empty. After a transition fired, the source places are emptied and the destination places are filled. Thus, arrows represent a relative ordering condition on the transitions.

STGs allow a better decomposition of the minimum propagation and cycle times. In Figure 3, the dashed arrow shows the shortest path from R_0+ to R_1+. Thus, T_p is the sum of δ_0 and the propagation time of the rising edge of L_1 to R_1. This time depends on the controller implementation. Similarly, the shortest cycle gives that $T_c = 2\delta_0 + T_{\mathrm{int}}$. Where the delays on the acknowledgment wires were neglected and T_{int} is the sum of 6 internal transitions inside the controller.

978-1-5090-2737-8/16 $31.00 © 2016 IEEE

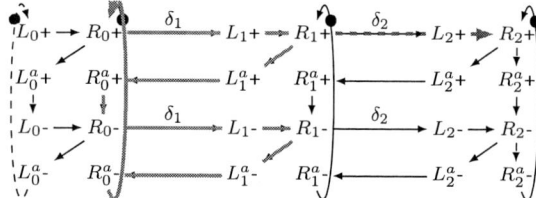

Figure 3. STG of 3 stages of a pipeline with WCHB controller. The critical cycle is highlighted and the bold dashed arrow indicates the propagation path.

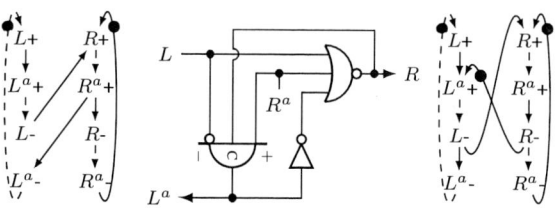

(a) Burst-mode STG [3] (b) Burst-mode controller. See Figure 6 for gate details. (c) Early ack. STG [11]

Figure 4. Existing Phase-3 protocols

B. Increasing the decoupling: other Phase-1 protocols

In WCHB, the first phase in the handshake ($L+$) indicates the arrival of a token. Hence, we call WCHB a Phase-1 protocol. After the data has been transmitted to the next channel ($R+$), the only thing to do for the channels is to return to zero. This RZ phase can be carried out independently by each channel. For example, $R-$ does not need $L-$ to happen before taking place itself. Thus, WCHB uses unnecessary synchronization between the channel. Decoupled protocols introduce more concurrency between the channels but require more complex controllers. Lines [5] and Yahya [9] proposed and compared Phase-1 protocols for QDI pipelines, while Furber and Day [4] did it for bundled-data systems.

In all Phase-1 protocols, as the forwarding of the token happens just after the first phase, $T_p \approx \delta$. Assuming that the same delay applies to the falling edge of the request, it comes $T_c \geq 2\delta$. Hence, if the critical path of the stage is long, the cycle time will be twice as long:

$$T_c \approx 2T_{\text{BD}} \qquad (1)$$

C. Putting the Return-to-Zero to work: Phase-3 protocols

QDI circuits must use Phase-1 protocols because the rise of the request is actually the dual-rail encoded data arriving to the memory block. Then, the RZ phase is the reset of the multi-rail logic. To reduce the cycle time, one possibility is to use a 2-phase protocol with encoding such as LEDR [10].

In bundled-data circuits, the event $L+$ is not necessarily bound to the arrival of a new data item. Phase-3 protocols, as we call them, use the falling edge of the input request ($L-$) to indicate a new token *instead* of $L+$ (Phase-1). Phase-3 protocols are 4-phase protocols because one data item is transmitted every four transitions. This Phase-3 sequencing allows to go through the request delay twice before indicating new valid data: once during the rising transition and once during the falling transition ($T_p \approx 2\delta$). Thus, given a critical path of the combinational in the datapath T_{BD}, the size of the delay line in the controller is halved ($\delta \approx T_{\text{BD}}/2$). This allows to reduce the area of the control part. Phase-3 protocols also allow a better usage of the cycle time, as shown in Section III.

To the best of our knowledge, two Phase-3 protocols have been proposed: the burst-mode protocol by Yun *et al.* [3], and the early acknowledgment protocol by Mannakkara and Yoneda [11]. Figures 4a and 4c present their respective STG. The two protocols apply the same concurrency reduction on the input channel L.

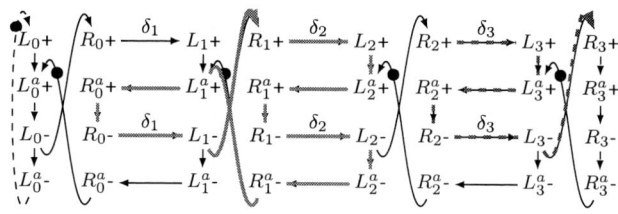

(a) STG of 4 stages of a pipeline of late capture controllers

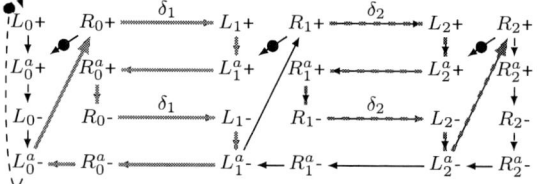

(b) STG of 3 stages of a pipeline of maximus controllers

Figure 5. STGs of pipelines of proposed controllers. The cycle path is highlighted and the dashed arrow shows the propagation path.

III. PROPOSED PROTOCOLS AND CONTROLLERS

The burst-mode and early acknowledgment differ in the moment when the input channel waits for the output channels. The synchronization of the channels in the early acknowledgment protocol is more intuitive: the controller waits until the output channel returns to zero before allowing a new cycle ($R- \rightarrow L^a+$). However, this sequencing requires a fast propagation of the falling edge of the request. Otherwise new data could be accepted before the next stage read the current one. The use of asymmetric delays forbids benefiting from the area reduction promised by Phase-3 protocols.

A. Late capture controller

The late capture protocol is a Phase-3 protocol that follows the same synchronization principle as the early acknowledgment protocol, but waits longer in order to allow the use of symmetrical delays: the input channel waits for R^a- instead of $R-$. According to Birtwistle's framework [7] the late capture protocol and the burst mode protocol are delay insensitive.

The shortest propagation paths for the two protocols contain two transitions on a request wire: for example $R_1+ \rightarrow L_2+$ and $R_1- \rightarrow L_2-$. Hence, the minimum propagation time is twice the length of the delay: $T_p \approx 2\delta_1$.

The shortest cycles of the protocols contain the propagation path plus one transition on a request wire: $R_2+ \rightarrow L_3+$ for

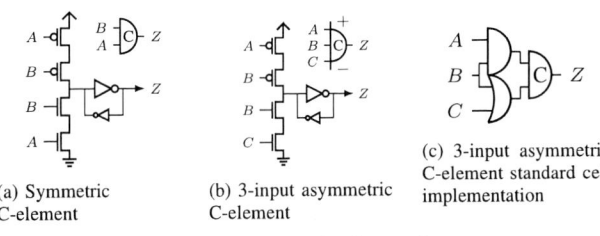

(a) Symmetric C-element

(b) 3-input asymmetric C-element

(c) 3-input asymmetric C-element standard cell implementation

Figure 6. C-element implementations.

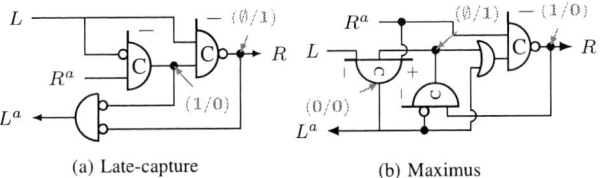

(a) Late-capture

(b) Maximus

Figure 7. Controller implementations. The bold net labels indicate the reset value of the reset values of the C-elements in the empty/full cases. $\emptyset$ means that the inputs are sufficient to reset the gate.

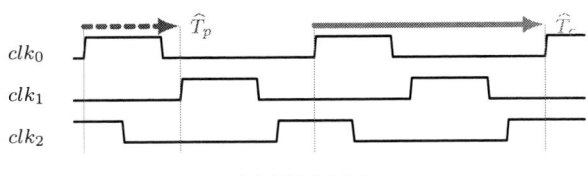

Figure 8. Simulation chronogram and measurements of the observed propagation time ($\widehat{T}_p$) and observed cycle time ($\widehat{T}_c$). clk_i is generated by the controller of the registers of stage i.

the burst mode, and $R_0- \to L_1-$ for the late capture. Thus, assuming that all stages have the same critical path length T_{BD} and that consequently the stage delays are minimized ($\delta_i = T_{\mathrm{BD}}/2$ so that $T_p = T_{\mathrm{BD}}$) then minimum cycle time is:

$$T_c \approx \frac{3}{2} T_{\mathrm{BD}} \qquad (2)$$

B. Maximus protocol

The maximus protocol was designed to maximize the pipeline throughput by having the ratio T_c/T_{BD} close to 1, compared to 2 for Phase-1 protocols (1) and $3/2$ for Phase-3 protocols (2). Figure 5b shows the STG of the protocol. The input and output channels only synchronize at the end of their respective handshake.

The additional $R+ \to L^a+$ synchronization formally reduces the protocol concurrency. However, in practice, the path $L^a- \to R+ \to L^a+$ is only made of internal transitions, and thus, takes less time than the path $L^a- \to L+ \to L^a+$, which goes through a request delay. Therefore, the additional synchronization does not reduce performances. Thanks to the concurrency reduction, only one internal state variable is needed to complete the state encoding.

This protocol is called maximus because it has states that do not exist in the "protocol of maximal concurrency" defined by Birtwistle and Stevens, LC_{max} [7]. Even more concurrent states can be found but they will require two additional state variable, and thus, will probably have a much higher hardware cost.

C. Implementation of the controllers

The implementation of the controllers uses C-elements, also known as Muller gates or rendezvous[2] gates. Figure 6a shows the gate symbol and one of the possible transistor-level implementations of the symmetric C-element. The controllers also use asymmetric C-elements. Figure 6b shows the gate symbol and one of the possible transistor-level implementations of the 3-input asymmetric C-element. 2-input asymmetric gates are similar except that a wire replaces the top transistor associated to the input A. Since our target technology library had only symmetric C-elements, we used a symmetric gate and one (resp. two) standard gate(s) to implement the 2-input (resp. 3-input) asymmetric C-element (see Figure 6c).

The speed-independent controllers were synthesized using the tool Petrify [12], and a mapping of the resulting equations into gates. Figure 7 shows the controller schematics.

[2]"a meeting at an appointed place and time" (Meriam-Webster)

IV. BENCHMARK

We compare the two proposed protocols and controllers to the burst-mode and WCHB protocols and controllers. The first goal of the simulations is to verify the previously described effects of the bundled-data constraint (T_{BD}) on the pipeline throughtput (at most one data item per T_c) and area.

For each controller, we simulate a ring of 20 identical stages. The circular configuration allows to keep a constant number of tokens in the pipeline. The *stage occupancy* of the pipeline is the ratio of the number of token over the number of stages. For the WCHB, each stage of the pipeline contains two controllers so that the token capacity of the pipeline is equal to the number of stages minus one as for the other protocols.

We simulate the pipelines with different values of request delay δ, which represent the delay of the combinational logic in the data-paths associated with the stages. All stages have the same delay length. This correspond to worse case for asynchronous pipelines [13].

The benchmark uses inverter-based delays: chaisn of minimal-size inverters. Indeed, using the same technology in the combinational logic and the associated delay correlates the delay under PVT variation and allows reducing margins when computing the required delay [14]. This delay is also symmetric: the propagation of a rising edge or a falling edge are similar. Only this kind of delay is in the scope of the present comparison.

The simulation is done at gate-level with retro-annotated delays. As shown in Figure 8, in each case, the testbench measures the minimum observed propagation time $T_p = \min(\widehat{T}_p)$ and the average observed cycle time $T_c = \mathrm{mean}(\widehat{T}_c)$.

V. ANALYSIS OF SIMULATION RESULTS

A. Propagation time and area

As expected, Figure 9a shows that the propagation time of Phase-3 protocols is two times longer than for WCHB for

978-1-5090-2737-8/16 $31.00 © 2016 IEEE

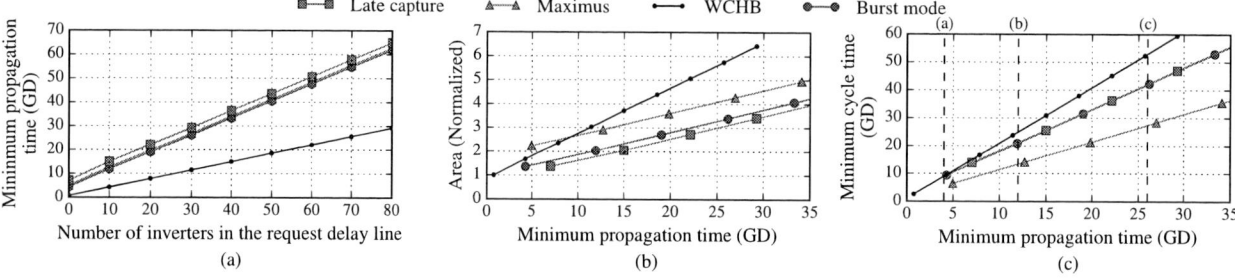

Figure 9. Area and cycle time in function of the length of the delay lines. One gate delay (GD) is the switching time of a F03 minimum size inverter. The propagation time separates the capture of data by two successive stages; the cycle time separates successive captures of data by one stage.

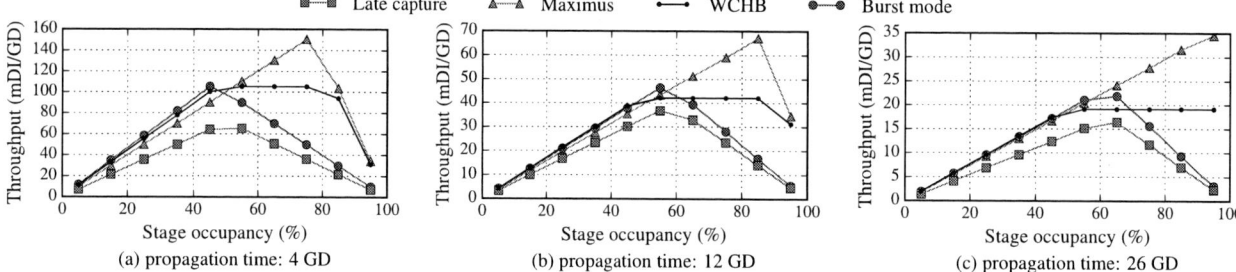

Figure 10. Throughput versus occupancy comparison of the pipelines. Depending on the protocol and the delay-line length, the maximum throughput is obtained for different occupancies. One gate delay (GD) is the switching time of a F03 minimum size inverter. One data item (DI) is the number of bits of the output stage of the pipeline. If the inverter delay is 100 ps and the pipeline output is 8 bits wide then 1 mDI/GD = 80 Mb/s

the same length of delay line. The time unit, the gate delay (GD), corresponds to the switching time of one minimum-size inverter connected to three other ones (FO3).

As a correlation, given a bundled-data constraint, Phase-3 protocols require to insert less inverters to meet the constraint. However, as Figure 9b shows, the (double) WCHB controller is smaller than the others: the late-capture and burst-mode controllers are as large as the WCHB with a 6-inverters chain, and the maximus controller is as large as the WCHB with a 14-inverters chain.

B. Cycle time

Figure 9c shows the cycle time in function of the propagation time. The slope of the lines are: 2 for the WCHB, 3/2 for the burst mode and the late capture, and 1 for the maximus protocol. This is consistent with our analysis of the critical cycle in the STGs. Only WCHB allows to have propagation times and cycle times smaller than 2ns. This is due to the computation overhead of the other protocols. In the following, the minimum propagation time is fixed to 3 values indicated by the vertical dashed lines in Figure 9c.

C. Throughput versus pipeline occupancy

The minimum cycle time computations assume that the transitions that are not on the critical STG path arrive in time. For example, in the critical cycle of WCHB (Figure 3), the transition R_1- happens right after L_1- because the transition R_1^a+ is assumed to have already happened. In practice, the cycle time is not equal to T_c because the controller either waits for new tokens (in cases of low occupancy) or bubbles (high occupancy).

For fair comparison, we compare the protocols with delay lines such that the protocols have matching propagation delay. Thus, all protocols have similar throughputs in the token-limited region (i.e. left part of Figure 10a), where the mean cycle time is dominated by the propagation of the few tokens in the pipeline.

Figure 10a shows the throughput of the pipeline as a function of the occupancy when the propagation time is short. The transition between token- and bubble-limited regimes happens around the occupancy of 50% for burst-mode and late-capture protocols because the propagation speed of tokens (available data) is similar to the propagation speed of bubbles (available registers). The transition happens later – 70% – for the others protocols because the bubbles propagate faster.

As shown in Figures 10b and 10c, for a longer propagation time, the regime transition is unchanged for WCHB because the bubbles are slowed as much as the token. For burst-mode and late-capture, the bubble slowdown is half the token slowdown, hence the transition shift. Finally for maximus, we see that the throughput at 95% of occupancy is constant. Indeed, the bubble propagation speed does not depend on the length of the delays. Thus, the regime transition is not observable in the graph. The phenomenon would reappear if a longer pipeline is used.

D. Figure of merit

In order to choose among the controllers, we define the following figure of merit (FOM) as a function of the minimum cycle time (T_c), minimum propagation time (T_p), and the stage area (A):

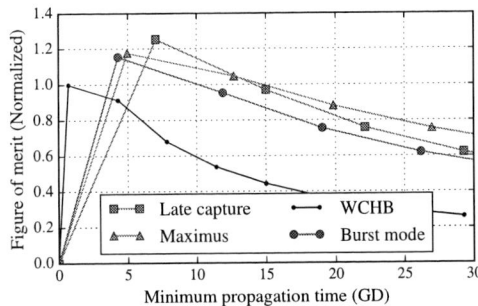

Figure 11. Figure of merit as a function of the minimum propagation time.

$$FOM = \frac{T_p}{T_c A}$$

Figure 11 plots the FOM of each controller as a function of the minimum propagation time. For short critical paths (less than 4 GD), Phase-3 protocols are too slow efficiently match the paths. The study of other Phase-1 protocols and the inclusion of asymmetric delays could balance Phase-1 and Phase-3 protocols for critical paths in the short to mid-length range because the asymmetric delays can shorten the RZ phase. A typical implementation uses a specific gate to combine the input and the output of the delay line. In order to avoid glitches, this specific gate cannot be a standard AND or OR gate. We found a safe semi-custom design that is out of the scope of this paper but has a significant area (16 transistors). Moreover, though the asymmetric delays are effective to reduce the cycle time of Phase-1 controllers, they have a negative impact on the area and power consumption.

For medium length path (between 4 and 15 GD), the late-capture protocol stands out because it is slow with a small area. Finally, for long critical path (more than 15 GD), the maximus protocol provides the best results because of its capacity to maintain a high throughput.

VI. CONCLUSION AND FUTURE WORKS

Phase-3 protocols are 4-phase protocols specific to bundled-data pipelines because the falling edge of the request indicates data validity. This feature allows to reduce the number of inserted delays in order to meet the bundled-data constraint, and also to reduce the duration of the return-to-zero phase. Hence, Phase-3 protocols can provide a significant area reduction and a throughput advantage. The new late capture protocol is as concurrent as the other Phase-3 protocols in the literature. However, it occupies less area at equal cycle time. The concurrency added in the maximus protocol allows to keep high throughput even in the bubble-limited regime.

We see three possible refinements to this study. Firstly, improving the delay model – especially including the Charlie effect [15] – would increase the accuracy of the throughput measure. Secondly, the factor of merit could consider the power consumption. Finally, other Phase-1 protocols with

and without asymmetrical delays could be compared to our protocols.

The maximus protocol is very effective in high-occupancy circuits. Such circuits appear especially interesting in the context of desynchronization. In Cortadella's desynchronization [16], the replacement of all the flip-flops of the synchronous circuits by a pair of latches is required in order to function near the optimal occupancy point (around 50%). Therefore, the maximus protocol could enable flip-flop based desynchronization frameworks.

ACKNOWLEDGEMENT

This work has been partially supported by the LabEx PERSYVAL-Lab (ANR-11-LABX-0025-01).

REFERENCES

[1] A. J. Martin, "The limitations to delay-insensitivity in asynchronous circuits," in *6th MIT Conference on Advanced Research in VLSI*, ser. AUSCRYPT '90. Cambridge, MA, USA: MIT Press, 1990, pp. 263–278.

[2] I. E. Sutherland, "Micropipelines," *Communications of the ACM*, vol. 32, no. 6, pp. 720–738, 1989.

[3] K. Yun, P. Beerel, and J. Arceo, "High-performance asynchronous pipeline circuits," in *2nd International Symposium on Advanced Research in Asynchronous Circuits and Systems (ASYNC)*, 1996, pp. 17–28.

[4] S. Furber and P. Day, "Four-phase micropipeline latch control circuits," *IEEE Transactions on Very Large Scale Integration (VLSI) Systems*, vol. 4, no. 2, pp. 247–253, 1996.

[5] A. M. Lines, "Pipelined asynchronous circuits," California Institute of Technology, Tech. Rep., 1998.

[6] P. B. McGee and S. M. Nowick, "A lattice-based framework for the classification and design of asynchronous pipelines," in *42nd annual Design Automation Conference (DAC)*. ACM, 2005, pp. 491–496.

[7] G. Birtwistle and K. S. Stevens, "The family of 4-phase latch protocols," in *14th IEEE International Symposium on Asynchronous Circuits and Systems (ASYNC)*, 2008, pp. 71–82.

[8] T.-A. Chu, "On the models for designing vlsi asynchronous digital systems," *INTEGRATION, the VLSI journal*, vol. 4, no. 2, pp. 99–113, 1986.

[9] E. Yahya, "Performace modeling, analysis and optimization of multi-protocol asynchronous circuits," Ph.D. dissertation, Institut National Polytechnique de Grenoble - INPG, Dec 2009.

[10] A. Mitra, W. F. McLaughlin, and S. M. Nowick, "Efficient asynchronous protocol converters for two-phase delay-insensitive global communication," in *13th IEEE International Symposium on Asynchronous Circuits and Systems (ASYNC)*, 2007, pp. 186–195.

[11] C. Mannakkara and T. Yoneda, "Asynchronous pipeline controller based on early acknowledgement protocol," *IEICE Transactions on Information and Systems*, vol. 93, no. 8, pp. 2145–2161, 2010.

[12] J. Cortadella, M. Kishinevsky, A. Kondratyev, L. Lavagno, E. Pastor, and A. Yakovlev, "Petrify: a tool for synthesis of petri nets and asynchronous circuits," Software, http://www.cs.upc.edu/~jordicf/petrify/ [accessed 2016-04-11].

[13] L. Wang, Z. y. Wang, and K. Dai, "An approximate method for performance evaluation of asynchronous pipeline rings," in *6th IEEE International Conference on Computer and Information Technology (CIT)*, 2006, pp. 244–244.

[14] J. Cortadella, L. Lavagno, D. Amiri, J. Casanova, C. Macián, F. Martorell, J. A. Moya, L. Necchi, D. Sokolov, and E. Tuncer, "Narrowing the margins with elastic clocks," in *IEEE International Conference on Integrated Circuit Design and Technology*, 2010, pp. 146–150.

[15] J. C. Ebergen, S. Fairbanks, and I. E. Sutherland, "Predicting performance of micropipelines using charlie diagrams," in *4th International Symposium on Advanced Research in Asynchronous Circuits and Systems (ASYNC)*, 1998, pp. 238–246.

[16] J. Cortadella, A. Kondratyev, L. Lavagno, and C. Sotiriou, "Desynchronization: Synthesis of asynchronous circuits from synchronous specifications," *IEEE Transactions on Computer-Aided Design of Integrated Circuits and Systems*, vol. 25, no. 10, pp. 1904–1921, Oct 2006.

A Design Methodology for Low-Noise CMOS Transimpedance Amplifiers Based on Shunt-Shunt Feedback Topology

A. F. Ponchet, E. M. Bastida, C. A. Finardi, R. R. Panepucci, S. Tenenbaum and S. Finco
DCSH Design House
Center for Information Technology Renato Archer
Campinas-SP, Brazil
aponchet@cti.gov.br; bastemar_e@hotmail.com;
celio.finardi@cti.gov.br; roberto.panepucci@cti.gov.br;
stefan.tenenbaum@cti.gov.br; saulo.finco@cti.gov.br

J. W. Swart
School of Electrical and
Computer Engineering
State University of Campinas
Campinas-SP, Brazil
jacobus@fee.unicamp.br

Abstract—The regulated cascode (RGC) is a widely used topology in transimpedance amplifiers projects. Although it is efficient with respect to power consumption and bandwidth, the regulated cascode presents high input-referred current noise levels that hinder the use of this topology in long distance optical networks. In this paper is proposed a design methodology for CMOS transimpedance amplifiers based on shunt-shunt feedback topology to achieve a broad-band response with low input referred current noise. A complete analysis of the shunt-shunt amplifier is presented. Experimental results shown a $51\,\mathrm{dB}\Omega$ transimpedance gain and a $10.54\,GHz$ bandwidth. The achieved average input referred current noise equals to $6.8\,\mathrm{pA}/\sqrt{(\mathrm{Hz})}$, the lowest one between other state-of-art CMOS designs. The circuit was manufactured in 130 nm RF CMOS technology.

I. INTRODUCTION

The huge growth in communication systems due to multimedia systems, high speed data-centers and other applications requires a corresponding bandwidth. Optical interconnects technology can overcome traditional limitations of electrical interconnects and keep the necessary bandwidth growth for the emerging applications [1], [2].

The first electronic device in an optical receiver after the photodetector is the transimpedance amplifier (TIA), which is responsible to convert the photodetector current into a voltage signal. The TIA must have a low input referred current noise to maximize the overall receiver sensitivity and mitigate the noise from succeeding stages [3]. In addition, the bandwidth of the TIA needs to be large enough for the required bit rate in order to minimize the intersymbol interference (ISI) [4]. In this paper, we present a low-noise transimpedance amplifier implemented in 130 nm CMOS technology with 10.5 GHz bandwidth and $51\,\mathrm{dB}\Omega$ transimpedance gain suitable for 10 Gbps optical receiver applications. A design methodology is proposed in order to minimize the input referred current noise and maximize the bandwidth.

II. THE SHUNT-SHUNT TOPOLOGY

The shunt-shunt TIA topology is is shown in figure 1 [4]. This topology is composed by a voltage amplifier with a transfer function given by $A(s)$ and a feedback resistor R_f connected between its input and the output. This configuration is also known by "voltage-current" feedback, where a negative feedback network (the feedback resistor) senses the voltage at the output and returns a proportional current to the input. This type of feedback is chosen for a TIA circuit because it lowers both the input and output resistances, thus increasing the input pole magnitude and allowing the amplifier to absorb the photodetector current, and also yielding a better drive capability [4].

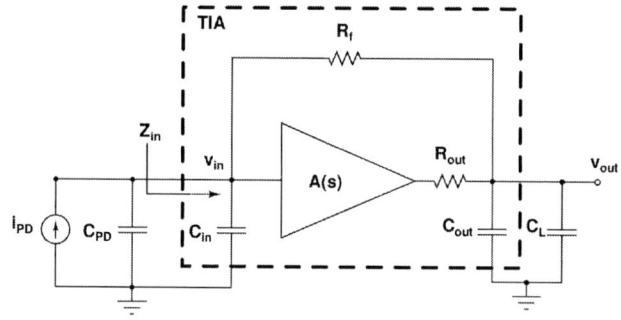

Fig. 1: Shunt-shunt feedback TIA.

The transimpedance equation and some design equations can be obtained if we apply the KCL in the input node of the voltage amplifier in figure 1. The transimpedance Z_T can be obtained as follows. Suppose that the voltage amplifier has a transfer function given by the equation 1 and $A_0 \gg 1$:

$$A(s) = \frac{A_0}{1 + \dfrac{s}{\omega_0}} \qquad (1)$$

978-1-5090-2737-8/16 $31.00 © 2016 IEEE

If we apply the KCL at the input node in figure 1, we have:

$$i_{PD}(s) = (sC_{in} + \frac{1}{R_f})v_{in}(s) + (-\frac{1}{R_f})v_{out}(s) \quad (2)$$

where i_{PD} is the photodetector current. Since $v_{in}(s) = \frac{v_{out}(s)}{A(s)}$, from equation 1 we obtain the following expression for the transimpedance:

$$Z_T(s) = \frac{v_{out}(s)}{i_{PD}(s)}$$
$$= \frac{-A_0 R_f}{\left(\frac{R_f C_{in}}{\omega_0}\right)s^2 + \left(R_f C_{in} + \frac{1}{\omega_0}\right)s + (1 + A_0)} \quad (3)$$

Th input impedance is obtained in a similar procedure by substituting $v_{out}(s) = A(s)v_{in}(s)$ into the equation 1, according to the equation 4:

$$Z_{in}(s) = \frac{R_f \omega_0 + R_f s}{R_f C_{in} s^2 + (R_f C_{in}\omega_0 + 1)s + \omega_0(1 + A_0)} \quad (4)$$

From equations 3 and 4, we can evaluate the transimpedance gain and the input impedance for low frequencies ($s \approx 0$) according to the equations 5 and 6:

$$Z_{T,DC} = -\frac{A_0}{(1 + A_0)}R_f \approx -R_f \quad (5)$$

$$Z_{in,DC} = \frac{R_f}{(1 + A_0)} \quad (6)$$

Equation 5 reveals that the feedback resistor R_f] sets the transimpedance value and must be maximized to minimize its noise contribution at the TIA input. Equation 6 relates R_f and the voltage amplifier DC gain A_0. This equation suggests that there is a trade-off between the DC transimpedance value and bandwidth, since the input pole sets the overall TIA bandwidth.

A first approximation for the TIA poles can be evaluated if we write the denominator of equation 3 like a second order system transfer function, expressed by the equation 7:

$$s^2 + 2\zeta\omega_n s + \omega_n^2 \quad (7)$$

where ζ is the "damping factor" and ω_n is the natural frequency of the system. For a critical damping, ζ must be equal to $\sqrt{2}/2$. If $\zeta < \sqrt{2}/2$ the step response exhibits ringing, creating ISI and corrupting the high and low levels of the transmitted data. Rewriting equation 4, we have

$$Z_T(s) = \frac{v_{out}(s)}{i_{PD}(s)}$$
$$= -\frac{\left(\frac{A_0\omega_0}{C_{in}}\right)}{\left(s^2 + \frac{R_f C_{in} + 1/\omega_0}{R_f C_{in}/\omega_0}s + \frac{(A_0 + 1)\omega_0}{R_{in}C_{in}}\right)}, \quad (8)$$

concluding that

$$\omega_n = \sqrt{\frac{(A_0 + 1)\omega_0}{R_f C_{in}}} \approx \frac{\sqrt{2}A_0}{R_f C_{in}} \quad (9)$$

$$\zeta = \frac{1}{2}\frac{R_f C_{in}\omega_0 + 1}{\sqrt{(A_0 + 1)\omega_0 R_f C_{in}}} \quad (10)$$

For a critical damping condition, the dominant pole of the voltage amplifier is given by equation 11:

$$\omega_0 = \frac{A_0 \pm \sqrt{A_0^2 - 1}}{R_f C_{in}} \approx \frac{2A_0}{R_f C_{in}} \quad (11)$$

The -3dB bandwidth of the TIA is finally obtained by setting the magnitude of equation 8 to $\sqrt{2}/2$ times its low-frequency value [4]. If we rewrite equation 8 in a more general form,

$$Z_T(s) = \frac{v_{out}(s)}{i_{PD}(s)} = -\frac{R_0\omega_n^2}{s^2 + 2\zeta\omega_n s + \omega_n^2}, \quad (12)$$

where $R_0 = \frac{A_0}{A_0 + 1}R_f$. Thus, we have:

$$\left|\frac{R_0\omega_n^2}{-\omega_{3dB}^2 + 2\zeta\omega_n s + \omega_{3dB} + \omega_n^2}\right| = \frac{\sqrt{2}}{2}R_0 \quad (13)$$

$$(\omega_n^2 - \omega_{3dB}^2)^2 + 4\zeta^2\omega_n^2\omega_{3dB}^2 = 2\omega_n^4 \quad (14)$$

$$\implies \omega_{3dB,TIA} = \omega_n \quad$$

Thus, the -3dB bandwidth of the second-order TIA is given by equation 15:

$$f_{3dB,TIA} = \frac{1}{2\pi}\frac{\sqrt{2}A_0}{R_f C_{in}} \quad (15)$$

III. THE PROPOSED CIRCUIT AND DESIGN METHODOLOGY

A. Circuit Topology

The proposed topology is a double cascode TIA with negative shunt-shunt feedback and inductive peaking techniques, optimized for very low noise performance [5-8]. Figure 2 shows the TIA schematic:

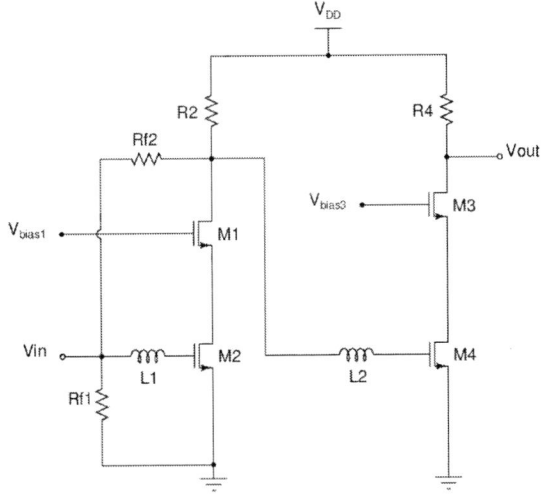

Fig. 2: Two stage cascode schematics (bias circuits not shown).

978-1-5090-2737-8/16 $31.00 © 2016 IEEE

The bias point of each transistor was carefully selected in order to guarantee a low noise performance. The inductors were designed for high Q performance. The microphotographs of each inductor are shown in figures 3(a) and 3(b):

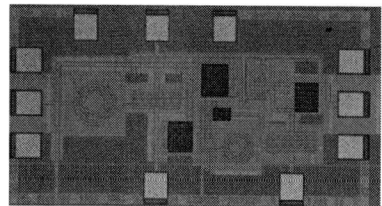

(a) $L_1 = 700\,\text{pH}$. (b) $L_2 = 1.7\,\text{nH}$.

Fig. 3: Microfotograph of the circular inductors.

Each inductor was designed if the top metal layer (metal 8) in order to minimize the parasitic capacitance to the substrate. Figure 4 shows the double cascode microphotograph:

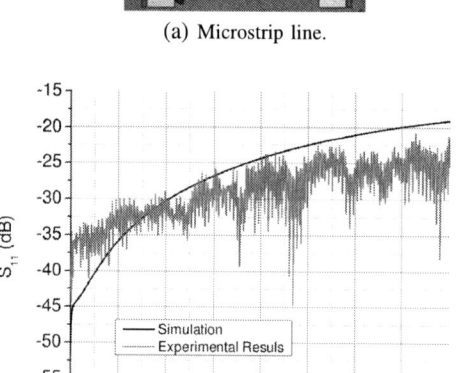

Fig. 4: TIA microphotograph.

All the layout lines were designed as microstrip transmission lines in order to minimize the losses and improve the overall performance. Figure 5 shows the microprotograph of the manufactured microstrip line, the simulated and measured S-parameters:

(a) Microstrip line.

(b) Measured S_{11} parameter.

Fig. 5: Microfotograph and S_{11} parameter of the microstrip line.

B. Biasing Optimization for Low Noise Performance

The biasing point of the TIA transistors was carefully defined by the characterization of the nMOS transistor. Some

simulations were performed with the testbench shown in figure 6 in order to find the best biasing operation point.

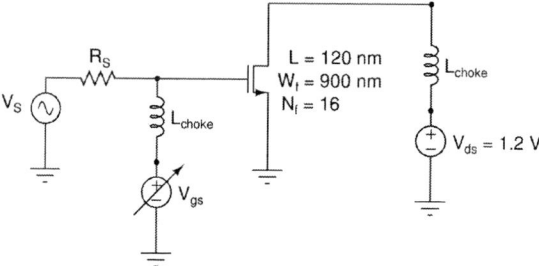

Fig. 6: nMOS transistor testbench.

The transit frequency (f_T), minimum noise figure (NF_{min}), intrinsic gain ($g_m r_o$) and current density (I_{Dens}) plots, as a function of the gate-source voltage (V_{gs}), were obtained from the nMOS testbench shown in figure 6. These plots are shown in figure 7(a) in the same horizontal axis. The g_m/I_{ds} plot was also obtained and it is shown in figure 7(b).

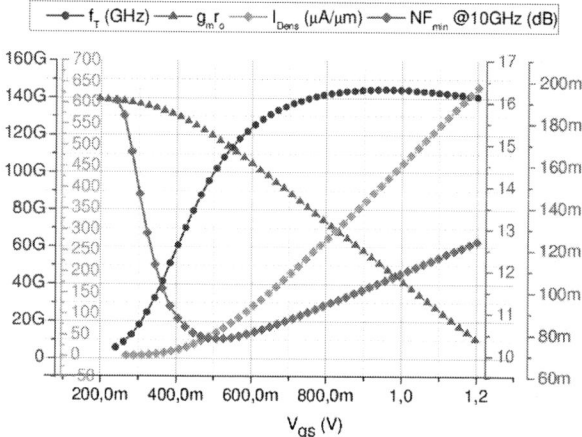

(a) nMOS plots for biasing optimization.

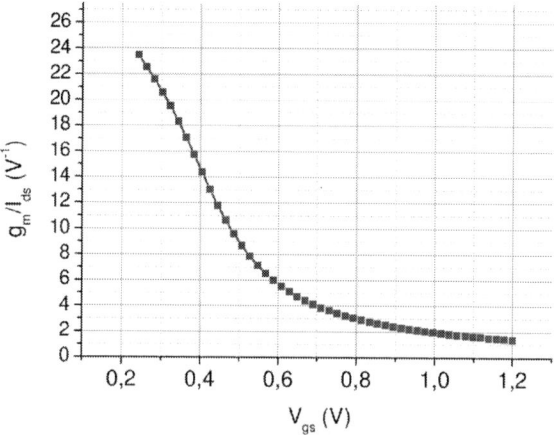

(b) g_m/I_{ds} plot for the nMOS transistor

Fig. 7: nMOS transistor plots for biasing optimization.

978-1-5090-2737-8/16 $31.00 © 2016 IEEE

From these plots a set of constraints for the circuit biasing point operation was defined:

$$400\,\text{mV} \leq V_{gs} \leq 600\,\text{mV} \tag{16}$$

$$6 \leq \left(\frac{g_m}{I_{ds}}\right) \leq 15 \tag{17}$$

$$25\,\mu\text{A}/\mu\text{m} \leq \left(\frac{I_{ds}}{W}\right) \leq 125\,\mu\text{A}/\mu\text{m} \tag{18}$$

Equations 16, 17 and 18 define the target biasing operation point for the TIA. The gate-source voltage (V_{gs}) defined by equation 16 assures that the transistors operates near the noise figure minimum and equations 17 and 18 help to define the transistors dimensions.

C. Noise Analysis

Figure 8 shows the double cascode diagram with the noise sources:

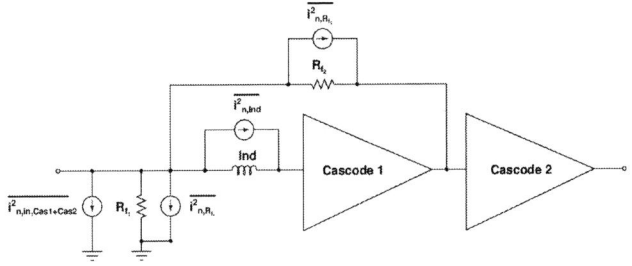

Fig. 8: TIA diagram with noise sources.

In the diagram shown in figure 8, the current sources $\overline{i_{n,in,Cas1}^2}$ e $\overline{i_{n,in,Cas2}^2}$ represent the input referred noise of the first and second stages respectively. Since each noise source is considered statistically independent from each other, the total TIA input referred noise is given by equation 19 [4]:

$$\overline{i_{n,in,TIA}^2} \approx \overline{i_{n,R_{f_1}}^2} + \overline{i_{n,R_{f_2}}^2} + \overline{i_{n,ind}^2} + \overline{i_{n,in,Cas_1+Cas_2}^2} \tag{19}$$

The noise generators of each cascode stage is obtained by the small signal analysis of the circuit in figure 9:

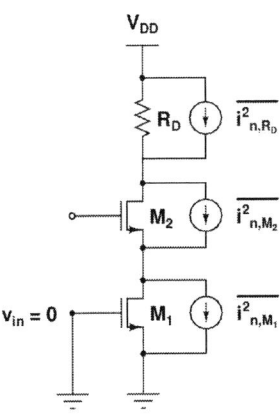

Fig. 9: Cascode schematic to evaluate $\overline{v_{n,out,Cas}^2}$.

The input referred noise voltage of the TIA is given by equation 20,

$$\overline{v_{n,in,Cas}^2} = \frac{\overline{v_{n,out,Cas}^2}}{A_{v,Cas}^2} \tag{20}$$

where $\overline{v_{n,out,Cas}^2}$ is the output noise voltage and $A_{v,Cas}^2$ is the cascode voltage gain. Since the input signal of the TIA is the photodetector current, the noise at the input must be represented by a current source. This noise source is calculated as follows.

In equation 20, $\overline{v_{n,out,Cas}^2}$ is given by equation 21,

$$\overline{v_{n,out,Cas}^2} \approx (\overline{i_{n,out,Cas}^2}) \times R_D^2$$

$$\approx 4kT\{\gamma(g_{m_1} + g_{m_2}) + \frac{1}{R_D}\} \times R_D^2 \tag{21}$$

where R_D^2 is the cascode stage transresistance (Z_T for $s \approx 0$). The substitution of equation 21 into equation 20, give us the TIA input referred noise voltage:

$$\overline{v_{n,in,Cas}^2} \approx \frac{4kT\left[\gamma(g_{m_1} + g_{m_2}) + \frac{1}{R_D}\right] \times R_D^2}{(g_{m_1}R_{out,Cas})^2} \tag{22}$$

The small signal analysis of the cascode schematic results in equation 23 that relates the TIA output noise voltage with the input noise current [9]:

$$\overline{v_{n,out,Cas}^2} \approx \underbrace{\overline{i_{n,in,Cas}^2}\left(\frac{1}{C_{in}\omega}\right)^2}_{v_{n,in,Cas}^2}(g_{m_1}R_{out,Cas})^2 \tag{23}$$

Finally, the comparison between equation 22 and the term in brackets in equation 23, give an approximation for the total input referred noise current:

$$\overline{i_{n,in,Cas}^2} \approx (C_{in}\omega)^2 \times$$

$$\left\{\frac{4kTR_D[\gamma R_D(g_{m_1} + g_{m_2}) + 1]}{(g_{m_1}R_{out,Cas})^2}\right\} \tag{24}$$

Equation 24 reveals that the TIA input referred noise current has a quadratic behavior with frequency.

D. Proposed Design Flow

The proposed design flow for the low noise TIA is given by the following steps:

1) *Define the resistor R_{f_2} value by the DC transimpedance:*

$$Z_{T(DC)} \geq 48\,\text{dB}\Omega$$

$$\geq 10^{48/20}$$

$$\geq 251,2\,\Omega \implies R_{f_2} \geq 251.2\,\Omega \tag{25}$$

2) *Plot the curves of f_T, NF_{min}, I_{ds}/W and g_m/I_{ds} for the nMOS transistor as a function of V_{gs}.*

3) *Define an interval for V_{gs} from the plots obtained in the previous step that ensures a DC operation near the NF_{min} minimum.*

978-1-5090-2737-8/16 $31.00 © 2016 IEEE

4) *Find an initial approximation for I_{ds}/W e g_m/I_{ds} from V_{gs} obtained in the previous step.*

5) *Define an initial value for the transconductance and output resistance of each cascode stage from the voltage gain and power consumption requirements. From these assumptions and the g_m/I_{ds} plot a initial value for the W/L relation can be obtained.*

6) *Optimize the input capacitance C_{in} of the cascode from the model defined by equation 24.*

7) *Repeat previous steps until the specifications are met.*

IV. EXPERIMENTAL RESULTS

The circuit was tested via on-wafer probing and mounted onto a RF microwave. Figures 10(a) and 10(b) show the chip mounted onto the PCB with connectors and the chip detail with bondwires connections, respectively.

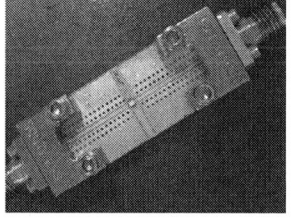

(a) Chip mounted onto a RF PCB with connectors.

(b) Chip with the bondwires.

Fig. 10: Chip mounted onto a RF PCB board.

The RF PCB was designed and manufactured in our lab. The transmission lines were designed with the aid of electromagnetic simulations.

Figures 11 and 12 show the voltage gain (S_{21}) and the transimpedance (Z_T) obtained from simulations and by experimental measurements, respectively:

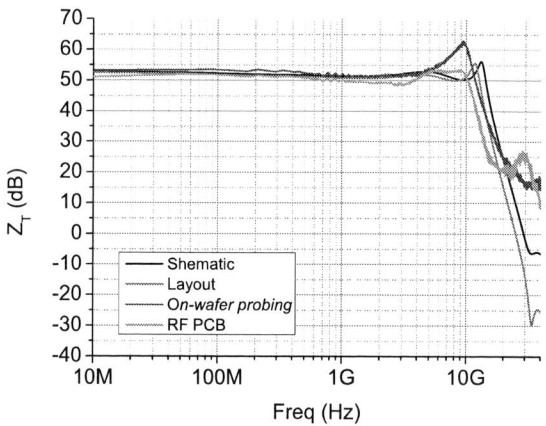

Fig. 12: Simulated and measured Z_T.

The results presented on figure 11 show us that the TIA has a 14 dB voltage gain with a 7.4 GHz bandwidth. The circuit draws a bias current equal to 35 mA from a 1.2 V DC supply. The simulations were performed considering a 35 fF photodetector capacitance. Figure 12 shows the results for the transimpedance gain that were obtained by the equation 26 from the S-parameters simulations:

$$Z_T = 50 \frac{S_{21}}{1 - S_{11}} \qquad (26)$$

The circuit has a $51\,\mathrm{dB\Omega}$ transimpedance gain and a 10.54 GHz bandwidth for the experimental results obtained with the chip mounted onto the PCB. From figure 11 it is possible to see that there is a reduction on the bandwidth, from the schematic to the PCB measurements, approximately equal to 3.5 GHz. This reduction is due the board and bondwires parasitics. The circuit has enough bandwidth to operate in 10 Gbps bit rates. As a rule of thumb, the TIA must have at least a bandwidth equal to 70 % of the bit rate in order to keep the ISI on acceptable levels [4]. Figures 13 and 14 show the DC values and the bandwidth for Z_T and S_{21} obtained from simulations and experimental results:

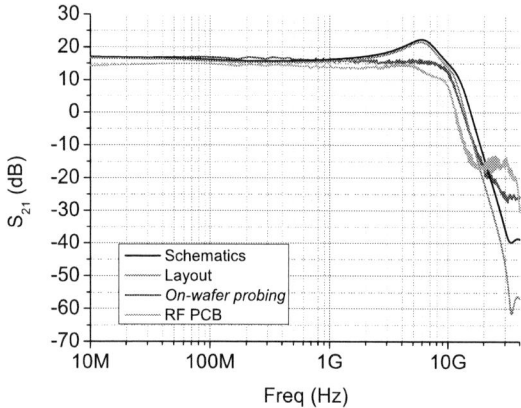

Fig. 11: Simulated and measured S_{21}.

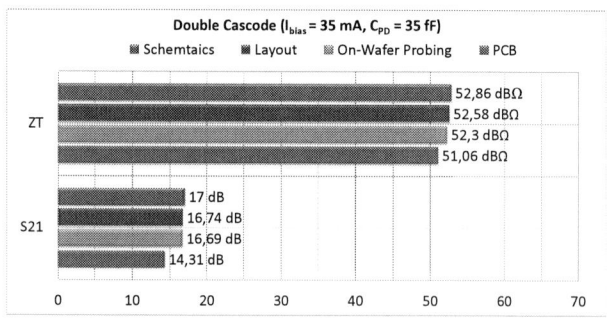

Fig. 13: Z_T and S_{21} DC values.

978-1-5090-2737-8/16 $31.00 © 2016 IEEE

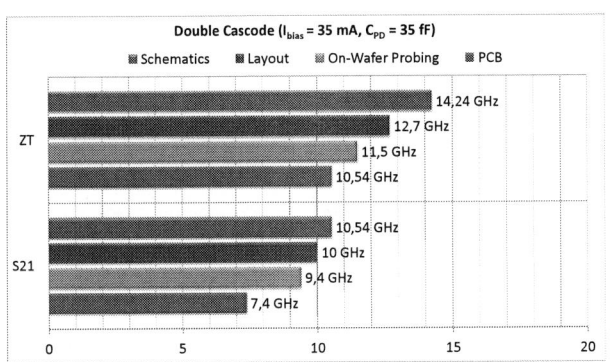

Fig. 14: Z_T and S_{21} bandwidth.

Table 1 shows the bias point obtained after the proposed design flow was completed:

TABLE I: nMOS transistors biasing point.

Trans.	Parameters					
	(W/L)	I_{DS}	V_{GS}	gm	$\dfrac{I_{DS}}{W}$	$\dfrac{gm}{I_{DS}}$
	$\mu m/nm$	mA	mV	mS	$\mu A/\mu m$	V^{-1}
M1	(280)/(120)	14.5	544.4	127.3	51.6	8.8
M2	(280)/(120)	14.5	526.4	134.8	51.6	9.31
M3	(280)/(120)	17.0	551.5	142.5	61.0	8.33
M4	(280)/(120)	17.2	546.3	141.0	61.5	8.19

From table 1 we can conclude that the biasing point of the double cascode TIA obey the set of constrains given by the equations 16, 17 and 18, necessary to keep the transistors working near the minimum of noise figure plot. Table 2 shows a complete summary and comparison with other state-of-art CMOS TIAs for 10 Gbps:

TABLE II: Summary and comparison with other 10 Gbps TIAs.

Reference/year	[11]	[12]	[13]	[14]	[15]	This work
	2013	2015	2015	2015	2004	2015
Tecnologia	CMOS	CMOS	CMOS	CMOS	CMOS	CMOS
	130 nm	130 nm	130 nm	180 nm	90 nm	130 nm
Topology	RGC	RGC	RGC	RGC	RGC	Double Casc.
Z_T (dBΩ)	52.9	50.1	59	57	54	**51**
V_{dd} (V)	1.8	1.5	1.3	1.8	1.0	**1.2**
BW (GHz)	14.3	7.0	6.9	8.2	13.4	**11.5**
Noise (pA/$\sqrt{\text{Hz}}$)	39	31.3	20	21	50	**6.8**
Sensitivity (dBm)	-16.9	-17.8	-19.8	-19.6	-15.8	**-24.5**
Power (mW)	2.7	7.5	16.9	22	2.2	**30**
Area (mm^2)	0.002	0.02	0.051	0.6	0.01	**0.16**

The proposed TIA has the lowest input referred noise current when compared to other state-of-art designs. The outstanding performance was achieved by applying the design flow method and the noise modeling proposed in this work.

V. CONCLUSION

A complete design flow and modeling procedure for low-noise CMOS transimpedance amplifiers were proposed in this work. The proposed circuit is based on a conventional cascode topology and the shunt-shunt feedback topology. The experimental results show that the proposed architecture has the lowest input referred current noise when compared with other state of art designs. The proposed circuit was manufactured and tested onto a RF PCB. The circuit was manufactured in 130 nm CMOS technology and the obtained transimpedance bandwidth ensure a 10 Gbps bit rate operation.

ACKNOWLEDGMENT

The authors would like to thank the technical staff of CTI, the Brazilian Research National Council (CNPq), FAPESP, CITAR Project and FINEP for the support and contributions during this work.

REFERENCES

[1] M. Asghari and A. V. Krishnamoorthy, "Silicon photonics: Energy-efficient communication," *Nature Photonics*, vol. 5, no.5, pp. 268-270, 2011.

[2] M. Hochberg et al., "Silicon Photonics: The Next Fabless Semiconductor Industry," *Solid State Cicruits Magazine, IEEE*, vol. 5, no. 1, pp.48-58, 2013.

[3] B. Razavi, *RF Microelectronics*, Prentice Hall, 2nd Edition, Oct. 2011.

[4] B. Razavi, *Design of Integrated Circuits for Optical Communications*, Wiley, 2nd Edition, Aug. 2012.

[5] A. F. Ponchet et al., "Design and Characterization of 0.13 μm CMOS and BiCMOS Low Noise Transimpedance Amplifiers for High Speed Optical Interconnects," 2015 SBMO/IEEE MTT-S International Microwave and Optoelectronics Conference, IMOC, 3-6 Novembro 2015, Porto de Galinhas, Pernambuco, Brazil.

[6] A. F. Ponchet et al., "Design and Optimization of High Sensitivity Transimpedance Amplifiers in 130nm CMOS and BiCMOS Technologies for High Speed Optical Receivers," *28th Symposium on Integrated Circuits and Systems Design, SBCCI*, 31 Aug-4 Sept. 2015, Salvador, Brazil.

[7] A. F. Ponchet et al., "SiGe HBT mm-wave DC Coupled Ultra-Wide-Band Low Noise Monolithic Transimpedance Amplifiers," *27th Symposium on Integrated Circuits and Systems Design*, SBCCI, 1-4 Sept. 2014, Aracaju, Brazil.

[8] A. F. Ponchet et al., "Low Noise Si Based Monolithic Transimpedance Amplifiers for 10 Gbps, 40 gbps and 100 Gbps Applications," *in Proc. of the IX ICCDCS Conference*, Playa del Carmen, Mexico, April, 2014, pp. 94-99.

[9] B. Razavi, *Design of Analog CMOS Integrated Circuits*, McGraw-Hill International Edition, 2001.

[10] J. Gao, "Optoelectronic Integrated Circuit Design and Device Modeling," John Wiley and Sons, 2011.

[11] D. Kim et al., "12.5-Gb/s Analog Front-End of an Optical Transceiver in 0.13 μm CMOS," IEEE International Symposium on Circuits and Systems (ISCAS), pp 1115-1118, May 2013.

[12] M. H. Taghavi et al., "10-Gb/s 0.13 μm CMOS inductorless Modified-RGC Transimpedance Amplifier," *IEEE Transactions on Circuits and Systems-I: Regular Papers*, Vol. 62, No. 8, August, 2015.

[13] J.ăSangirov et al., "Design and fabrication of a 10 Gbps transimpedance amplifier-receiver for optical interconnects," *Journal of Computational Electronics*, vol. 14, no. 3, pp. 669-674, May. 2015.

[14] S. Qiwei et al.,"Novel-pre Equalization Transimpedance Amplifier for 10 Gb/s Optical Interconnects," *Journal of Semiconductors (Chinese Institute of Electronics)*, vol.36, no. 7, DOI:10.1088/1674-4926/36/7/075002, July, 2015.

[15] C. Kromer , G. Sialm , T. Morf , M. L. Schmatz , F. Ellinger , D. Erni and H. Jackel "A low-power 20-GHz 52-dBΩ transimpedance amplifier in 80-nm CMOS", *IEEE J. Solid-State Circuits*, vol. 39. no. 6, pp. 885 -894, June 2004.

A 450 mV Supply Self-biased Wideband Inductorless Balun LNA for sub-GHz Applications

Arthur Liraneto Torres Costa*, Hamilton Klimach[†] and Sergio Bampi*

*PGMicro - Informatics Institute
[†]PGMicro - Electrical Engineering Department
Federal University of Rio Grande do Sul, Porto Alegre-RS
Email: {altcosta, bampi}@inf.ufrgs.br, hamilton.klimach@ufrgs.br

Abstract—This paper presents a CMOS wideband LNA topology operating under a 450 mV voltage supply in the analog TV white spaces frequency band from 54 MHz to 862 MHz. It could be used in a wideband RFID or energy harvesting communication network applications. The proposed circuit is self-biased, uses no inductors and it has a cascaded amplifier in the noise canceling branch for a better trade-off of noise figure (NF), S11 and IP3. The 450 mV operation is achieved by a proper choice of the basic amplifiers that compose the noise canceling topology. This paper demonstrates the use of low-VT PMOS transistors and zero-VT NMOS transistors in a RF circuit, in order to be self-biased in a 130 nm CMOS technology PDK from Global Foundries. The post-layout simulations included bondwire inductances and pad capacitances parasitics for more realistic results. Under such a low supply, the LNA circuit is capable of Voltage Gain > 17 dB, NF < 6.2 dB and S11 < -10.3 dB, in the 54 MHz - 862 MHz range. The overall LNA power consumption is only 2 mW.

I. INTRODUCTION

The wideband operation of radio receivers has become an important research topic since the publication of the noise-canceling technique for low-noise amplifiers (LNA) by [1] and the release of the IEEE 802.22 WRAN standard, which operates in the 54 MHz to 862 MHz frequency range. The reduction of chip area/costs and power consumption provided by the use of a single radio instead of multiple narrowband front-ends to cover a wide range of frequencies has made the wideband receiver a very attractive topology [2]. The toughest challenge in designing a wideband receiver is the reception of all in-band interferers along with the desired signal in the receiver. This increases the IP2 and IP3 requirements of the radio, making it difficult to maintain a low noise figure (NF) and a low intermodulation distortion altogether. The main receiver block responsible for most of the NF-IP3 trade-off is the LNA. Moreover, in CMOS wideband designs, the LNA is the bottleneck for the IP2, as its own second order intermodulation can down-convert the interferers into the desired band of a direct-conversion receiver [2]. Thus, once the LNA design can assure a low NF and a reasonable IP2 and IP3, all other blocks in the receiver front-end chain have less stringent and more feasible specifications.

Recently, digital applications have gone towards a decrease in voltage supplies in order to reduce power consumption and to allow CMOS FETs below 32 nm minimum channel lengths. The RF front-end has been left behind namely by operating at higher voltages, which in turn taxes power consumption

and also the compatibility to digital circuits voltage supplies. The usual approach in a transceiver design is to add a separate higher voltage supply for the RF front-end (F/E). While in fact the supply rails for the F/E have mandatory physical separation to the digital supplies, the diversity of supplies poses a system-level concern for the power delivery network. In order to solve this problem, some recent works have tried to design low-voltage narrowband receivers such as in [3]. There has been some improvement in ultra low-voltage (ULV) wideband receivers targeting energy harvesting applications such as in [4].

In wideband designs, the challenge is increased, as the trade-off of NF and IP2/IP3 is already at its limits. A voltage supply reduction in a wideband RF circuit such as an LNA implies in a limited achievable gain, giving the design a degraded NF to start with. When one tries to increase the gain for a better NF (by increasing power consumption) it usually degrades the IP2 or IP3. Not to mention, operating in sub-1V power supplies requires the transistor to enter the weak inversion region in most cases. For digital CMOS there is a wealth of demonstrations [5] that the maximum energy efficiency occurs at near-VT supplies around 300 mV at room temperature. The voltage supply decrease, in fact, increases the LNA non-linearity, which also degrades IP2 and IP3. There are some proposals of designs in the literature that implemented wideband low voltage and ultra-low voltage (ULV) LNAs such as [6]–[8] and [9]–[12]. Most of them have either used inductors, do not behave as a balun, need an external bias circuit, or do not mention the IP2.

In this paper we propose an inductorless wideband balun LNA topology which is self-biased and capable of providing a high voltage gain, consuming only 2 mW under a 450m V power supply. The absence of inductors in our new design also minimizes the area of the CMOS circuit and this circuit does not require a separate bias circuit, as it is self-biased. This LNA is suited for the 54 MHz - 862 MHz wideband where many communication standards and white spaces from analog TV coexist.

This paper is organized as follows: in section II the circuit design is presented. In section III, the post-layout simulation results including bondwire inductances and pad capacitances parasitics are shown and compared to other designs at the state of the art. Section IV summarizes the main results and presents the key conclusions.

978-1-5090-2737-8/16 $31.00 © 2016 IEEE

II. TOPOLOGY DESCRIPTION AND DESIGN

The design of every noise-canceling LNA is composed by, at least, two branches of signal path. One where the main amplifier is responsible for input impedance matching and another containing the auxiliary amplifier, which is able to cancel the main amplifier noise at the output. In [13], a cascaded auxiliary amplifier was proposed, in order to break the NF-IP3 trade-off and improve the LNA overall parameters. In this paper, we propose a different cascaded auxiliary amplifier, suited for ULV operation and self-biased.

When choosing the basic amplifiers to insert in the main and auxiliary branches, the designer must have in mind the role of each block, exceptionally when operating under a 450 mV supply of a CMOS 130 nm process with 1.2V of nominal voltage supply. If a smaller node process is available, such as 90 nm or 45 nm, the results can be even better or the voltage supply could be decreased below 450 mV, using the same strategy.

The main amplifier in our proposed circuit is responsible for input impedance matching and half the gain at the differential output, as its noise is going to be canceled by the auxiliary branch. Among the basic amplifiers, the ones that provide a good input impedance matching are the Common-Gate (CG) and the resistive-feedback inverter. The latter is a preferable choice, as only half the 20 mS g_m value in each transistor is needed to achieve a 50 Ω matching, which is an extremely important feature under ULV gain limitations. Also, the resistive-feedback can be self-biased.

In the auxiliary branch, the challenge of ULV operation is greater, since its noise is not going to be canceled at the output. The first stage must have the least influence possible in the input impedance matching and have a low gain, delivering the least possible distortion to be amplified by the second stage. The least value for NF in this first stage is also required, since its NF dominates the auxiliary branch. In this situation, a simple CS amplifier is good enough for the NF requirement, but its gain is too high to be inserted as the first stage of the auxiliary branch, causing too much distortion. Lowering its gain by loading it with a small impedance would imply in a large transistor size, to compensate its NF loss, degrading the bandwidth of the input impedance matching due to the large size of the transistor parasitic capacitances. However, if the first stage load could be set by the input impedance of the second stage of the cascade, it would give more degrees of freedom to size the transistor of the first stage, in order not to disturb the bandwidth of the LNA input impedance matching. In order to implement this idea, the second stage was chosen to be another resistive-feedback CS amplifier to support the first stage being a simple CS amplifier. In this auxiliary branch setup, the first stage gain is set by the g_m of the transistors in the second stage, which in turn, has its gain adjustable by the feedback resistor. All independently adjustable, closing the auxiliary branch topology design with a small gain, small transistors in the first stage and a high gain in the second stage. The proposed LNA topology is shown in Fig. 1.

The transistor M_3 is a load diode that helps the linearity improvement of the main amplifier. Its bias current is 1-2% of the current of M_1, in order not to interfere significantly in the gain of the resistive-feedback amplifier ($g_{m3} << (g_{m1}+g_{m2})$).

The purpose of M_3 is to cancel non-linear terms of M_1 current and to improve the IP2 and IP3 as used in [13].

All small signal analysis were done neglecting the parasitic capacitances. Care must be taken with the transistor size, which can decrease the bandwidth. The purpose of this analysis is to give an insight for the designer of how the circuit behaves and which variables are most important for the electrical design. As in modern nanometer CMOS technologies the transistor is very difficult to model, simulation models such as BSIM, which contains hundreds of parameters, make electrical simulations a good approach to achieve the desired g_m under the specified transistor bias.

The input resistance of the LNA is dominated by the resistive-feedback inverter in the main branch, which is given by (considering the effect of the diode load):

$$R_{IN} = \frac{1 + \frac{R_F}{r_{ds_1}//r_{ds_2}} + g_{m3}R_F}{g_{m1} + g_{m2} + g_{m3}} \approx \frac{1}{g_{m1} + g_{m2}} \quad (1)$$

Initially, the first order estimation for the R_{IN} value is not going to consider the effect of the transistors r_{ds}, but they were included in the equation above in order to be able to verify if they could actually be neglected. The purpose of R_F is to reduce the NF of the LNA, which will be discussed further in this section. As the value of g_{m3} is set to be intentionally very small (as discussed later), it can already be neglected. The input impedance R_{IN} must match the signal source impedance R_S, leading to a first order estimation of (considering $r_{ds_1}//r_{ds_2} >> R_F$ initially): $g_{m1} + g_{m2} = \frac{1}{R_S}$. The small signal voltage gain of resistive-feedback inverter with the diode load is given by:

$$A_{V_1} = \frac{1 - (g_{m_1} + g_{m_2})R_F}{1 + (g_{m_1} + g_{m_2})R_S + g_{m3}(R_F + R_S)} \quad (2)$$

If g_{m3} is neglected, since it is too small compared to g_{m1}, assuming $R_F >> R_S$ and input impedance matching ($R_{IN} = R_S$), it yields:

$$A_{V_1} \approx -\frac{R_F}{2R_S}. \quad (3)$$

The value of R_F was set to 416 Ω, leading to a first order estimation of $A_{V_1} = 4.2$ V/V. As two of the three amplifiers in this circuit are self-biased, the simple CS first stage in the auxiliary cascade was biased by the second stage V_{GS} through R_{BIAS}. All transistors are biased around moderate or strong inversion to get the best IP2-IP3 possible. This is done by analyzing the second and third derivatives of the transistor current.

The auxiliary amplifier is composed by M_4 to M_6. These transistors are organized in two cascaded amplifiers: a CS and a resistive-feedback inverter. The CS amplifier has a resistor as a load just to provide a DC path to the transistor, given its load will heavily depend on the next stage $1/g_m$ input impedance (as a first order estimation). The second stage is a self-biased resistive-feedback inverter to provide a low input impedance as

978-1-5090-2737-8/16 $31.00 © 2016 IEEE

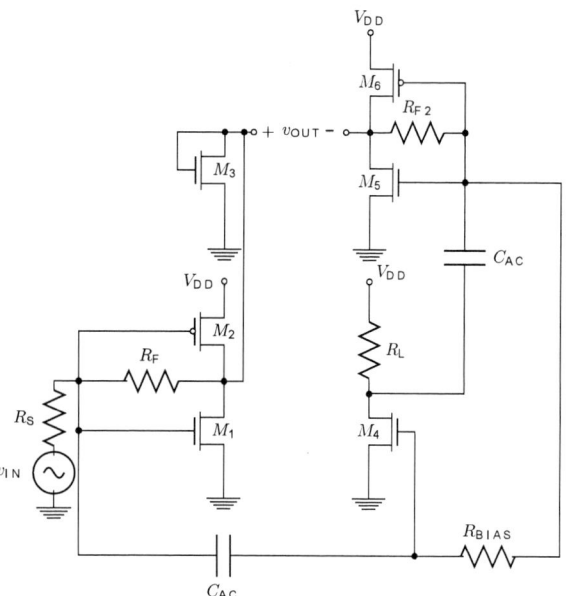

Fig. 1. The proposed LNA topology.

a load to the first stage. The transistors r_{ds} were also included in the input impedance equation of the second stage, in order to be able to verify if they could be neglected again, which is given by:

$$R_{IN_{AUX2}} = \frac{\frac{R_{F2}}{r_{ds5}//r_{ds6}} + 1}{g_{m5} + g_{m6}} \qquad (4)$$

As R_{F2} was set to 384 Ω, and $r_{ds5}//r_{ds6} = 278.4\ \Omega$, so they could not be neglected in this case, yielding $R_{IN_{AUX2}} = 65\ \Omega$. The small signal voltage gain of the first CS amplifier is:

$$A_{V_{AUX1}} = -g_{m4}(R_L//R_{IN_{AUX2}}) \qquad (5)$$

in which, R_L is dominated by $R_{IN_{AUX2}}$. The small signal expression for the voltage gain of the second stage is similar to the expression of A_{V_1}, except that R_S is replaced by the output impedance of the previous stage R_L (assuming $R_L << r_{ds4}$) and there is no g_{m3}. Thus:

$$A_{V_{AUX2}} = \frac{1 - (g_{m5} + g_{m6})R_{F2}}{1 + (g_{m5} + g_{m6})R_L} \approx -\frac{R_{F2}}{R_L} \qquad (6)$$

As R_L is only intended to provide a DC path to the first stage, its value can be small to increase the next stage voltage gain. Finally, the overall LNA voltage gain at the differential output is:

$$A_{V_{TOTAL}} = -\left(\frac{R_F}{2R_S} + \frac{g_{m4}R_{IN_{AUX2}}R_{F2}}{R_L} \right) \qquad (7)$$

As this circuit behaves as a balun, both output branches must have the same voltage gain, in order to be balanced. Each branch in this design was designed to have at least 3.6 V/V gain (at the cutoff frequency), in order to achieve a minimum differential voltage gain of 7.2 V/V or 17 dB. The main amplifier was designed to have a 4.2 V/V gain and the auxiliary branch, the first stage has a voltage gain of 1.16 V/V and the resistive-feedback inverter has a voltage gain of 3.6 V/V, yielding a voltage gain of 4.18 V/V in the auxiliary branch.

The resistive-feedback inverter has a good IP2 due to its bias point at strong inversion, near moderate inversion, and due to its feedback. The second stage on a cascade has to have a better IP2 than the first stage, so as to dominate the resulting IP2 of this branch.

The main amplifier diode load has very low current when compared to the transistor which provides gain, the effect is the non-linearity subtraction in the currents with no significant noise addition nor gain decrease.

The NF of a generic amplifier is given by [14]:

$$NF_{amp} = 1 + \frac{\overline{V_{n_{OUT}}^2}}{|\alpha|^2 A_V^2} \frac{1}{\overline{V_{n_{RS}}^2}} \qquad (8)$$

where $\overline{V_{n_{OUT}}^2}$ is the amplifier noise at the output, α is 1/2 for perfect input matching, $\overline{V_{n_{RS}}}$ is the source impedance noise and A_V is the amplifier voltage gain.

In the resistive-feedback amplifier, the squared output voltage noise is given by:

$$\overline{V_{n_{OUT}}^2} = 4kTR_F + 4kT\gamma(g_{m1} + g_{m2})\frac{(R_F + R_S)^2}{4} \qquad (9)$$

whose two terms are the output noise that results from the feedback resistor and from the transistors when considering only thermal noise. Hence, for the resistive-feedback amplifier $NF_{MAIN} = 1 + \frac{4R_S}{R_F} + \gamma(g_{m1} + g_{m2})R_S$, where the last term we seek to cancel at the output by virtue of the cascaded amplifiers. Noise canceling technique was first addressed by [1] using a single transistor amplifier in the auxiliary branch. Adding the noise of the cascaded amplifiers, the total NF (considering only thermal noise) for the LNA is:

$$NF = 1 + \frac{4R_S}{RF} + \frac{A_{V_{AUXtotal}}^2 v_{n_{AUX}}^2 R_S}{kTR_F^2} \qquad (10)$$

where k is the boltzmann constant, T is the absolute temperature and $v_{n_{AUX}}^2$ is the input referred noise of the cascaded amplifiers given by:

$$v_{n_{AUX}}^2 = \left(\frac{4KTR_L^2}{g_{m4}^2 R_{IN_{AUX2}}^2 R_{F2}^2} \right)(R_{F2} + \gamma g_{m4}R_{F2}^2 + \qquad (11)$$

$$+ \frac{\gamma(g_{m5} + g_{m6})(R_{F2}^2 + R_L^2)}{4} + \frac{R_{F2}^2}{R_L}) \qquad (12)$$

978-1-5090-2737-8/16 $31.00 © 2016 IEEE

The IP2 optimization was made by changing the ratio of M_3-M_1. This IP2 adjustment was carried out via electrical simulations of post-layout with the BSIM4 model for the transistors provided by the foundry. The final sizing for the complete LNA is shown in Table I for the 130 nm technology. Transistors M_1, M_3-M_5 are zero-VT NMOS FETs available in this mixed-signal CMOS technology. All PMOS have low-VT values. The value of R_L was designed to be just 96 Ω just to provide a DC path for the first stage of the auxiliary branch. The g_{m4} was set to give just a unite gain for the first stage. The main concern here is the input impedance matching bandwidth. The R_{BIAS} value is 33 kΩ and all C_{AC} values are 44 pF. The values for the R and C of this high-pass filter is designed to have the best linearity possible. If one chooses to increase R_{BIAS}, reducing C_{AC} in order to save area, it will degrade the IP2, because the dual MIM capacitors used in this design are more linear than the poly resistors. Thus, there is a trade-off between the RC sizes and the IP2.

TABLE I.   DEVICE SIZING.

MOSFETs	SIZING		
	W(μm)	L(μm)	g_m(mS)
M_1	250	0.42	15
M_2	400	0.19	15.4
M_3	8	0.42	0.2
M_4	350	0.42	30
M_5	300	0.42	16.2
M_6	1000	0.12	20.4

III.   RESULTS AND COMPARISON

The LNA was fully designed, laid out in CMOS 130 nm technology, and fully extracted, using Cadence Virtuoso^{TM} and Cadence Spectre^{TM} for post-layout simulations. The simulation test-bench is shown in Fig. 3 and includes estimated bondwire inductances provided by the foundry (L_{BOND} = 2 nH), PAD capacitances (C_{PAD} = 60 fF extracted from simulation of PDK library cell) and external decoupling output capacitors ($C_{AC_{IN}}$ = 120 pF and $C_{AC_{OUT}}$ = 7 pF). A load composed by R_L = 1.5 kΩ and C_L = 100 fF was also considered. This load was provided from a differential mixer designer in our group which is going to be used as the load for this LNA.

These package parasitics L_{BOND} and C_{PAD} were taken into account for the simulations, degrading the performance of the LNA. In case the LNA is designed to be used as an internal block of an application circuit, like in an RF front-end, results can be improved further without them.

The layout of the LNA is shown in Fig. 2 and its simulation results post-extraction are shown in Figs. 4, 5, 6. The gain and NF are nicely flat in the 54 MHz - 862 MHz frequency band. The S_{11} is below -10.5 in the entire band.

The topology presented a voltage gain above 17 dB and NF 5.3-6.2 dB in the entire frequency band as can be seen in Fig. 4. The highest IP2 is 15.5 dBm as can be seen in Fig. 6. The worst case of the output gain imbalance is 1.7 dB, which is showed in Fig. 5, with both output gains separately showed.

A.   Comparison with recently published ULV Wideband LNAs

Table II presents the performance parameters used to compare our design with 7 other ULV wideband LNA circuits

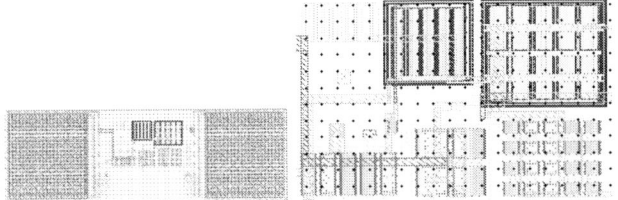

Fig. 2.   Layout of the LNA (left) with decoupling capacitors. Zoom of the LNA core (right).

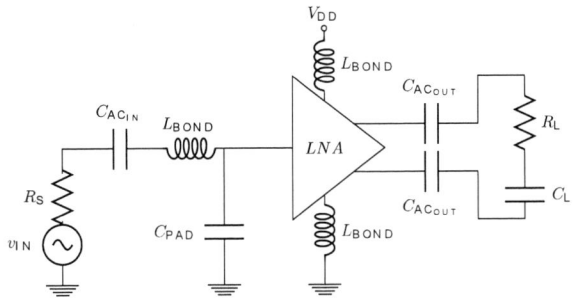

Fig. 3.   The test-bench used to simulate the layout with extracted parasitics.

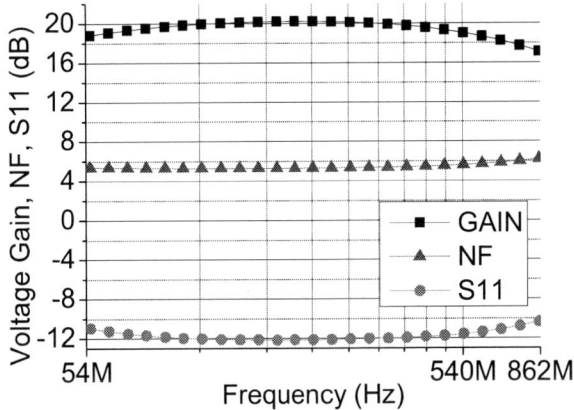

Fig. 4.   Results for voltage gain, NF and S_{11}.

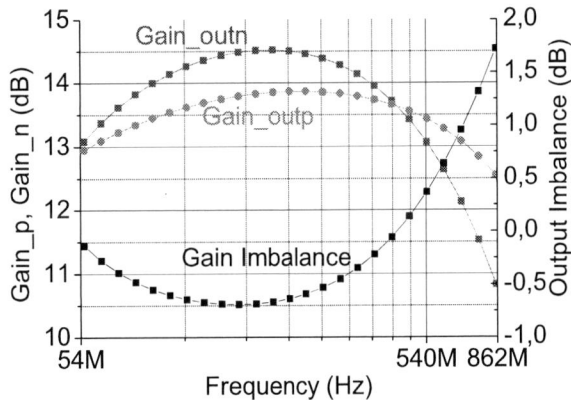

Fig. 5.   Results for gain imbalance in this balun.

published in the literature, for processes from 90 nm to 250 nm.

TABLE II.   Comparison table with recently published wideband LNAs. The * means power gain values. Index 1 are for measured and 2 for simulated designs.

Ref.	BW (GHz)	Gain (dB)	NF (dB)	S_{11} (dB)	IIP2 (dBm)	IIP3 (dBm)	Power (mW)	Area (mm^2)	Tech (nm)	V_{DD} (V)	BALUN
[3][1]	3.4~6.9	10*	4.5	-	-	-1	3.5	0.72	250	1	NO
[4][1]	0.2~3.8	19	2.8~3.4	<-9	-	-4.2	5.7	0.025	130	1	YES
[5][2]	2~10.1	10.2*(max)	3.7(min)	<-9.7	-	-1@6GHz	7.2	1.2	180	1	NO
[6][1]	0.4~0.9	20	2.95	-	-	-	0.36	0.07	180	0.5	NO
[7][2]	3.1~10.6	20*	1.2~2.6	<-10	-	-8	12.6	-	90	0.6	NO
[8][1]	0.1~7	12.6*(max)	5.5~6.5	<-10	-	-9~-6	0.75	0.23	90	0.5	NO
[9][1]	3~8.1	12*	2.8~4.7	<-12.5	-	4.2	4.2	0.87	180	0.6	NO
This work	**0.054~0.86**	**17~20.2**	**5.3~6.2**	**<-10.5**	**13~15.5**	**-14.6~-12.5**	**2**	**0.046**	130	0.45	**YES**

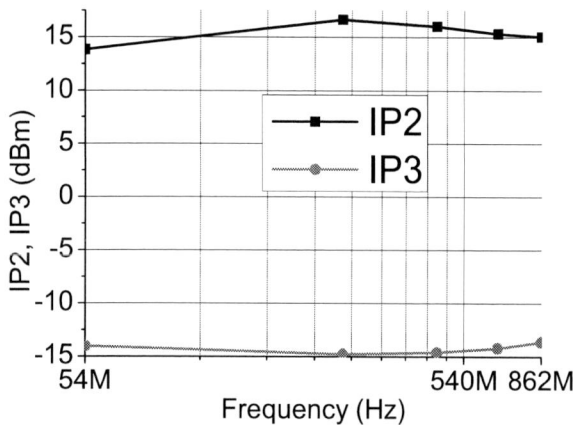

Fig. 6.   Results for the IP2 and IP3 over frequency in the whole band in the right.

Our design is the only LNA in Table II serving as a balun below 1-V power supply and also the only one focusing on IP2 for direct conversion receivers. Our design is also the only one targeting frequencies as low as 54 MHz (where the worst IP2 value is expected). The input impedance matching S_{11} of this design has a worst case of -10.5 dB in the frequency band limits. Despite the fact that high decoupling capacitors are needed in our design, its area in silicon is the 3rd best (including decoupling capacitors). One smaller design had no IP2 reported, and have a worse S_{11}.

IV.   CONCLUSION

An ultra-low voltage wideband LNA topology for balun use with no inductors, for sub-GHz applications was presented. A high gain (>17 dB) in the entire band was achieved under a 450 mV power supply targeting direct conversion receiver applications. This topology was designed in a 130 nm CMOS technology for the 54 MHz - 862 MHz frequency band. Post-layout simulations were done including packaging bondwire inductances and PAD capacitances parasitics. The LNA featured a maximum IP2 of 15.5 dBm, a NF = 5.3-6.2 dB and a voltage gain > 17 dB for the entire band. The circuit consumes only 2 mW of power, which could be used in energy harvesting wideband receivers.

ACKNOWLEDGMENT

The authors gratefully acknowledge the financial support from CAPES and CNPq agencies, the MOSIS for silicon prototyping and NSCAD group for EDA support.

REFERENCES

[1]   F. Bruccoleri, E. Klumperink, and B. Nauta, "Wide-band cmos low-noise amplifier exploiting thermal noise canceling," *Solid-State Circuits, IEEE Journal of*, vol. 39, no. 2, pp. 275–282, Feb 2004.

[2]   B. Razavi, "Cognitive radio design challenges and techniques," *Solid-State Circuits, IEEE Journal of*, vol. 45, no. 8, pp. 1542–1553, Aug 2010.

[3]   A. Balankutty and P. Kinget, "An ultra-low voltage, low-noise, high linearity 900-mhz receiver with digitally calibrated in-band feed-forward interferer cancellation in 65-nm cmos," *Solid-State Circuits, IEEE Journal of*, vol. 46, no. 10, pp. 2268–2283, Oct 2011.

[4]   J. Correia, N. Mancelos, J. Oliveira, and L. Oliveira, "A low-voltage lna and current mode mixer design for energy harvesting sensor node," in *Mixed Design of Integrated Circuits Systems (MIXDES), 2014 Proceedings of the 21st International Conference*, June 2014, pp. 523–528.

[5]   A. L. R. Rosa, L. B. Soares, K. H. Stangherlin, and S. Bampi, "Designing cmos for near-threshold minimum-energy operation and extremely wide v-f scaling," in *Proceedings of the 28th Symposium on Integrated Circuits and Systems Design*, ser. SBCCI '15. New York, NY, USA: ACM, 2015, pp. 1:1–1:6. [Online]. Available: http://doi.acm.org/10.1145/2800986.2801004

[6]   D. Barras, F. Ellinger, H. Jackel, and W. Hirt, "A low supply voltage sige lna for ultra-wideband frontends," *Microwave and Wireless Components Letters, IEEE*, vol. 14, no. 10, pp. 469–471, Oct 2004.

[7]   H. Wang, L. Zhang, and Z. Yu, "A wideband inductorless lna with local feedback and noise cancelling for low-power low-voltage applications," *Circuits and Systems I: Regular Papers, IEEE Transactions on*, vol. 57, no. 8, pp. 1993–2005, Aug 2010.

[8]   B.-Y. Chang and C. Jou, "Design of a 3.1-10.6ghz low-voltage, low-power cmos low-noise amplifier for ultra-wideband receivers," in *Microwave Conference Proceedings, 2005. APMC 2005. Asia-Pacific Conference Proceedings*, vol. 2, Dec 2005, pp. 4 pp.–.

[9]   J. Liu, H. Liao, and R. Huang, "0.5 v ultra-low power wideband lna with forward body bias technique," *Electronics Letters*, vol. 45, no. 6, pp. 289–290, March 2009.

[10]   S. Pandey and J. Singh, "A 0.6 v, low-power and high-gain ultra-wideband low-noise amplifier with forward-body-bias technique for low-voltage operations," *Microwaves, Antennas Propagation, IET*, vol. 9, no. 8, pp. 728–734, 2015.

[11]   M. Parvizi, K. Allidina, and M. El-Gamal, "A sub-mw, ultra-low-voltage, wideband low-noise amplifier design technique," *Very Large Scale Integration (VLSI) Systems, IEEE Transactions on*, vol. 23, no. 6, pp. 1111–1122, June 2015.

978-1-5090-2737-8/16 $31.00 © 2016 IEEE

[12] C.-S. Chang and J.-C. Guo, "Ultra-low voltage and low power uwb cmos lna using forward body biases," in *Radio Frequency Integrated Circuits Symposium (RFIC), 2013 IEEE*, June 2013, pp. 173–176.

[13] A. Costa, H. Klimach, and S. Bampi, "High linearity 24 db gain wideband inductorless balun low-noise amplifier for ieee 802.22 band," *Analog Integrated Circuits and Signal Processing*, vol. 83, no. 2, pp. 187–194, 2015. [Online]. Available: http://dx.doi.org/10.1007/s10470-015-0531-1

[14] B. Razavi, *RF Microeletronics*, 2nd ed., T. S. Rappaport, Ed. Upper Sadle River, NJ-US: Prentice Hall, 2011.

A Hardware Accelerator for the Alignment of Multiple DNA Sequences

Antonyus P. A. Ferreira*, João G. M. Silva,
Jefferson R. L. Anjos, Luiz H. A. Figueiroa,
Edna N. S. Barros and Manoel E. Lima
Informatics Center, Federal University of Pernambuco
Recife, Pernambuco, Brazil
Email: *apaf@cin.ufpe.br

Victor W. C. Medeiros
Department of Statistics and Informatics
Federal Rural University of Pernambuco
Recife, Pernambuco, Brazil

Abstract—The comparison of DNA sequences is a classic problem in molecular biology. Forensic applications uses this comparison for personal identication. For instance, in the USA, the CODIS system has today 14.9 million DNA proles stored on its database. To accelerate the recurrent task to query into similar databases, this work presents a hardware acclerator for the parallel alignment of multiple DNA sequences, aiming for the maximum throughput. The proposed accelerator architecture optimizes the use of hardware resources, the data access strategy and, as a result, memory bandwidth. The experiments were conducted using a DNA database with 8 million individuals, in which, each of them is represented using a set of 15 sequences with a length of 256 nucleotides. In this case study, a prototype of the proposed hardware accelerator using a single Stratix IV FPGA and running at the frequency of 250MHz outperforms by tens of times consolidated software applications like SWIPE and FASTA which are running in a GPP platform, as well as an optimized GPU implementation in OpenCL.

I. INTRODUCTION

The DNA sequencing process extracts the biologic information encapsulated at DNA chain and translate it into a linear sequence of symbols. Four symbols compose the alphabet of the DNA (a = adenine, c = cytosine, t = thymine, g = guanine), they are also called nitrogenous bases or nucleotides.

Since the first DNA sequencing in 1970, the sequencing cost per individual is becoming cheaper. Consequently, the number of species completely sequenced is growing [1]. For instance, in 2002 the cost of sequencing an entire genome was U$ 100 million. In 2015, it reached near U$ 1 thousand. Until 2007, the cost evolved year by year decreasing but, with the diffusion of the NGS (Next Generation Sequencers) [2], this tendency had a more intense falling rate.

The need to process all this information so it can be used in scientific advances has created entirely new problems. For example, databases must store all the information generated and the understanding of molecular sequences requires sophisticated techniques of pattern recognition. The classic example of a problem in molecular biology solvable by an algorithm is sequence comparison: given two sequences representing bio-molecules, one wants to know how similar they are. Moreover, there are other applications such as taxonomy, forensic analysis, personal identification, genetic engineering, and many others. In these applications, this problem must be solved thousands of times every day. Sequence comparison is the most used operation in computational biology, serving as a basis for many other, more complex, manipulations [1].

Applying such concepts, the USA and the UK have forensic DNA databases to help the authorities in crimes investigations. In the USA, the system called CODIS (Combined DNA Index System) was created in 2007 by the FBI to support criminal justice DNA analysis. Presently, CODIS databases count with more than 14 million DNA profiles [3]. In this context, the problem of verifying the identity of criminal consists in comparing the collected material in the crime scene with all the existing profiles stored in the database, in a 1-by-n query style. Every individual is identified by p short DNA sequences. In the UKs DNA Database $p = 15$, and in CODIS $p = 13$. Nevertheless, DNA sequences comparisons implies to perform the pairwise global alignment between them.

Aiming to accelerate this growing number of sequences comparisons, many works have proposed solutions to speed up this task, ranging from digital hardware implementations [4], passing through GPUs (Graphic Processing Units) [5][6], and improvements in conventional GPPs (General Purpose Processors) based implementations. This work presents an optimized hardware architecture targeted at FPGA devices and it explores the temporal and spatial parallelism through a full-pipelined data path. For this purpose, the database is organized according to the data dependency to reduce the memory accesses, allowing internal memory reuse. The reported results are better when compared with well-established software like FASTA[7] and SWIPE[8] running in dual GPP and also better than own GPU implementation.

This paper is organized as follows: Section II presents the concepts concerning to sequence alignment, after that in Section III it is exposed the related works. Section IV presents the main strategies used for the hardware architecture. In Section V, the experiments are detailed and the results are discussed. Section VI presents conclusions and future works.

II. COMPARING DNA CHAINS

Given two chains $A = (a_1, a_2, ..., a_n)$ and $B = (b_1, b_2, ..., b_m)$ with length n and m, and alphabet $\Sigma = \{a, c, g, t\}$, a pairwise alignment between the two sequences is the operation that, conceptually, consists of the matching of similar regions of A and B. Then, the resultant alignment between A and B is the tuple $R = < A', B' >$, on which $A' = (a'_1, a'_2, ..., a'_l)$, $B' = (b'_1, b'_2, ..., b'_l)$, and l is between $max(m, n) \leq l \leq (n + m)$. Where R has the alphabet $\Sigma' = \{a, c, g, t\} \cup \{-\}$ that corresponds to the nucleotides set including the blank symbol $'-'$. This way, the alignment of two characters a_i and b_i may result one of the following outcomes showed in **Equation 1** .

$$\mathbf{R} < a_i, b_i >= \begin{cases} a\ match, & if\ a_i = b_i \\ a\ mismatch, & if\ a_i \neq b_i \\ an\ insertion, & if\ a_i = '-' \\ a\ deletion, & if\ b_i = '-' \end{cases} \quad (1)$$

The algorithm that solves this problem uses a computational technique known as dynamic programming. Dynamic programming, as I. Parberry says: is a fancy name for recursion with a table. Instead of solving sub-problems recursively, solve them sequentially and store partial results in a table [9].

These operations (match, mismatch, insertion and deletion) have evolutionary meanings: a match is a preservation of the genetic information, and a mismatch represents a swap in a single nucleotide without to affect the chain length. The deletion and insertion events are less likely because they insert gaps in the DNA chain.

The solution to the global alignment problem was proposed by Needleman [6]. The Needleman-Wunsch (NW) algorithm reduces the complexity of the problem from exponential to quadratic $O(nm)$, for two sequences of length n and m. The main bottleneck to speed-up this algorithm is its intrinsic data dependency. **Equation 2** shows iterative rule to build the matrix $M_{n \times m}$. It uses the nucleotides positions, of each sequence, as the matrix indexes (see **Fig. 1**).

$$\mathbf{M}(i, j) = max \begin{cases} \mathbf{M}(i - 1, j - 1) + \mathbf{S}(a_i, b_i) \\ \mathbf{M}(i - 1, j) + g_1 \\ \mathbf{M}(i, j - 1) + g_2 \end{cases} \quad (2)$$

$$\mathbf{S}(a_i, b_i) = \begin{cases} \delta, if\ a_i = b_i \\ -\delta, if\ a_i \neq b_i \end{cases} \quad (3)$$

In this equation, $M(i, j)$ represents the $n \times m$ score matrix; $S(a_i, b_i)$ (**Equation 3**) is the substitution score for the nucleotides with positions i and j; g_1 and g_2 are called the gap penalty values; and δ is the match/mismatch cost. Where g are normally negative cost associated with the deletions and insertions of nucleotides.

The Needleman-Wunsch algorithm follows three steps:

1) Initialize the matrices.
2) Compute the scores and the traceback matrices
3) Obtain the alignment from the traceback matrix

In Step 1 the borders of the matrix are filled up following the rules: $M(0, 0) = 0$, $M(i, 0) = i * g_1$, $M(0, j) = j * g_2$. It is necessary to compute the neighbors cells (the left, the upper left, and the upper cells) before the computing of each cell. An example of these parameters could define, for instance, $g_1 = g_2 = -1$ and $\delta = 1$ and the Step 1 results in the filled

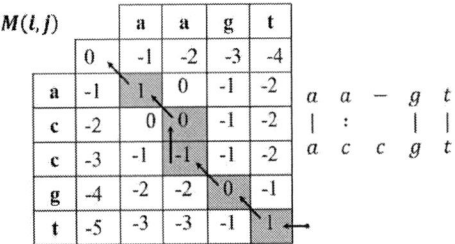

Fig. 1: Alignment result for query **accgt** and sequence **aagt**

borders shown in **Fig. 1**. The sequence in the vertical direction is called the query sequence $accgt$ and the second the original sequence $aagt$.

Starting the Step 2 for the position $\mathbf{M}(1, 1)$, a match happens ($\mathbf{R} < a, a >$) so $\mathbf{S}(a, a) = 1$ and $\mathbf{M}(1, 1) = max(0 + 1, -1 - 1, -1 - 1) = max(1, -2, -2) = 1$. The iteration of the Step 2 of the algorithm continues until the matrix shown in **Fig. 1** is filled.

At this point, the Step 3 can start. The highlighted positions of **Fig. 1** depict the path chosen in the traceback step. It starts from $M(5, 4)$ (the bottom-right corner), and it goes cell-by-cell, selecting the local maximum score until to get at position $M(0, 0)$ (upper-left corner).

The **Fig. 1** (right) also shows the optimal global alignment obtained. First, a match $(a-a)$ occurs, after a mismatch $(a-c)$, then an insertion of (c) in the original sequence, and at the end, two matches $(gt-gt)$. Once obtained the alignment, a scoring strategy, to measure the similarity between the two sequences, may now be chosen. An integer weight is defined for each one of the four events possible (for instance $match = 3$, $mismatch = -1$, $gap\ open = -5$ and $gap\ extension = -2$). A gap open is the first occurrence of a gap (insertion or a deletion) in a row. Then the gap extension is the second occurrence onward of a gap, after a gap open. So the score computation for the resultant alignment of **Fig. 1** results in $3 - 1 - 5 + 3 + 3 = 3$.

III. RELATED WORKS

The utilization of FPGAs in many application areas is growning and being consolidated [10]. The FPGAs are broadly used as a rapid way of prototyping digital systems, which aim ASICs (Application Specific Integrated Circuits) designs.

Nonetheless, in many other cases, the FPGA implementation may achieve substantial performance gains, in comparison with the corresponding application in a GPP [10]. In such situations, the FPGA is used as a target platform. In applications like image processing and 2D and 3D matrix processing, the GPUs competes in a higher league than the FPGAs because of its customized architecture for vector processing, optimized memory hierarchy and large memory bandwidth[1].

In the real world, even in applications in which GPUs provide better performance than FPGAs, one may also be looking for favorable niches to the FPGAs. A recurrently presented point is a metric performance per watt that may

---

[1] http://www.nvidia.com/gtx-700-graphics-cards/gtx-780ti/

benefit the FPGAs, considering their lower power consumption when compared with GPUs. In fact, the power requirements have been an important constraint on the performance growing of the GPUs that limit their frequencies around 1 GHz, while the FPGAs maximum frequencies are around 300 - 400 MHz. In the long term, the performance/watt metric may also pay off the higher investment and project costs of an FPGA solution.

Isa et. al. [11] propose an area efficient, architecture to process the alignment of biologic sequences. The proposed architecture was prototyped in an FPGA platform. The implemented processing elements (PEs) follow the systolic array configuration. The results showed a minimum speedup of 15x, over the FASTA[7] software, running at 195 MHZ.

Benkrid et al. [12] bring a comparative analysis of FPGAs performance in a scenario of the Smith-Waterman application. The FPGAs are compared with GPUs, IBMs CELL BE, taking as base the GPP. The comparison criteria include processing speed, energy consumption, overall cost, and project time.

Again, the PEs distribution was the systolic array configuration. The authors do not compute the traceback step inside the FPGA, and this phase must be computed at the host. The internal hardware resources, apparently, limit the number of PEs instantiated. For example, they could instantiate 500 PEs.

The speed-up achieved with the FPGA implementation has the best performance, with two orders of magnitude (228x at 80 MHz) higher than the GPP application followed by The Cell BE (45x) and GPU (14x) implementations. The authors measure the performance in GCUPS (Giga Cell Updates per Second) per dollar spent. Again, the FPGA solution overcame the GPU 1.9, and the Cell BE 2.3 solutions. The results showed the benefit of an FPGA solution, counterbalancing the larger development time in an FPGA design.

Sebastiao [13] proposed a ASIC for the Smith-Waterman algorithm with 512 PEs running at 250MHz that achieved 278x over conventional software application.

Differently of the works in the literature, this work proposes to optimize the alignment of hundreds of relatively small sequences in parallel (instead of aligning one large sequence) to address the forensic problem of personal identification.

IV. PROPOSED HARDWARE ARCHITECTURE

The proposed FPGA implementation used a Gidel's Stratix IV FPGA board[2] that is composed of four FPGAs. Each one possesses three DDR2 memories attached to it (two SODIMM 4GB modules, plus an onboard 512MB module). The PCIe interface is used as Hardware/Software interface between the PC host application and the FPGA core of the proposed architecture. The software does the partition of the sequences database between the two larger memories and sends the data to them. The smaller memory module is used to store the query sequences and the processing results to be read later by the software application.

The FPGA processing starts after the database transfer into the DDR2 memories. After that, the hardware processing core waits for software requisitions to query the database.

[2]Gidel PROCStar IV board - http://www.gidel.com/PROCStar IV.htm

It was possible to fit a DNA database with 8 million individuals, each one with 15 sequences yielding a total of 120 million sequences of 256 nucleotides. This amount of information could be stored in the 2x4GB available memory due to the 2-bits nucleotides encoding.

A. External Memory Organization

As mentioned before, the platform used has two DDR2 memories. These external memories are accessed from the FPGA design through the Gidel's PROCMultiport module that manages the memory transfers making them available to the user design.

The bank of DNA sequences, with m length, was organized in the DDR2 memories as describes **Fig. 2a**. This scheme allows data access for multiple processing elements (PEs) per alignment, multiple alignments per individual, and multiple individuals processed in parallel.

In **Fig. 2a**, each one of the sub-matrices $\{\{S_{1,1}, S_{1,2}, \cdots, S_{1,k}\}; \{S_{2,1}, S_{2,2}, \cdots, S_{2,k}\}; \cdots \{S_{n,1}, S_{n,2}, \cdots, S_{n,k}\}\}$ represents a single m length DNA sequence, divided into $n = m/k$ slices to be computed by k PEs. The number of p sequences compound each individual in the database, then processing i individuals in parallel, inside the FPGA, needs to read a word of $k * p * i$ nucleotides of 2 bits each. The entire database with b individuals has $(b/i) * n$ lines of $k * p * i * 2$ bits. Using this configuration, the database can be split vertically and stored in several memories.

Similarly to the database distribution, the query memory organization is configured to make the data accessible to the FPGA computation. **Fig. 2b** shows the p sequences, with m size, of the individual query concatenated vertically.

B. Computing the Scores and the Directions

The scores computation is the step 2 of NW algorithm, where each cell is computed by a PE. The PEs are organized in a systolic array configuration[14], and they are phase-shifted from each other by one clock cycle, as shown **Fig. 3**. The query is disposed vertically and the bank sequence in the horizontal. This way only one nucleotide of the query is read per cycle and carried out through the PEs, from PE_1 to PE_4. Once per slice, the bank sequence is read one nucleotide per PE in parallel, so per alignment are read $k * 2$bits, with k PEs.

The score matrix computation is divided into processing slices whose width is defined by the number of PEs. **Fig. 3** represents four PEs that calculate, each one, a single column independently. Herein it is used a single one dimension score vector with the length of the sequence instead of the two dimensions score matrix. This score vector is implemented as a FIFO (First in First Out) module inside the FPGA. The score FIFO stores only the last PE output (PE_4 in **Fig. 3**) to allow the computation of the next slice.

Instead of storing the score matrix, it is used a directions matrix that indicates which event produced the current score. An example of a direction encoding is shown in **Equation 4**, so each element of directions matrix is encoded in only 2 bits.

978-1-5090-2737-8/16 $31.00 © 2016 IEEE

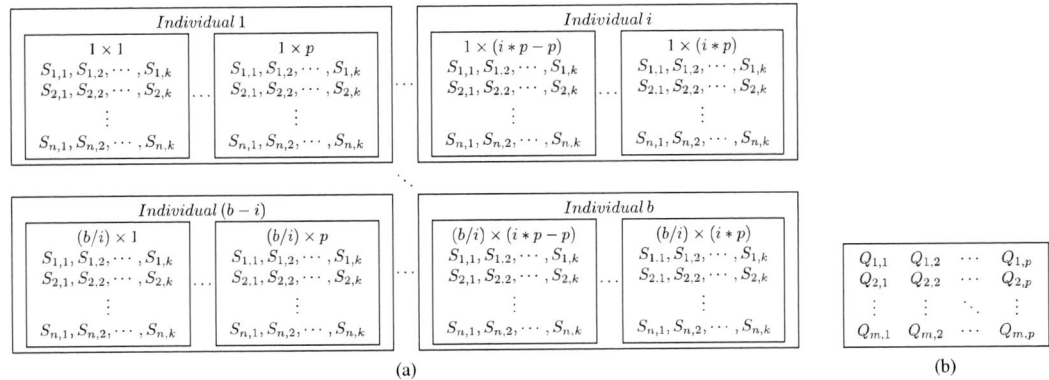

(a)

(b)

Fig. 2: (a) Memory organization for a bank with m length sequences, k processing elements, p sequences per individual, $n = m/k$ slices, i individuals processed in parallel in the FPGA, and b individuals in the database. (b) Memory organization for a query with m length and p sequences per individual.

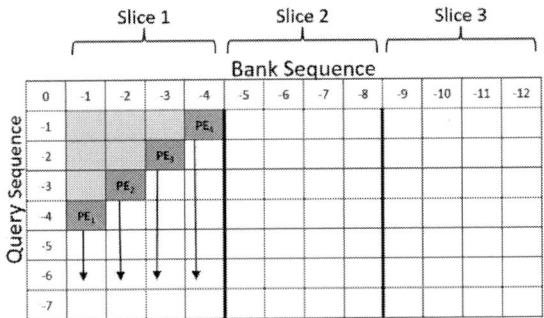

Fig. 3: Computation of the scores and the PEs arrangement

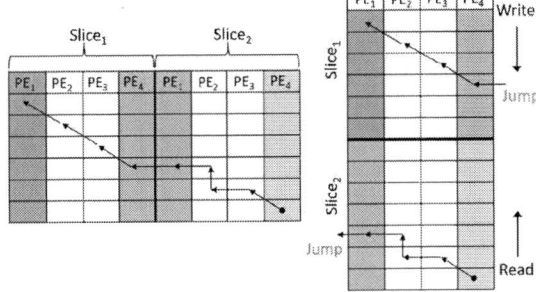

Fig. 4: Directions Buffer

$$Directions(a_i, b_i) = \begin{cases} 0, & a_i = b_i, a\ match \\ 1, & a_i \neq b_i, a\ mismatch \\ 2, & b_i = -, an\ insertion \\ 3, & a_i = -, a\ deletion \end{cases} \quad (4)$$

C. The Tracing Back

After the completion of the forward phase, the directions matrix is complete and the trace back operation may start. The directions buffer stores the directions in an internal FPGA RAM memory. The slices are disposed vertically as shown in **Fig. 4**. The trace back process starts at the last written address of the directions buffer, and reads must start in bottom-up direction. **Fig. 4** also shows this trace back reads strategy from the directions buffer. When the left side of the memory is reached (PE_1), it is necessary to jump to the correspondent position in the previous slice to continue the trace back process. The process ends at the origin $M(0,0)$ of the matrix. It simulates, inside the FPGA, what would be easily implemented in software navigating in a 2D matrix.

D. Hardware Modules

The top module of the hardware architecture, shown in **Fig. 5a**, is called the Top_arch. It includes the control module

that accesses the multiple external memories. It instantiates i Arch_align_groups modules which are related to individuals in the bank. Each Arch_align_group has p Arch_align modules, that perform the alignment itself (computation of scores and trace back), query and score FIFOs.

Following the hierarchy of modules, the Arch_align module (**Fig. 5b**) makes the computation of the scores in the Forward_align module and the trace back in the Backward_align module. The Forward_align module instantiates k PEs and its directionals are stored in the Directions_buffer inside the Backward_align module. The result of the Backward_align module is the score number of the alignment which measures the similarity between the DNA chains.

The **Fig. 5c** presents the internal PE architecture. The core of the computation of the scores, the PE module, initiates the processing making a comparison to define the function $S(a_i, b_i)$, then two sums of the diagonal value with S cost and left value with G costs. After this, it is conducted a comparison to obtain the maximum result between them. Then another turn of sum and comparisons are performed, this time with the previously computed value. At the end, one cell is computed according to **Equation 2**. Notice that the left score is received from the left PE and the output score becomes the left score to the next PE. The diagonal score is the left score delayed,

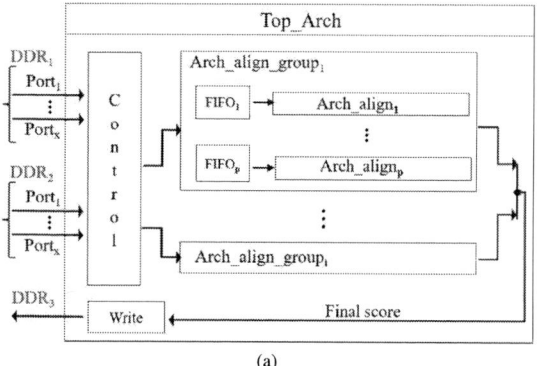

(a)

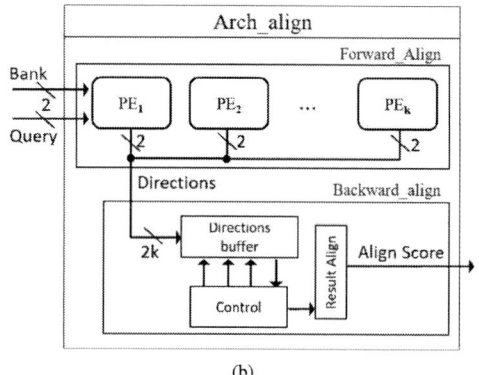

(b)

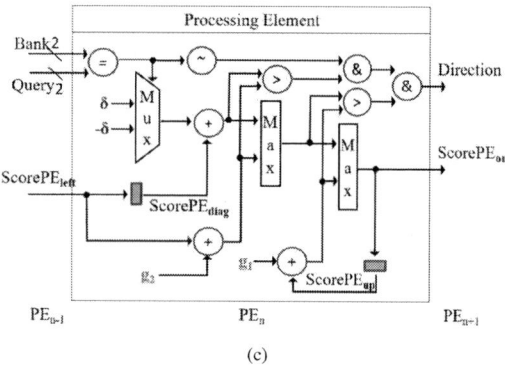

(c)

Fig. 5: (a) Top_arch Module. (b) Arch_align Module. (c) PE Module internal architecture.

and the up score is the previously computed result.

E. Performance Considerations

The external memory access organization proposed in this work and the query sequences storage in an internal FIFO optimized the access for data reuse. This way, the query data is reused to compare with all individuals in the database. The second point is that a single line of the database is needed to be read for all alignments in parallel, during the time of an entire slice processing. A slice corresponds to the computation, in an alignment, of the scores with the PEs modules for the all m nucleotides of a query (see **Fig. 3**). For instance, in a

single alignment, with k PEs and a query with m size, only k nucleotides of the memory bank are read within m clock cycles. So the memory has m clock cycles after a data request before a new read. This fact makes the alignment a processing bound application, instead of a memory bound application.

Another way to avoid external memory access is the use of the directions buffer, using the compressed 2 bits representation of the directions, allows the internal storage of this matrix. Thereby the performance limiter becomes the internal operation frequency and the FPGA resources.

For instance, a single alignment ($k = 16$, $m = 256$, and $f_{max} = 250MHz$) demands the memory bandwidth of $(16PEs * 2bits * 250MHz)/(8bits * 256nucleotides) = 3.9MB/s$. A DDR2 800 (PC2-6400) RAM memory has ideal bandwidth of $6400MB/s$ and real bandwidth of $70\%*6400 = 4480MB/s$. So, with one memory, it is possible to feed $1,148$ alignments. In opposition, writing the directions in the DDR memory, the bandwidth consumed is: $3.9MB/s + (16PEs * 2bits * 250MHz)/(8bits) = 1003.9MB/s$.

V. RESULTS

The experiments were conducted using a random generated DNA bank with 8 million individuals. Each of these individuals is represented by 15 DNA sequences with the length of 256 nucleotides. This accounts for a total of 120 million sequences. All the presented implementations consist of the three steps of the NW algorithm (initialization, computation of the scores, and the traceback phase). Additionally, the experiments only compute processing times. The initial FPGA data transfer penalty is neglected since the 8 million individuals database transfer is made only once for an unlimited number of queries.

The proposed FPGA architecture was implemented using SystemVerilog HDL language and prototyped in a Gidel's Stratix IV FPGA board. Only one FPGA, out of four, was used. The PC host application partitions and sends the database to the board memories via the PCIe bus. After the transfer,the processing begins, so this database transfers are made offline.

In the experiments, the platform was instantiated for several number of individuals (i), $p = 15$ sequences per individual, $k = 16$ PEs, and $m = 256$ nucleotides. **Fig. 6** shows the resources utilization for $i = 1, ..., 7$ including the modules of the platform to access the external memories. The maximum quantity of individuals that could fit the FPGA was $i = 7$, and the frequency of operation was $250MHz$. Then the performance measures were taken using the size of $i = 7$.

To compare FPGA performance with other architectures, it was also implemented an optimized GPU version of the algorithm in OpenCL. This implementation runs on a GPU AMD Radeon HD7770 with 1 GHz core frequency, 1 GB GDDR5 memory with 72 GB/s of bandwidth and 640 stream processors. We also compared the results with two consolidated third-party software SWIPE [8] and FASTA [7] running in a high-performance Supermicro server with dual Intel Xeon E5645 (24 threads) 2.4GHz, and with 48GB of DDR3 RAM.

The GCUPS metric (Giga Cells Updates per Second) is herein used to compare the performance between the

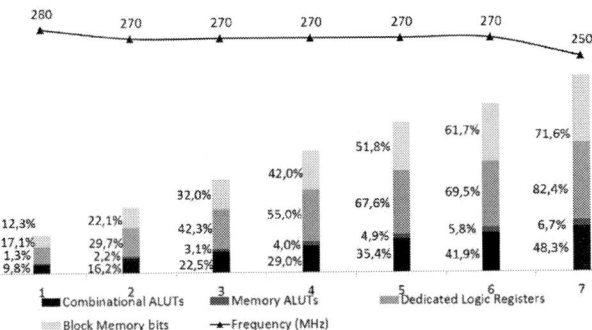

Fig. 6: Resources utilization and frequency of operation with the increasing of the number of individuals

TABLE I: Processing Times, GCUPS, and the Speed-ups

Implementation	Processing Time	CGUPS	Speed-up
FASTA[7]	63.41 min	2.07	1x
SWIPE[8]	40.20 min	3.26	1.58x
GPU	26.79 min	4.89	2.37x
FPGA	**0.557 min**	**235.46**	**113.90x**

implementations. GCUPS is computed using the equation: $GCUPS = (b * m^2 * p)/(PT * 10^9)$. It corresponds to the celerity in which all individuals in the database are aligned with the query. Where, each individual has a time complexity of $O(m^2)$ to compute the scores, each individual has p sequences that represent him, and b stands for the number of individuals in the database and PT for the processing time.

The equation covers only the computation of the scores but the measured time, in all experiments, also includes the trace back stage. It could be seen in **Table I**, that the FPGA implementation in a single FPGA achieved the highest performance of 235.46 GCUPS, which represents a speed-up of 113.90x over the FASTA software and 48.13x over the GPU. It is important to remark that the PCI-e data transfers did not penalize the GPU performance. Although the database size was bigger than the GPU memory, only the processing time was computed. The GPU implementation runs an entire alignment per thread in an independent way. This way, it avoids memory fences and synchronization points that may degrade the overall speed-up. The GPU performance was limited by the number of the stream processors and its frequency of operation.

In comparison with related works the presented work demonstrates higher performance as displays **Table II**, that also lists the frequency of operation, the performance (GCUPS metric), and, for comparison purposes, the device where each accelerator was implemented. The comparison shows that the proposed work was one order of magnitude faster than other works in FPGA and in GPU.

VI. CONCLUSIONS AND FUTURE WORKS

This work presents an optimized FPGA-based architecture to solve the problem of searching in a large DNA database

TABLE II: Performance Comparison Against other Implementations

Work	Device (IC Technology)	F_{max}	GCUPS
Benkrid [12]	Virtex-4 160 (90nm)	80MHz	19.4
ISA [11]	Virtex-5 110 (65nm)	200MHz	39.0 (Peak)
Liu [6]	GPU GTX680 (28nm)	1GHz	80.0
Proposed	**Stratix IV 530 (40nm)**	**250MHz**	**235.46**

as happens in the forensic databases. Each query, to this database, consists in a global alignment one against all the other sequences. Its performance was compared with GPP and GPU platforms, and also with the software FASTA and SWIPE as benchmarks. The FPGA implementation achieved better performance demonstrated using GCUPS metric.

The main contribution of this work is the design of an architecture to align many sequences in parallel inside the FPGA to best fit the problem.

As future works, all the current implementation may be extended to be applied in other areas such as diseases diagnosis. It is also planned to exploit the OpenCL for FPGAs.

ACKNOWLEDGMENT

The authors would like to thank the support of Brazilian National Counsel of Technical and Scientific Development - CNPQ and the Petrobras - PADMEC group.

REFERENCES

[1] J. Setubal and J. Meidanis, *Introduction to computational molecular biology*. PWS Publishing, 1997.
[2] J. Xu, *Next Generation Sequencing: Current Technologies and Applications*. Caister Academic Press, 2014.
[3] F. , "CODIS - NDIS Statistics." [Online]. Available: https://www.fbi.gov/about-us/lab/biometric-analysis/codis/ndis-statistics
[4] S. Aluru and N. Jammula, "A Review of Hardware Acceleration for Computational Genomics," *IEEE Design Test*, vol. 31, no. 1, pp. 19–30, Feb. 2014.
[5] I. Savran, Y. Gao, and J. D. Bakos, "Large-Scale Pairwise Sequence Alignments on a Large-Scale GPU Cluster," *IEEE Design Test*, vol. 31, no. 1, pp. 51–61, Feb. 2014.
[6] Y. Liu, A. Wirawan, and B. Schmidt, "CUDASW++ 3.0: accelerating Smith-Waterman protein database search by coupling CPU and GPU SIMD instructions," *BMC Bioinformatics*, vol. 14, no. 1, pp. 1–10, 2013.
[7] W. R. Pearson, "Effective protein sequence comparison," in *Methods in Enzymology*. Academic Press, 1996, vol. Volume 266, pp. 227–258.
[8] T. Rognes, "Faster Smith-Waterman database searches with inter-sequence SIMD parallelisation," *BMC Bioinformatics*, vol. 12, no. 1, pp. 1–11, 2011.
[9] I. Parberry, *Problems on Algorithms*, 1st ed. Prentice Hall, 1995.
[10] M. B. Gokhale and P. S. Graham, *Reconfigurable Computing: Accelerating Computation with Field-Programmable Gate Arrays*, 1st ed. Springer Publishing Company, Incorporated, 2010.
[11] M. N. Isa, S. A. Z. Murad, R. C. Ismail, M. I. Ahmad, A. B. Jambek, and M. K. Md Kamil, "An efficient processing element architecture for pairwise sequence alignment," in *Electronic Design (ICED), 2014 2nd International Conference on*. IEEE, 2014, pp. 461–464.
[12] K. Benkrid, A. Akoglu, C. Ling, Y. Song, Y. Liu, and X. Tian, "High Performance Biological Pairwise Sequence Alignment: FPGA versus GPU versus Cell BE versus GPP," *International Journal of Reconfigurable Computing*, vol. 2012, pp. 1–15, 2012.
[13] N. Sebastiao, N. Roma, and P. Flores, "Configurable and scalable class of high performance hardware accelerators for simultaneous dna sequence alignment," *Concurrency and Computation: Practice and Experience*, 2013.
[14] T. Oliver, B. Schmidt, and D. Maskell, "Hyper customized processors for bio- sequence database scanning on fpgas," *Proc. 13th Int. Symp. Field- Programmable Gate Arrays*, pp. 229–237, 2005.

Inserting permanent fault input dependence on PTM to improve robustness evaluation

Rafael B. Schivittz[1], Rafaél Fritz[1], Denis T. Franco[1,2], Lirida Naviner[3], Cristina Meinhardt[1], Paulo F. Butzen[1]

[1]Programa de Pós-Graduação em Engenharia de Computação – PPGComp – Universidade Federal do Rio Grande – FURG
[2]Universidade Federal de Pelotas - UFPEL
[3]Institut TELECOM, Télécom-ParisTech, LTCI-CNRS, COMELEC Paris, France

Abstract— **Many of the nanometer CMOS challenges are seriously compromising the gains attained with technology scaling, mainly impacting the yield and the circuit reliability. To cope with these problems, new design methodologies are necessary to improve the robustness of the circuits. Given the overheads associated with the traditional fault-tolerant approaches, alternative solutions, based on partial fault tolerance and fault avoidance, are also being considered as possible solutions to the reliability problem. These approaches are based on the application of fault tolerance to a restricted part of the circuits or hardening of individual cells, allowing reliability improvements and limiting the associated overheads. In this context, a fast and accurate evaluation of circuit's reliability is fundamental, to allow a reliability-aware automated design flow, where the synthesis tool could rapidly cycle through several circuit configurations to assess the best option. This work presents a methodology to calculate circuit reliability, based on the Probabilistic Transfer Matrix (PTM) method, and using a probabilistic model for stuck-on faults that considers a fault probability for each input vector. The work shows that considering the same error probability for all input vectors underestimates the input influence on the overall circuit reliability. The proposed model of gate reliability associated with the PTM method can provide results that are more accurate in terms of circuit reliability. Results obtained with the proposed approach show a difference up to 15% when compared with the traditional application of the PTM method with equal input vector probabilities, when applied to a set of combinational circuits.**

Keywords— CMOS, Stuck-On faults, PTM, EDA.

I. Introduction

Device scaling has been the main strategy adopted by the semiconductor industry to increase the performance of integrated circuits (ICs). The continuous scaling has emphasized several aspects neglected in earlier technologies nodes. The circuit reliability has been pointed out as one of the major challenges in deep sub-micron CMOS circuits [1].

In nanoscale designs, many factors associated with technology scaling, like manufacturing precision limitations, supply voltage reduction, higher operation frequency and power dissipation have influenced the need for reliability. These factors increase the circuit fault probability, mainly permanent faults generated during the fabrication process steps. One of these faults is the Stuck-On Fault (SOnF). SOnF is a permanent fault that occurs in transistors. The transistor

with a Stuck-On fault will drive current, independently of the signal applied in the gate terminal.

Many techniques are proposed to mitigate the problems generated by the continuous scaling. These techniques are usually based on redundancy in time, hardware and/or information [2][3]. However, these techniques present penalties in original circuit characteristics. For example, hardware redundancy techniques increase the circuit area. The Triple Modular Redundancy (TMR) is one of the most adopted technique, producing a circuit robust to any fault in one of the three modules and have more than 3X of penalty in area. Thus, TMR adoption should be carefully explored.

The main challenge is to identify the moment when advantages in reliability are bigger than disadvantages in other circuit characteristics. In some cases, these techniques reduce the scaling gains due to the higher complexity of reliability approaches. To deal with this balance, probabilistic methods are one of the better options [4]. At circuit level, this method can be modeled by a matrix to reproduce the logic gates behavior. One of the most adopted technique is the probabilistic transfer matrices (PTM) that represents the expected circuit output for each input combination. In PTM, a matrix M is composed by i rows and j columns, where the $(i, j)th$ input represents the probability of an occurrence of the output j given the input i, denoted by $p(i \mid j)$.

Although most of PTM implementations are presented considering the same probability q for all input vectors of a gate, there is a different fault probability for each input combination and also a different fault probability for a different kind of fault. A simple q value can underestimate the fault probability in PTM implementation. Assuming a same value q for all input combination can mask the influence of the input in different kinds of faults. In this context, the main contribution of this work is to show the input influence on robustness of Stuck-On faults and also to present a methodology to model SOnF over a PTM making the robustness evaluation of a circuit more accurate.

This work is organized as follows. Section II presents a background about SOnF and the probabilistic transfer matrices. In Section III is presented the proposed PTM model considering SOnF and input dependence. In Section IV the results of this work are presented, including a library of PTM modified for gates and the application of this library to compare robustness of a circuit with the traditional PTM. Finally, in section V is presented the final remarks.

II. BACKGROUND

This section presents an introduction about the Stuck-On faults behavior. It is also shown how the probabilistic transfer matrices represent the expected output of the gates and the reliability computation of a circuit.

A. Stuck-On faults

In a circuit, problems such as manufacturing issues, aging effects or even single events can make the transistors remain permanently in the on state, which characterizes the Stuck-On faults. The transistor that remains permanently stuck-on can affects the expected logic gate behavior, competing with its complementary transistors to control the output in some input combinations.

This work focuses on Stuck-On faults effects on combinational circuits. To exemplify the SOnF behavior, Fig. 1 shows the circuit of an Inverter logic gate and its outputs in the correct operation (OUT) and the output considering that a SOnF occurred in the transistor N_1 in the pull-down network (OUT'). The effects of this fault are visible when the input vector A=0 is applied, creating a short circuit between VDD and GND and causing an error state in the output with SOnF (OUT'). In other words, when the input is A=1, the circuit behavior with SOnF is equal to the circuit expected, because there is no fault in the pull up network, i.e., the transistor with SOnF is the transistor responsible to make a path between *GND* and *OUT*.

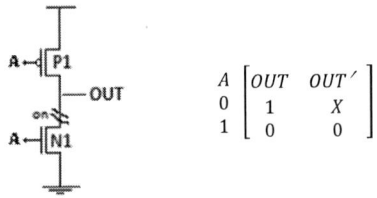

Fig. 1. Inverter with SOnF in transistor N_1 and its truth table

B. Probabilistic Transfer Matrices

In an error-free operation system, the function of a combinational logic circuit can be represented by a truth table, which is a deterministic mapping of input values to output values. Another way to represent this function is using the ITM (Ideal Transfer Matrix). In ITM, row indices represent all input combinations, and column indices represent possible output values. The ITM models the function inputs to outputs in a system with no error. The value in a cell of the matrix is 1, that represents 100% of chance of the input to be in that state, or 0, that represents no chances of the output to be in that state.

The probabilistic transfer matrix (PTM) method presents a matrix that lists the inputs and outputs of a logic gate considering its topology and individual reliability of its gates. In the presence of soft errors, permanent faults or manufacturing defects sometimes an input can produce a wrong output value. If we know how often this is likely to happen, we can model this behavior using a PTM (Probabilistic Transfer Matrix). In PTM, as well as in ITM, row indices represent all input combinations, and column indices represent possible output values.

For example, let's consider the possible probabilistic transfer matrices shown in Fig. 2. In these examples, the gates PTM provide the correct output value with probability q, where q represents the reliability factor of the gate. The error probability of the gate is represented by $p=1-q$. In general, it is possible to use any fixed probability distribution for the rows of the matrices depending of the reliability factor of the technology.

The difference between ITM and PTM is that ITM represents a gate with error-free and PTM represents the probability of error in the gate. In other words, the ITM is a PTM matrix with reliability factor q equal to 1.

$$ITM_{INV} = \begin{bmatrix} 0 & 1 \\ 1 & 0 \end{bmatrix} \quad ITM_{NAND2} = \begin{bmatrix} 0 & 1 \\ 0 & 1 \\ 0 & 1 \\ 1 & 0 \end{bmatrix} \quad ITM_{NOR2} = \begin{bmatrix} 0 & 1 \\ 1 & 0 \\ 1 & 0 \\ 1 & 0 \end{bmatrix}$$

$$PTM_{INV} = \begin{bmatrix} p & q \\ q & p \end{bmatrix} \quad PTM_{NAND2} = \begin{bmatrix} p & q \\ p & q \\ p & q \\ q & p \end{bmatrix} \quad PTM_{NOR2} = \begin{bmatrix} p & q \\ q & p \\ q & p \\ q & p \end{bmatrix}$$

Fig. 2. ITMs and PTMs for three different logic gates

The generation of the global circuit PTM involves the combination of gates and interconnections. With the information about reliability of each gate, it is possible to compute the reliability of the whole circuit. The global PTM is obtained dividing the circuit in levels and combining the PTM of each gate presented in the circuit. To compute the global PTM of a circuit, two rules have to be respected:

- The multiplication between elements of the same level occurs using tensor product or (Kronecker product ⊗).

- The multiplication between levels occurs multiplying the PTMs of each level.

The circuit shown in Fig. 3 is separated into two levels. The first level (L1) is formed by a *NAND2* and an *INV*. To start the computation is necessary to know the PTM of the logic gates adopted in the circuit. The next step is to compute the PTM for each logic level in the circuit. The PTM of logic level 1 (L1), called PTM_{L1}, is calculated with the tensor product between PTM_{NAND} and PTM_{INV} as shown in Eq. 1. The PTM of the logic level 2 (L2), called PTM_{L2} is the PTM of the NOR gate, as shown in Eq. 2.

$$ITM_{L1} = ITM_{NAND} \otimes ITM_{INV}$$
$$PTM_{L1} = PTM_{NAND} \otimes PTM_{INV}$$
(1)

$$ITM_{L2} = ITM_{NOR}$$
$$PTM_{L2} = PTM_{NOR}$$
(2)

The PTMs of all levels are multiplied to determine the circuit PTM. In the example, the circuit PTM is determined by the multiplication of PTM_{L1} and PTM_{L2} as shown in Eq. 3. To compute the gate reliability is necessary to multiply the ITM_{CIR} and the PTM_{CIR} considering the input signal probabilities. A specific condition is observed when subsequent levels have different number/order of fanins and fanouts. In this case an ITM of interconnection is used to map the behavior [5].

$$ITM_{CIR} = ITM_{L1} * ITM_{L2}$$

$$PTM_{CIR} = PTM_{L1} * PTM_{L2} \tag{3}$$

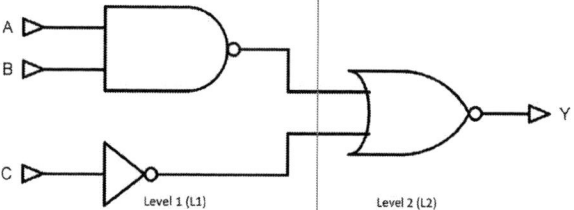

Fig. 3. Example of circuit level PTM computation.

III. PROPOSED METHOD TO INSERT STUCK-ON FAULT INPUT DEPENDENCE ON PTM

This section presents the method proposed in this paper to evaluate SOnF and input vector dependence in different transistor arrangements. It is also shown that different input vectors and transistor arrangement have different probability to produce a correct output.

The traditional PTM models use the reliability of the gate, as the reliability of the output to be correct for each input combination [6][7][8][9]. However, the use of PTM can be modified to explore the circuit and the different possibility of an error taking into account the input vectors [10][11]. In this context, this work proposes a new methodology to insert the input influence on the traditional robustness evaluation with PTM models when considering Stuck-On fault effect on combinational circuits. The new method can be applied in all traditional CMOS gates composed by single stages. Gates that present multiple stages have to be evaluated using the traditional circuit PTM computation method described in Section II.

The main goal of this work is to model the circuit reliability considering the SOnF probability of each input combination. The methodology starts defining a Stuck-On model based on the transistor arrangement and the input dependence. Next Subsection details the modelling of Stuck-On Faults. Then, the input dependence is transposed to the PTM, implying in a new PTM model that take into account the input dependence for SOnF. Finally, this Section presents a case study showing the difference of the traditional PTM methodology and the method proposed when applied to a *NAND2* gate.

A. Modelling Stuck-On Faults

Let's consider a CMOS gate, where C is the circuit description composed by t transistors and I inputs. The inputs generate 2^I combinations and each combination is represented as X_i. The set of all transistors in the netlist is named T. The circuit is described by two graphs, named *G1* and *G0*. The graph *G1* is the one that sets the output to '1' (pull-up network) and the graph in *G0* sets the output to '0' (pull-down network).

In the ITM, an input combination X_i has its normal output already defined, as shown in the examples presented in Fig. 2. To define the error probability under SOnF for each

combination in X, this work models the circuit considering the ITM for each input. If the expected output is '1', only faults present in pull-down network can affect the output of the gate. Otherwise, if the expected output is '0', only faults presents in transistors of the pull-up network can affect the output of the gate, and the analysis to compute the reliability considers the graph *G1*.

Furthermore, in each input combination there are transistors that are already in conductive state. These transistors are not present in the reliability computation because a SOnF in a transistor that is already conducting does not affect the circuit output. These situations allow simplifying the graph in order to remove the transistors in a conductive state and, then, to compute the reliability only with the critical transistors to the related input combination. To do it, this work presents a probabilistic SOnF model that determines the input fault error probability considering the transistor arrangements in the circuit. This model determines the behavior for series, parallel and combined series/parallel transistors arrangements.

1) Serial arrangements

Serial arrangements consist in more than one transistor connected in serial condition. To represent statistically the probability of a SOnF between two terminals in this arrangement is necessary that all transistors present a SOnF. This occurs because the stuck-on fault is only propagated to the output if there is a path able to propagate the signal between the terminals. For example, considering Fig. 4, the fault will just be propagated if both transistors T_1 AND T_2 are Stuck-On, creating a conductive path.

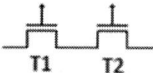

Fig. 4. Serial arrangement of transistors

To compute the error probability in this transistor arrangement, let's consider that, for each transistor t_j with $[j=1...n]$ where n is the number of transistor in serial arrangement, the error probability $P(e)$ is modeled as the fault probability in t_j and t_{j+1} through T_n, as shown in Eq. (1). The probability of the transistor failing is given by $P(t_j)$. The statistics operation to represent the *and* relation is the intersection, as presented in Eq. (2). Eq. (3) shows the error probability of the serial arrangement presented in Fig. 4.

$$P(e) = P(t_1) \ and \ P(t_2) \ and \ ... and \ P(t_n) \tag{1}$$

$$P(e) = P(t_1) \cap P(t_2) \cap ... \cap P(t_n) \tag{2}$$

$$P(e) = P(T_1) \cap P(T_2) \tag{3}$$

2) Parallel arrangements

This arrangement consists in more than one transistor connected in parallel condition. To represent statistically the probability of a SOnF between two terminals in this arrangement is necessary that at least one transistor present a SOnF. This occurs because the stuck-on fault is only propagated to the output if there is a path able to propagate the signal between the terminals. For example, considering Fig. 5, the error is propagated if transistor T_1 or T_2 are Stuck-On, creating a conductive path.

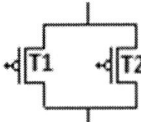

Fig. 5.   Parallel arrangement of transistors

To compute the error probability in this transistor arrangement, let's consider that, for each transistor t_j with $[j=1...n]$ where n is the number of transistor in parallel arrangement, the error probability $P(o)$ is modeled as the fault probability in t_j or t_j+1 through T_n, as showed in Eq. (4). The probability of the transistor failing is given by $P(t_j)$. The statistics operation to represent the *or* relation is represented in Eq. (5). Eq. (6) shows the error probability of the parallel arrangement presented in Fig.5.

$$P(o) = P(t_1) \; or \; P(t_2) \; or \; ... \; or \; P(t_n) \qquad (4)$$

$$P(o) = P(t_1) \cup P(t_2) \cup ... \cup P(t_n) \qquad (5)$$

$$P(o) = P(T_1) \cup P(T_2) \qquad (6)$$

3) Serial/Parallel arrangements

In this arrangement condition, the methodology needs to simplify the graph, in order to obtain the error probability of the arrangement. Let's consider Fig. 6(a) as an example, in this case, the first simplification occurs in parallel arrangements and the circuit becomes as in Fig. 6(b). In the arrangement shown in Fig. 6(b) is just necessary to compute the reliability of transistors in serial condition, as explained before.

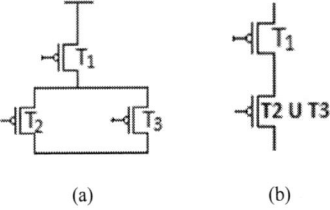

(a)   (b)

Fig. 6.   Serial and parallel arrangement of transistors (a); Parallel simplification in transistor arrangement (b).

4) NAND2 Case Study

The traditional PTM of a NAND2 is independent of the transistor arrangement and obtained just considering the NAND2 function. The traditional PTM is shown in Fig. 7(b). To exemplify the proposed methodology, let's consider a NAND2 gate described in Fig. 7(a). Different of the traditional PTM, the PTM considering SOnF and input dependence has relation with the transistor arrangement and it implies in the new PTM showed in Fig. 7(b). It is possible to note that, for each input combination, there is a different value of q due to the transistor arrangement in the circuit, which causes a different possibility to propagate a wrong output value. The error propagation condition is shown in Fig. 7(d) for each input combination.

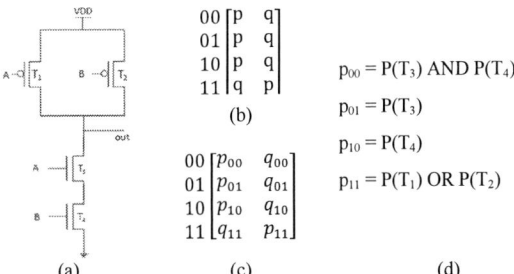

(a)   (c)   (d)

Fig. 7.   NAND2 gate schematic (a); Traditional PTM (b); PTM considering SOnF and input dependence (c); Probability of error propagation (d).

B. PTM considering SOnF model

To validate and allow the adoption of the method proposed this work applied the traditional PTM method and the method proposed in a set of the most frequently used combinational standard cells from a commercial library.

The generation of a library with all information about SOnF in a set of combinational logic gates is made using an automatic generator of PTM considering SOnF and input dependence. The entire process to compute the reliability of a gate under Stuck-On faults is described in Algorithm 1. In the rest of this work, the PTM generated considering the inputs influence is called as PTM modified. The algorithm inputs are the circuit description (C); the list of inputs (I) and the reliability value of the technology (q). The algorithm output is the PTM modified of the gate.

The algorithm starts organizing the circuit description in the two graphs ($G0$ and $G1$) adopted in the methodology (line 1). Then, the list of inputs (I) is used to compute the number of input combinations of the circuit (line 2). For each combination in I, is realized the simplification of the graph eliminating the transistors in conductive state (line 3). After that (line 4), it is computed the serial and parallel arrangements of transistors based on the graph information in order to obtain the total reliability of the graph. After establishing all arrangements and computing the whole equation representing reliability for the input combination X_i, the circuit reliability is computed and the PTM modified is generated.

Algorithm: PTM_Creator (C, I, q)

Input data:
        I // List of Inputs
        C // Circuit description Netlist
        q // Gate Reliability

```
1    G = create_graphs(C);
2    For each input combination {
3       G = graph_simplification(I);
4       reliability_computation(G, q); }
```

Output data:
        PTM matrix of circuit C

Algorithm 1. Methodology to create the PTM of a gate.

The main step in the Algorithm is in line 4. In this step, the methodology creates the PTM considering the reliability of the arrangement found at the graph analyzed. This work uses graph simplification to produce the equation to be solved by combining the serial and parallel arrangements. This method searches for parallel arrangements and simplifies the transistors in parallel, computing its reliability. After removing the parallel arrangements, the next step of this method is to simplify the serial arrangements that are possible to simplify. This method will be repeated while there are removable nodes in the graph. At the end of this process, the PTM modified is finished and available to the user.

IV. EXPERIMENTAL RESULTS

This section presents the experimental results separated in three parts. The first one shows the error probability computed for all single stage logic gates from a Nangate FreePDK45 Open Cell Library [12]. The second part explores logic gates that present its reliability equals to the used technology reliability. In other words, this value could be misunderstanding as equal to the traditional PTM approach that consider the same value of q for all input vectors. A deeper analysis presents this coincidence and shows the difference in several input vectors. Finally, the constructed library of logic gates PTM is applied into two different circuits to compare our method and the traditional PTM values in related works.

A. Standard cell modified PTM

Fig. 8 shows the error probability obtained from proposed method (red columns) and the error probability considering the traditional PTM (blue line) considering an arbitrary technology reliability $q = 0.99$ for evaluated logic gates. The gates are organized by the number of transistors, from left to right in figure, starting with 2 transistors of the INV to the 12 transistors of the OAI33. In Fig. 8 is possible to notice that the complementary gates, i.e., AOI-OAI, NAND-NOR, have the same error probability. Another observation is that NAND2, NOR2, INV, AOI211 and OAI211 have the same error probability computed in both methods. Otherwise, NAND4 and NOR4 considering the proposed method presents the smaller error probability compared to the traditional method. It occurs because they present more transistors in serial condition and this kind of arrangement makes the error probability smaller.

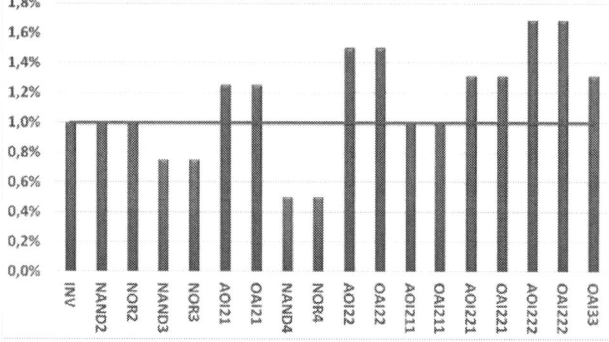

Fig. 8. Error probability for a set of logic gates validated considering SOnF.

The main difference in the gates reliability considering input dependence is that when there is some transistors in series condition in many of the input combinations, the reliability is increased. On the other hand, circuits with many parallel arrangements present a smaller reliability because this arrangement is extremely sensitive to a Stuck-On fault, propagating if any of the transistors present a fault. In NAND and NOR arrangements, only one network presents parallel arrangement, in the case of AOI and OAI they can present parallel arrangements in both networks, what reduces the gate reliability.

Analyzing the gates in Fig. 8 is possible to notice that some gates have the same error probability considering both methods. To explore these gates let consider the NAND2 gate. Fig. 7(a) presents the transistor arrangement, Fig. 7(c) and Fig. 7(d) presents the difference in both methods between PTM considering input vector dependence and the PTM traditional. The same arbitrary value of reliability $q=0.99$ is applied and Fig. 9(a) and Fig. 9(c) show the values computed for NAND2 gate for PTM with input dependence and PTM traditional, respectively. If you compare each row of the PTMs is possible to notice that the rows '00' and '11' present different reliabilities. However, if you compute the gate reliability (PTM*ITM), the value is the same and equal to 0.99. The same comparison is done in an AOI21 in Fig. 9(b) for PTM modified and Fig. 9(d) for PTM traditional. In AOI21, only three in eight values of reliability are the same than the traditional PTM. This difference does not cause difference in logic gate reliability, by as explored in next section, they are observed in circuit analysis. The PTM modified of other gates are not present in the figure because they have more input combination and it would be impractical due to space limitations.

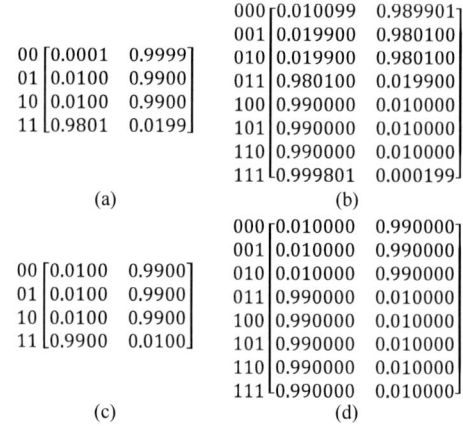

Fig. 9. NAND2 gate PTM considering SOnF (a); AOI21 PTM considering SOnF (b); NAND2 traditional PTM (c); AOI21 traditional PTM (d).

B. Impact on the Circuit Reliability

Using the library created by our method, the information can be used to compute circuit reliability for any circuit composed by the cells present in our analysis. Then, two circuits are selected in order to explore the difference in circuit error probability and reproduce our analysis: a C17 using NAND2 gates and a multiplexer. Both circuits are mapped using only the logic gates that present the same reliability in traditional PTM and from proposed method. This choice will allow the isolation of the input vector influence in the circuit reliability.

The C17 circuit presented in Fig. 10 is composed of six *NAND2* gates. In this analysis is considered that all inputs vectors have the same probability of occurrence. The interesting point is that *NAND2* presents the same error probability in both methods, but different error probability for some input vectors. This difference in circuit error probability considering both methods is up to 15% in C17. This is the consequence of the difference in the reliability of some input vectors of the *NAND2* gate. This difference is propagated during the PTM operations to compute the circuit reliability.

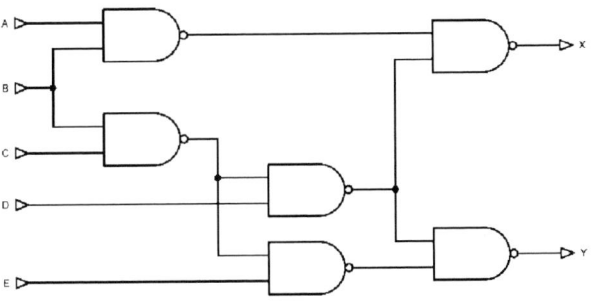

Fig. 10. ISCAS C17 circuit using *NAND2* gates.

The multiplexer shown in Fig. 11 is composed by *NAND2*, *NOR2*, and *Inverters*. As the circuit if formed by cells with the same error probability in both methods, the difference in circuit error probability is caused because the gates present different error probabilities for some input vectors and this difference causes an error probability different when the PTM operations are applied in circuit arrangement.

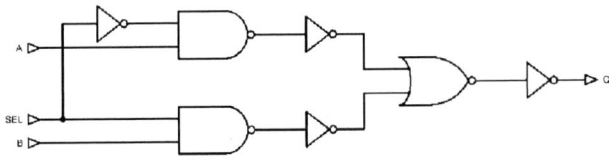

Fig. 11. Multiplexer circuit.

The results obtained for both circuits are presented in Table I. A different reliability of the technology *q=0.95* is also used to complement the analysis and verify if the difference in both methods remain the same [9].

Considering *q=0.99* the difference in circuit C17 using the traditional PTM and the PTM modified, considering error probability is 18%. This represents that the PTM modified increases the error probability in 18%. The multiplexer error probability using PTM modified is 13% higher than PTM using traditional values.

Considering *q=0.95* the difference in circuit C17 using the PTM modified in error probability is 23% higher. Analyzing

the multiplexer error probability using PTM modified is 10% higher than PTM using traditional values.

TABLE I. CIRCUITS ERROR PROBABILITY CONSIDERING TRADITIONAL PTM AND PTM WITH INPUT VECTORS DEPENDENCE

Circuit	Error probability using Traditional PTM (%)		Error probability using PTM modified (%)	
	q=0.99	q=0.95	q=0.99	q=0.95
C17	4,80	21.61	5,71	24.97
Multiplexer	5,13	21.39	5,81	23.54

V. FINAL REMARKS

Circuit reliability estimation is an important aspect that has to be considered in modern circuit design to avoid the use of fault tolerance techniques that can mitigate the gains achieved by the technology scaling. This work presents a method to compute logic gate reliability using a probabilistic model for stuck-on faults that explores the transistor arrangement and the input vector influence to compute fault probability for each input combination. The results show that considering the same error probability for all input vectors underestimates the input influence on the overall circuit reliability. The proposed method can provide results that are more accurate in terms of circuit reliability and guarantee more accurate results in reliability circuit analysis.

ACKNOWLEDGMENT

This work is supported by *Coordenação de Aperfeiçoamento de Pessoal de Ensino Superior* – CAPES – Brazil.

REFERENCES

[1] Borkar, S. et al. "Design and reliability challenges in nanometer technologies". DAC 2004. pp. 75.

[2] Fang, L.; Hsiao, M. S. Bilateral testing of nano-scale fault-tolerant circuits. Journal of Electronic Testing, Springer, v. 24, n. 1-3, p. 285–296, 2008.

[3] Vial, J. et al. Using TMR architectures for yield improvement. DFTVS, 2008. p. 7–15.

[4] Naviner, L. A. et al. Efficient computation of logic circuits reliability based on probabilistic transfer matrix. DTIS 2008.

[5] Beg, Azam, and Walid Ibrahim. "On teaching circuit reliability." FIE 2008. 38th Annual. IEEE, 2008.

[6] Ketan N. Patel, Igor L. Markov, and John P. Hayes. Evaluating circuit reliability under probabilistic gate-level fault models. IWLS 2003.

[7] S. Krishnaswamy, G.F. Viamontes, I.L. Markov, and J.P. Hayes. Accurate reliability evaluation and enhancement via probabilistic transfer matrices. DATE, 2005.

[8] Xiao, J., et al. A method of gate-level circuit reliability estimation based on iterative PTM model. *IEEE PRDC*, 2011.

[9] Singh, N. S. S., et al. "Sensitivity analysis of probability transfer matrix (PTM) on same functionality circuit architectures."(CSPA), 2012 IEEE 8th International Colloquium on. IEEE, 2012.

[10] Grandhi, S., Spagnol, C., & Popovici, E. Reliability analysis of logic circuits using probabilistic techniques. PRIME. 2014

[11] Krishnaswamy, Smita, et al. "Probabilistic transfer matrices in symbolic reliability analysis of logic circuits." TODAES. 2008.

[12] NanGate FreePDK45 Open Cell Library, available at: http://nangate.com

An FPGA-based accelerator for multiple real-time template matching

Erika S. Albuquerque, Antonyus P. A. Ferreira, João G. M. Silva, João P. F Barbosa,
Renato L. M. Carlos, Djeefther S. Albuquerque and Edna N. S. Barros
Informatics Center
Federal University of Pernambuco
Recife, Pernambuco, Brazil
Email: esa3@cin.ufpe.br

Abstract—Object tracking with multiple dense templates is a challenging problem in the context of continuous monitoring video cameras. Applications of the multiple match template technique range from tracking of multiple independent objects to several pose variations of a single object. Due to its high accuracy and tolerance to brightness and contrast changes, the zero mean normalized cross-correlation(ZNCC) was selected as similarity measure. This paper proposes an FPGA architecture that explores a full pipeline implementation and maximizes the internal data reuse to calculate several ZNCC-based template matching in an efficient approach. Experimental results shows that the proposed implementation achieves the real-time performance reaching up to 30fps running ten parallel ZNCCs in a single Stratix IV FPGA.

I. Introduction

Real time tracking systems have to combine high accuracy with short processing time [8]. A traditional approach to this class of problems is to use template matching to track the target. This method proves itself useful, although a non-optimized implementation can deliver a large processing time, which increases with frame dimensions.

The working principle of template matching in tracking algorithms is to search for the window in the region of interest in each frame which is the most similar to the template [11]. The region of interest(ROI) of a frame is the region that includes the last template position and its surroundings. The basic idea is to consider continuous moves to find the searched object. Aiming to carry out this search, we use a similarity measure between the template and each image window where the template is superimposed. The maximum similarity score is selected as the match location.

The detection of an object translation is a kindly easy problem to be solved with image correlation. The trouble in tracking 3D objects is that real objects, besides the translation, may rotate and change its pose. In this scenario, the captured template is outdated, and the algorithm can not provide an accurate result.

A common solution to this problem is to store, in a temporal buffer, pose variations of the 3D object intended to track. Although this strategy increases algorithm accuracy, it also greatly increases the computational cost of this application.

Another frequent problem in object tracking is the multiple independent objects tracking. In this case, multiple separate

Fig. 1. (a) Template image sliding over source image. (b) Result of ZNCC similarity measure for each template position.

objects are continuously tracked. The solution of this problem is quite similar of the multi-pose tracking, on which real-time comparisons are made between a set of templates against each video frame.

To tackle this problem, the present work proposes a hardware architecture to compute ZNCC (Zero Mean Normalized Cross Correlation) with multiple templates in a more efficient way. In next sections more details about this work are presented. Section *II* presents the object tracking algorithm using multiple ZNCC-based template matching. Section *III* presents some important related works, Section *IV* describes the proposed architecture to optimize the ZNCC calculus. Section *V* shows the implementation details of the proposed hardware architecture, Section *VI* depicts the experiments and the results obtained, at last, Section *VII* shows conclusions and future works.

II. Multiple Template Object Tracking

A. ZNCC-based template matching

Template matching is an image processing technique to looking for areas of a source image that match to a template image. The search consists in sliding a template image over a source image and calculates a similarity measure between template and each image window. The Result matrix (R) numerically represents how well the template matches in each image position, as we can see in Figure 1.

The Zero Mean Normalized Cross Correlation (ZNCC) is a similarity measure that presents as an advantage the robustness against the linear intensity distortions and image contrast variations[6]. It also supports the use of a similarity threshold since it gives normalized results.

Due to its robustness, the ZNCC-based template matching is widely used in a large variety of applications in image and video processing, and in computer vision applications such as fingerprint recognition[10], object detection[3], face recognition[4], defect detection[14], stereo vision[13] and object tracking[12].

The computation of ZNCC could be accelerated using fast Fourier transform(FFT) and integral images[9] or using multi-resolution search[17]. Nonetheless, software approaches are hard to fit in real time applications.

Equation 1 shows how result values $R(x,y)$ are calculated from image $I(x,y)$ pixels and template $T(x,y)$ pixels.

$$R(x,y) = \frac{\sum_{i,j}^{N}(T_{i,j} - \overline{T}) \cdot (I_{i',j'} - \overline{I}_{x,y})}{\sqrt{\sum_{i,j}^{N}(T_{i,j} - \overline{T})^2 \cdot \sum_{i,j}^{N}(I_{i',j'} - \overline{I}_{x,y})^2}} \quad (1)$$

where i', j', $\overline{T}$, and $\overline{I}_{x,y}$ are defined as:

$$i' = x + i \text{ and } j' = y + j$$

$$\overline{T} = \frac{1}{N} \cdot \sum_{i,j}^{N} T_{i,j} \quad (2)$$

$$\overline{I}_{x,y} = \frac{1}{N} \cdot \sum_{i,j}^{N} I_{x+i,y+j} \quad (3)$$

B. Object tracking using multiple templates

Tracking can be defined as the problem of estimating the trajectory of an object in the image plane as it moves around a scene. A tracker assigns consistent labels to the tracked objects in different video frames [16]. The object tracker algorithm implemented in this work assumes that the object location and appearance do not change abruptly.

The algorithm of object tracking using multiple templates searches for a bank of templates in a region of interest of the frame. The bank of templates is composed of 'm' images of the same object in slightly different poses. The bank is updated with time, as we can see in Figure 2. This algorithm is presented in Table 1. The resulting matrices, one for each template, have values $R(x,y)$ between -1 and 1. If the maximum result value is smaller than a minimal threshold it is considered that object is occluded and template position is not updated.

When there is at least one result greater than the minimal threshold, it is considered that the object is in the scene and not occluded. The template position is updated. If, in addition, this value is smaller than another required threshold, it is considered that the object changed its pose. In this case, a new template is extracted from the current frame and it replaces the template that produced the smallest maximum result value for this frame.

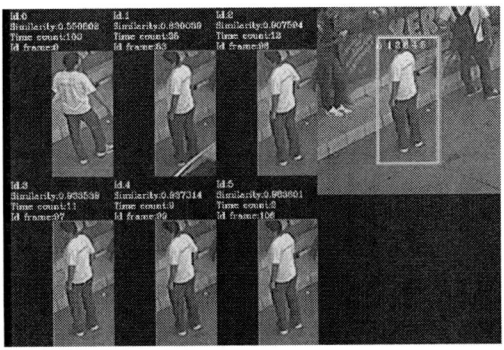

Fig. 2. Templates of a pedestrian with pose variations.

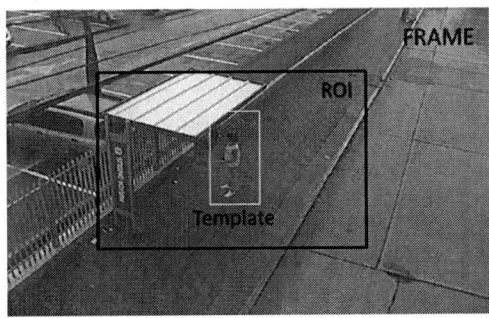

Fig. 3. Region of interest, around last template position.

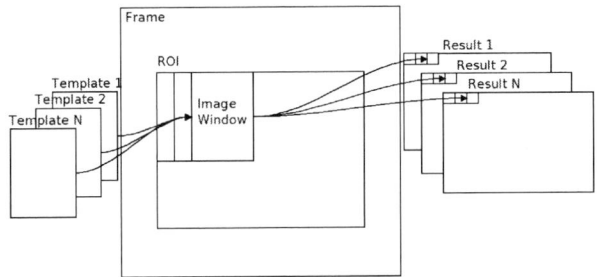

Fig. 4. Multiple Templates generating multiple Results matrix.

Figure 3 illustrates typical sizes of frames, ROI and templates for our application. Figure 4 illustrates how the resulting similarity matrices are calculated from ROI and templates matrices.

III. RELATED WORKS

Chen et al [5] proposed a hardware architecture using a simplified measure based on ZNCC. The simplification avoids the square root operation but it also implies that the results are no more bounded between -1 and 1, an undesired by-product. In order to accelerate the computation, the work proposes a multiresolution search, which can be a good idea to improve time efficiency but decreases the result accuracy.

J. P. Lewis [9] presented a way to accelerate ZNCC computation, which is the base of matlab's function "normxcorr2". He proposes to calculate the correlation in frequency domain and sums it through running sum, also called integral image.

Algorithm 1: Object tracking using multiple template matching

Step 1: Get a frame.

Step 2: User chooses the template from current frame.

Step 3: Initialize all templates from template bank with 'T1' copy.

Step 4: Get region of interest(ROI), around last template position from new video frame.

Step 6: Convert ROI to gray scale.

Step 7: Compute ZNCC matrices between ROI and each template.

Step 8: Take max point values and positions from result matrices.

Step 9: If there is at least one max value greater than minimal similarity threshold, template position is updated.

Step 10: If the max value is greater than minimal similarity threshold but smaller than a required threshold, the current template go to template bank, replacing the template which resulted in the smallest max value.

Step 11: Go to step 4

This work is a reference in ZNCC computation. To calculate correlation in frequency domain improves algorithm performance, especially when the size of the template is near to the size of the image. The matlab implementation of this work does not support real time processing. The paper presents good ideas for software implementation but for hardware implementation the frequency domain conversion can consume too many resources [2] and the conversion time can reduce the algorithm performance.

Hashimoto et al. [7] implemented a hardware architecture to compute ZNCC. The paper has good architecture insights but they did not implement the complete system with hardware and software integration, and the parallelism they implemented is only possible because the template is really small (4x4 pixels).

IV. THE PROPOSED ARCHITECTURE FOR ZNCC CALCULATION

The computation of ZNCC between ROI and multiple templates is the critical step in the object tracking algorithm discussed in this work. For the dimensions used in this work, ZNCC calculation time represents 97% of total loop time, therefore this work proposes a hardware architecture to accelerate this calculation step while keeping other steps executing by software.

The ZNCC computation, Eq. 1, is hard to implement due to the need of pre-computing image window sums to be able to calculate the final result $R(x, y)$. The formula was algebraically manipulated to simplify hardware implementation.

Expanding numerator of equation 1 we have

$$\sum_{i,j}^{N}(T_{i,j} \cdot I_{i',j'} - \overline{T} \cdot I_{i',j'} - T_{i,j} \cdot \overline{I}_{x,y} + \overline{T} \cdot \overline{I}_{x,y}) \quad (4)$$

distributing summations and using the property

$$\sum_{i,j}^{N} A_{i,j}\overline{B} = \overline{B}\sum_{i,j}^{N} A_{i,j} = N\overline{A}.\overline{B} \quad (5)$$

we get

$$\sum_{i,j}^{N}(T_{i,j} \cdot I_{i',j'}) - N\overline{T} \cdot \overline{I}_{x,y} . \quad (6)$$

Now that we simplified the numerator, we will also simplify the denominator, first expanding to get

$$\sqrt{\sum_{i,j}^{N}(T_{i,j}^2 - 2N \cdot \overline{T} \cdot T_{i,j} + \overline{T}^2) \cdot \sum_{i,j}^{N}(I_{i',j'}^2 - 2N \cdot I_{i',j'}\overline{I}_{x,y} + \overline{I}_{x,y}^2)} \quad (7)$$

Distributing summations and using the property below

$$2\sum_{i,j}^{N} A_{i,j}\overline{A} = 2\overline{A}\sum_{i,j}^{N} A_{i,j} = 2N\overline{A}^2 \quad (8)$$

it follows

$$\sqrt{\sum_{i,j}^{N} T_{i,j}^2 - N\overline{T}^2} \cdot \sqrt{\sum_{i,j}^{N} I_{i',j'}^2 - N\overline{I}_{x,y}^2} \quad (9)$$

and combining numerator and denominator, we get

$$\frac{\sum_{i,j}^{N}(T_{i,j} \cdot I_{i',j'}) - N\overline{T} \cdot \overline{I}_{x,y}}{\sqrt{\sum_{i,j}^{N} T_{i,j}^2 - N\overline{T}^2} \cdot \sqrt{\sum_{i,j}^{N} I_{i',j'}^2 - N\overline{I}_{x,y}^2}} . \quad (10)$$

Replacing averages by its expressions, the following equation is obtained

$$\frac{N\sum_{i,j}^{N}(T_{i,j} \cdot I_{i',j'}) - \sum_{i,j}^{N} T \cdot \sum_{i,j}^{N} I_{i',j'}}{\sqrt{N\sum_{i,j}^{N} T_{i,j}^2 - (\sum_{i,j}^{N} T)^2} \cdot \sqrt{N\sum_{i,j}^{N} I_{i',j'}^2 - (\sum_{i,j}^{N} I_{x,y})^2}} . \quad (11)$$

Now the complete computation includes only five summations that can be computed in parallel, which is much simpler for a hardware implementation than the original formula. Besides that, the terms with only template dependency do not change during the ZNCC computation. Thus, these terms can be provided by software. The constant terms during computation are given below.

$$\sum_{i,j}^{N} T \quad (12)$$

$$\sqrt{N\sum_{i,j}^{N} T_{i,j}^2 - (\sum_{i,j}^{N} T)^2} \quad (13)$$

The proposed hardware needs to compute three summations and needs to do some arithmetic reductions to obtain $R(x, y)$. The reduction can be implemented in pipeline as sketched in Figure 5.

It remains to design a strategy to compute correlation and summations. The proposed technique is based on a bottom up approach where we first implement a correlation cell. The cell receives the image pixel and the templates pixels and its

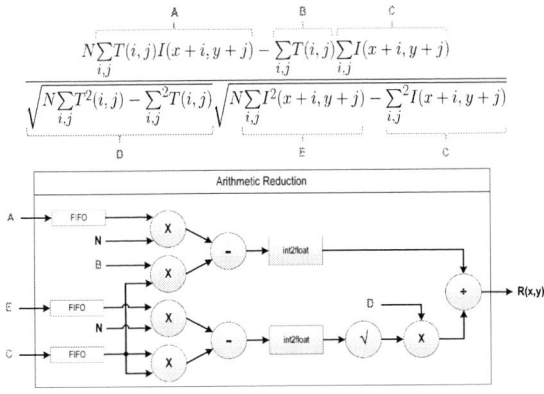

$$\frac{N\sum_{i,j}T(i,j)I(x+i,y+j) - \sum_{i,j}T(i,j)\sum_{i,j}I(x+i,y+j)}{\sqrt{N\sum_{i,j}T^2(i,j) - \sum_{i,j}^2 T(i,j)}\sqrt{N\sum_{i,j}I^2(x+i,y+j) - \sum_{i,j}^2 I(x+i,y+j)}}$$

Fig. 5. Illustration of arithmetic reduction blocks.

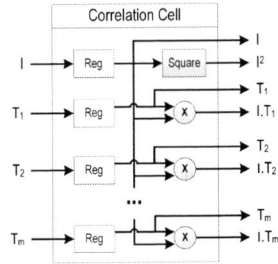

Fig. 6. Correlation cell for 'm' templates.

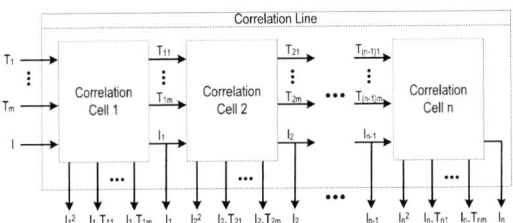

Fig. 7. Parallelism and pipeline diagram of correlation line for 'm' templates.

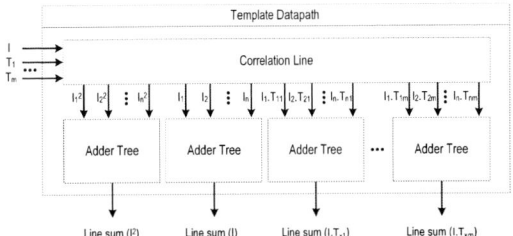

Fig. 8. Template datapath composed of correlation line and adder trees.

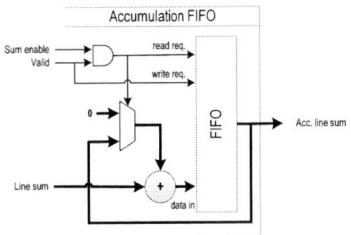

Fig. 9. Correlation window computation using feedback FIFOs.

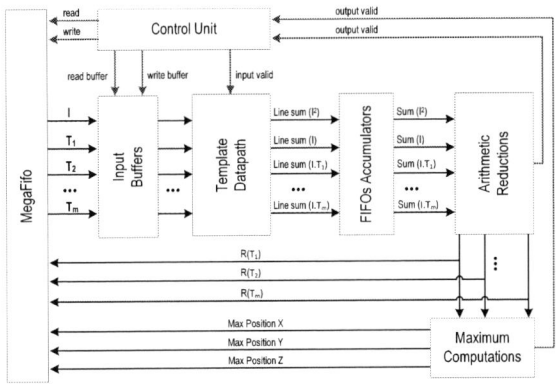

Fig. 10. ZNCC processing architecture overview.

outputs are the square of image pixel, the image pixel single value aligned in time and the products between the image pixel and each template pixel. The inputs 'T' and the 'IxT' outputs are unpacked verilog arrays, therefore, the number of templates 'm' is a cell parameter, see Fig. 6.

The cell is instantiated multiple times in a correlation line block. The correlation line computes I, I^2 and IT_1 to IT_m values of a complete template line in parallel. The template line size is a parameter for correlation line generation, see 7.

The template datapath is composed of a correlation line and some trees of adders in the pipeline to compute the summations $\sum I$, $\sum I^2$ and $\sum IT_1$ to $\sum IT_m$. Figure 8 shows template datapath.

The computations needed for a complete window are serialized with lines results being accumulated in FIFOs, as can be seen in Figure 9. At the beginning of computations we fill in the FIFOs with first line results. From this step forward, we read the accumulated results and sum it to the new line results.

When the last line is being computed, the FIFOs are filled in with the summations $\sum I$, $\sum I^2$ and $\sum I * T_1$ to $\sum I * T_m$, that will pass by arithmetic reductions of Figure 5 and then become the results R_1 to R_m

This section proposed an architecture to efficiently compute ZNCC between multiple templates and a single image. We added to the architecture a maximum computation block to reduce communication time of the result sending step. The input of the object tracking algorithm are the maximum values and their positions, so the similarity matrices are not necessarily read by the host.

The input buffers stores input data to reuse it in serialized computation. The control unit controls input reading, data processing and output writing. The megafifo protocol is used to improve communication and we will see more about it in the next section. The architecture overview is shown in Figure 10.

The platform that contains the ZNCC module will be described in the next section as well as the communication between the hardware and the host software.

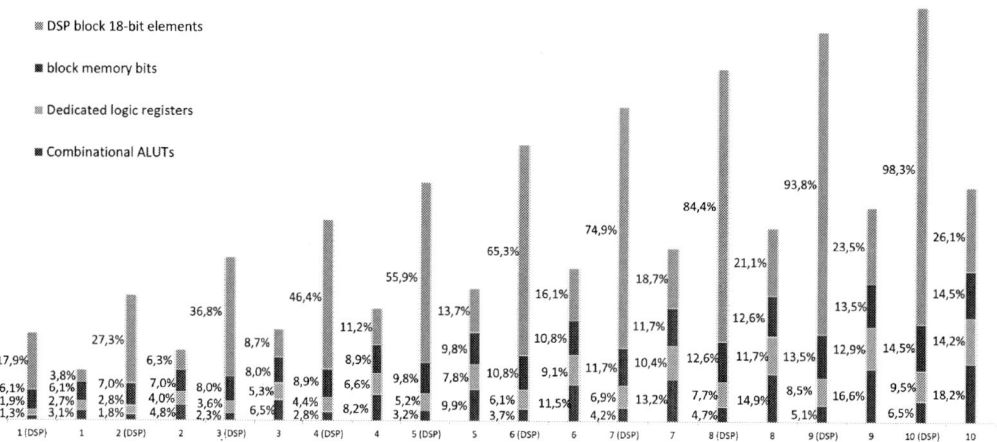

Fig. 11. FPGA resources usage for complete system by design type(multipliers built using DSP blocks or ALUTs) and the number of templates

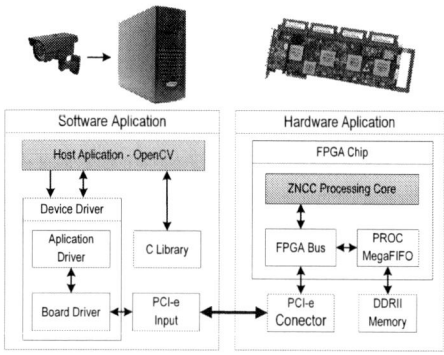

Fig. 12. Overview of the object tracking system.

V. FPGA IMPLEMENTATION

This work was developed using the Gidel Proc-Star IV prototyping platform, which includes four Stratix IV FPGAs (EP4SE530H35C2). However, the developed system requires only one of these four FPGAs. Communication between the platform and the host is via the PCI-express bus using DDR2 memories which are connected to each FPGA.

The Gidel PROCWizard library provides some IP cores to facilitates the hardware/software integration: PROCMegaFIFO, PROCMegaDelay and PROCMultiPort. The PROCMegaFIFO IP is designed for stream applications, so it is well fitted to our real time tracking application. The PROCMegaFIFO uses the DDR2 memory. Figure 12 shows the system block diagrams.

The computer host used to running the application has an i7 core processor, 3.3Ghz, 16GB of RAM, it could be replaced by an embedded processor without significant performance losses as it runs the less critical steps of the algorithm. The only requirement for this host processor is a PCIe port.

The resource usage of megaFIFO IP is less than 2% of FPGA available resources. Although the ProcMegaFIFO occupies some of the FPGA area, its use is highly recommended

since developers may focus their efforts in implementing processing architecture without need to develop modules that will read and write directly onto the DDR2 memory.

8bits integer multipliers are basic components of the correlation cell. Such a multiplier can be built using either 80 ALUTs blocks or an 18bit DSP block. Since the FPGA board can be used for other tasks, we did two versions of cell implementation: using or not using the DSP on cell multipliers. Figure 11 shows the resource usage according to implementation type and template quantity.

The average communication time measured at run-time is lower than 0,05ms. It is very small compared to processing time.

VI. RESULTS

The hardware parameters are the quantity of templates and template and ROI dimensions. We target it to our application dimensions, that are 480x240 image ROI and 72x144 template. The template dimension is an average pedestrian dimensions in our videos. The region of interest (ROI) dimensions were chosen observing that pedestrians normaly moves faster horizontaly than verticaly in image.

Increasing the quantity of templates enhances algorithm accuracy and leads to zero or almost zero performance degradation, due to hardware parallelism. However, it increases the hardware resources usage, so we performed the tests to a maximum of ten templates, see Figure 11.

Aiming to compare the performance of the proposed system in FPGA, we implemented a reference model in software, using the OpenCV template matching function and a custom optimized GPU implementation using CUDA. The GPU implementation runs in a GTX760 GPU with frequency of 1GHz with 2GB of GDDR5 of dedicated video memory.

The OpenCV implementation runs in a desktop with i7 core processor with 3.3Ghz, 16GB of DDR3 memory.

The FPGA device, as aforementioned, is a stratix IV running 200MHz of clock frequency. We compare FPGA frames per

978-1-5090-2737-8/16 $31.00 © 2016 IEEE

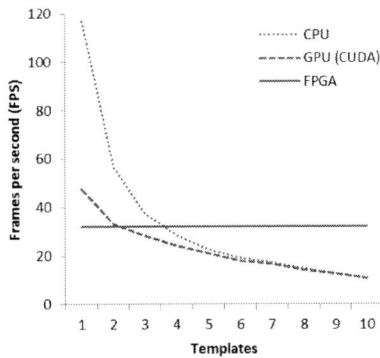

Fig. 13. Performance in frames per second of CPU, GPU and FPGA implementations according to templates quantity.

second rate with software implementations running in CPU and GPU.

As one can see in Figure 13, the CPU implementations surpasses the GPU performance. This could be explained the overhead of launching the GPU kernels, which decreases the GPU performance. The performance gap decreases as the template number increases, corroborating the overhead justify.

Others works achieve better GPU performance in different contexts [1] and [15], a direct comparison can not be done as the works used others CPU and GPU devices and even other template matching formula, but they point that a carefull GPU implementation can be usefull in template matching solutions.

It seems unfair to compare performances of an embedded device as FPGA running 200MHz with such a powerful processor, but as we can see in Figure 13, when the number of templates is bigger than 4, the FPGA is faster than both CPU and GPU. This is because the FPGA computes all templates in parallel. The proposed architecture is three times faster than CPU implementation when we use ten templates in the object tracker algorithm. Note that the FPGA execution time is 31.13ms. This results in a frames per second rate of 32.12FPS regardless of the number of templates.

VII. CONCLUSION AND FUTURE WORKS

This work has presented the development of a parametric hardware architecture that compute the multiple templates tracking algorithm. The architecture has a pre-synthesis option for usage optimization of DSP blocks or combination ALUTs.

A hybrid system composed by CPU + FPGA was developed for real time implementation of the algorithm, this system reached up to 30 FPS even when analyzing 10 templates in parallel. The object tracking algorithm was also implemented using CPU and GPU. The results obtained indicate that the system composed of CPU + FPGA reaches a speed-up of 3x when compared to the CPU and GPU implementations, even with at most one fifth of the frequency of operation.

Future works may explore the integration of this accelerator in a system with the HTTP + Ethernet infrastructure to send the directions to the camera while tracking a subject. Due to the small resource utilization, the proposed system can also

be integrated with a broad list of video processing algorithms inside the same FPGA. Another possible development of this work is to implement this design in ASIC technology to easily embarking it on camera.

ACKNOWLEDGMENT

We gratefully acknowledge the financial support of CNPq and Capes.

REFERENCES

[1] Robert Finis Anderson, J Steven Kirtzic, and Ovidiu Daescu. Applying parallel design techniques to template matching with gpus. In *International Conference on High Performance Computing for Computational Science*, pages 456–468. Springer, 2010.

[2] João PF Barbosa, Antonyus PA Ferreira, Rodrigo CF Rocha, Erika S Albuquerque, Josivan R Reis, Djeefther S Albuquerque, and Edna NS Barros. A high performance hardware accelerator for dynamic texture segmentation. *Journal of Systems Architecture*, 61(10):639–645, 2015.

[3] Mohammed Bennamoun and George J Mamic. *Object recognition: fundamentals and case studies*. Springer Science & Business Media, 2012.

[4] Roberto Brunelli and Tomaso Poggio. Face recognition: Features versus templates. *IEEE Transactions on Pattern Analysis & Machine Intelligence*, (10):1042–1052, 1993.

[5] Jiun-Yan Chen, Kuo-Feng Hung, Hsin-Yi Lin, Yen-Chung Chang, Yin-Tsung Hwang, Ciao-Kai Yu, Cheng-Ru Hong, Chin-Chia Wu, and Yung-Jung Chang. Real-time fpga-based template matching module for visual inspection application. In *Advanced Intelligent Mechatronics (AIM), 2012 IEEE/ASME International Conference on*, pages 1072–1076. IEEE, 2012.

[6] Luigi Di Stefano, Stefano Mattoccia, and Federico Tombari. Zncc-based template matching using bounded partial correlation. *Pattern recognition letters*, 26(14):2129–2134, 2005.

[7] Koji Hashimoto, Yu Ito, and Kaoru Nakano. Template matching using dsp slices on the fpga. In *Computing and Networking (CANDAR), 2013 First International Symposium on*, pages 338–344. IEEE, 2013.

[8] Gayithri Kuruppu, Chandrabose Manoj, SR Kodituwakku, and UAJ Pinidiyaarachchi. Comparison of different template matching algorithms in high speed sports motion tracking. In *Industrial and Information Systems (ICIIS), 2013 8th IEEE International Conference on*, pages 445–448. IEEE, 2013.

[9] JP Lewis. Fast normalized cross-correlation. In *Vision interface*, volume 10, pages 120–123, 1995.

[10] Almudena Lindoso and Luis Entrena. High performance fpga-based image correlation. *J. Real-Time Image Processing*, 2(4):223–233, 2007.

[11] Arif Mahmood and Sohaib Khan. Correlation-coefficient-based fast template matching through partial elimination. *Image Processing, IEEE Transactions on*, 21(4):2099–2108, 2012.

[12] P Mishra, JK Kishore, R Shetty, Ahana Malhotra, and Ratandeep Kukreja. Robust template matching based obstacle tracking for autonomous rovers. In *Electronics, Computing and Communication Technologies (CONECCT), 2013 IEEE International Conference on*, pages 1–5. IEEE, 2013.

[13] A Qayyuma, AS Malik, MNM Saad, M Iqbal, F Abdullah, W Rahseed, TARBT Abdullah, and AQ Ramlib. Vegetation height estimation near power transmission poles via satellite stereo images using 3d depth estimation algorithms. *International Archives of the Photogrammetry, Remote Sensing & Spatial Information Sciences*, 2015.

[14] Du-Ming Tsai and Chien-Ta Lin. Fast normalized cross correlation for defect detection. *Pattern Recognition Letters*, 24(15):2625–2631, 2003.

[15] Nicholas A Vandal and Marios Savvides. Cuda accelerated iris template matching on graphics processing units (gpus). In *Biometrics: Theory Applications and Systems (BTAS), 2010 Fourth IEEE International Conference on*, pages 1–7. IEEE, 2010.

[16] Alper Yilmaz, Omar Javed, and Mubarak Shah. Object tracking: A survey. *Acm computing surveys (CSUR)*, 38(4):13, 2006.

[17] Shinichi Yoshimura and Takeo Kanade. Fast template matching based on the normalized correlation by using multiresolution eigen images. In *Intelligent Robots and Systems' 94.'Advanced Robotic Systems and the Real World', IROS'94. Proceedings of the IEEE/RSJ/GI International Conference on*, volume 3, pages 2086–2093. IEEE, 1994.

978-1-5090-2737-8/16 $31.00 © 2016 IEEE

Successful Prototyping of Complex Integrated Circuits with Focused Ion Beam

E. Petitprez, D. M. Colombo, F. M. Henes, L. Courcelle, R. Tararam,
S. Jacobsen, R. Soares, C. Krug and M. Lubaszewski

CEITEC S.A.
Porto Alegre, RS, Brazil
emmanuel.petitprez@ceitec-sa.com

Abstract—Focused ion beam is a tool that allows to perform microsurgery of the interconnect network of an integrated circuit in order to change its functionality, without need of a metal tweak. This paper reports on the application of this technique to complex CMOS integrated circuits. Two case studies, conducted in Brazil, are presented. Electrical results show the focused ion beam circuit prototyping were successful. To our knowledge, this is the first time such result is reported in Latin America.

Keywords—focused ion beam, integrated circuit, interconnects

I. INTRODUCTION

In the past decade, significant efforts have been devoted to improvements in the integrated circuit (IC) design verification flow in order to fix eventual design errors prior to IC tape-out [1]. However, despite verification, fault-free IC implementation remains very difficult and some layout corrections may still be required during the early IC development phase. These corrections usually include design fix, new tape-out, lithography mask reorder and silicon wafers manufacturing. Because of the increasing complexity in IC technology, and inherent increasing mask cost and silicon manufacturing cycle time, such corrective loops, when repeated several times, can result in severe financial impact for the whole project [2].

In this context, Focused Ion Beam (FIB) has proven to be very useful, as it offers the possibility to reconfigure an IC through physical modification of its metal interconnects [3][4]. In this way, IC rapid prototyping can be performed on some samples to verify the projected layout fix produces the expected functional result. For complex ICs, a significant amount of time and money can be saved with the use of FIB during the IC prototyping phase [5].

Up to recently, FIB circuit edit service was not available in Latin America (LATAM). Therefore, LATAM IC designers had to outsource the service from abroad providers, and deal with additional logistics, cost and delay issues. However, the implementation of a FIB service offering is currently gaining momentum at CEITEC, Brazil. Reports of FIB circuit reconfigurations performed in Brazil, including IC edit methodology and electrical characteristics, have already been presented earlier [6][7].

In this paper we report on complex IC reconfiguration performed in Brazil. We present two case studies from two different ICs in which FIB was used to modify parts of the

interconnect network in order to change or improve the IC functionality. For each case we first describe the functional issue, then we present the layout modification required to fix it, as well as its execution by FIB, and end up showing electrical results that support the successfulness of the FIB IC reconfigurations we performed. Based on these two examples, we finally provide some recommendations to circuit designers in order to produce FIB-friendly IC layout.

II. CASE STUDY #1: CHIP A

A. Functional issue and suggested fix

In Chip A, a battery supplied temperature measurement subsystem (TSS, depicted in Fig. 1) is composed by a regulator, a control logic for the subsystem, a temperature sensor and an analog-to-digital converter (ADC, not shown on figure 1).

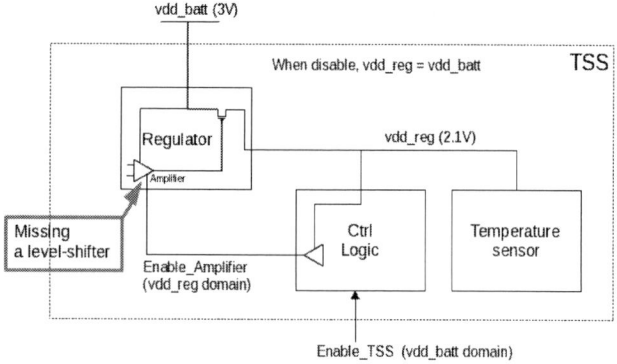

Fig. 1. Chip A TSS functional issue description.

The TSS subsystem is only turned on when a temperature measurement is required. When the TSS is off, the regulator output is set at the battery supply (3V). When the measurement process is started, the regulator output provides a 2.1V nominal supply to the whole TSS. The only block inside the TSS that remains powered by the battery (3V) is the regulator amplifier. This one is also turned on and off together with the whole TSS. The enable signal that enters the regulator amplifier (*amplifier_enable*) comes from the 2.1V voltage domain provided by the regulator and goes to the amplifier which is powered by a 3V supply.

A level-shifter (2.1V to 3V) should have been inserted in the enable signal path to ensure the supply domain transition. Indeed, without the level-shifter, the enable/disable switch transistors are not put in OFF state when needed. This creates a parasitic current which degrades the amplifier gain and prevents the regulator from operating properly. Consequently, the temperature sensor is not tightly regulated and this directly impacts the temperature measurement error.

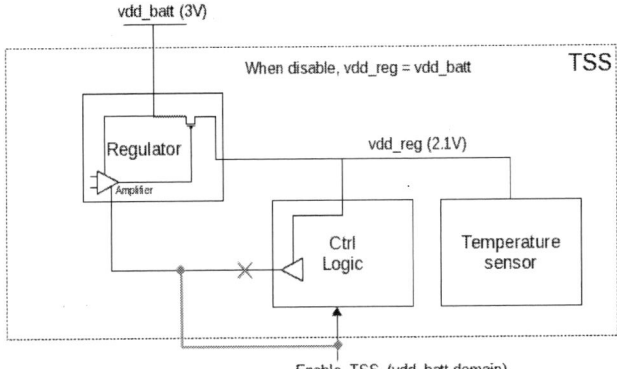

Fig. 2. Suggested fix to solve chip A TSS functional issue.

To overcome the absence of a level-shifter, it has been identified the *amplifier_enable* signal could be removed and substituted by the global *TSS_enable* signal taken from the battery supply domain (3V). This would allow the amplifier enable/disable switch transistor to remain turned off without parasitic current, and hence the regulator to work properly (Fig. 2).

B. Fix implementation with FIB

Chip A is fabricated in CMOS 0.35µm technology and its interconnect network has 4 metallization levels. The circuit routing was carefully analyzed to determine the best location to perform the reconfiguration. A region was found where both signals (*amplifier_enable* signal and *TSS_enable* signal) were passing quite close one to the other. In this region, *amplifier_enable* signal is located at M3 and M4 levels, and *TSS_enable* signal at M3 level. Both metal lines are 0.6µm wide (Fig. 3).

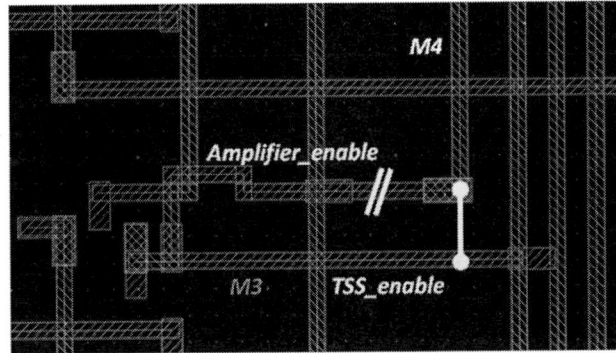

Fig. 3. Chip A layout.modification required for fix implementation

Step-by-step IC reconfiguration execution is shown on Fig. 4a. The first step of the reconfiguration process is to open an electrical connection to *amplifier_enable* by milling a 1.5x1.5 µm² hole through the circuit passivation down to M4, i.e. IC topmost metal level.

In the second reconfiguration step (Fig. 4b), the same process is repeated above the M3 buried line (*TSS_enable*). In this case the FIB milling box positioning is somehow tricky as M3 line is not visible at all from the chip surface. The FIB box is therefore positioned with respect to the previously exposed M4 line and to additional M4 features on the chip surface.

The third step of the reconfiguration is to establish an electrical connection between the two previously exposed metal lines. This is achieved through ion-induced tungsten deposition [8][9]. The beam current was chosen to ensure a proper fill of the milled holes by the FIB-deposited metal [10], which in this case was tungsten.

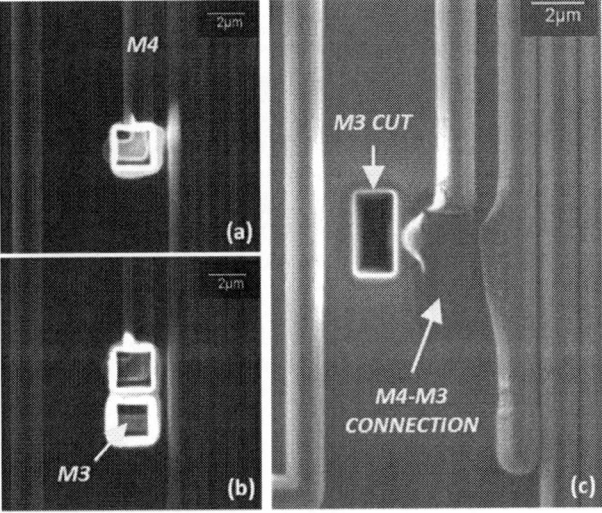

Fig. 4. FIB top views showing step-by-step IC reconfiguration process.

The last reconfiguration step is to disconnect the *amplifier_enable* by cutting the M3 connection to the control logic block. In this case the FIB milling pattern can easily be positioned thanks to the previously exposed M4 line, as well as to M4 topology on the chip surface. In order to cut the M3 line, the milling process is continued up to complete removal of the metal material. Fig. 4c shows the final result of the FIB reconfiguration process performed on chip A.

C. Electrical validation

The temperature sensor response was characterized for original samples (no FIB reconfiguration) and samples on which microsurgery had been performed by FIB. The results are shown on Fig. 5.

On the original samples, the temperature sensor response remains constant up to 2.8V, but exhibits significant variation when increasing supply voltage above this value. This illustrates the fact the regulator fails to operate properly due to parasitic current of the enable/disable switch transistors. The samples that have been submitted to the FIB reconfiguration described in the above section, however, show constant temperature sensor response on the whole power supply range, allowing for an accurate temperature measurement. This indicates that, as expected, the regulator is now working properly, and therefore demonstrates we successfully reconfigured the layout and functionality of these circuits with FIB operation.

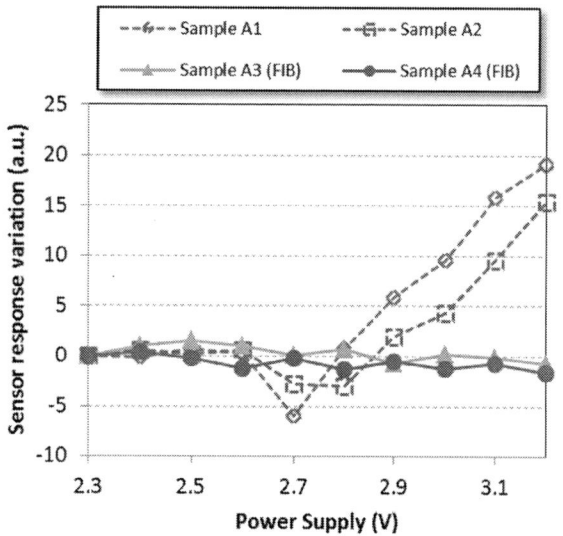

Fig. 5. Temperature sensor response for original and reconfigured samples.

III. CASE STUDY #2: CHIP Q

A. IC edit context

Our second case study deals with current references sub-circuits in Chip Q. Current references provide a stable output current that is weakly dependent on fabrication process and supply voltage. Moreover, this type of circuit is usually immune (on first order) to temperature variations.

Some applications have sub-circuits that require currents with opposite temperature coefficients (TC). As for instance, a current-based voltage oscillator that has its output voltage with intrinsically negative temperature coefficient requires a bias current with opposite temperature coefficient in order to achieve temperature compensation, and therefore produce an output frequency independent of the temperature of operation. In addition, a power-on-reset circuit may require a temperature independent bias current in order to reduce the spread of its output voltage over the temperature range of operation.

The implemented application in Chip Q has two current reference circuits in order to generate currents, $I_{REF,1}$, $I_{REF,2}$ and $I_{REF,3}$ with positive, negative and zero temperature coefficients, respectively.

Current reference $I_{REF,2}$ is generated by means of $I_{REF,1}$ as shown in Fig. 6. Current $I_{REF,1}$ with positive TC is injected in a diode-connected NMOS transistor array. The value of I2 and its temperature coefficient are (on first order) determined by the width and length of all transistors connected at node A and resistor R1.

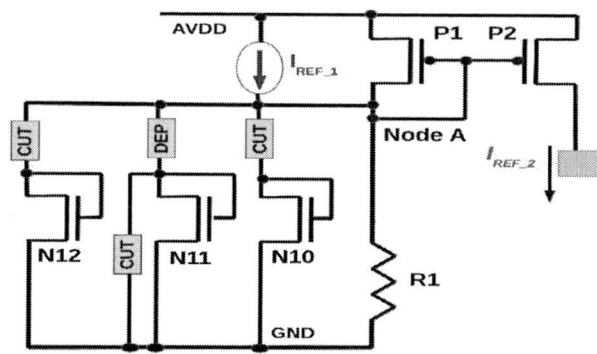

Fig. 6. Chip Q : Current reference $I_{REF,2}$

In order to achieve robustness regarding fabrication process variability, it is possible to correct an eventual deviation of the expected value of $I_{REF,2}$ and its TC by connecting or disconnecting transistors N10, N11 or N12 at node A. For this purpose, dedicated metal tweak cells that allow FIB cut or deposition have been placed into the circuit layout. Besides, these cells also give the possibility to adjust the value of $I_{REF,2}$ and its TC in case circuit specifications change according to future modifications of the application.

As shown in Fig. 7, current I3 is the sum of $I_{REF,1}$ and $I_{REF,2}$ and does not vary with temperature because $I_{REF,1}$ and $I_{REF,2}$ have exactly opposite temperature coefficients. As can be concluded, metal tweak cells presented in Fig. 6 are used to correct the temperature coefficient of $I_{REF,2}$, and as a consequence, to indirectly adjust $I_{REF,3}$ TC. In addition, current reference $I_{REF,3}$ has a second set of metal tweak cells in order to allow a fine adjust of $I_{REF,3}$ value through connection or disconnection of N3 and N4 transistors at output node. $I_{REF,3}$ value is given by the product of $I_{REF,A}$ multiplied by the aspect ratio of N1 and all transistors connected at output node.

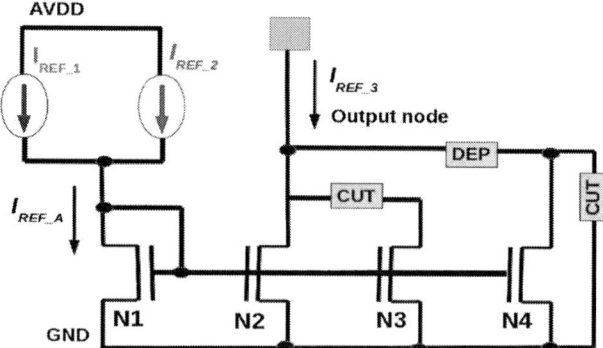

Fig. 7. Current reference $I_{REF,3}$

B. FIB reconfiguration process for Chip Q

The first characterization of Chip Q silicon wafers showed that $I_{REF,3}$ temperature coefficient was way too negative and should be corrected. The layout modification required to bring $I_{REF,3}$ TC close to zero consisted in cutting 7 metal wires from the interconnect network (Fig. 8a).

Chip Q is fabricated in CMOS 0.18µm technology and its interconnect network has 5 metallization levels. It is therefore significantly more complex than Chip A reported in the previous case study, and one can assume its reconfiguration process by FIB would also be much more challenging.

In fact, Chip Q reconfiguration process turned out to be much simpler than for Chip A thanks to the presence of metal tweak cells. Moreover, these cells have been implemented on last metal level, which facilitates their localization thanks to the topmost metal level topography visible in the FIB equipment[6].

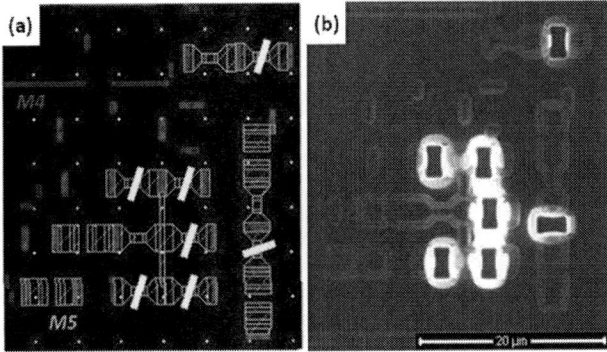

Fig. 8. Layout.modification required for fix implementation on Chip Q (a), and FIB top view after reconfiguration complete (b).

Thanks to these M5 metal tweak cells, the FIB milling pattern can easily be positioned directly on the M5 feature to be cut. Alike the previous case study, the milling process first removes the passivation to expose the metal, and then continues until the metal material is completely sputtered off. This process is repeated 7 times at the 7 locations specified on Fig. 8a to achieve the required circuit modification (Fig. 8b).

C. Electrical results

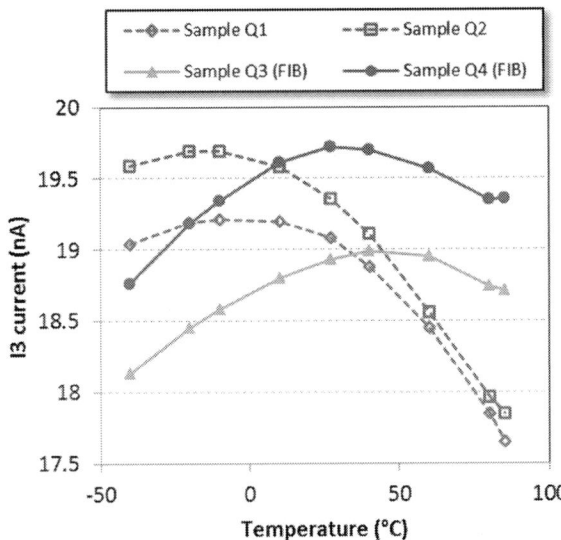

Fig. 9. $I_{REF,3}$ current temperature dependence, for default and reconfigured samples.

Fig. 9 presents temperature performance of I3 for 4 samples of the fabricated circuit. Samples Q1 and Q2 represent the

fabricated circuit with the default configuration (no action was taken in the metal tweak cells). On these samples, $I_{REF,3}$ exhibits a slightly negative temperature coefficient, with an overall variation of 1.9 nA in the entire temperature range. Samples Q3 and Q4 have been submitted to the FIB reconfiguration described above. As can be noticed, the temperature performance was enhanced. Indeed, the current variation decreased to 0.9 nA over the whole temperature range, demonstrating a 50% temperature coefficient improvement after FIB operation.

IV. DISCUSSION

These two case studies illustrate how the difficulty of performing FIB modifications on complex ICs can be significantly relieved if already anticipated during the physical design phase of the IC development. We would therefore like to suggest the circuit designers to implement metal tweak cells, or at least try to bring critical signals as close as possible to the topmost metal level, in order to guarantee

1. *better milling process control*: FIB milling process is monitored by secondary electrons (SE) generated when the ion beam hits the sample surface. Topmost metal level offers shallow milling cavities providing higher secondary electrons signal than deep cavities.

2. *better positioning accuracy*: topmost metal level offers easy localization thanks to its topography on the sample surface.

3. *faster operation* : exposing topmost metal level features requires short milling times because less oxide thickness needs to be removed.

4. *enhanced success rate*: dedicated metal tweak cells offer room to perform FIB modifications without risk of damage to the neighboring metal features

Finally, in the presence of metal tweak cells, one can wonder about the advantages of FIB reconfiguration compared to the widely used laser cut operation[11]. We see three main advantages supporting the FIB:

1. FIB offers a more accurate positioning since it is based on an electron beam with inherent higher precision than an optical beam used for laser cut.

2. FIB allows tight process control in all 3 directions, reducing the risk of damage for neighboring metal features, or even for active devices placed underneath the modification zone.

3. FIB offers the unique possibility to establish new electrical connections through metal deposition when laser cut technique does not.

V. CONCLUSION

In this paper we have presented two case studies of focused ion beam reconfiguration on complex CMOS integrated circuit. The FIB process has been entirely conducted in Brazil. For both cases we have shown electrical results demonstrating the success of the prototyping. Moreover, the comparison of these two cases evidences the usefulness of metal tweak cells for secure, reliable and rapid prototyping of complex integrated circuits. We believe the use of such metal tweak cells, combined with FIB prototyping, can provide significant cycle time saving during the IC development phase. To our knowledge, this is the first time such FIB prototyping results are reported in Latin America. These results are an important milestone on the path of implementing FIB reconfiguration service offer in Latin America.

ACKNOWLEDGMENTS

The authors are grateful to the financial support from Conselho Nacional de Desenvolvimento Científico e Tecnológico (CNPq, Bolsa Pesquisador Visitante Especial, Processo 314075/2013-5) as well as from Fundação de Amparo a Pesquisa do Rio Grande do Sul (FAPERGS).

REFERENCES

[1] K. Chang, I. Markov, V. Bertacco, Design Errors in Digital Circuits, Springer Lecture Notes in Electrical Engineering, Vol. 32 (2009)

[2] R. Collet & D. Pyle, McKinsey on Semiconductors, N. 3, p24 (2013)

[3] J. Melngailis, et al., "The focused ion beam as an integrated circuit restructuring tool" Journal of Vacuum Science and Technology B., 4, pp 176-180 (1986).

[4] S. Smith, A. Walton, S. Bond, A. Ross, J. Stevenson and A. Gundlach, "Test structures for the electrical characterisation of platinum deposited by focused ion beam" Proceedings of the 2002 IEEE International Conference on Microelectronic Test Structures V15 (2002)

[5] T. Mohiuddin, "FIB Circuit Edit Becomes Increasingly Valuable In Advanced Node Design", Electronic Design, Jan 24, 2014, available at http://electronicdesign.com/eda/fib-circuit-edit-becomes-increasingly-valuable-advanced-node-design

[6] E. Petitprez, S. Jacobsen, R. Tararam, C. Krug and M. Lubaszewski, "Repairing integrated circuits with focused ion beam", Proceedings of the XXX Simpósio Sul de Microeletrônica (2015)

[7] E. Petitprez, S. Jacobsen, R. Tararam, C. Krug and M. Lubaszewski, "Electrical characterization of integrated circuit interconnects processed with focused ion beam", Proceedings of the 2015 Workshop on Circuits and Systems Design (2015)

[8] S. Smith, A. Walton, S. Bond, A. Ross, J. Stevenson and A. Gundlach, "Electrical characterization of platinum deposited by focused ion beam", IEEE Trans. On Semiconductor Manufacturing, vol. 16, pp 199-206 (2003)

[9] M.M. da Silva, A.R. Vaz, S A. Moshkalev and J.W. Swart, "Electrical characterization of platinum thin films deposited by focused ion beam", ECS Transactions, vol. 9 pp 235-241 (2007)

[10] S. Jacobsen, E. Petitprez, R. Tararam, C. Krug and M. Lubaszewski, "Morphological characterization of metallic lines deposited by focused ion beam under different beam currents", Proceedings of the XIV SBPMat conference (2015)

[11] H. Yamaguchi, M. Hongo, T. Miyauchi and M. Mitani, "Laser cutting of aluminium stripes for debugging integrated circuits", Solid-State Circuits, Vol. 20, pp 1259-1264 (1985)

Automatic Layout Integration of Bulk Built-In Current Sensors for Detection of Soft Errors

Mário Vinícius Guimarães
School of Engineering
Universidade Federal de Minas Gerais
Belo Horizonte, Brazil

Frank Sill Torres
Department of Electric Engineering
Universidade Federal de Minas Gerais
Belo Horizonte, Brazil

Abstract— **Soft error resilience is of rising importance for the design of integrated circuits realized in CMOS nanometer technologies. Therefore, Bulk Built-In Current Sensors (BBICS) have been proposed as a fast and efficient technique for detecting transient faults that might lead to soft errors. An important requirement for application of these sensors in common designs is the automatic integration. The aim of this work is to present a methodology for automatic insertion of BBICS in common standard cell designs. Further, two different placement strategies are introduced and compared. Experiments demonstrate the feasibility of the approach and indicate requirements for future BBICS developments in order to reduce area offset.**

Keywords—Soft error, transient fault, EDA, Bulk Built-in Current Sensors

I. INTRODUCTION

Current nanometer scale CMOS technologies are facing an increasing amount of fault sources that can lead to serious reliability problems. Examples are effects like parameter variations [1], oxide breakdown [2], and radiation [3]. In case of the latter, high energetic particles inject electrical charge into sensitive regions of the semiconductor devices that can result in transient faults and soft errors [4]. For long time, researches on soft errors due to radiation focused mainly on memories and avionics and aerospace applications. However, with the uprising of nanometer scale technologies, soft error resilience is also a concern for applications on ground level. Several concurrent error detection and/or correction techniques have been presented to avoid the effects of radiation. This includes multiple clocking schemes [5], checker based arithmetic units [6], and selective redundancy [7]. In contrast to these gate and system level techniques, Bulk Built-In Current Sensors (BBICS) are an approach on transistor level, which enables the detection of radiation induced particle strikes immediately after its occurrence [8]. The advantages of these sensors are fast error detection and low power penalty with moderate costs in terms of area [9-11].

However, to the best of our knowledge, there is still no solution for automatic integration of BBICS in random designs published. Thus, this work presents a modified standard cell design flow that enables the automatic insertion of modular BBICS [9, 12]. Further, requirements for layout and placing algorithms will be introduced. It can be said that the results of this work are an essential step towards the consolidation of the BBICS approach for application in common designs.

The rest of the paper is organized as follows. Section 2 gives preliminary information, while section 3 details the proposed flow. Section 4 discusses the layout design of the standard cells and Section 5 presents experimental results. Finally, section 6 concludes this work.

II. BASICS

This section presents fundamental information that support the understanding of the paper.

A. Standard Cell Design

Standard cell design is a CMOS design methodology that applies cells, which are pre-designed and pre-verified on logic, schematic, and layout level. This enables the separation of design tasks into different abstraction layers, *e.g.*, logic synthesis, placement, and routing [13].

The layout of each standard cell has same height but different widths. Further, cells are designed such that they can be placed horizontally next to each other. This also includes the adequate positioning of supply wires as well as N- and P-wells. Thus, cells can be organized in rows during placement.

B. Transient Faults and Soft Errors

Strikes of high energetic ion particles, like alpha particles or neutrons, into a sensitive region of an integrated device can cause transient faults that might turn into soft errors [4].

The most susceptible regions to ionizing radiation events are reverse biased junctions, as for instance the drain of a transistor in off-state. In the event of a hit, the particle's path forms an electron-hole pair track. When this track traverses a depletion region, carriers are rapidly collected by the electric field creating a distortion of the potential into a funnel shape [14]. This carrier collection can be observed as a current and voltage transient on the affected node. The following phase is dominated by diffusion, in which additional charge diffuses into the depletion region over a longer period of time [4].

The described effect can be modeled by double exponential current pulse I_{strike} by following equation [4]:

$$I_{strike}(t) = \frac{Q_{coll}}{t_f - t_r}\left(e^{\frac{-t}{t_f}} - e^{\frac{-t}{t_r}}\right) \qquad (1)$$

with Q_{coll} is the total collected charge, t_r indicates the time constant for the funnel collection and t_f represents the second phase of carrier collection.

978-1-5090-2737-8/16 $31.00 © 2016 IEEE

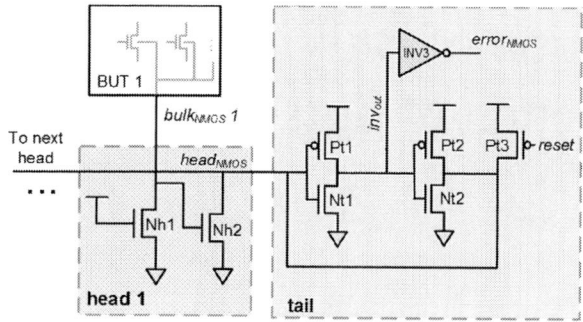

Fig. 1. Principal architecture of modular BBICS (mBBICS)

C. Modular Bulk Built-In Current Sensors

The principal idea of a Bulk Built-In Current Sensors (BBICS) is to measure the anomalous current in the transistor bulk in case of an ionizing particle strike [8, 15]. Thus, BBICS enable the detection of transient faults that might lead to soft errors. Amongst the several BBICS architectures, the modular BBICS (mBBICS) represents a promising trade-off between response time, robustness and costs in terms of area [9].

The mBBICS consists of two functional blocks: the head and the tail (see Fig. 1). The head circuits are connected to the bulk of the monitored transistors (BUT in Fig. 1) while the outputs of several heads are latched by the tail circuit. NMOS devices are monitored by the NMOS type mBBICS, while the bulks of the PMOS counterparts are connected to the PMOS type mBBICS

Considering the NMOS version of the mBBICS, the gate of transistor Nh1 is connected to VDD and its drain to the bulk of the monitored transistors (see Fig. 1). In normal operation, the drain of Nh1 acts as a virtual GND while the signal head$_{NMOS}$ is at V_{DD} level. In case of a particle strike, the fault current is conducted through Nh1 resulting in a voltage drop over Nh1 which increases the gate-source voltage of Nh2. If this voltage exceeds the threshold voltage $v_{th,Nh2}$ of Nh2 the signal head$_{NMOS}$ is pull-down. Consequently, the connected tail circuit latches this signal and triggers the error flag. The circuit remains in this state until the reset transistor Pt3 is activated, causing the circuit to returning to its initial state [9].

The processing of a detection of a particle strike is realized on higher processing layers and not considered in this work. Further information can be found in [16-18].

III. FLOW FOR BBICS INSERTION

This section presents the proposed flow for automatic insertion of BBICS in a random design.

A. Flow

The proposed flow applies the modular BBICS presented in section II [9]. It should be noted, though, that the flow can be adapted with low effort for any other BBICS architecture. The purposed flow is also resilient to changes in the netlist during the place and route phase, since it uses relative cell position.

Each mBBICS head monitors a certain amount of transistors. These transistors can be placed in different or same

N- or P-wells. However, devices that are monitored by different sensors must be placed in different wells. In this case, wells have to be separated by a minimum distance, which easily can be as wide as a standard cell [19]. Consequently, it is desired to place all standard cells, whose transistors are monitored by the same mBBICS, in the same N- or P-well.

The insertion of mBBICS heads and tails on logic level is not recommended as there is still no information about the design layout available. Thus, we propose to include the mBBICS only after initial placement. This solution permits less control over which cells are monitored with same mBBICS, but it considerably reduces the area penalty.

Fig. 2 depicts the simplified version of the proposed flow for mBBICS insertion. After initial synthesis and floorplanning, the standard cells are placed by an automatic placement tool. During the next step, the cells of each row are clustered to be monitored by same heads. Then, cells are shifted in order to create space, the wells are separated, and NMOS and PMOS type heads are added to the row. In the final steps, tail circuits are added at the end and beginning of each row and the routing is executed by an automatic tool.

B. Clustering and head insertion strategies

The amount of transistors an mBBICS head can monitor is limited by the capacitive load due to the bulks of the monitored devices [20-22]. If this load is higher than the sensor's capability, the sensibility of the mBBICS will not be sufficient to detect all particle strikes that can lead to a soft error.

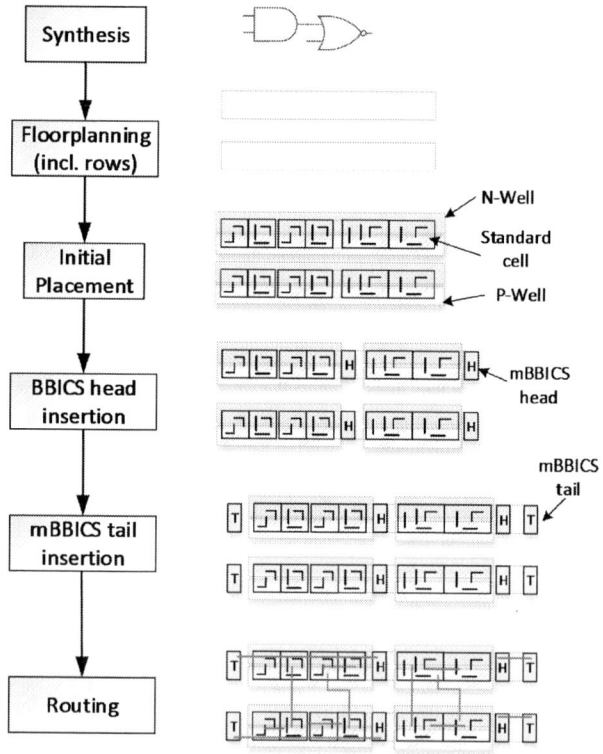

Fig. 2. Proposed flow for mBBICS insertion (simplified)

978-1-5090-2737-8/16 $31.00 © 2016 IEEE

Hence, we propose to define for each cell i the two weight values $\gamma_n(i)$ and $\gamma_p(i)$, which are related to the bulk capacitances of its NMOS and PMOS devices. During clustering, the total weights $\gamma_{cl,n}$ and $\gamma_{cl,p}$ of the cluster are incremented when a standard cell is added to the cluster. Further, for both mBBICS head types are estimated the maximum capacitive loads $\beta_{max,n}$ and $\beta_{max,p}$ for which they have still sufficient sensibility.

The proposed methodology offers two kinds of mBBICS head cells, which are detailed in the following.

1) Double head

A *Double head* mBBICS cell consists of a NMOS and a PMOS type head. The resulting insertion algorithm is listed in Fig. 3. The processing of each row of the design starts with the initialization of the cluster weights (line 3). Next, cells are added successively to the cluster starting from left side (lines 5 and 6). This is done until the value of at least one of the total cluster weights $\gamma_{cl,n}$ and $\gamma_{cl,p}$ would cross the corresponding maximum capacitive load (line 7). Then, a *Doubled head* is inserted (line 10) and a new cluster is formed (lines 8 and 9). In case of an insertion in the layout, the current row is extended and all right-hand cells are shifted rightwards by the width of the *Doubled head*, which is then added. It should be noted that a *Doubled head* realizes the separation of the wells. Further, the netlist is updated in order to enable the later routing of the mBBICS head cell.

2) Single head

The disadvantage of the *Double head* cells is its incapability to handle unbalanced cell weights, *i.e.*, clusters in which $\gamma_{cl,n}$ and $\gamma_{cl,p}$ strongly differ. In these cases, one of the mBBICS heads in the *Double head* cell monitors less devices than it could. Hence, we propose the additional use of *Single head* cells, which consist of a single NMOS or PMOS type mBBICS head.

The modified algorithm is listed in Fig. 4. Similar to the previous algorithm it starts with an initialization of the cluster weights (line 3). This is followed again by successive addition of cells to the cluster (lines 5 and 6). In contrast to the algorithm presented in Fig. 3, a *Double head* is only inserted if at least one of the cluster weights is higher than the corresponding maximum load, while the opposite cluster weights crosses a predefined fraction α of the corresponding

---

```
 1:   Do for all rows {
 2:       Get n, row[1], …, row[n]   // array with all cells of the row
 3:       γcl,n = γcl,p = 0
 4:       Do for i = 1 to n {
 5:           γcl,n = γcl,n + γn(row[i])
 6:           γcl,p = γcl,p + γp(row[i])
 7:           // 0 ≤ α ≤ 1
 8:           if [(γcl,n > βmax,n) and (γcl,p > α * βmax,p)]
                 or [(γcl,n > α * βmax,n) and (γcl,p > βmax,p)] {
 9:               γcl,n = γn(row[i])
10:               γcl,p = γp(row[i])
11:               insert Double head before row[i]
12:           } else if (γcl,n > βmax,n) {
13:               γcl,n = γn(row[i])
14:               insert Single head NMOS before row[i]
15:           } else if (γcl,p > βmax,p) {
16:               γcl,p = γp(row[i])
17:               insert Single head PMOS before row[i]
18:           }
19:       }
20:       Insert Double head
21:   }
22:   End
```

Fig. 4.   Algorithm for insertion of mBBICS *Double* and *Single head* cells

maximum load (line 8). The value of α should be determined in empiric studies.

If only one of the cluster weights reaches the maximum load while the opposite weight is lower than the predefined fraction, a corresponding *Single head* is inserted (lines 14 and 17) and only the related cluster is reformed (lines 13 and 16). The addition of a *Single head* is realized in the same way as for the *Double head* described above.

C. Final Steps

After the head cells have been inserted in all rows, tail cells are added. Thereby, PMOS type tail cells are added on the right side of a row and NMOS type ones on the left side. The number of tail cells in each row depends on the amount of heads each mBBICS tail can monitor [9]. However, each row has at least one tail of each type. Additionally, the netlist of the design is updated.

Finally, the whole design is routed by an automatic routing tool. Thereby, also the bulk connections of each standard cell are routed with its corresponding head cell.

IV. CELL IMPLEMENTATION

This section presents the applied technology and the implemented cells.

A.   Technology

All designs have been realized in a commercial 180 nm technology with Triple-well option. The nominal voltage is 1.8 V, the minimum transistor length is 180 nm and the minimum width is 220 nm.

B.   Standard cells

The BBICS approach requires that all monitored transistors are located in a P- or N-well. Hence, it was necessary to create

---

```
 1:   Do for all rows {
 2:       Get n, row[1], …, row[n]   // array with all cells of the row
 3:       γcl,n = γcl,p = 0
 4:       Do for i = 1 to n {
 5:           γcl,n = γcl,n + γn(row[i])
 6:           γcl,p = γcl,p + γp(row[i])
 7:           if (γcl,n > βmax,n) or (γcl,p > βmax,p) {
 8:               γcl,n = γn(row[i])
 9:               γcl,p = γp(row[i])
10:               insert Double head before row[i]
11:           }
12:       }
13:       Insert Double head
14:   }
15:   End
```

Fig. 3.   Algorithm for insertion of mBBICS *Double head* cells

a standard cell library that complies with this requirement. Fig. 5 depicts the layout of a NAND2 and an INV standard cell which are realized in a Triple-well process. The standard cell height was defined with 16.5 μm.

As can be identified in Fig. 5, the Triple-well structure of the NMOS devices as well as the N-well of the PMOS devices are not closed within the cell. Hence, this closing has to be realized by filler cells or mBBICS head cells. Further, the bulk contacts are pins that can be connected during the routing process.

All cells have been characterized by the tool Synopsys SiliconSmart [23].

C. Layout of mBBICS head cells

This subsection presents the layouts of the head cells of the mBBICS.

1) Double head

The *Double Head*, introduced in section III, contains both PMOS and NMOS heads in the very same cell. Fig. 5 depicts the layout of this cell. The size of the cell is dominated by the required minimum distances between wells of same ($MinD_{eq}$) and different potential ($MinD_{dif}$). Further, its NMOS devices are not realized in Triple-well. The connections (pink wires) between the input pins $bulk_{NMOS}$ and $bulk_{PMOS}$ and the corresponding bulk connections of the monitored cells NAND2 and INV had been added only during routing. The width of the cell is 12.6 μm.

As mentioned in the previous subsection, the *Double head* also contains structures to close the N-wells and Triple-wells of the neighbor standard cells (see left and right corners of the

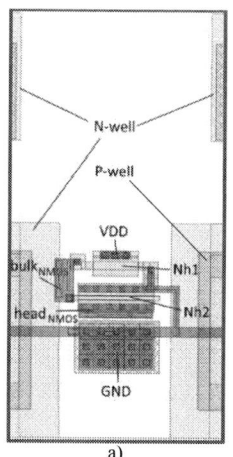
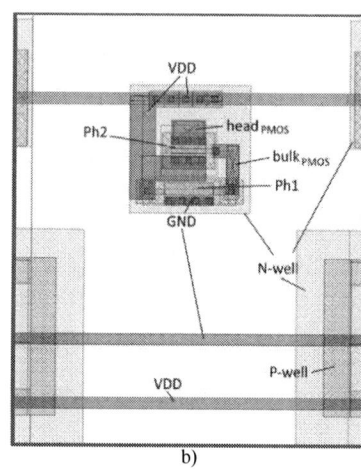

Fig. 6. Layout of a) NMOS *Single head* cell, b) PMOS *Single head* cell

Double head cell in Fig. 5).

2) Single head

Fig. 6 depicts the NMOS and PMOS *Single head* cells. The width of the NMOS version is 8.2 μm, while its PMOS counterpart has the same width as a Double head, *i.e.*, 12.6 μm. This difference in size follows from the required minimum distances of N-wells with same and different potential, as mentioned in the previous subsection. Given that the NMOS head applies no PMOS devices and no Triple-well, it can be realized with smaller area.

Further, both cells contain again structures to close the N-wells and Triple-wells of neighbor cells.

D. mBBICS Sensibility estimation

The sensibility of the BBICS was estimated by using inverters. Initially, a chain of two inverters was created and a double exponential current pulse with $t_r = 1ps$ and $t_f = 20\ ps$ (see equation 1) was added to the drain of the 1st inverter in order to simulate the effects of a particle strike. Next, the value for the collected charge Q_{coll} was increased until an error occurred at the output of the 2nd inverter. Thus, the critical collected charge for particle strikes in NMOS and PMOS devices could be determined.

In the following step, the NMOS and PMOS mBBICS cells were connected with the bulks of the corresponding inverter's transistor and the estimated values for the critical charge were applied. Then, the amount of monitored inverters was increased in order to determine the maximum amount of device and thus, the maximum capacitive load the mBBICS can monitor.

The weight of each standard cell (see also section III) was directly related to the gate-bulk capacitances of its devices.

V. ANALYSIS

This section presents and discusses the results for the exploration of the proposed flow.

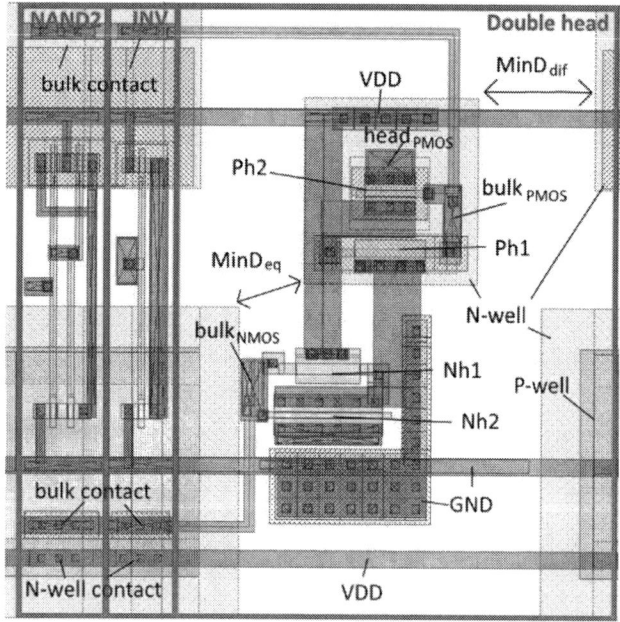

Fig. 5. Layout of *Double head* cell and NAND2 and INV standard cells. The bulk connections (pink wires) had been added during routing. $MinD_{dif}$ and $MinD_{eq}$ indicate minimum distances between wells with same and different potential.

Fig. 7. Exemplary layout of the circuit c499 with *Double head* mBBICS

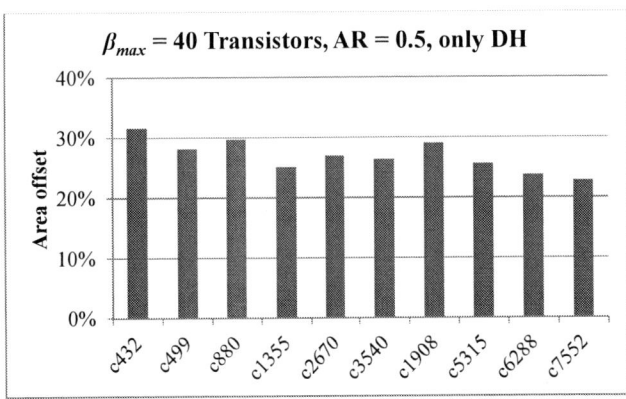

Fig. 9. Area increase for layout aspect ratio AR = 0.5, solely application of *Double heads*, and max. load β_{max} of 40 min. sized transistors.

A. Environment

Automatic placing and routing was realized by the tool Cadence Encounter [24]. All scripts have been implemented in TCL. The designs are taken from the ISCAS benchmark suite [25] and were verified after application of the proposed flow. Fig. 7 depicts an exemplary result for the circuit c499 with inserted *Double head* and tail mBBICS cells.

B. Comparison of head insertion strategies

In a first attempt, the application of solely *Double heads* and mixed *Single* and *Double heads* was compared. Therefore, the maximum load was set to β_{max} = 40 minimum sized transistors for both mBBICS types and a layout aspect ratio of AR = 0.5 was chosen. During this analysis, the algorithm for mixed application of *Single* and *Double heads* was applied, whereas α varied between 0 and 1 (see algorithm in Fig. 4). It should be noted that the results for α = 0 are identical to an application of solely *Double heads*.

The results shown in Fig. 8 indicate that the area offset improves for decreasing α, with achieving lowest offset for α < 0.8. Further, for α values below 0.8 only *Double head* cells have been applied. This can be explained by the rather balanced NMOS and PMOS weights of the applied standard cells. Consequently, the number of required mBBICS for NMOS and PMOS devices is very uniform in each row. Thus, there is no gain in using cells with only one head. It should be

noted, that these results might differ for other standard cell libraries.

Following from the observed results, all further experiments were based on the solely application of *Double head* cells.

C. Impact of design

In the following analysis, the variation of the area offset for different designs was compared. The aspect ratio was set to AR = 0.5, β_{max} was defined with 40 minimum transistors, and only *Double heads* were inserted. The results are shown in Fig. 9.

The average area increase is 26.9 %, while the results vary between 22.8 % (c7552) and 31.7 % (c432). It follows that, as expected, the design has average impact on the area offset.

D. Maximum load

The next analysis focused on the reduction of the area offset if the maximum load of the mBBICS could be improved. The results might indicate guidelines for further optimizations of the mBBICS.

Fig. 10 shows that the area offset saturates for maximum

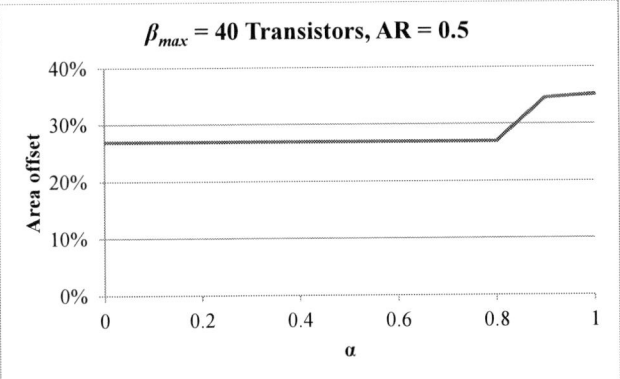

Fig. 8. Average increase of area for varying α, max. load of β_{max} = 40 (in terms of min. sized transistors), layout aspect ratio AR = 0.5 and mixed application of *Single* and *Double heads*.

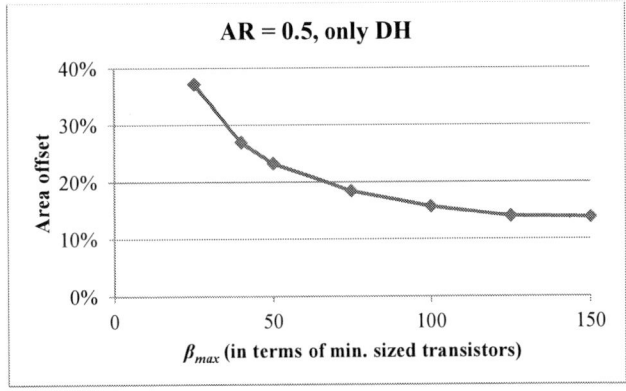

Fig. 10. Average increase of area for aspect ratio AR = 0.5, solely application of *Double heads*, and varying max. load β_{max} of the mBBICS heads.

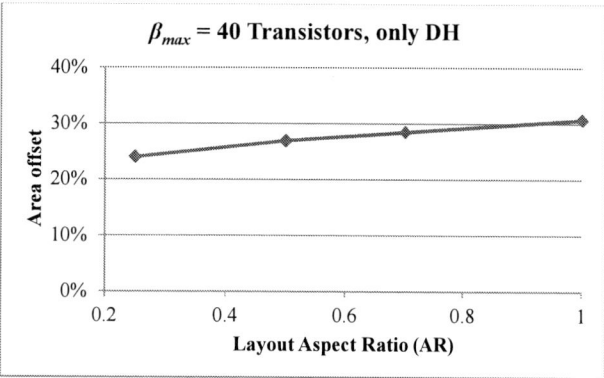

Fig. 11. Average increase of area for different Aspect Ratio (AR) with max. load of $\beta_{max} = 40$ (in terms of min. sized transistors) and only DH cells.

loads above 100 minimum sized transistors at around 13 %. This is expected as each row applies at least 2 tail circuits, resulting in a minimum area offset. Even though the minimal achievable area offset differs amongst the designs, *e.g.* 9.0 % for c6288 and 18.7 % for c432, the observed tendency is the same for all analyzed circuits.

Hence, it is recommended to develop mBBICS that can monitor at least 100 minimum sized transistors.

E. Aspect Ratio

The final analysis concentrated on the impact of the aspect ratio on the results. Therefore, each design was implemented with solely *Double head* mBBICS and a maximum load of $\beta_{max} = 40$ minimum sized transistors.

The results depicted in Fig. 11 show that the area offset varied from 24.0 % for AR = 0.25 to 30.7 % for AR = 1. Thus, there is only a low impact of the AR value on the results.

VI. CONCLUSION

Radiation-induced soft errors are a rising concern in designs realized in current nanometer CMOS technologies. mBBICS are a promising approach to circumvent these problems. This work proposes a flow that enables the automatic integration of the sensors into common designs, which is a mandatory step towards the application of this approach. It could be shown, that it is recommendable to use of cells that combine NMOS and PMOS head sensors. Further results indicate that mBBICS heads should be able to monitor at least 100 min. sized inverters in order to achieve an area increase of 13 % or lower.

REFERENCES

[1] O. S. Unsal, J. W. Tschanz, K. Bowman, V. De, X. Vera, A. Gonzalez, and O. Ergin, "Impact of parameter variations on circuits and microarchitecture," *IEEE Micro,* vol. 26, pp. 30-39, Nov-Dec 2006.

[2] J. Srinivasan, S. V. Adve, P. Bose, and J. A. Rivers, "The impact of technology scaling on lifetime reliability," in *Dependable Systems and Networks, 2004 International Conference on,* 2004, pp. 177-186.

[3] T. Karnik, P. Hazucha, and J. Patel, "Characterization of soft errors caused by single event upsets in CMOS processes," *IEEE Transactions on Dependable and Secure Computing,* vol. 1, pp. 128-143, Apr-Jun 2004.

[4] R. C. Baumann, "Radiation-induced soft errors in advanced semiconductor technologies," *IEEE Transactions on Device and Materials Reliability,* vol. 5, pp. 305-316, Sep 2005.

[5] N. D. P. Avirneni and A. K. Somani, "Low Overhead Soft Error Mitigation Techniques for High-Performance and Aggressive Designs," *Ieee Transactions on Computers,* vol. 61, pp. 488-501, Apr 2012.

[6] S. Pontarelli, P. Reviriego, C. J. Bleakley, and J. A. Maestro, "Low Complexity Concurrent Error Detection for Complex Multiplication," *Ieee Transactions on Computers,* vol. 62, pp. 1899-1903, Sep 2013.

[7] S. N. Pagliarini, G. G. dos Santos, L. A. D. Naviner, and J. F. Naviner, "Exploring the feasibility of selective hardening for combinational logic," *Microelectronics Reliability,* vol. 52, pp. 1843-1847, Sep-Oct 2012.

[8] E. H. Neto, I. Ribeiro, M. Vieira, G. Wirth, and F. L. Kastensmidt, "Using bulk built-in current sensors to detect soft errors," *IEEE Micro,* vol. 26, pp. 10-18, Sep-Oct 2006.

[9] F. Sill Torres and R. Possamai Bastos, "Detection of Transient Faults in Nanometer Technologies by using Modular Built-In Current Sensors," *Integrated Circuits and Systems, Journal of* vol. 8, pp. 89-97, 2013.

[10] A. Simionovski and G. Wirth, "Simulation Evaluation of an Implemented Set of Complementary Bulk Built-In Current Sensors With Dynamic Storage Cell," *IEEE Transactions on Device and Materials Reliability,* vol. 14, pp. 255-261, Mar 2014.

[11] Z. Zhang, T. Wang, L. Chen, and J. Yang, "A new Bulk Built-In Current Sensing circuit for single-event transient detection," in *Electrical and Computer Engineering (CCECE), 2010 23rd Canadian Conference on,* 2010, pp. 1-4.

[12] F. Sill Torres and R. P. Bastos, "Robust modular Bulk Built-in Current Sensors for detection of transient faults," in *Integrated Circuits and Systems Design (SBCCI), 2012 25th Symposium on,* 2012, pp. 1-6.

[13] N. H. E. Weste and D. M. Harris, *CMOS VLSI design : a circuits and systems perspective,* 4th ed. Boston: Addison Wesley, 2011.

[14] C. M. Hsieh, P. C. Murley, and R. R. O'Brien, "A field-funneling effect on the collection of alpha-particle-generated carriers in silicon devices," *Electron Device Letters, IEEE,* vol. 2, pp. 103-105, 1981.

[15] E. H. Neto, F. L. Kastensmidt, and G. Wirth, "Tbulk-BICS: A Built-In Current Sensor Robust to Process and Temperature Variations for Soft Error Detection," *IEEE Transactions on Nuclear Science,* vol. 55, pp. 2281-2288, Aug 2008.

[16] R. P. Bastos, G. Di Natale, M. L. Flottes, and B. Rouzeyre, "How to sample results of concurrent error detection schemes in transient fault scenarios?," in *Radiation and Its Effects on Components and Systems (RADECS), 2011 12th European Conference on,* 2011, pp. 635-642.

[17] F. Leite, T. Balen, M. Herve, M. Lubaszewski, and G. Wirth, "Using Bulk Built-In Current Sensors and Recomputing Techniques to Mitigate Transient Faults in Microprocessors," *Latw: 2009 10th Latin American Test Workshop,* pp. 147-152, 2009.

[18] R. P. Bastos, G. Di Natale, M. L. Flottes, F. Lu, and B. Rouzeyre, "A New Recovery Scheme Against Short-to-Long Duration Transient Faults in Combinational Logic," *Journal of Electronic Testing-Theory and Applications,* vol. 29, pp. 331-340, Jun 2013.

[19] R. J. Baker, *CMOS Circuit Design, Layout, and Simulation:* Wiley-IEEE Press, 2010.

[20] J. G. M. Melo, F. Sill Torres, and R. P. Bastos, "Exploration of Noise Robustness and Sensitivity of Bulk Current Sensors for Soft Error Detection," presented at the CMOS Variability, 6th International Workshop on, Salvador, Brazil, 2015.

[21] R. P. Bastos, J. M. Dutertre, and F. S. Torres, "Comparison of bulk built-in current sensors in terms of transient-fault detection sensitivity," in *CMOS Variability (VARI), 2014 5th European Workshop on,* 2014, pp. 1-6.

[22] J. G. M. Melo and F. Sill Torres, "Exploration of Noise Impact on Integrated Bulk Current Sensors," *Journal of Electronic Testing,* vol. 32, pp. 163-173, 2016.

[23] "SiliconSmart ACE User Guide," Version 2014.09 ed: Synopsys, Inc, 2014.

[24] "EDI System Menu Reference," ed: Cadence Design Systems, Inc, 2012, p. 1168.

[25] J. P. Hayes, M. C. Hansen, and H. Yalcin, "Unveiling the ISCAS-85 benchmarks: A case study in reverse engineering," *IEEE Design & Test of Computers,* vol. 16, pp. 72-80, Jul-Sep 1999.

Analytic Boundaries for 6T-SRAM Design in Standby Mode

Fabián Olivera and Antonio Petraglia

Federal University of Rio de Janeiro - EPOLI/PEE/COPPE - Rio de Janeiro, RJ Brazil

Emails: {folivera, antonio}@pads.ufrj.br

Abstract—In this paper an analytic model is presented, whose purpose is to determine boundaries inside which supply voltage is minimized and leakage current is limited under manufacturing process variations for 6T-SRAM cells operating in standby mode. Drain-induced barrier lowering (DIBL) effects are considered in view of the fact that their influences become more critical in nanometer (below 90 nm) CMOS nodes, when devices are dimensioned with minimum sizes and operate in sub-threshold region. The analytic expressions were verified by comparisons with electrical simulation results obtained with a 28 nm FDSOI CMOS process.

Index Terms—DIBL, minimum-supply-voltage, SNM, ULP, ULV,

I. INTRODUCTION

In the last years, ultra low power (ULP) has gained more importance in SoC design by the emergence of several circuits that need to prolong battery life, such as mobile phones, personal computers, sensor networks and implantable circuits [1], [2].

According to the International Technology Roadmap for Semiconductors (ITRS) forecast [3], SRAM arrays currently occupy above 90% of the SoC area, and their leakage currents are one of the major reasons for static power dissipation. An effective way to reduce it is by decreasing the supply voltage which has an exponential influence on the leakage current [4], [5]. Therefore, the minimum supply voltage is a performance parameter of memory cells that determines the static power dissipation of the SRAM array.

The aggressive scaling of CMOS technology below 90nm nodes creates a crucial challenge for designers, since manufacturing process variability is not scaling at the same rate as does transistor dimensions [3]. Usually, the smallest possible dimensions are used to realize a high-density SRAM arrays, and consequently the large variability of these transistors has a dramatic impact on the minimum supply voltage that ensures the correct operation of the memory cells, with no loss of stored data. One of the most commonly metrics to quantify SRAM robustness is the static noise margin (SNM) [6], [7], which is used to estimate the data retention voltage (DRV) [4], since SNM positive values allow safe operation of each memory cell. When the minimum transistor dimensions are used, DIBL effects reduce the SNM and hence such effect should be taken into account in the memory cell designs.

This paper presents analytic boundaries for 6T-SRAM cell design, which are employed with the purpose of enabling the designer to minimize the supply voltage and limit leakage current under the variability of the manufacturing process.

This paper is organized as follows. Section II presents models for SNM and leakage current in standby mode that strongly consider the DIBL effect influences. Section III describes the use of analytic boundaries in the design of 6T-SRAM cells. Section IV shows the simulation results of an example design. Conclusion remarks are made in Section V.

II. 6T-SRAM IN STANDBY MODE

The 6T-SRAM cell of Fig. 1 is formed by two crossed coupled inverters (M_1-M_3 and M_2-M_4) and two access transistors (M_5 and M_6). In standby mode the cells only retains stored bits and the objective of designers is to minimize the supply voltage, without data loss. Usually this is obtained by careful modeling of the devices operating in sub-threshold region.

A. Transistor Model at Sub-Threshold Region

The sub-threshold current of MOS transistor at low supply voltage can be approximated by the well-known expression [7], [8]

$$I_{DSi} = \beta_i \exp\left(\frac{V_{GS} + \lambda_i V_{DS} + \gamma_i V_{BS}}{n_i U_T}\right)\left[1 - \exp\left(\frac{-V_{DS}}{U_T}\right)\right] \quad (1)$$

where,

$$\beta_i = I_{oi}\frac{W_i}{L_i} \cdot \exp\left(-\frac{V_{Toi}}{n_i U_T}\right), \quad (2)$$

λ_i and γ_i model the DIBL and body effects, respectively, U_T is the thermal voltage, V_{Toi} is the threshold voltage, n_i is the slope factor, I_{oi} is a technology dependence current factor parameter, and W_i and L_i are the transistor width and length respectively.

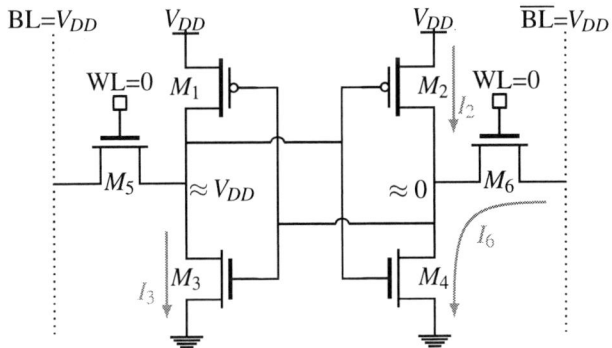

Fig. 1. Schematic of the 6T-SRAM circuit in standby mode.

978-1-5090-2737-8/16 $31.00 © 2016 IEEE

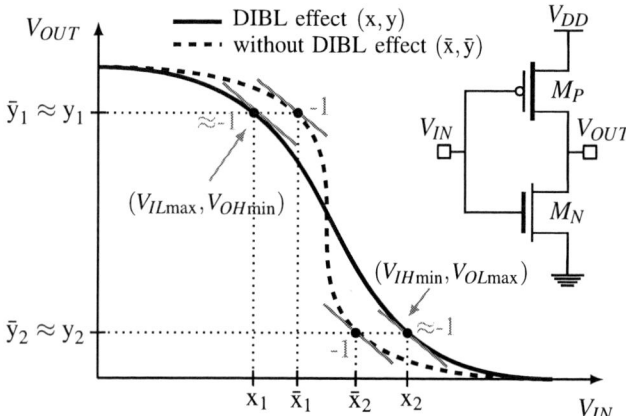

Fig. 2. VTC illustrative diagram of the inverter circuit.

B. Static Model of the Inverter

The voltage transfer curve (VTC) [9] of the inverter circuit illustrated in Fig.2, considering DIBL effects, can be expressed as

$$V_{IN} = C_o + \frac{nU_T}{2} \log\left[\frac{1 - \exp\left(\frac{V_{OUT} - V_{DD}}{U_T}\right)}{1 - \exp\left(\frac{-V_{OUT}}{U_T}\right)} \right] - \lambda V_{OUT} \quad (3)$$

where,

$$C_o = \frac{nV_{DD}(1 - \lambda_p)}{2n_p} + \frac{nU_T}{2} \log\left(\frac{\beta_p}{\beta_n}\right) \quad (4)$$

$$n = 2 \cdot \frac{n_p n_n}{n_p + n_n} \quad (5)$$

$$\lambda = \frac{n}{2} \cdot \left(\frac{\lambda_p}{n_p} + \frac{\lambda_n}{n_n}\right) \quad (6)$$

It is important to note that when $\lambda = \lambda_n = \lambda_p = 0$, Eq. (3) describes an inverter without DIBL effects, and hence

$$\overline{V_{IN}} = \overline{C_o} + \frac{nU_T}{2} \log\left[\frac{1 - \exp\left(\frac{\overline{V_{OUT}} - V_{DD}}{U_T}\right)}{1 - \exp\left(\frac{-V_{OUT}}{U_T}\right)} \right] \quad (7)$$

where,

$$\overline{C_o} = \frac{nV_{DD}}{2n_p} + \frac{nU_T}{2} \log\left(\frac{\beta_p}{\beta_n}\right) \quad (8)$$

C. Standby SNM Considering DIBL effects

As long as the analysis is concerned in standby mode, it is possible to neglect the influence of access the transistors to derive the SNM, and hence the circuit of Fig. 1 can be approximated as two crossed coupled inverters. According with [7], the noise margins low (NM_L) and high (NM_H) of the inverter are defined as

$$NM_L = V_{ILmax} - V_{OLmax} \quad (9)$$

$$NM_H = V_{OHmin} - V_{IHmin} \quad (10)$$

where, V_{OH} and V_{IL} denote, respectively, the high and low nominal output voltages obtained with full-swing input, whereas the pairs (V_{ILmax}, V_{OHmin}) and (V_{IHmin}, V_{OLmax}) denote the unity-gain points, as illustrated in Fig. 2. In order to derive the noise margins including DIBL effects we propose to approximate the vertical axis of the unity-gain points (y_1 and y_2) by the other ones without DIBL effects ($\bar{y}_1$ and $\bar{y}_2$), which can be established by Eq. (7). Therefore, V_{OHmin} is found by equaling the derivative of Eq. (7) to -1 and using the approximation $\exp(-V_{OUT}/U_T) \ll 1$ for high V_{OUT}, yielding

$$V_{OHmin} = y_1 \approx \bar{y}_1 = V_{DD} - U_T \log\left(\frac{n+2}{2}\right) \quad (11)$$

On the other hand, by equaling the derivative of Eq. (7) to -1 and using $\exp((V_{DD} - V_{OUT})/U_T) \gg 1$ for low V_{OUT}, we obtain

$$V_{OLmax} = y_2 \approx \bar{y}_2 = U_T \log\left(\frac{n+2}{2}\right), \quad (12)$$

which agrees with [7]. Now, by evaluating Eq. (3) at $V_{OUT} = V_{OHmin}$ we obtain, using Eq. (11),

$$V_{ILmax} = x_1 = \frac{nV_{DD}}{2}\left(\frac{1}{n_p} - \frac{\lambda_n}{n_n}\right) + U_T \lambda \log\left(\frac{n+2}{2}\right) \\ - \frac{nU_T}{2} \cdot \left[\log\left(\frac{\beta_n}{\beta_p}\right) + \log\left(\frac{n+2}{n}\right)\right] \quad (13)$$

and by evaluating Eq. (3) at $V_{OUT} = V_{OLmax}$, it follows that

$$V_{IHmin} = x_2 = \frac{nV_{DD}}{2}\left(\frac{1}{n_p} + \frac{\lambda_p}{n_p}\right) - U_T \lambda \log\left(\frac{n+2}{2}\right) \\ - \frac{nU_T}{2} \cdot \left[\log\left(\frac{\beta_n}{\beta_p}\right) - \log\left(\frac{n+2}{n}\right)\right] \quad (14)$$

Finally, the NM_L and NM_H defined by Eqs. (9) and (10) can be obtained as

$$NM_L = \frac{nV_{DD}}{2}\left(\frac{1}{n_p} - \frac{\lambda_n}{n_n}\right) + (\lambda - 1)U_T \log\left(\frac{n+2}{2}\right) \\ - \frac{nU_T}{2} \cdot \left[\log\left(\frac{\beta_n}{\beta_p}\right) + \log\left(\frac{n+2}{n}\right)\right] \quad (15)$$

$$NM_H = \frac{nV_{DD}}{2}\left(\frac{2}{n} - \frac{\lambda_p + 1}{n_p}\right) + (\lambda - 1)U_T \log\left(\frac{n+2}{2}\right) \\ - \frac{nU_T}{2} \cdot \left[\log\left(\frac{\beta_p}{\beta_n}\right) + \log\left(\frac{n+2}{n}\right)\right] \quad (16)$$

As can be observed from the above expressions, the influence of supply voltage on the NM_L and NM_H is reduced by the DIBL effects. This property can also be observed by differentiating Eqs. (15) and (16) with respect to V_{DD}, that is

$$\frac{\partial NM_L}{\partial V_{DD}} = \frac{nV_{DD}}{2}\left(\frac{1}{n_p} - \frac{\lambda_n}{n_n}\right) \quad (17)$$

$$\frac{\partial NM_H}{\partial V_{DD}} = \frac{nV_{DD}}{2}\left(\frac{2}{n} - \frac{\lambda_p + 1}{n_p}\right) \quad (18)$$

which presents a similar result as that reported in [10]. The behavior of Eqs. (15) and (16) are shown in Fig. 3(a). This curves indicate low relative error along the simulated supply voltage range, as can be seen in Fig. 3(b).

978-1-5090-2737-8/16 $31.00 © 2016 IEEE

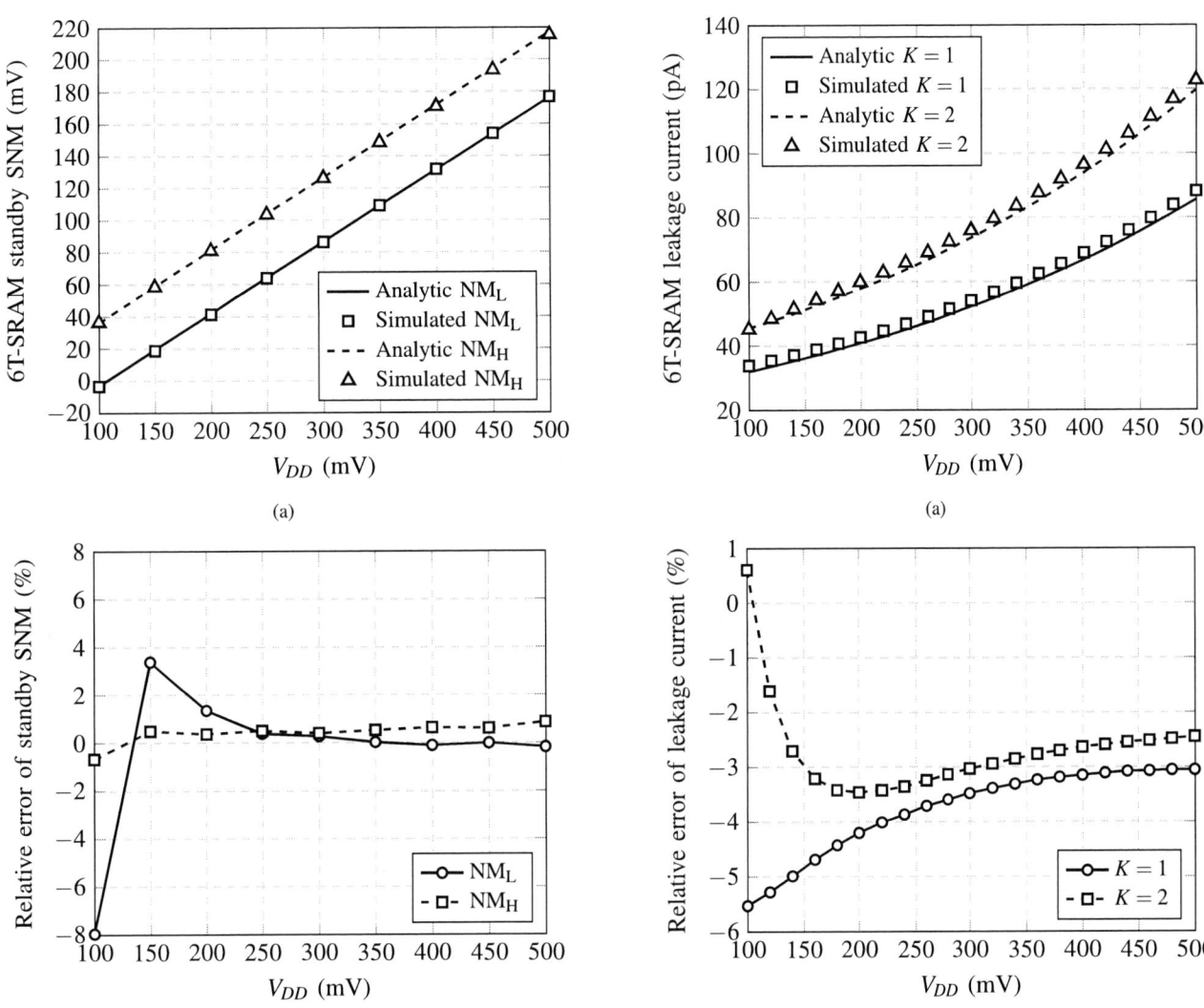

Fig. 3. Analytic noise margin *versus* simulations results: (a) noise margin behavior; (b) relative error (E_r=[Analytic-Simulated]/Simulated).

Fig. 4. Analytic leakage current *versus* simulations results: (a) leakage current behavior; (b) relative error (E_r=[Analytic-Simulated]/Simulated).

D. Leakage Current

According to [11], the total standby leakage current of the 6T-SRAM cell depicted in Fig. 1 can be derived by adding the leakage currents of M_2 (M_1), M_3 (M_4) and M_6 (M_5) when a logic "1"("0") is stored, which leads to

$$I_{Leak} = \beta_2 \exp\left(\frac{\lambda_2 V_{DD}}{n_p U_T}\right) + \beta_3 \exp\left(\frac{\lambda_3 V_{DD}}{n_n U_T}\right) + \\ \beta_6 \exp\left(\frac{\lambda_6 V_{DD}}{n_n U_T}\right) \quad (19)$$

Considering that the width of access transistor M_6 (M_5) is K times of M_3 (M_4), $\beta_2 = \beta_p$ and $\beta_3 = \beta_6 = \beta_n$, Eq. (19) can be written as

$$I_{Leak} = \beta_p \exp\left(\frac{\lambda_p V_{DD}}{n_p U_T}\right) + \beta_n (1+K) \exp\left(\frac{\lambda_n V_{DD}}{n_n U_T}\right) \quad (20)$$

Fig. 4(a) shows the behavior of Eq. (20) with respect to simulations varying V_{DD} from 100mV to 500mV in cases that

$K = 1$ and $K = 2$. The absolute value of relative error (see Fig. 4(b)) is less than 6% along the simulated V_{DD} range.

III. BOUNDARIES BASED ON β PARAMETER

The parameter β defined by Eq. (2) is strongly affected by the manufacturing process, mainly due to threshold voltage and mobility variations. Exploiting the fact that Eqs. (15), (16) and (20) depend on β_n and β_p, we propose design worst-case inequalities by defining values for NM_L, NM_H and I_{Leak} as

$$NM_L \geqslant 0: \quad \beta_p \geqslant \beta_n \cdot C_o^{-1} \cdot \exp\left(-\frac{V_{DD}}{U_T}\left[\frac{1}{n_p} - \frac{\lambda_n}{n_n}\right]\right) \quad (21)$$

$$NM_H \geqslant 0: \quad \beta_p \geqslant \beta_n \cdot C_o \cdot \exp\left(\frac{V_{DD}}{U_T}\left[\frac{2}{n} - \frac{\lambda_p + 1}{n_p}\right]\right) \quad (22)$$

where

$$C_o = \frac{n}{2}\left[1 + \frac{n}{2}\right]^{\frac{2}{n}(\lambda - 1) - 1} \quad (23)$$

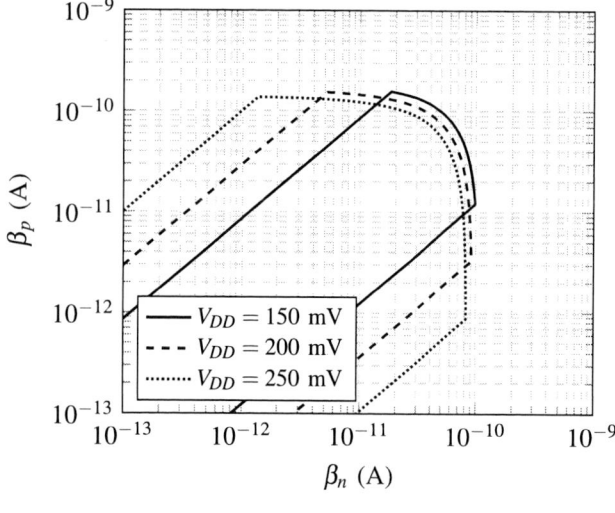

(a)

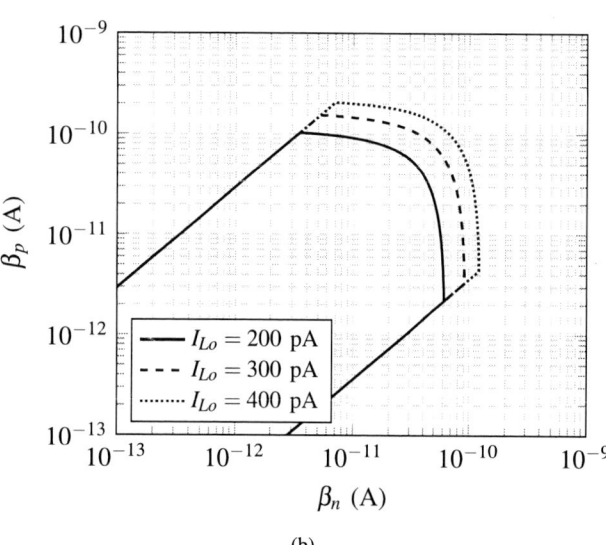

(b)

Fig. 5. Boundaries at different configurations: (a) varying V_{DD} at $I_{Lo} = 300$nA; (b) varying I_{Lo} at $V_{DD} = 0.2$V.

and

$$I_{Leak} \leqslant I_{Lo}: \quad \beta_p \leqslant I_{Lo} \exp\left(-\frac{V_{DD}}{U_T}\frac{\lambda_p}{n_p}\right) - (1+K)\beta_n \exp\left(\frac{V_{DD}}{U_T}\left[\frac{\lambda_n}{n_n} - \frac{\lambda_p}{n_p}\right]\right) \quad (24)$$

These inequalities create boundaries over the β_n-β_p plane in which the variations should be confined. Similar concept was introduced in [12], [13], in which V_{Tn} and V_{Tp} were used as control parameters.

Eqs. (21), (22) and (24) depend on the parameters λ_i and n_i, but their variations are slightly affected by the manufacturing process. Therefore it is possible to adjust the boundaries by varying V_{DD} and I_{Lo} as shown in Figs. 5(a) and 5(b), respectively. Note that the region inside the boundaries, which determines the feasible values of β_n and β_p, increases at the cost of higher values for supply voltage and leakage current.

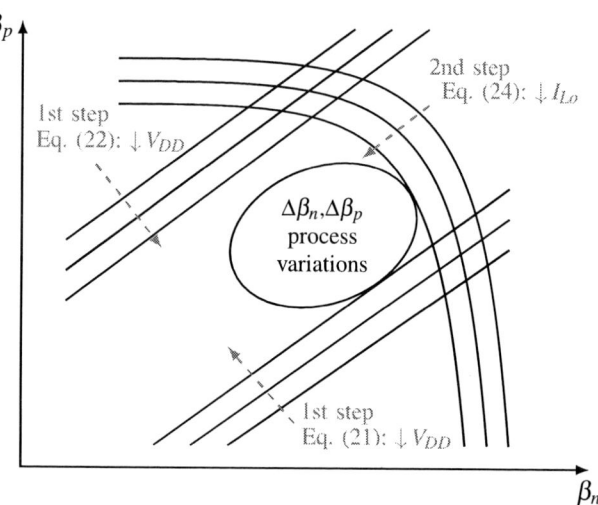

Fig. 6. Illustrative diagram of the design based on boundary equations.

In order to estimate the minimum V_{DD} and maximum leakage current of memory cells, we use the boundaries as indicated in Fig 6. First, by decreasing V_{DD} so does the distance between the boundaries described by Eqs. (21) and (22), which can be adjusted at the limit imposed by $\Delta\beta$ process variations obtained either by modeling or simulation results. Generally, the boundary established by Eq. (21) is the more critical than that of Eq. (22) due to the difference between n-type and p-type mobilities. Once the minimum supply voltage is determined, Eq. (24) is also adjusted by decreasing I_{Lo} down to the limit of $\Delta\beta$ variations, hence the maximum leakage current is obtained.

IV. A DESIGN EXAMPLE

A design example was developed in 28 nm FDSOI CMOS process and validated through simulations using HSPICE, which incorporates a model of intrinsic floating body effect. Minimum transistor lengths were chosen to realize the 6T-SRAM memory cell. The corresponding extracted parameters are listed in Table I at all n-type/p-type corners (TT, FF, SS, FS and SF), being the substrates connected to V_{SS} and V_{DD} for n-type and p-type transistors, respectively. As previously mentioned, and which can also be observed in Table I, the parameters β_n and β_n are the most sensitive to corner variations, in contrast with n_n, n_p λ_n and λ_p, whose values at the TT corner were used to model the boundaries. Regarding the threshold voltage, we used the available regular-V_T (RVT) transistors. Their extracted TT values are $V_{Tn}=447.6$ mV and $V_{Tp}=509.7$ mV (see Eq. (2)).

The boundaries were adjusted at the smallest as possible supply voltage and leakage current, producing the analytic values of 0.2V and 320pA, respectively. Fig. 7 shows the minimum region over 1000 samples of β_n and β_p. It should be noted that the influence of local variations has significant impact, and hence must be taken into account. According to [7] and Fig. 7, FS (fast n-type and slow p-type) corner and its local variations are decisive to determine the minimum supply

978-1-5090-2737-8/16 $31.00 © 2016 IEEE

TABLE I
EXTRACTION RESULT OF TRANSISTOR PARAMETERS ($L_n=L_p=30$ nm, $W_n=80$ nm AND $W_p=240$ nm).

	TT	FF	SS	FS	SF
β_n [pA]	10.67	27.11	4.18	16.2	7.06
β_p [pA]	3.69	9.98	1.37	2.68	5.09
n_n	1.395	1.425	1.368	1.390	1.401
n_p	1.443	1.497	1.401	1.450	1.44
λ_n [m]	84.0	95.0	74.3	84.0	83.7
λ_p [m]	114.3	127.2	103.2	114.2	114.5

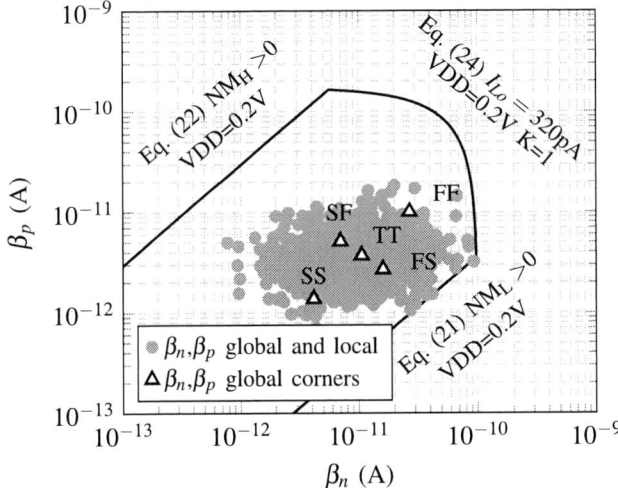

Fig. 7. Adjusted boundaries for β variations with manufacturing process (over 1000 samples).

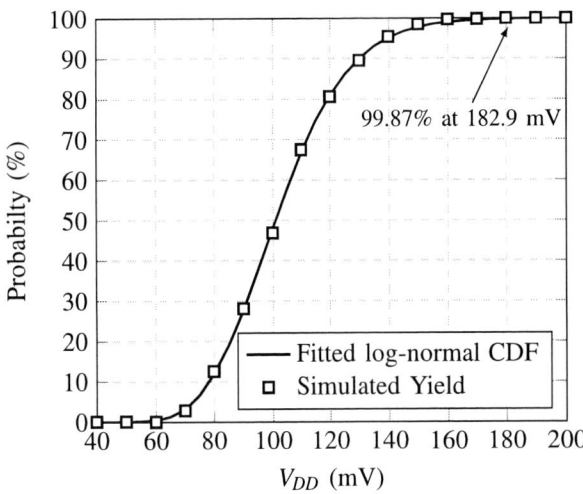

Fig. 8. Yield Monte Carlo simulation results considering global and local variations (over 1000 samples) on the 6T-SRAM at 27°C.

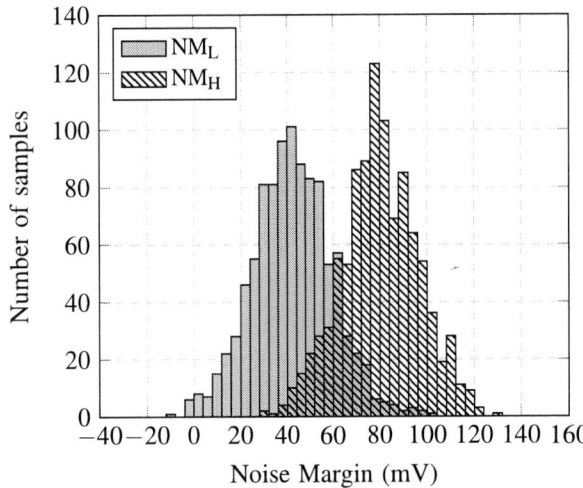

Fig. 9. Noise margin Monte Carlo simulation results considering global and local variations (over 1000 samples) on the 6T-SRAM at 27°C.

voltage. However, Fig. 7 indicates that FF (fast n-type and fast p-type) corner is also important and has influence on the leakage current since it determines the limit of the boundary described by Eq. (24).

Fig. 9 presents the SNM results, and as previously mentioned, NM_L is more critical than NM_H to determine the minimum V_{DD}, as negative SNM values do not ensure the correct operation of the cells. Nevertheless the main simulation to validate the minimum supply is the yield of 6T-SRAM memory cells when the voltage is decreased. Indicated in Fig. 8, 99.87% at $V_{DD}=182.9$ mV was obtained by fitting the yield result with a log-normal cumulative distribution function (CDF). This result is slightly overestimated by the analytic V_{DD} of 200 mV, which was found by adjusting the boundary on the worst-case of β_n and β_p. The difference between the analytic and simulated values is due to the fact that SNM=0 is an approximation to determine the DRV. The leakage current is presented in Fig. 10, which shows a mean value of 56.1 pA and a maximum value of 319.2 pA, that is in close agreement with the maximum analytically predicted value of 320 pA. The performance summary of this design example is shown in Table II.

V. CONCLUSIONS

In this paper, boundaries were advanced by using analytic models that determine a region inside which minimum supply voltage and maximum leakage current of 6T-SRAM memory cells can be predicted. Simulation results were in close agreement with those obtained by using the analytic models, and verified that the proposed boundaries can ensure stable operation of 6T-SRAM under the variability of manufacturing process.

ACKNOWLEDGMENT

This work was supported in part by CNPq and FAPERJ.

REFERENCES

[1] O. Andersson, K. Chon, L. Sornmo, and J. Rodrigues, "A 290 mv sub-vt asic for real-time atrial fibrillation detection," *Biomedical Circuits and Systems, IEEE Transactions on*, vol. 9, no. 3, pp. 377–386, June 2015.

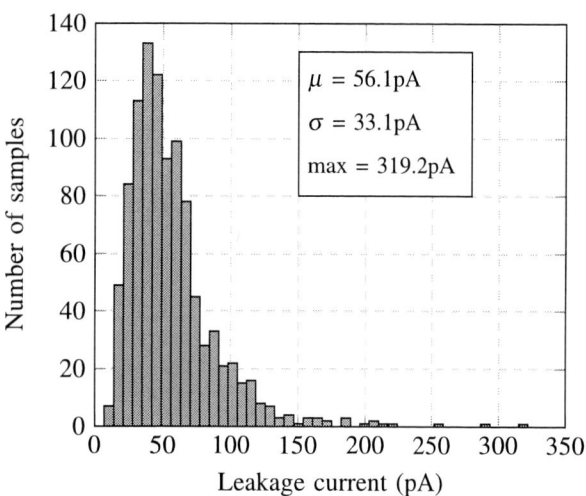

Fig. 10. Leakage current Monte Carlo simulation results considering global and local variations (over 1000 samples) on the 6T-SRAM at 27°C.

TABLE II
PERFORMANCE SUMMARY OF THE 6T-SRAM MEMORY CELL IN STANDBY MODE.

	Analytic	Simulated
Minimum supply	200 mV	182.9 mV
Maximum leakage	320 pA	319.2 pA

[2] Y.-P. Chen, D. Jeon, Y. Lee, Y. Kim, Z. Foo, I. Lee, N. Langhals, G. Kruger, H. Oral, O. Berenfeld, Z. Zhang, D. Blaauw, and D. Sylvester, "An injectable 64 nw ecg mixed-signal soc in 65 nm for arrhythmia monitoring," *Solid-State Circuits, IEEE Journal of*, vol. 50, no. 1, pp. 375–390, Jan 2015.

[3] M. H. Abu-Rahma and M. Anis, *Nanometer Variation-Tolerant SRAM*. New York: Springer New York, 2013.

[4] H. Qin, Y. Cao, D. Markovic, A. Vladimirescu, and J. Rabaey, "Sram leakage suppression by minimizing standby supply voltage," in *Quality Electronic Design, 2004. Proceedings. 5th International Symposium on*, 2004, pp. 55–60.

[5] G. Huang, L. Qian, S. Saibua, D. Zhou, and X. Zeng, "An efficient optimization based method to evaluate the drv of sram cells," *IEEE Transactions on Circuits and Systems I: Regular Papers*, vol. 60, no. 6, pp. 1511–1520, June 2013.

[6] E. Seevinck, F. J. List, and J. Lohstroh, "Static-noise margin analysis of mos sram cells," *IEEE Journal of Solid-State Circuits*, vol. 22, no. 5, pp. 748–754, Oct 1987.

[7] M. Alioto, "Understanding dc behavior of subthreshold cmos logic through closed-form analysis," *IEEE Transactions on Circuits and Systems I: Regular Papers*, vol. 57, no. 7, pp. 1597–1607, July 2010.

[8] Y. Tsividis, *Operation and Modeling of the MOS Transistor*, 2nd ed. Oxford University Press, USA, 2003.

[9] E. A. . Vittoz, "Weak inversion for ultimate low-power logic," in *Low-Power Electronics Design*, C. Piguet, Ed. CRC Press, 2004.

[10] A. Tajalli and Y. Leblebici, "Design trade-offs in ultra-low-power digital nanoscale cmos," *IEEE Transactions on Circuits and Systems I: Regular Papers*, vol. 58, no. 9, pp. 2189–2200, Sept 2011.

[11] H. Qin, "Deep sub-micron sram design for ultra-low leakage standby operation," Ph.D. dissertation, University of California, Berkeley, 2007.

[12] M. Yamaoka, K. Osada, R. Tsuchiya, M. Horiuchi, S. Kimura, and T. Kawahara, "Low power sram menu for soc application using yin-yang-feedback memory cell technology," in *VLSI Circuits, 2004. Digest of Technical Papers. 2004 Symposium on*, June 2004, pp. 288–291.

[13] M. Yamaoka, N. Maeda, Y. Shinozaki, Y. Shimazaki, K. Nii, S. Shimada, K. Yanagisawa, and T. Kawahara, "Low-power embedded sram modules

with expanded margins for writing," in *Solid-State Circuits Conference, 2005. Digest of Technical Papers. ISSCC. 2005 IEEE International*, Feb 2005, pp. 480–611 Vol. 1.

A 0.7V Fully Differential First Order GZTC-C Filter

Pedro Toledo[1,2], Renê Timbo[1,2], David Cordova[1,2], Hamilton Klimach[2], Sergio Bampi[2] and Eric Fabris[1,2]

[1] NSCAD Microeletronica [2] Graduate Program on Microelectronics - UFRGS - Porto Alegre, RS, Brazil

(toledo, rene.timbo, david)@nscad.org.br, hamilton.klimach@ufrgs.br, (bampi, fabris)@inf.ufrgs.br

Abstract—A 0.7 V supply voltage fully differential first order GZTC-C filter is herein proposed. The GZTC-C filter definition is used in this paper as a Transconductance-Capacitor filter (Gm-C) in which its gm stage is biased exactly on transconductance zero-temperature (GZTC) bias condition. This special bias point has all necessary conditions to design MOSFET transconductors with low temperature dependence. Additionally, two calibrate-points with three-bit resolution have been added to mitigate process variations which may cause bias shift errors on sensitive nodes, driving the circuit out of its normal GZTC operation region. The final circuit has been designed in a 130 nm CMOS process generating a 5 MHz cutoff frequency. The filter occupies around 0.01 mm^2 of silicon area while consuming just 21 μW. Monte Carlo (MC) post-layout and post-calibrated simulation has presented an Effective Cutoff Frequency Temperature Coefficient (TC_{eff}) average of 73.4 ppm/$^\circ$C with 30.6 ppm/$^\circ$C standard deviation for a temperature range from -40 to $+120^\circ$C and has estimated a cutoff frequency fabrication sensitivity of $\sigma/\mu = 0.87\%$, including average process and local mismatch variability.

Index Terms—CMOS, analog integrated circuits, Low Temperature Sensitivity Transconductors, GZTC Condition.

I. INTRODUCTION

MOSFET Transconductor (gm) is essential building block for analog, mixed-signal and RF designs. They are used in a large variety of analog/RF circuits, such as filters, multipliers, oscillators, and amplifiers. Consequently, the gm is a key element in determining various performance parameters, such as gain, bandwidth, and noise. A robust transconductance design under temperature and process variations perspective is a must in such applications and recently several works have been addressed such topic [1]–[6] including its impact on blocks such as LNA [7] [8], OTA [9], readout circuit for capacitive MEMS sensors [10], capacitance to frequency converter for MEMS sensors [11], 8-bit digital to analog converter [12] and sigma-delta modulator [13].

These proposed techniques for robust gm generation [1]–[6] have in common the use of some kind of a modified beta multiplier-based Gm-R bias circuit in order to compensate gm transconductance over temperature and/or process variations. For instance, in [1] a novel beta multiplier-based Gm-R bias circuit for implementing a PT-invariant transconductor using a MOSFET in triode region, a Proportional To Absolute Temperature (PTAT) supply generator and process tracking circuit has been proposed. Or as in [2] in which a beta multiplier-based Gm-R bias circuit, which has its resistor that is tuned with a fully integrated CMOS Phase Locked Loop (PLL)

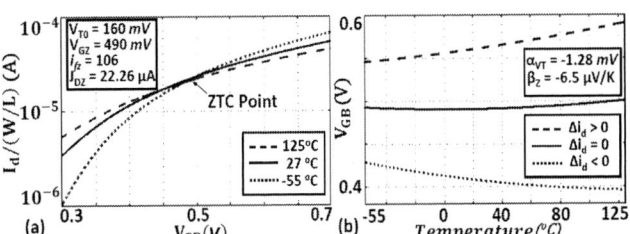

Fig. 1. (a) ZTC condition for an NMOS transistor in a 130 nm process, and (b) $V_{GB}(T)$ for $\Delta I_d > 0$, $\Delta I_d = 0$ and $\Delta I_d < 0$.

locked to an external frequency reference (normally present in most systems), has been used as well. By comparison, all these implementations always have been designed to have both: a suitable bias temperature dependence such that it compensates temperature variation in mobility and/or any kind of the feedback loop in order to mitigate process variations (note that trimming or calibration does not cease to be a kind of feedback).

Recently in [14], following a similar idea from [15], the transconductance zero-temperature (GZTC) bias condition has been delved showing as a good approach to design gm with low temperature dependence. Further, according to [14], to mitigate the gm temperature dependence it is necessary to bias the gm cell with a PTAT bias. That statement is in close agreement to all techniques applied before in the literature [1]–[5].

Main idea of this paper is to design a 0.7 V Fully Differential First Order Gm-C Filter biasing the gm stage exactly on transconductance zero-temperature (GZTC) bias condition. The Gm-C filter biased on GZTC condition is defined here as GZTC-C filter. To support the design flow, a brief review of ZTC and GZTC analysis and qualitative analysis of GZTC vicinity behavior are both introduced.

The paper is organized as follows: Section II presents a brief review of ZTC and GZTC analysis for the MOSFET transistor, based on a continuous MOSFET model that can predict its behavior from weak to strong inversion [16], and a qualitative investigation of GZTC vicinity behavior. In Section III, the 0.7 V supply voltage fully differential first order GZTC-C filter is presented including its design flow. Simulation results are shown in Section IV and Section V presents the conclusions.

978-1-5090-2737-8/16 $31.00 © 2016 IEEE

II. A BRIEF REVIEW OF MOSFET ZTC AND GZTC CONDITION

A. MOSFET ZTC Condition Review

The MOSFET bias ZTC condition derives from the mobility and threshold voltage dependencies on temperature mutual cancellation, which happens at a particular transistor gate voltage bias resulting a drain current that barely depends on the temperature, as can be seen in Fig. 1(a). Shoucair [17] and Prijic [18] were the first to report the bias ZTC point for bulk CMOS in both the linear and the saturation regions. All investigations have been done for a temperature ranged between 27℃ and 200℃.

Few years later, in [19], Osman investigated the drain current ZTC point in Partially Depleted (PD) Silicon-on-Isolator (SOI) MOSFET. In his analysis it also has been included the thermal dependence of degradation of low-field mobility with respect to the applied transverse electric field. All these previous ZTC modeling [17]–[19] were aiming to provide an accuracy model against the CMOS or SOI-MOS technology of that time. Nevertheless, it visible that all these proposed modeling so far were no longer "user-friendly" for analog design purpose.

Then in 2001, Fylanovisky made this analysis more intuitive for MOSFET operating in strong inversion and saturation regime [15]. From [15], the ZTC operating point is given by Eqs. (1) and (2)

$$V_{GZ} = V_{T0}(T_0) + nV_{SB} - \alpha_{V_{T0}}T_0 \qquad (1)$$

$$J_{DZ} = \frac{I_{DZ}}{(W/L)} = \frac{\mu_n(T_0)T_0^2 C'_{ox}}{2n}\alpha_{V_{T0}}^2 \qquad (2)$$

where T_0 is the room temperature, $V_{T0}(T_0)$ is the threshold voltage at room temperature, n is the slope factor, V_{SB} is the source-bulk voltage, $\alpha_{V_{T0}}$ is the thermal coefficient of the threshold voltage (stressing that V_T decreases with T), $\mu_n(T_0)$ is the low field mobility at room temperature, C'_{ox} is the oxide capacitance per unit of area and $\frac{W}{L}$ is the transistor aspect ratio. J_{DZ} can be defined as ZTC normalized drain-current and one can readily conclude that V_{GZ} and J_{DZ} are only dependent on device fabrication processes.

Fig. 1(a) shows the drain current (in a log scale) as a function of the gate-bulk voltage (V_{GB}) of a saturated long-channel NMOSFET, simulated under temperatures ranging from -55^oC to $+125^oC$, for a regular transistor in a commercial 130nm CMOS process. The ZTC operation point can be seen around $V_{GB} \approx 490mV$ for a transistor with $V_T = 160mV$, resulting that the ZTC point occurs for an overdrive voltage around 330mV, meaning the transistor operates in strong inversion.

Following this context, in [20], Toledo et al. have realized that the ZTC condition does not only happens in strong inversion regime. In other words, a more complete MOSFET model should have been used to model this condition, such as the one presented in [16], that describes continuously the transistor behavior at any inversion level. From this more complete

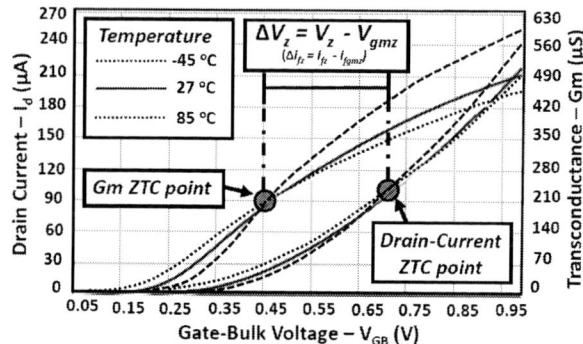

Fig. 2. ZTC and GZTC condition for a PMOS transistor in a 130 nm process with $V_{T0} = 250$ mV.

analysis, the condition for which the drain current has its temperature dependence negligible $((\partial I_D)/(\partial T)|_{T=T_1} = 0)$ is given by Eq. (3).

$$\frac{|\alpha_{V_{T0}}|q}{nk} = \left(\frac{\alpha_\mu + 2}{2}\right)\left(\frac{-i_{fz}}{\sqrt{1+i_{fz}}+1}\right) + \left[\sqrt{1+i_{fz}} - 2 + ln\left(\sqrt{1+i_{fz}}-1\right)\right] \qquad (3)$$

where k is the Boltzmann constant, q is the elementary electric charge, α_μ is the temperature dependence power coefficient for the mobility, and i_{fz} is defined as the ZTC forward inversion level. Eq. (3) means that when a saturated transistor is biased in this inversion level at the source i_{fz}, the drain current results insensitive to temperature.

In parallel, the ZTC vicinity condition has also been investigated. Firstly defined in [15] and subsequently widespread in [20], Eq. (4), (5) and (6) show how the transistor operates if it is biased in ZTC vicinity.

$$V_{GB}(T) \approx V_{GZ} - \frac{\alpha_{V_{T0}}\Delta i_f}{2f(i_{fz})(\sqrt{1+i_{fz}}-1)}T = V_{GZ} - \frac{\beta_z\Delta I_d}{I_{SQ}\frac{W}{L}}T \qquad (4)$$

$$\beta_z = \frac{\alpha_{V_{T0}}}{2f(i_{fz})(\sqrt{1+i_{fz}}-1)} \qquad (5)$$

$$f(i_{f(r)}) = \left[\sqrt{1+i_{f(r)}} - 2 + ln\left(\sqrt{1+i_{f(r)}}-1\right)\right] \qquad (6)$$

where $\Delta I_d = I_D - I_{DZ}$ indicates how far the transistor is biased from the ZTC operating point and β_z is defined as the ZTC slope. Eq. (4) shows that V_{GB} presents a linear temperature dependence in the vicinity of V_{GZ}, and that this dependence can be positive or negative, depending on the ΔI_d chosen, as shown in Fig. 1 (b).

B. MOSFET GZTC Condition Review

Similarly to the last subsection II-A, an inversion level (i_{fgz}), which has a transconductance with low temperature dependence, can be found and defined as well [14], [21]. This condition is called the transconductance zero temperature coefficient, or GZTC, and its definition is important since in any analog signal processing block the gain is fundamentally determined by the transistors transconductance. If the design is developed with the GZTC point in mind, the gain results less sensitive to temperature.

From [14], [21], applying the condition $(\partial g_{mg})/(\partial T)|_{T=T_1} = 0$, where g_{mg} is the gate-bulk transconductance, Eq. (7) is achieved.

$$0 = \alpha_\mu(\sqrt{1 + i_{fgz}} - 1) - 2 +$$
$$\frac{\sqrt{1 + i_{fgz}} + 1}{\sqrt{1 + i_{fgz}}} \left(\frac{|\alpha_{V_{T0}}|q}{nk} + 2 - ln\left(\sqrt{1 + i_{fgz}} - 1\right) \right) \quad (7)$$

where i_{fgz} is defined as GZTC forward inversion level.

One can note that, as in the case of the bias current I_D, the condition GZTC derives from the mutual cancellation of the mobility and threshold voltage dependencies on temperature, which happens for a particular bias condition, $V_{GB}(i_{fgz}) = V_{GGZ}$. Another interesting conclusion from [14], [21] is that ZTC will be aways above the GZTC bias point, i. e., GZTC is working in ZTC vicinity with a Complementary To Absolute Temperature (CTAT) behavior (Fig. 1 (b)), as can be also seen in Fig. 2 for a PMOS transistor in the same 130 nm technology.

Finally in [21], since the GZTC is working in ZTC vicinity with a CTAT behavior it is evaluated the necessary positive temperature coefficient (TC_I) to maintain i_{gz} stable over temperature variations in such manner that the transconductance (gm_g) becomes temperature independent. In this case, the requested Current Bias GZTC Temperature Coefficient (TC_{IGZ}) is given by

$$\frac{|\alpha_{V_{T0}}|q}{nk} = \left[\frac{TC_{IGZ}T_0}{2} - \left(\frac{\alpha_\mu + 2}{2} \right) \right] \left(\frac{i_{fgz}}{\sqrt{1 + i_{fgz}} + 1} \right) +$$
$$\left[\sqrt{1 + i_{fgz}} - 2 + ln\left(\sqrt{1 + i_{fgz}} - 1\right) \right]$$

Values for TC_{IGZ} between 3000 and 9500 ppm/°C have been calculated, which comply with CMOS PTAT current reference found in literature, where TC_I between 1000 and 10000 are readily achieved [22]. As final result, combination of Eq. (7) with Eq. (8) means that if the MOSFET transductors are biased in GZTC condition along with right amount of PTAT current defined by TC_{IGZ} coefficient, the transconductance will present very low temperature sensitivity.

C. GZTC Vicinity Qualitative analysis

A hefty share of the problem when a specific bias point is used in the design is that after fabrication this expected designed biasing or inversion level likely will not be the measured one. Once the variability is intrinsic to the CMOS technology, the GZTC bias condition is also sensitive to process variations. After fabrication, the GZTC bias condition can be driven out of its normal operation, i.e., the bias condition can be working into the GZTC vicinity. A qualitative knowledge of the GZTC vicinity behavior gives to the designer how transconductance varies with temperature and it is here presented.

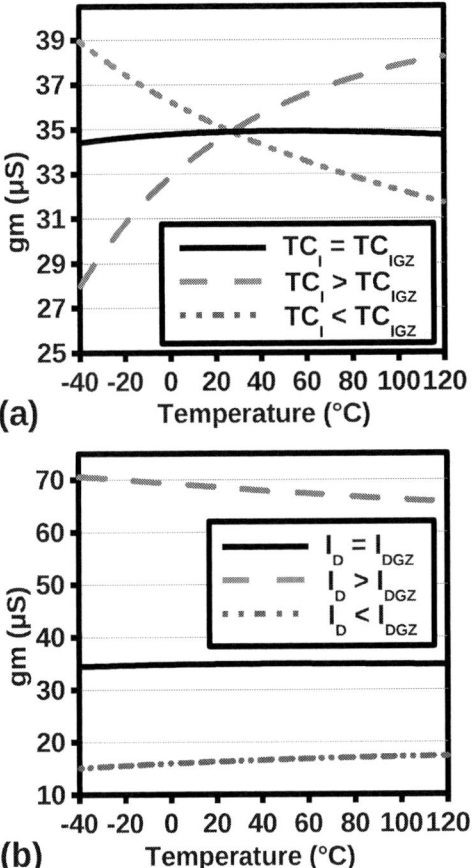

Fig. 3. (a) GZTC vicinity for different TC_I (b) GZTC vicinity for different I_D for an 130 nm process.

Figs 3 (a) and (b) show the transconductance versus temperature for different TC_I and I_D around the GZTC bias condition (TC_{IGZ}, I_{DGZ}), respectively. Figs 3 (a) shows that gm presents a linear temperature dependence in the vicinity of GZTC and that this dependence can be positive or negative, depending on if the TC_I is larger or not than TC_{IGZ}. In similar way, Figs 3 (b) also shows that gm presents a linear temperature dependence which can be positive or negative, depending on how far the MOS transistor is biased from I_{DGZ}.

III. A 0.7 V FULLY DIFFERENTIAL FIRST ORDER GZTC-C FILTER

The Figs. 4 (a), (b) and (c) show the proposed 0.7 V fully differential first order GZTC-C filter. It is composed by two symmetric balanced fully differential CMOS operational transconductance amplifier (OTA) [23] and C_p capacitors. These OTAs have in common an inherent common-mode detection [23] that can be used to economically implement the common-mode feedback circuit. The former one, $GZTC_1$, is formed by regular transistors NM_1 to NM_9, PM_1 to PM_6 and a pole TC_{eff} calibration cell. The latter one is just comprised by the transistors inside of dotted red box.

The pole TC_{eff} calibration cell is implemented using diode-connected high threshold voltage (NM_{H1} to NM_{H7})

978-1-5090-2737-8/16 $31.00 © 2016 IEEE

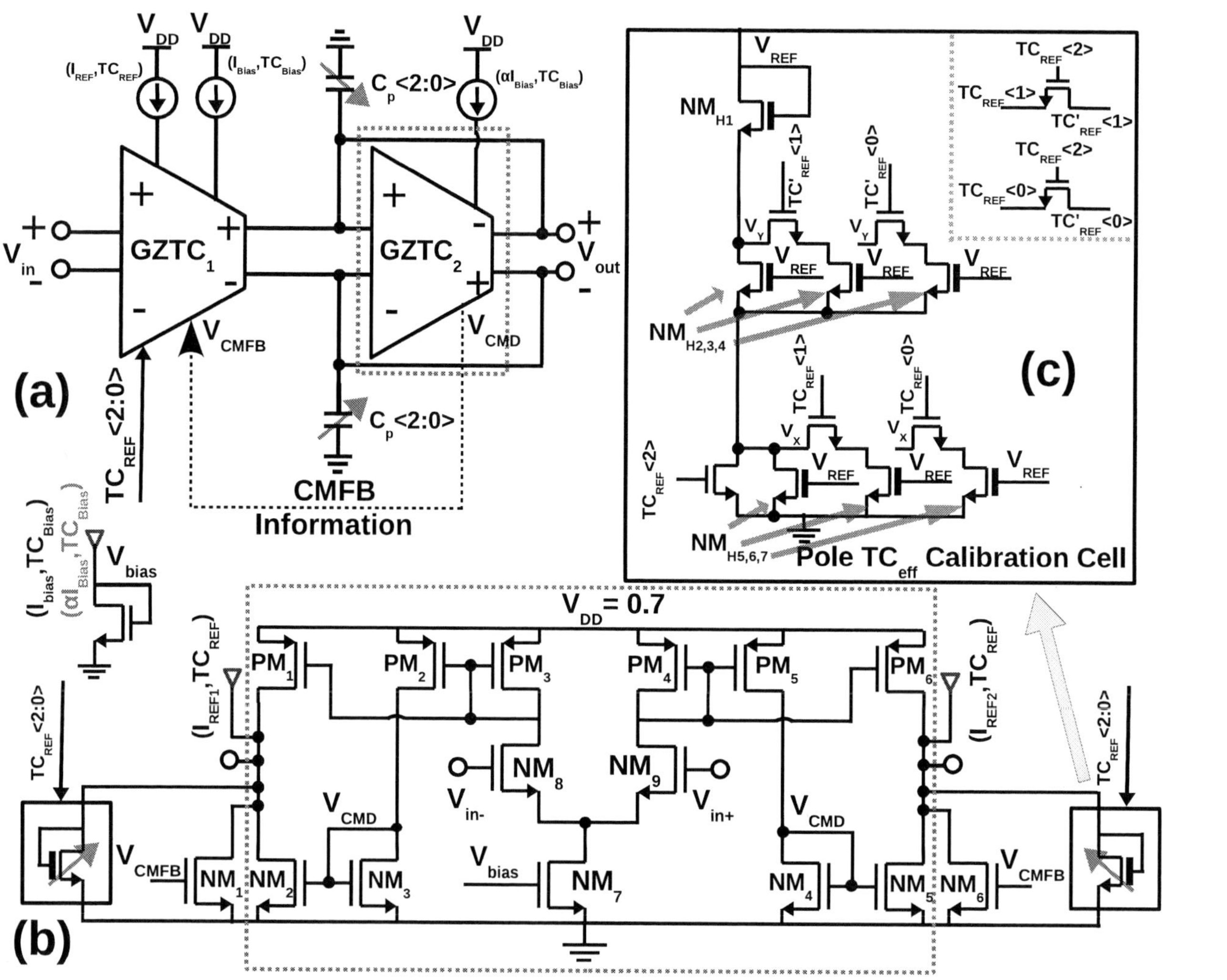

Fig. 4. (a) Fully Differential First Order GZTC-C Filter (b) symmetric balanced fully differential CMOS operational transconductance amplifier [23] (c) The pole TC_{eff} calibration cell.

transistors in which there are regular transistors inside working as switches to trim their final widths (W) and lengths (L) values. This cell is in charge to calibrate the filter transconductance back to GZTC condition, which can be deviated from its normal operation due to process variations, as explained at subsection II-C. In addition, there is also a extra trimming point with three bit resolution in parallel to C_p capacitor, as shown in Fig. 4 (a).

A. Circuit Analysis and Design

The proposed filter low frequency gain and dominant pole frequency are defined by transconductance parameters as indicated in Eq. (9).

$$\frac{V_{out}(s)}{V_{in}(s)} = \frac{g_{m8,9_{GZ1}}}{(g_{m8,9_{GZ2}} + g_{mH}) + sC_p} \quad (9)$$

where $g_{m8,9_{GZ1}}$ is the transconductance of the first OTA, $g_{m8,9_{GZ2}}$ is the transconductance of the second OTA and g_{mH} is the transconductance provided by the pole TC_{eff} calibration cell.

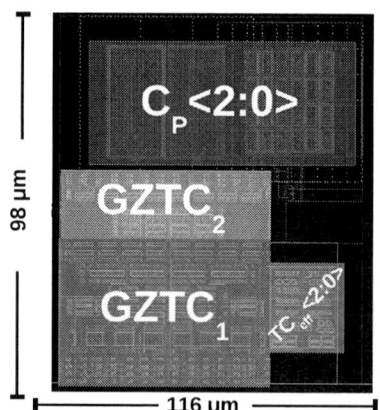

Fig. 5. Layout - $116\mu m$ X $98\mu m$.

Considering the three fixed bias currents (I_{REF}, I_{BIAS} and αI_{BIAS}), the $NM_{8,9}$ for both $GZTC_1$ and $GZTC_2$ OTAs aspect ratios (W/L) must be defined so that their inversion

978-1-5090-2737-8/16 $31.00 © 2016 IEEE

TABLE I
COMPARISON WITH OTHER ROBUST TRASNCONDUCTANCE CIRCUITS

Reference	CMOS Process	Fabrication Sensitivity (σ/μ)	Average Current/V_{DD}	TC_{eff} (ppm/^{o}C)	Temperature Range (^{o}C)	On/Off Chip Trimming
This Work*†	0.13 μm	0.87%	29 μA/0.7V	73.4	-40 to 120	on-chip digital trimming
TVLSI 2015$^{\oplus}$ [1]	0.18 μm	2.24%	55 μA/2.5V	16,000	0 to 100	None
JSSCC 2001$^{\oplus}$ [2]	0.35 μm	3%	378μA/3.3V	22.000	0 to 60	on-chip digital trimming + External Frequency
Spring J. of AICSP, 2003* [3]	0.18 μm	3.84%	270μA/1.8V	130	25 to 125	None
Spring J. of AICSP, 2011* [6]	0.5 μm	NA	46 nA /1.8V	5000	0-100	Precise External Resistor
ISCAS 2006* [4]	0.18 μm	NA	850 μA/1.5V	15000	0 to 60	Master-Slave tuning
NWSCAS 2005* [5]	0.18 μm	0.5%	NA/1.3V	5000	0 to 100	Precise External Resistor

* Simulation Results †Process and Mismatch (100 runs) $^{\oplus}$Measurement

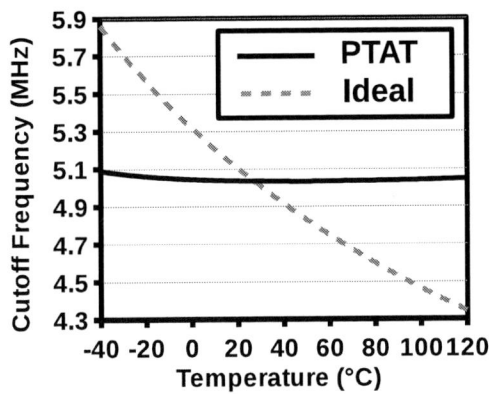

Fig. 6. Filter biased on GZTC Bias condition with and without required TC_{IGZ}.

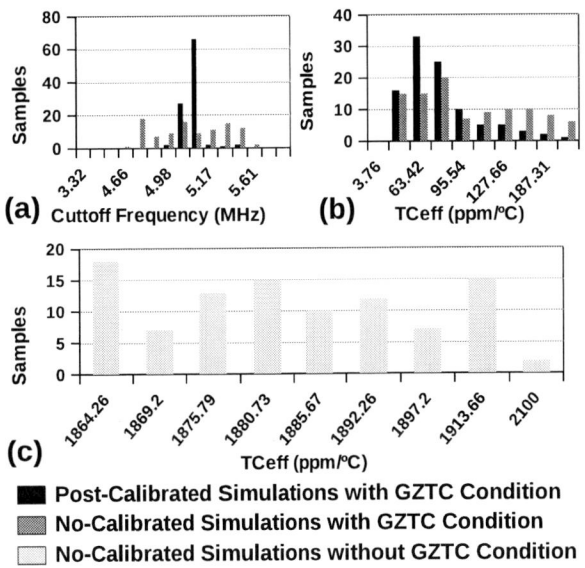

Fig. 7. MC simulation for (a) cutoff frequency and (b) (c) TC_{eff}.

levels are i_{fgz} and then it is necessary to ensure that the GZTC bias condition remains stable over temperature. As this condition is located in ZTC bias vicinity (below the bias ZTC point), it is necessary to cancel its CTAT bias behavior (Fig. 1 (b)) by applying a small amount of a PTAT bias current (TC_{REF} and TC_{BIAS}), as shown in Fig. 4 (a) and described by Eq. (8). It deserves to note that the TC_{REF} stabilizes the GZTC condition for the high voltage transistor within the pole TC_{eff} calibration cell and the TC_{BIAS} for the regular transistors in the OTAs.

B. Layout

The layout is small, occupying around 0.01 mm^2, as shown in Fig. 5. The devices placement has been performed taking all precautions to minimize the global mismatch effects.

IV. SIMULATION RESULTS

The proposed filter has been designed and simulated in a 130 nm CMOS commercial process. The calculated bias currents I_{REF}, I_{BIAS} and αI_{BIAS} are 3, 3 and 1.5 μA for the transconductors, with a TC_{REF} and TC_{BIAS} equal to 2800 and 3000 ppm/^{o}C, respectively. The capacitor C_p = 500 fF has been used to set the cutoff frequency equals to 5 MHz along with extra parallel small capacitors for trimming issues.

Fig. 6 presents the cutoff frequency versus temperature for both case: with required PTAT current behaviour and using a ideal current source. The black continuous curve shows that the cutoff frequency presents low temperature sensitivity,

978-1-5090-2737-8/16 $31.00 © 2016 IEEE

having around a TC_{eff} = 58 ppm/^{o}C under V_{DD} = 0.7V. The resulting coefficient is comparable to TCs obtained in some CMOS voltage and current reference circuits found in the literature [21]. On the other hand, if it is not applied the right amount of PTAT current the cutoff frequency will be more susceptible to temperature variations, as also shown in dashed red line in Fig. 6 resulting a TC_{eff} = 1885 ppm/^{o}C.

To predict vulnerability to manufacturing process variations (average process variations + mismatch), 100 simulations samples have been performed for the proposed first order filter. Both cases have been accomplished: Ideal current source and PTAT current source. Fig. 7 (a) shows the Monte Carlo simulations of the cutoff frequency for post-calibrated and no-calibrated scenario, presenting a fabrication sensitivity of $\sigma/\mu = 0.87\%$ and 3.79%, respectively.

Fig. 7 (b) shows the Monte Carlo simulations for the TC_{eff} in the post-calibrated and no-calibrated configuration. From that, one can note that the TC_{eff} spreading has been reduced with the aid of the pole TC_{eff} calibration cell decreasing average value from 95.5 to 73.4 ppm/^{o}C and the standard deviation from 45.8 to 30.6 ppm/^{o}C. Also, Fig. 7 (c) further reinforces that without the right amount of PTAT current the cutoff frequency will be more susceptible to temperature variations.

Finally, Table I presents a comparison of recently published robust transconductance circuits generator and techniques. Clearly the main advantages of our technique is the competitive low temperature coefficient TC_{eff} under wide temperature range and the low fabrication sensitivity even using low resolution trimming (In our design only 3 bits).

V. CONCLUSIONS

A fully differential first order GZTC-C filter has been presented. The ZTC and GZTC conditions have been both reviewed and a new design technique has been also proposed. The circuit has been designed in a 0.13 μm CMOS process, generating an average cutoff frequency of 5 MHz at room temperature under a power supply higher than 0.7 V. Post-layout and post-calibrated simulation for typical device parameters has resulted an effective temperature coefficient of 58 ppm/^{o}C from -40 to 120 ^{o}C, and a maximum of 217 ppm/^{o}C for the same temperature range including process and mismatch variability effects. Monte Carlo simulations show a spread of $\sigma/\mu = 0.87\%$ for average process variation and mismatch while consuming a maximum power consumption of 21 μW and the silicon area of $0.010mm^2$.

REFERENCES

[1] A. Amaravati, M. Dave, M. S. Baghini, and D. K. Sharma, "A fully on-chip pt-invariant transconductor," *IEEE Transactions on Very Large Scale Integration (VLSI) Systems*, vol. 23, no. 9, pp. 1961–1964, Sept 2015.

[2] A. McLaren and K. Martin, "Generation of accurate on-chip time constants and stable transconductances," *IEEE Journal of Solid-State Circuits*, vol. 36, no. 4, pp. 691–695, Apr 2001.

[3] J. Chen and B. Shi, "Circuit design of an on-chip temperature-compensated constant transconductance reference," *Analog Integrated Circuits and Signal Processing*, vol. 37, no. 3, pp. 215–222. [Online]. Available: http://dx.doi.org/10.1023/A:1026221809719

[4] N. Talebbeydokhti, P. K. Hanumolu, P. Kurahashi, and U.-K. Moon, "Constant transconductance bias circuit with an on-chip resistor," in *Circuits and Systems, 2006. ISCAS 2006. Proceedings. 2006 IEEE International Symposium on*, May 2006, pp. 4 pp.–2860.

[5] M. Danaie and R. Lotfi, "A low-voltage high-psrr cmos ptat amp; constant-g/sub m/ reference circuit," in *Circuits and Systems, 2005. 48th Midwest Symposium on*, Aug 2005, pp. 1807–1810 Vol. 2.

[6] V. Agarwal and S. Sonkusale, "Ultra low power pvt independent sub-threshold gm-c filters for low frequency biomedical applications," *Analog Integr. Circuits Signal Process.*, vol. 66, no. 2, pp. 285–291, Feb. 2011. [Online]. Available: http://dx.doi.org/10.1007/s10470-010-9546-9

[7] M. El Kaamouchi, M. Moussa, J. Raskin, and D. Vanhoenacker-Janvier, "Zero-temperature-coefficient biasing point of 2.4-ghz lna in pd soi cmos technology," in *Microwave Conference, 2007. European*, Oct 2007, pp. 1101–1104.

[8] D. Gomez, M. Sroka, and J. L. G. Jimenez, "Process and Temperature Compensation for RF Low-Noise Amplifiers and Mixers," *IEEE Transactions on Circuits and Systems I: Regular Papers*, vol. 57, no. 6, pp. 1204–1211, jun 2010.

[9] Y. Wang and V. P. Chodavarapu, "High-temperature general purpose operational amplifier in IBM 0.13 µm CMOS process," in *2014 IEEE International Conference on Electron Devices and Solid-State Circuits*. IEEE, jun 2014, pp. 1–2.

[10] ——, "Design of a CMOS readout circuit for wide-temperature range capacitive MEMS sensors," in *Fifteenth International Symposium on Quality Electronic Design*, vol. 9. IEEE, mar 2014, pp. 738–742.

[11] ——, "Design of CMOS capacitance to frequency converter for high-temperature MEMS sensors," in *2013 IEEE SENSORS*. IEEE, nov 2013, pp. 1–4.

[12] K. S. Greig and V. P. Chodavarapu, "Extreme wide-temperature range 8-bit digital to analog converter in bulk CMOS process," in *2014 IEEE 27th Canadian Conference on Electrical and Computer Engineering (CCECE)*, no. 2. IEEE, may 2014, pp. 1–5.

[13] Y. Wang and V. P. Chodavarapu, "Design of a sigma-delta modulator in standard CMOS process for wide-temperature applications," in *Sixteenth International Symposium on Quality Electronic Design*. IEEE, mar 2015, pp. 107–111.

[14] P. Toledo, H. Klimach, D. Cordova, S. Bampi, and E. Fabris, "MOSFET ZTC Condition Analysis for a Self-biased Current Reference design," *Journal of Integrated Circuits and Systems*, vol. 10, no. 2, pp. 103–112, December 2015.

[15] I. Filanovsky and A. Allam, ""mutual compensation of mobility and threshold voltage temperature effects with applications in cmos circuits"," *IEEE Trans. Circuits Syst. I, Fundam. Theory Appl.*, vol. 48, no. 7, pp. 876–883, Jul. 2005.

[16] C. Schneider and C. Galup-Montoro, *CMOS Analog Design Using All-Region MOSFET Modeling*, 1st ed. Cambridge University Press, 2010.

[17] F. Shoucair, "Analytical and experimental methods for zero-temperature-coefficient biasing of mos transistors," *Electronics Letters*, vol. 25, no. 17, pp. 1196–1198, Aug 1989.

[18] Z. Prijic, S. S. Dimitrijev, and N. Stojadinovic, ""the determination of zero temperature coefficient point in cmos transistor"," *Microelectron. Reliab.*, vol. 32, no. 6, pp. 769–773, 1992.

[19] A. Osman, M. Osman, N. Dogan, and M. Imam, "Zero-temperature-coefficient biasing point of partially depleted soi mosfet's," *Electron Devices, IEEE Transactions on*, vol. 42, no. 9, pp. 1709–1711, Sep 1995.

[20] p. Toledo, H. Klimach, D. Cordova, S. Bampi, and E. Fabris, ""mosfet ztc condition analysis for a self-biased current reference design"," *The Journal of Integrated Circuits and Systems*, vol. 10, no. 2, pp. 103–112, 2015.

[21] P. Toledo, "Mosfet zero-temperature-coefficient (ztc) effect modeling and analysis for low thermal sensitivity analog applications," Master's thesis, UNIVERSIDADE FEDERAL DO RIO GRANDE DO SUL, DOI: 10.13140/RG.2.1.2965.9286, sep 2015.

[22] F. Serra-Graells and J. Huertas, "Sub-1-v cmos proportional-to-absolute temperature references," *Solid-State Circuits, IEEE Journal of*, vol. 38, no. 1, pp. 84–88, Jan 2003.

[23] A. N. Mohieldin, E. Sanchez-Sinencio, and J. Silva-Martinez, "A low-voltage fully balanced ota with common mode feedforward and inherent common mode feedback detector," in *Solid-State Circuits Conference, 2002. ESSCIRC 2002. Proceedings of the 28th European*, Sept 2002, pp. 191–194.

Software-Defined Radio Design based on GALS Architecture for FPGAs

Eduardo Lussari[1,2], Duarte L. Oliveira[1], Lester A. Faria[1], Orlando Verducci[1]

[1]Instituto Tecnológico de Aeronáutica – Divisão de Engenharia Eletrônica – SJC – SP – Brazil
[2]Oldebrecht Defesa e Tecnologia – SJC – SP – Brazil
e-mail: lussari@gmail.com, [duarte,lester]@ita.br, verducci@gmail.com

Abstract— The design, implementation and comparison of Software-Defined Radio (SDR) based on GALS architectures were focused on. GALS port controllers, previously proposed for implementation in ASIC, have been redesigned for use in conventional FPGAs, eliminating the need for hard macros. A GALS architecture for SDR was proposed and validated, comprising a wrapper described in VHDL. In addition, some guidelines for its implementation in synchronous designs are shown. Whereas asynchronous wrappers were designed to be robust and to dispense timing verification, the verification effort is negligible, since the same test benches can be used for both designs. When compared with a synchronous design, a reduction of 39% in the dynamic consumption was obtained, what may be even greater if the idle clock periods are considered. We saw that for the same device, it is possible to earn 53% in the data flow (throughput), signifying that the radio is able to process a broader band, or that the cost of equipment can be reduced, once it is possible to adopt a cheaper FPGA version with lower performance.

Keywords— asynchronous wapper; XBM specification; port controller; GALS pipeline

I. INTRODUCTION

Historically, common radio equipment is designed to perform specific functions in a fixed and well-defined environment, facilitating its optimization in terms of consumption and performance. On the other hand, radios that are able to perform different functions and to adapt to various environments can offer significant advantages in effectiveness and in terms of logistics [1].

Software-defined radio (SDR) are able to withstand numerous waveforms (set of parameters such as modulation, operating frequency, coding and encryption of the link), which requires signal digital processing resources with high performance and the possibility to be reprogrammable by the user. However, problems of consumption, performance and development costs still appear as major challenges [1].

Digital systems with SDR typically use a global clock signal to synchronize operations. However, the use of a global clock imposes limitations to the circuit design, such as: *a)* reduced performance, since the minimum period of the global clock is limited to the longest path of the combinatorial logic; *b)* Power consumption, because all parts of the circuit use the same clock, making even an idle sub circuit to consume power; *c)* High noise generated by the global clock, causing spikes in circuit power lines and high-frequency electromagnetic emissions in the same frequency of the global clock and its upper harmonics; and *d)*

Modularity and reuse, because if a component present a low performance when compared to the other sub circuits, the global circuit will have its performance degraded [2,3].

Any design style that does not work under a single global control signal is considered as being asynchronous. Asynchronous circuits have the advantages of not having a clock signal to synchronize their operations. They operate by events and their operations are synchronized by a handshake protocol. However, asynchronous circuits have, as disadvantages, the lack of design tools, difficulties in hazard-free design and the fact that the asynchronous logic is not familiar to designers [3].

A solution that significantly reduces the problems related to the global clock is the Globally Asynchronous Locally Synchronous methodology (GALS), proposed initially by Chapiro [4]. A GALS system consists of functional synchronous modules, which presents individual clocks with non-related frequencies. They communicate to each other asynchronously. This paradigm is a combination of synchronous and asynchronous paradigms, where each one of the modules is synchronous. To handle the asynchronous communication between modules, an interface circuit must be added around each synchronous module, creating an asynchronous wrapper (see Fig. 1) [5]. A local clock, a FIFO, or an asynchronous communication controller (inputs ports, output ports) may compose this interface, for example. Techan et al. [6] shows different styles of asynchronous interfaces focusing on GALS systems. Among all of them, the use of "communication ports" show to be very interesting, once they allow removing the handshake of synchronous modules. Therefore, the synchronous modules can be designed through standard and well-known techniques of synchronous design (see Fig. 2).

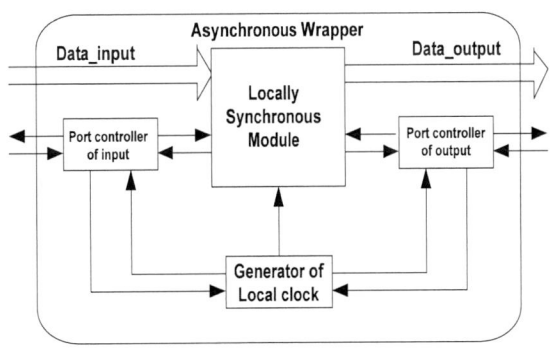

Fig.1. Asynchronous wrapper.

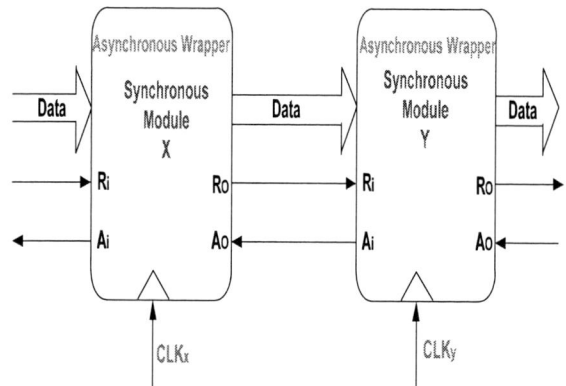

Fig.2. Data communication: synchronous modules.

This approach aims to join the advantages of synchronous design (easy to design and availability of tools), with the advantages of asynchronous design (low power and sub circuits modularity) [6]. The GALS approach has been successfully implemented in different ASICs, such as in the cryptographic processors Acacia [7] and SAFER GALS [8]. There are also studies on GALS design for FPGAs, such as a proposal for a reprogrammable chip called GAPLA [9], and GALS architectures synthesizable in commercial FPGAs, such as [10].

Among the various existing technologies for embedded processing, Field Programmable Gate Arrays (FPGAs) present great importance because they allow to implement a dedicated hardware for high efficiency processing, as well as in an Application Specific Integrated Circuits (ASIC), besides having the possibility to be reprogrammed at any time. For purposes of SDR designs in FPGAs, the use of Globally Asynchronous Locally Synchronous (GALS) circuits present interesting features, allowing implementing circuits with reduced power consumption, reusable sub circuits and performance improvements in processing rate.

Concerning to SDR, there is implementation in different technologies [11,12]. Several studies have implemented SDR based on FPGAs, such as the works of [13] and [14]. However, these proposals are based on synchronous designs without exploring the advantages that the GALS approach can offer. The authors propose the use of GALS for SDR implemented through digital circuits, but the work does not consider the use of FPGAs. Nyogi et al [15] propose the use of GALS to applications of SDR to be implemented by hardware, but the paper does not consider the use of FPGA devices. Finally, the authors [16] propose the use of GALS in an SDR implemented in FPGA, which proves to be advantageous by preserving the benefits of synchronous design and by exploring the asynchronous routing advantages, but does not provide a GALS architecture for the application, either demonstrate its robustness.

This paper proposes a robust GALS architecture for SDR applications on FPGAs, evaluating the performance of this architecture in a broadband QPSK receiver, such that it is possible to demonstrate the suitability of the proposed architecture for SDR applications. When compared to the synchronous SDR design, the SDR_GALS design obtained a 39% reduction in dynamic power dissipated and an increase in throughput of 53%.

II. SDR ADVANTAGES & WAVEFORM

According to Grayver et al. [17], the six main advantages of SDR are:

Interoperability: an SDR is able to communicate with multiple incompatible radios, and act as a bridge between them.

Efficient use of resources under different conditions: an SDR is able to adapt its waveform to maximize an important metric. For example, the radio can opt for a low-power waveform if the radio is in low battery level.

Frequency reuse based on opportunity: an SDR can obtain an advantage in the use of available spectrum. If the owner of a given spectrum band is not using it at the time, the radio can "borrow" this band until its owner again use it.

Reduced obsolescence: an SDR can be field upgraded to support the latest telecommunication standards.

Cost reduction by scale: A single radio may be used for different markets and for different applications.

Research and development: developing a waveform is much faster with SDR, it is not necessary to build an entire hardware for this purpose.

A. Defining a Waveform

The design of an SDR can be regarded as the specification of a waveform. Figure 3 shows the waveform defined by IEEE, which is widely used in wireless networks of personal computers (WLAN or Wi-Fi). Its physical layer is shown in Fig. 3 is based on OFDM (Orthogonal Frequency-Division Multiplexing) [18]. This paper proposes a waveform to SDR, which will be implemented in FPGA and is described in Fig. 4. This waveform is composed by an enlace radio receiver of UHF broadband (2.5 Mbps on a channel of approximately 1.6 MHz), encrypted and modulated in QPSK. Both the portion of transmission and the reception were developed, but only the portion of the reception will be object of study for this paper.

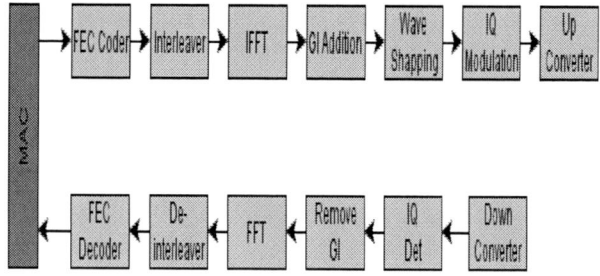

Fig.3.Functional diagram of the physical layer 802.11a [18].

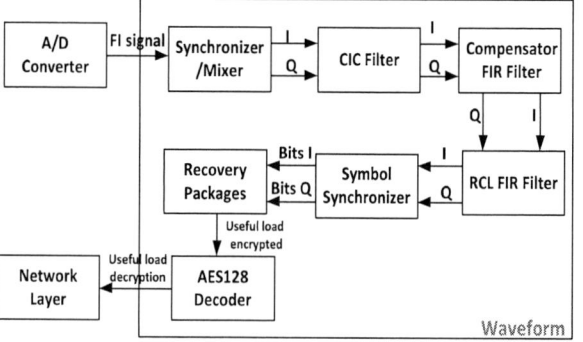

Fig.4. Waveform proposal for SDR.

III. GALS ARCHITECTURE FOR SDR

In this paper, we propose a GALS architecture focused on FPGAs (Field Programmable Gated Array) that will be used in the SDR project. Asynchronous interfaces, in this case the communication ports, when implemented in FPGAs are subject to essential hazard because of delays between the macrocells, as well as for logical hazard due to mapping, where the decomposition of a Boolean function is carried out in the conventional way [19]. Figure 5 shows the proposed asynchronous wrapper based on gated-clock. The asynchronous wrapper is composed of two communication ports. The communication ports are described in the extended burst-mode specification of [20] and are asynchronous finite state machines (XBM_AFSMs). The proposals ports of input and output are based on the ports of [8] (see Fig. 6a, b). The A_{P1}, R_{P1} and R_{I11} signals are inserted in the specification of the two ports to make them robust to essential hazard. The two ports meet the essential signal property [19] and are implemented by direct mapping using the method proposed in [20]. The resulting circuits are robust, supporting any kind of mapping in an FPGA, dispensing the use of hard macros [21], and so being free of logical hazard and essential hazard.

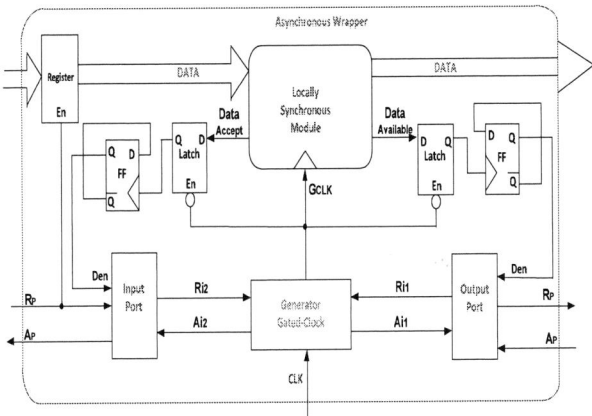

Fig.5. Proposed GALS architecture: asynchronous wrapper

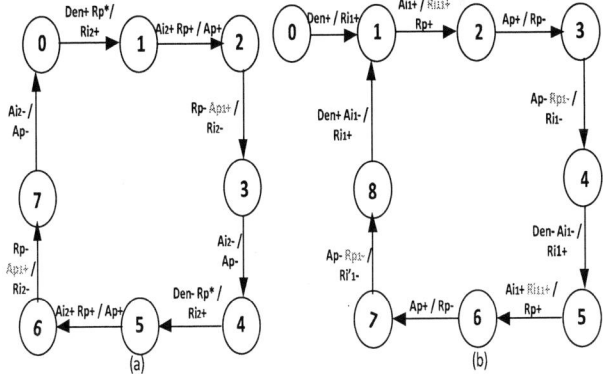

Fig. 6. BM/XBM specification: a) input port; b) output port.

B. Clock-Gated Generator for Architecture Wrapper

In this paper, it is also proposed a clock-gated generator (CGG) that is free of metastability. Figure 7 shows the timing diagram of the CGG, highlighting the stopping and activation of the clock signal. The stopping (pause) of the GCLK signal occurs when *Ri2* or *Ri1* switches 0→1. Figure 8 shows the architecture of the CGG that is based in the latch RS and a control based on basic gates.

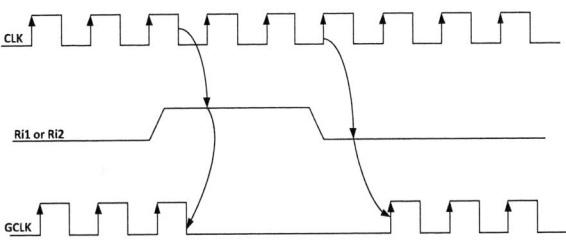

Fig.7. Timing diagram: gated-clock generator.

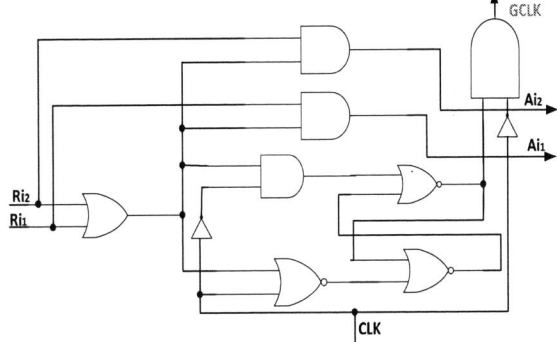

Fig.8. Proposed gated-clock generator.

C. Communication Ports for Architecture Wrapper

The input and output ports were implemented by based method in direct mapping [22]. The method uses for each state of the XBM specification a memory element shown in Fig. 9. Figure 10 shows the logic circuit of input port.

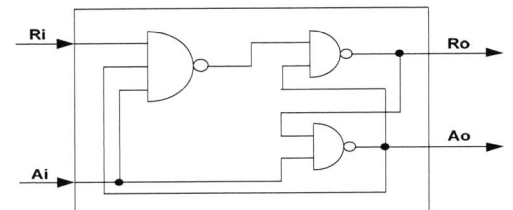

Fig.9. Element of memory [22].

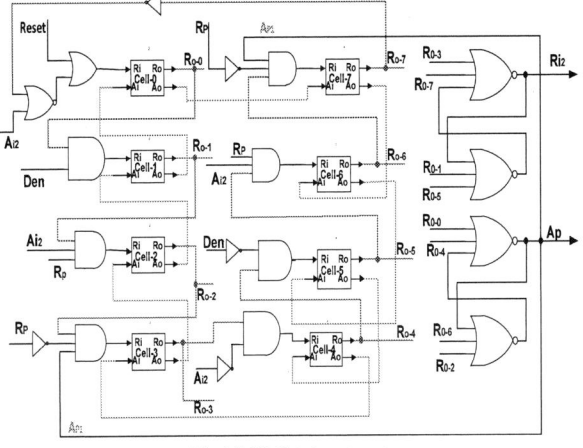

Fig.10. Netlist: input port.

D. GALS_SDR Design: Partitioning

Partitioning is a creative process, and to derive directly an algorithm is not something feasible. Still, it is important to strengthen some aspects of the designer in the form of six guidelines that will help the partitioning:

1. If a region processes data of sources with very different rates, this region is a candidate to be partitioned;
2. Circuit regions that presents critical paths, that is, combinatorial logic whose delay circuit limits the maximum frequency should be considered as candidates for operating in a domain of different clock of the remainder;
3. For this architecture, each transaction between synchronous blocks implies pause the clock by up to 1.5 cycles. This means that frequent transactions penalize the circuit flow rate when compared with its globally synchronous version;
4. Each partition created involves the addition of a GALS wrapper, which in turn consumes so much area with power;
5. Each partition created consumes a dedicated line to clocks the FPGA. Although current devices present a reasonable number of clock lines (at least 20, even for the low-cost lines, this is a factor that must be considered (Altera – [23]);
6. Clock skew affects the performance of a synchronous circuit. Locally synchronous modules very large, or that for some reason need to be spread over a broad region in the FPGA, should be considered as candidates for partitioning;

The synchronous SDR receiver design was realized with seven modules, as seen in Fig. 4. Using the partitioning guidelines for GALS architecture, the waveform was partitioned into four partitions, as shown in Fig. 11.

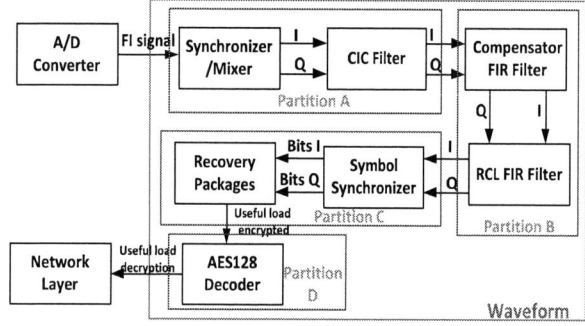

Fig.11. Waveform of SDR: proposed partitioning for GALS.

IV. DESIGN & SIMULATIONS & TESTS

The chosen waveform implements both the physical and datalink layers of a broadband QPSK receiver (2.5 Mbps of raw data). The project is able to: *a)* receiving a digitalized intermediate frequency (IF) signal type of 10 MHz; *b)* tune the IF signal and down converting it, generating a signal in baseband; *c)* extracting the information signal, modulated by quadrature phase shift (QPSK), and the matched filter of type raised root cosine; *d)* retrieve the symbol synchronization; *e)* Retrieve the frame structure and extract the payload, being the framework composed of a word of 32 bits of preamble (0x55555555), 1 bit of alignment (0xAA7C1DD5), and 16 bits of data; and *f)* decrypt the data of the payload with AES-128 algorithm.

As interface requirements, the waveform must (Fig. 12): *a)* receive, at the RF front-end, a signal of 80 MSPS, amplitude quantized by 12 bits and by 8 samples per carrier period; and *b)* delivering for the network layer 128 bits data packages.

Fig. 12. Waveform interfaces.

As a mean of comparison, and functional check of the GALS architecture designed for SDR, it was decided to implement the previously described requirements in two different ways: the first as a fully synchronous design and the second in the target GALS topology. The results are compared for different parameters.

In the fully synchronous implementation, a set of codes in VHDL was developed to describe the digital hardware, similar to the designed waveform. In addition to the developed codes, it also used a previously developed block for the AES decoder, which adds 2276 lines of code to the project. The verification of the description behaviour is made observing the project's internal signals. The main evidence of the correct operation of the circuit is in the output of the block, which corresponds to the data of the payload of incoming packages. Figure 13 shows a result of the simulation.

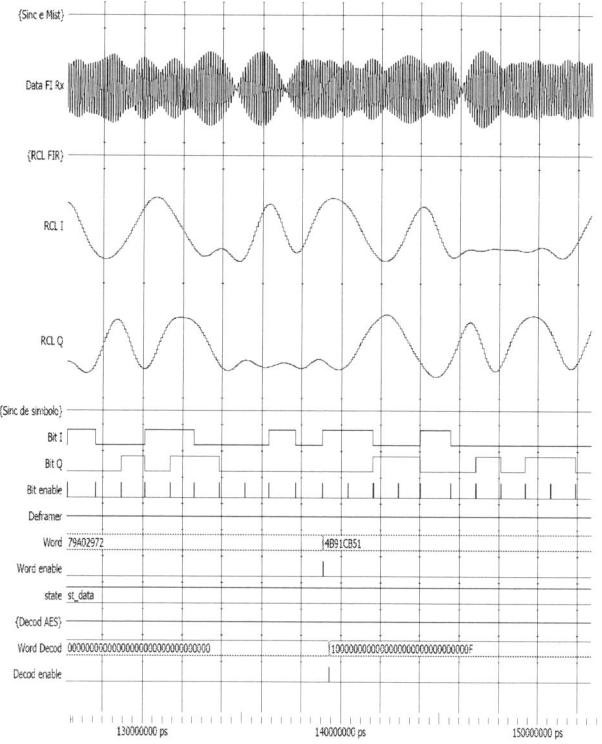

Fig. 13. Example of a receptor post layout simulation: synchronous version.

The Altera tool Quartus II Web Edition 14.0 was used to synthesize the synchronous design for the 5CSEMA5F31C6 device of the Cyclone V family [23]. The compilation was made using the factory default settings, which parameterize the synthesis and layout of the project for a result balanced for area, speed, and compilation time. As objective for Fmax (maximum frequency circuit operation), it was set a value of 100 MHz. Although it supports higher rates, the circuit

978-1-5090-2737-8/16 $31.00 © 2016 IEEE   232

operated at 50 MHz in this test, since the A/D converter of the hardware supports a maximum rate 60 MSPS. The test was run until 800,000 data packages were transmitted, which is equivalent to 410 million bits. No error (package either loss or bit error) was recorded in the test run, which shows an error rate (BER, bit error rate) of less than 10^{-9}. It is important to say that this kind of test does not aim to characterize the rate of failures, leading only to verification, by demonstration, that the circuit implemented in hardware operates exactly as the functional specification of the previously described waveform.

The synchronous version was partitioned into seven modules and an analysis of the maximum frequency and processing time was done on the eight modules, which led to a new partitioning to GALS architecture. Then, the project was divided into locally synchronous blocks, and we apply wrappers to these blocks, resulting in a GALS design for the defined waveform, which was made for four distinct partitions. Once defined the partitions, the VHDL code that instantiates the highest level blocks had to be modified so that the other blocks work with a local clock, and have interfaces between them. This process comes down to instantiate the wrapper components in the code for each region, to connect the request and acknowledge signals between the wrappers, and to replace the input of the clocks to each synchronous component. Enable signals to the wrappers were also implemented, as seen in Figure 14.

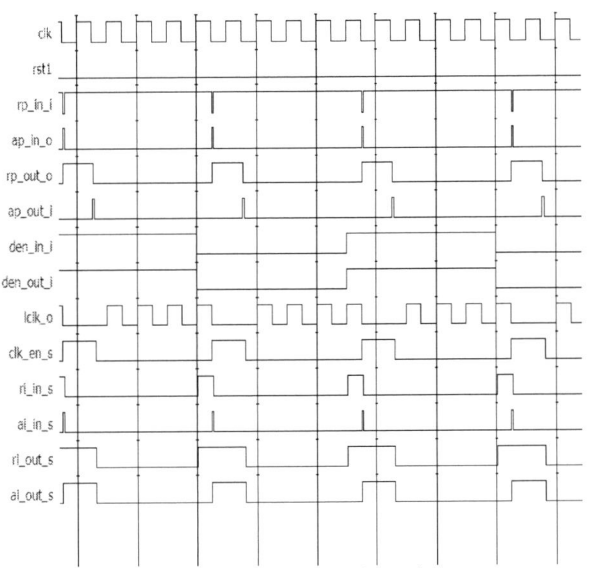

Fig.14. Post layout simulation: proposal asynchronous wrapper.

The testbench previously built for synchronous design was reused, requiring just a few modifications. The simulation performed in the Mentor Graphics Modelsim Altera Starter Edition tool presented outputs that were consistent with the waveform specification, showing the correct application of the GALS wrappers. Figure 15 shows an example of a functional verification by simulation, highlighting the wrapper vectors ({Comp RCL Wrp}), as well as the inputs and outputs of a locally synchronous component ({Comp+RCL FIR}).

The test assembly previously used in the synchronous design was used again and the GALS design was loaded in

the FPGA receiving board. Despite being able to operate at higher rates, the circuit was tested with a 50 MHz clock due to hardware performance limitation, as in the test of the synchronous design. The test was run until 800,000 data packages were transmitted, which is equivalent to 410 million bits. No error (package either loss or bit error) was recorded in the test run, which leads to an error rate (BER, bit error rate) of less than 10^{-9}.

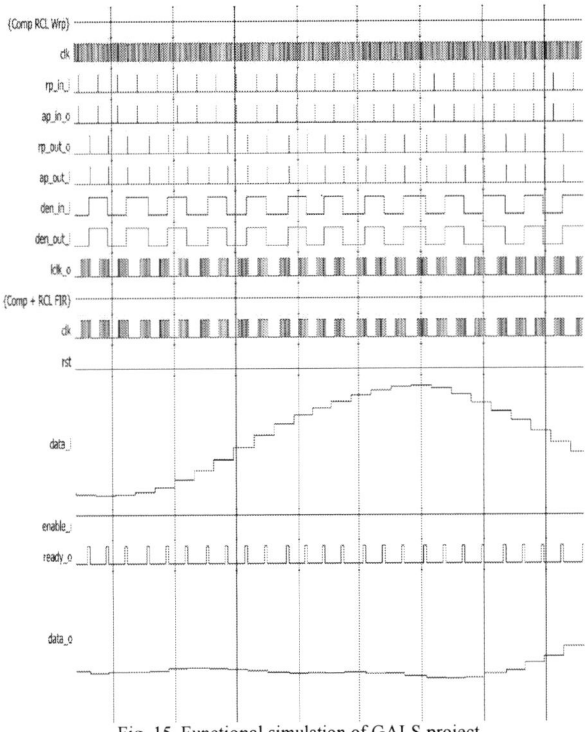

Fig. 15. Functional simulation of GALS project.

V. RESULTS & ANALYSIS

Concerning to resources, the inclusion of the GALS wrappers resulted in a penalty of 32 registers and 167 ALM (*Adaptative Logic Element*), as shown in the Table I. Naturally other metrics remained the same, since the locally synchronous blocks are the same in both designs. Although there is a penalty, it was only 4% for the use of ALM, and 1% for registers.

TABLE I COMPARISON ON RESOURCES

Parameter	Synchronous	GALS	Ratio
Logical Use (in AKM)	3894	4061	4%
Registers	5281	5313	1%
Pins	125	125	0%
Block memories (bits)	11264	11264	0%
DSP blocks	56	56	0%
PLLs	1	1	0%

In what concerns to the data flow, the synchronous design was able to operate at 95.92 MHz in the worst case, which means that the circuit is capable of processing 95.92 million samples per second. Once the circuit needs 18,432 samples for each package (or 512 bit of payload), the synchronous design is capable of processing about 2,664 kbit/s while the GALS design is able to process approximately 4,079 kbit/s (to see Table II). In a comparison between the two projects,

978-1-5090-2737-8/16 $31.00 © 2016 IEEE   233

the GALS architecture allowed an improvement of 53% in the data flow performance and a maximum flow 4,079,979.84 bits/s while the synchronous project led to 2,664,535.04 bits /s.

TABLE II DATA FLOW FOR GALS DESIGN

Partition	Fmax (MHz)	Cycles/ package	Transf/ package	Bits/sec
A	147.56	18432	0	4,098,831.01
B	100.98	9216	3456	4,079,979.84
C	203.05	2304	432	37,996,853.39
D	66.40	128	4	257,538,535.21

Finally, the dynamic power consumption was approximately 113 mW for the synchronous design, and approximately 69 mW for the GALS design, as shown in Table III. The application of GALS architecture thus represented a reduction of 39% in the dynamic consumption, which is the ratio of the total consumption that can be reduced by design [23].

TABLE III COMSUMPTION COMPARISON

Project	Dynamic (mW)	Total (mW)
Synchronous	113.83	543.83
GALS	69.44	499.44
Ratio	-39%	-8%

If we consider the total consumption, which includes the static consumption of 415 mW and the I/O consumption of 15 mW for both cases, the consumption reduction decreased to 8%. It should be noted that the circuit occupies a small area on the user FPGA (only 12%), so the static consumption is dominant.

VI. CONCLUSION

In the present study, the design, implementation and comparison of Software Defined Radio (SDR) based on GALS architectures were focused on GALS port controllers, previously proposed by [5,8] for implementation in ASIC, have been redesigned for use in conventional FPGAs. Unlike other controllers proposals found in the literature, such as the one presented by [21], our proposal eliminates the need for hard macros, having been demonstrated in hardware.

A GALS architecture for SDR was proposed and validated. This architecture comprises a wrapper described in VHDL, and also some guidelines for its implementation in synchronous designs. It has been shown that the adoption of GALS paradigm is suitable for SDR applications, providing the reuse of functional blocks, improvement in performance of the processing rate, and also a power consumption reduction. Furthermore, the proposed architecture has proved to be advantageous, while not inserting efforts of design and verification, allowing data exchange between blocks and, in general, by ensuring performance improvement.

When comparing the power consumption, a reduction of 39% in the dynamic consumption was obtained, and may be even greater if the idle clock periods are considered. Whereas asynchronous wrappers were designed to be robust and to dispense timing verification, the verification effort is negligible, since the same testbenches can be used for both designs. Finally, it became clear that the performance improvement is possible in SDR projects without the need for modification of the original modules. We saw that for the same device, you can earn 53% in the data flow (throughput),

signifying that the radio is able to process a broader band, or that the cost of equipment can be reduced, once it is possible to adopt a cheaper FPGA version with lower performance.

REFERENCES

[1] T. Ulversoy, "Software Defined Radio: Challenges and Opportunities," IEEE Communications Surveys & Tutorials, v. 12, n. 4, p. 531-550, October 2010.

[2] C. H. V. Berkel, et al., "Applications of asynchronous circuits," Proceedings of IEEE, Vol. 87, No. 2, p. 223-233, 1999.

[3] S. Hauck, "Asynchronous Design Methodologies: An Overview," Proceedings of IEEE, v. 83, n. 1, p. 69-93, January 1995.

[4] D. M. Chapiro, Globally-Asynchronous Locally-Synchronous Systems, PhD thesis, Stanford University, October 1984.

[5] D. S. Bormann and P. Y. K., "Asynchronous Wrappers for Heterogeneous Systems," Proc. Int. Conf. Computer Design (ICCD), pp.307-314, October 1997.

[6] P. Techan, M. Greenstreet, and G. Lemieux, "A Survey and Taxonomy of GALS Design Styles," IEEE Design & Test of Computers, vol. 24, pp.418-428, September-October 2007.

[7] F. Gurkaynak, et al. "Improving DPA security by using globally-asynchronous locally-synchronous systems," Solid-State Circuits Conference, ESSCIRC 2005. Proceedings of the 31st European, 12-16 September pp. 407-410,2005.

[8] J. Muttersbach, "Globally-Asynchronous Locally-Asynchronous Architecture for VLSI Systems," Zurich, Suiça, Dissertation submited to the Swiss Federal Institute of Technology for the degree of Doctor of Technical Sciences, p. 138, 2001.

[9] X. Jia, "The GAPLA: A Globally Asynchronous Locally Synchronous FPGA Architecture," Field-Programmable Custom Computing Machines, FCCM 2005. 13th Annual IEEE Symposium on, Cincinnati, pp. 291-292, 18-20 April 2005.

[10] R. I. Soares, et al., "A Robust Architectural Approach for Cryptographic Algorithms Using GALS Pipelines," IEEE Design & Test of Computers, vol.28, Issue: 5, pp.62-71, 2011.

[11] H. Lee, "A Baseband Processor for Software Defined Radio Terminals," PhD Thesis, University of Michigan, 2007.

[12] L. R. Soriano, "Design and Implementation of a Reconfigurable Radio Platform," M. Eng. Sc Thesis, National University of Ireland, 2007.

[13] J. Degroat, et al., "Synthesizing FPGA Digital Modules for Software Defined Radio," Aerospace and Electronics Conference, NAECON 2008., Dayton, OH, pp. 358-362,16-18 July 2008.

[14] M. Gautier, et al., "Design Space Exploration in an FPGA-Based Software Defined Radio," Digital System Design (DSD), 17th Euromicro Conference on, Verona, pp. 22-27, 27-29 August 2014.

[15] K. Nyogi, D. Marculescu, "System Level Power and Performance Modeling of GALS Point-to-point Communication Interfaces," IEEE Proc. Int. Low Power Electronics and Design, pp. 381-386, 2005.

[16] M. Hentati, et al., "Software Defined Radio Equipment: What's the Best Design Approach to Reduce Power Consumption and Increase Reconfigurability?," International Journal of Computer Applications, IJCA, v. 45, n. 14, p. 26-32, December 2012.

[17] E. Grayver, "Implementing Software Defined Radio," 1st. ed. New York: Springer Science Business, 2013.

[18] IEEE Computer Society. IEEE Std 802.11™-2012. New York, USA, IEEE Standard for Information Technology, 2012.

[19] D. L. Oliveira, et al., "Burst-Mode Asynchronous Controllers on FPGA", Int. Journal of Reconfigurable Computing, Vol. 2008, pp. 1-10.

[20] K. Y. Yun and D. L. Dill, "Automatic Synthesis of Extended Burst-Mode Circuits: Part I (Specification and Hazard-Free Implementation) and Part II (Automatic Synthesis)," IEEE Trans. on CAD of Integrated Circuit and Systems, vol. 18:2, pp. 101-132, February 1999.

[21] J. Pontes, et al., "SCAFFI: an Intrachip FPGA asynchronous interface based on hard macros", 25th Int. Conf. on Computer Design, pp. 541-546, October 2007.

[22] D. L. Oliveira, L. A. Faria and E. Lussari, "Design of an Improved and Robust Asynchronous Wrapper (AW) for FPGA Applications," Journal of Integrated Circuits and Systems, v. 8, n. 1, p. 54-63, 2013.

[23] ALTERA CORPORATION. PowerPlay Power Analysis. PowerPlay Power Analysis, 11 abril 2013. Disponivel em: <https://www.altera.com/en_US/pdfs/literature/hb/qts/qts_qii53013.pdf>. Acesso em: 19 abril 2015.

A Digital Offset Correction Method for High Speed Analog Front-Ends

Andres Amaya, Hector Gomez and Elkim Roa
UIS — Universidad Industrial de Santander, Bucaramanga, Colombia
{andres.amaya1,hector.gomez}@correo.uis.edu.co, efroa@uis.edu.co

Abstract—This paper presents an offset voltage correction technique for high-speed digital interfaces. Contrary to conventional way of measuring offset, the proposed technique is based on the phase measurement of a slicer output avoiding the input connection to a common mode voltage. A fully-digital implementation allows phase measurement maintaining offset accuracy. Proper operation of calibration technique is achieved when the input signal is comparable to the offset and sensitivity of the whole interface. Thus, the proposed method could be used during on-line operation, without breaking the communication link. The circuit has been implemented in a 130nm TSMC standard CMOS process, and simulation results show an offset reduction nearly 90% in the analog front-end with a low area overhead.

Index Terms—Offset correction, digital calibration, voltage comparator.

I. INTRODUCTION

In a high-speed serial interface, the front-end of the reception block (Rx) recovers the transmitted information performing three main tasks: equalization, amplification and sampling. After that, the pre-processed signal passes through a Clock-and-Data Recovery (CDR) circuit, which recognizes and separates the clock signal. Besides recovering the clock, the CDR synchronizes the system if the input signal is delivered with a proper quality by previous circuits. The front-end can include a couple of continuous-time linear equalizers (CTLEs), several amplifier stages, a slicer circuit or a sampler as shown Fig. 1 contributing to a large accumulated offset at the slicers input.

The CTLEs offset might saturate the signal at the front of the system making impossible the information recovery. Typically, offset correction techniques are applied in various stages in order to avoid loss of the information. Several offset reduction alternatives have been proposed. In [1], Kimura uses a low-pass RC filter to extract DC of the signal. Kimura [1] calibrates only the analog part by tapping the signal path to sense offset which increases loading. In [2], Redman-White uses a modified CDR to include robustness to offset instead of introducing correction techniques on the analog part of the front-end. The proposed CDR uses ten phases and a complex algorithm in order to extract the input symbols in a way that offset impact is reduced.

This paper presents a solution that reduces offset in high-speed digital interfaces. The reduction technique can calibrate analog and digital part from sampler and can be used to correct the overall offset of the front-end. The correction is performed by sensing the signal phase of the sampler output, generating a control signal with a quite-simple algorithm. The proposed technique works based on the fact that the amplitude of input signal should be less than the offset plus the sensitivity of slicer.

Some advantages of the proposed technique are its low-complexity and fully digital implementation —adding low hardware overhead—. Moreover, offset correction can be applied at any of the stages by sensing only the outputs of the sampler stages. Furthermore, the proposed technique avoids the input connection to a common mode signal as in classical schemes and does not need additional signal paths. The resulting implementation does not compromise high-speed operation and can be used without the need of interrupting data transfer.

II. PROPOSED OFFSET CORRECTION METHOD

Offset correction is typically done in each block of the analog front-end. Fig. 1 shows a classical reception circuit composed by one variable-gain amplifier (VGA), two continuous-time linear equalizer (CTLE), and a 2-tap decision-feedback equalizer using predictive-DFE (prDFE) as its first TAP. For each block an offset sensing and correction circuit is implemented: for the CTLEs is common to use auto-zero; for the comparator is more usual to introduce digital calibration algorithms [3]. In these techniques the offset reduction procedure begins with setting the input signal to a common mode voltage with the purpose of sensing offset only. As a consequence, it is necessary to use auxiliary reception circuits in order to keep the communication link, introducing additional power consumption and area.

The proposed method is shown in Fig. 2, which is based on sensing offset through the comparator (slicer) output signal phase and frequency. This sensing is done by a digital phase detector, whose outputs control the transitions of a Finite-State-Machine (FSM). Then, the FSM outputs are connected to a digital-to-analog converter (DAC), with the purpose of controlling the bias current of the first CTLE. This change in the bias current will induce an additional voltage contrary to the total offset.

When offset if present and its magnitude is too high so that any change of the input signal will not produce any change at the output data, one output of the slicer will remain constant at VDD, and the other will be oscillating for each clock cycle between VDD and ground i.e. from the reset to the comparison phase. As a consequence, the phase detector (PD) —which

978-1-5090-2737-8/16 $31.00 © 2016 IEEE

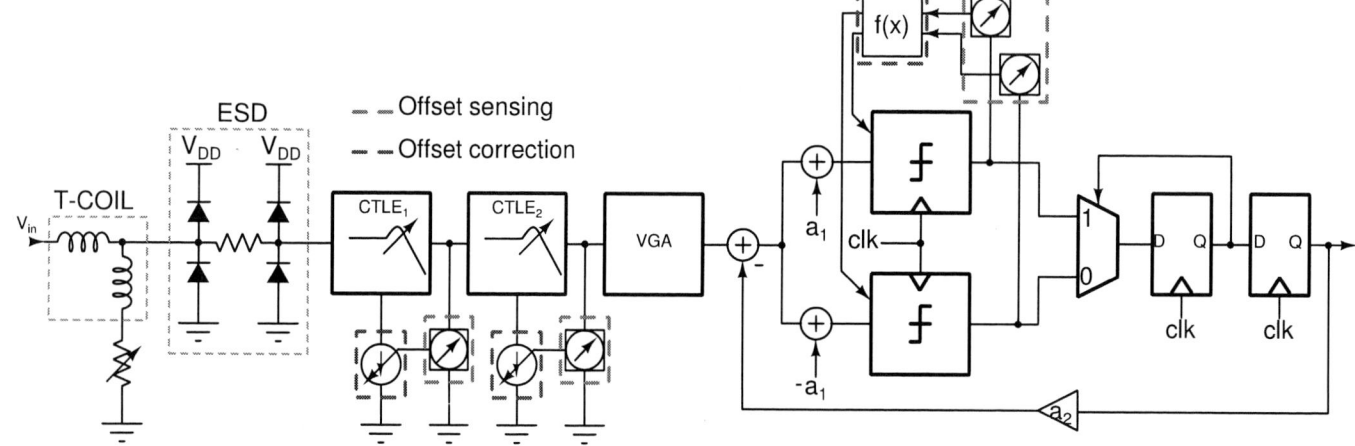

Fig. 1: Classical serial high speed interface with offset correction.

corresponds to a variation of the Bang-Bang circuit (Fig. 3)—senses that one slicer output has a different frequency than the other, producing a change in its UP and DOWN signals. For example, if V_{o1} is stacked to V_{DD} and V_{o2} is oscillating, the PD detects that V_{o2} has a frequency equal to the sampling rate, and that V_{o1} has a lower one; so, the DOWN output is high while the UP signal is low.

These two signals control the FSM transitions regarding the states diagram described in Fig. 4, such that the X_2 signal increases each clock cycle, while X_1 decreases. Then, X_1 and X_2 are converted to analog signals by a digital-to-analog converter, controlling the gate-to-source voltage of transistors M_x and M_y. As a consequence, the circuit modifies the bias current of the first amplifier, trying to compensate the total offset. The calibration finishes when the change in the bias current is able to produce that both V_{o1} and $Vo2$ oscillate between VDD and ground each clock cycle, causing that both phase detector outputs signals are always high.

One of the main advantage of the proposed technique is the fact that the calibration process can be done while the whole interface is still working. It means that, unlike classical offset reduction techniques, the input does not need to be set at a common mode voltage while the calibration is carried out. When offset is larger than the input signal, any change at the input does not produce a variation at the output. If calibration circuit is turned on the offset will be reduce until it is low enough to allow that the input signal can be sensed by the slicer. This characteristic is critical for high speed operation because no additional load is introduced at input of the system. Also, there is no need to execute high complexity algorithms that might introduce latency.

III. SIMULATION RESULTS

To simulate a more realistic situation, the input of Fig 2 was connected to the output signal of a second-order low-pass programmable filter and a Pseudo-Random Bit Generator PRBS, as Fig. 5 shows. The filter simulates the finite channel bandwidth, while the PRBS acts as the information source.

Also, the filter is based on the Gm-C topology, using Nauta circuits [4] as transconductance blocks. The input signal and clock frequency are 3GHz. At this frequency the filter attenuation is 35dB, so its output is 20mV, implying that the offset requirement of the whole system must be less than 5mV. It is important to highlight that all the digital circuits were fully synthesized and simulated at transistor level, allowing to include mismatch and process variations. The CTLE circuits are based on the degenerated common-source topology with resistive load (Fig. 6) which is very common in high-speed applications. The bias topology current is composed by two constant current sources and two additional transistors —M_x and M_Y— used for offset compensation. The comparator corresponds to the strong-arm circuit (Fig. 7) only for validation issues. This topology includes reset transistors at the output nodes for increasing speed [5]. Finally, the current mirrors are biased by the two 8bits R2R calibration DACs.

Figure 8 shows the output signals DAC_1 y DAC_2 of the DAC. Offset was simulated by MonteCarlo analysis in order to include mismatch; however, for explanation purposes, Fig. 8 presents only one iteration. The offset compensation process begins with the two DAC outputs from ground; at this point the comparator can not recover the information from a signal whose amplitude is 100mV. For that reason output signal V_{o2} is always high, while V_{o1} is oscillating between V_{dd} and ground. Some clock cycles later, signal DAC1 increases while signal DAC2 remains constant. After 400ns, these two signals reach a steady-state and the counters and phase detectors are disabled, thus keeping the last data. Once the calibration has been done the two comparator output signals oscillate each clock cycle, recovering the information without breaking the communication link. Moreover, Fig. 9 presents the calibration signals for several Monte Carlo iterations, achieving always a correct offset compensation.

Fig. 11 shows the slicer outputs while an online offset reduction is carried out; in this case, the input signal is provided by the system shown in Fig. 5. At the beginning the signal V_{o2} is always near to V_{DD}, and V_{o1} is oscillating

978-1-5090-2737-8/16 $31.00 © 2016 IEEE

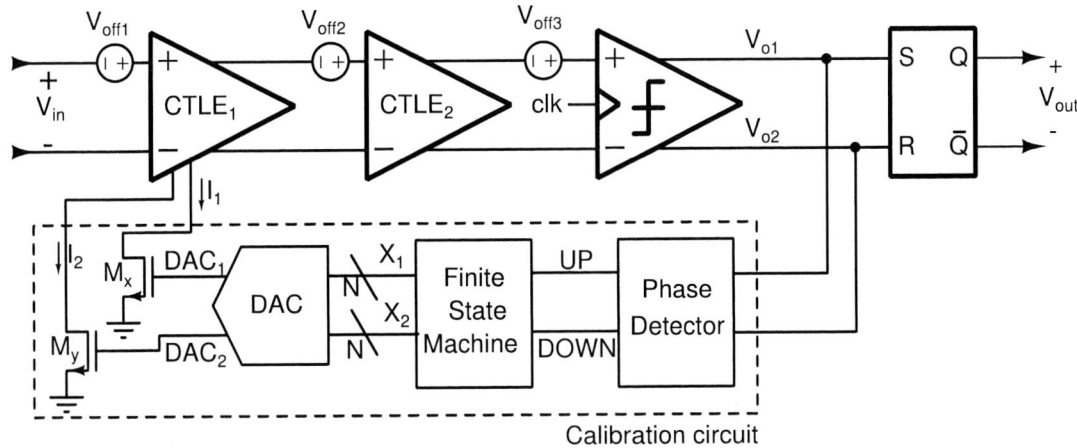

Fig. 2: Proposed offset compensation method.

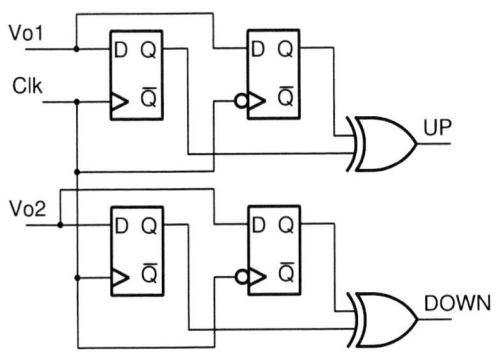

Fig. 3: Implemented Phase Detector.

	Typical	Worst Speed Case	Best Speed Case
Offset Before Cal.	100mV	80mV	110mV
Offset After Cal.	220μV	500μV	700μV
Preamp+Comp Power	3.5mW	2.8mW	4.1mW
Additional Power	580μW	430μW	650μW
Supply Voltaje	1.2V	1V	1.4V
Calibration Time	400ns*	400ns*	400ns*

* The clock frequency of the calibration circuit is 250MHz

TABLE I: Performance of the calibration technique

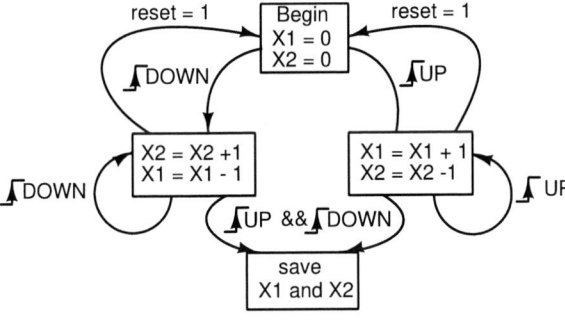

Fig. 4: State diagram of the calibration circuit FSM.

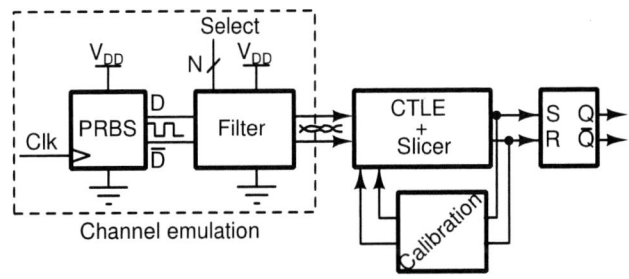

Fig. 5: PRBS and filter used for validating the calibration process.

than 1mV, implying a reduction of more than 100X.

between the power rails at the clock frequency, indicating that offset is greater than signal amplitude. Then, 400ns later, the calibration circuit adjusts the first CTLE bias current so that both slicer outputs can oscillate regarding the input signal.

In addition, Fig. 10 shows the DAC calibration signals for the worst operation condition: slow process corner, low supply voltage (1V) and low temperature (-40°C); and for the best case: fast process corners, high supply voltage (1.4V) and high temperature (120°C). Both present a successful calibration process in less than 500ns.

Also, table I summarizes the performance of the proposed technique. It is important to highlight that final offset is lower

IV. Summary

In this paper, a low-hardware overhead calibration technique for high-speed digital interfaces has been proposed. The proposed technique for offset measurement detecs the phase difference between the outputs of the slicer to adjust a bias current to reduce the offset to a permissible value. Calibration is triggered when the amplitude of the input signal at the slicer is less than the sum of the offset and sensitivity. The proposed technique avoids the input connection to a common mode signal enabling the possibility to perform on-line calibration. The calibration circuitry was fully synthesized in 130nm showing the potential to scale the technique to different fabrication process.

978-1-5090-2737-8/16 $31.00 © 2016 IEEE

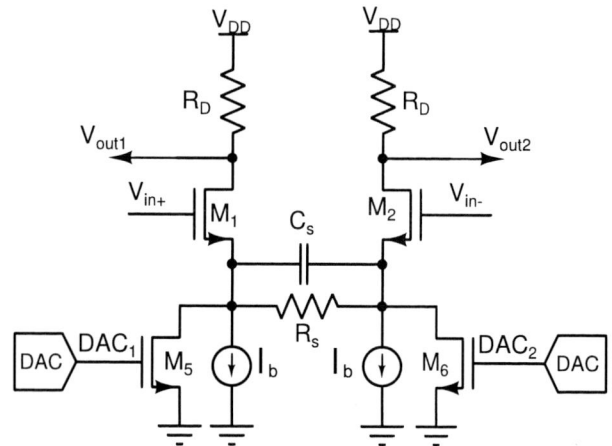

Fig. 6: Traditional Continuous-time Linear Equalizer circuit.

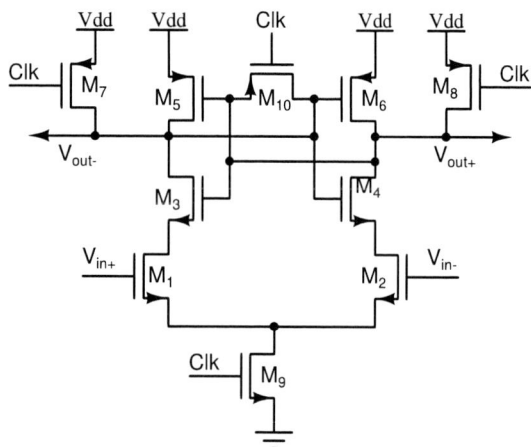

Fig. 7: Strong-arm topology used for the slicer circuit

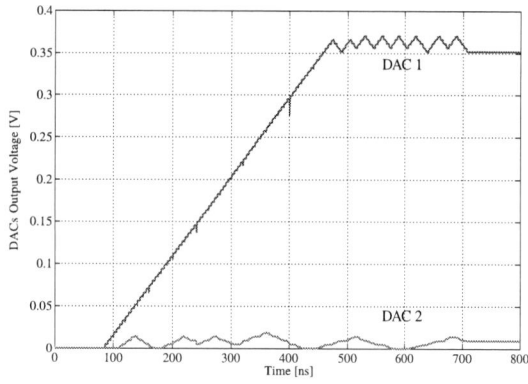

Fig. 8: DACs output signals while the calibration process is carried out.

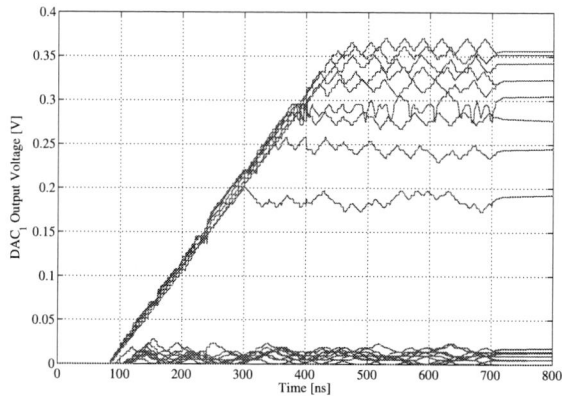

Fig. 9: First DAC output signal for Monte-Carlo samples.

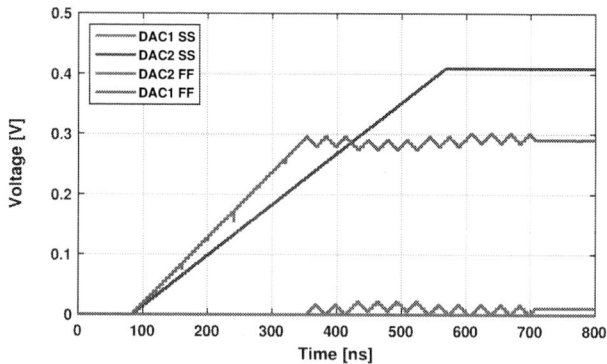

Fig. 10: DACs signals for the best and worst operating case.

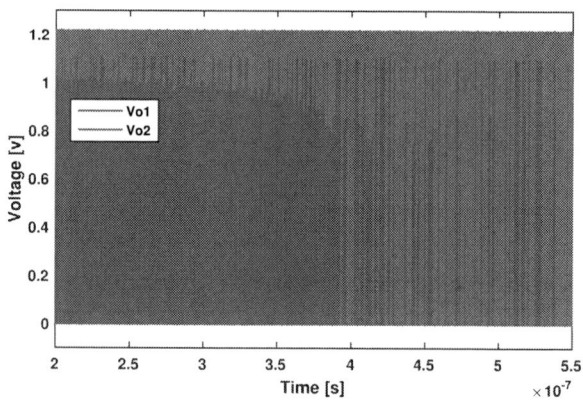

Fig. 11: Comparator output signals while calibration is carried out. The Fig. shows that the system can compensate an offset greater than 100mV.

REFERENCES

[1] H. K. et. al, "2.1 28Gb/s 560mW Multi-Standard SerDes with Single-Stage Analog Front-End and 14-tap Decision-Feedback Equalizer in 28nm CMOS," in *2014 IEEE International Solid-State Circuits Conference Digest of Technical Papers (ISSCC)*, Feb 2014, pp. 38–39.

[2] W. R.-W. et. al, "A Robust High Speed Serial PHY Architecture With Feed-Forward Correction Clock and Data Recovery," *IEEE Journal of Solid-State Circuits*, vol. 44, no. 7, pp. 1914–1926, July 2009.

[3] R. Wu, J. H. Huijsing, and K. A. A. Makinwa, *Precision Instrumentation Amplifiers and Read-Out Integrated Circuits*. Springer, 2013.

[4] S. Kumaravel, A. Gupta, and B. Venkataramani, "VLSI Implementation of Gm-C Filter using Modified Nauta OTA with Double CMOS Pair," in *Recent Advances in Intelligent Computational Systems (RAICS), 2011 IEEE*, Sept 2011, pp. 216–220.

[5] B. Razavi, "The StrongARM Latch [A Circuit for All Seasons]," *IEEE Solid-State Circuits Magazine*, vol. 7, no. 2, pp. 12–17, Spring 2015.

IEEE
445 Hoes Lane
Piscataway, NJ 08854-4141

ISBN 978-1-5090-2737-8